Chlorinated Dioxins and Dibenzofurans in Perspective

Edited by

Christoffer Rappe

Gangadhar Choudhary

Lawrence H. Keith

LEWIS PUBLISHERS, INC.

Library of Congress Cataloging-in-Publication Data

Chlorinated dioxins and dibenzofurans in perspective.

Based on the 189th National Meeting of the American
Chemical Society, held in Miami Beach, Fla. in Sept. 1985.
Includes bibliographies and index.
  1. Dioxins — Environmental aspects. 2. Dibenzofurans —
Environmental aspects. 3. Environmental chemistry.
4. Man — Influence of environment. I. Rappe, Christoffer.
II. Choudhary, Gangadhar, 1935- . III. Keith,
Lawrence H., 1938- . IV. American Chemical Society.
Meeting (189th : 1985 : Miami Beach, Fla.) [DNLM:
1. Benzofurans — analysis — congresses. 2. Dioxins —
analysis — congresses. 3. Environmental Exposure —
congresses. 4. Environmental Pollutants — analysis —
congresses. WA 671 C5435 1985]

TD196.C5C45   1986   628.5   86-18537
ISBN 0-87371-056-8

Second Printing 1987

LEWIS PUBLISHERS, INC.
121 South Main Street, P.O. Drawer 519, Chelsea, Michigan 48118

PRINTED IN THE UNITED STATES OF AMERICA

Preface

This volume was developed from the proceedings of a symposium held in Miami Beach, Florida at the 189th National Meeting of the American Chemical Society. It is the result of the combined efforts of many experts whose efforts have advanced our knowledge of the production, analysis, distribution, effects and control of chlorinated dioxins, dibenzofurans and related compounds. This is the third in a series of publications originating from current technology presented at national meetings of the American Chemical Society. Using this forum as a catalyst, researchers from all over the world came together to present and discuss their data and plan future work in this rapidly developing and sometimes highly emotional technical area.

This book incorporates a number of changes: the title, the publisher and the emphasis. The title and content reflect the maturation and evolution of this subject. The first two books, *Chlorinated Dioxins and Dibenzofurans in the Total Environment*, were indicative of the initial phases of research — developing the measurement process and defining the extent of the problem caused by these compounds. In contrast, this book emphasizes refining the measurement process and assessing the effects, measurable or potential, on humans and their environment. The first two books were published by Ann Arbor Science and Butterworth Publishers, respectively (the former was acquired by the latter). Although this book has yet a third publisher, there is a string of continuity in that Ed Lewis was president of Ann Arbor Science, then was associated with Butterworth, and is now president of Lewis Publishers.

The opening section is devoted to the concerns and manifestations of human exposure to chlorinated dioxins and dibenzofurans. The chapters in this section describe the analysis and distribution of dioxins and dibenzofurans in humans. This includes the special analytical techniques that are required and the levels and distributions of these compounds that are being found; one unique study involves the analysis of human samples from Vietnam.

The second section is comprised of chapters involving formation and emissions of these compounds from incineration sources. Increased interest in energy conversion from municipal incinerators and the massive decontamination costs that have resulted from accidental fires in buildings containing transformers with polychlorinated biphenyls (PCBs) have catapulted this source of chlorinated dioxins and dibenzofurans into local, national and international prominence. The concern revolves around the potential large-scale epidemiological exposure to these compounds when they are formed — either accidentally or intentionally — by fires.

The third section is concerned with the continuing problems of sampling, analysis, fate and destruction of chlorinated dioxins and dibenzofurans in soils and related matrices. It was the earlier problems associated with improper disposal of contaminated wastes onto and into soils that focused much of the initial attention of environmental contamination from these compounds in the USA. The emphasis now has moved from sampling and distribution to fate and, ultimately, to removal and destruction of these pollutants.

The fourth and fifth sections involve analytical techniques: the fourth is comprised of chapters describing the latest bioassays, and the fifth is a collection of the most advanced sampling and analytical methodologies. Together these sections describe state-of-the-art techniques for measuring the subject compounds or their effects in a wide variety of environmental matrices. New cleanup techniques and efficient automated fractionations are described, along with advances, refinements and in-depth comparisons of the latest analytical methods.

The final section describes current efforts aimed at synthesis (for analytical standards) and destruction of these highly complex and resistant compounds. As the science of understanding and coping with these compounds and their problems matures and evolves, so does the focus of our research efforts. The first of our books, published in 1983, contained two chapters on synthesis and none on destruction. This volume contains three chapters on synthesis and/or standards preparation and four devoted to destruction, removal, reclamation, and disposal of these pollutants.

Thus, throughout this volume we see an emphasis not only on describing the problems and dangers caused by chlorinated dioxins, dibenzofurans and associated compounds, but also on discussions of their relationship to man and the environment. As the science of the study of these compounds continues to mature, we can expect to see more emphasis placed on the consequences of exposure to these chemicals — health effects, financial costs, and environmental effects — in relation to the benefits and risk assessments of removing, containing, destroying, avoiding and legislating against them. The story isn't over yet.

C. Rappe
G. Choudhary
L. H. Keith

Christoffer Rappe is a Professor and Chairman of Organic Chemistry at the University of Umeå, Sweden. He was previously an associate professor at the University of Uppsala, Sweden, where he also received his PhD in 1965. Professor Rappe is Rapporteur in ad hoc groups on dioxin problems of the World Health Organization (WHO) and of the International Agency for Research on Cancer (IARC). An internationally recognized authority on chlorinated dioxins and dibenzofurans, Professor Rappe has been involved with their synthesis, occurrence, formation, degradation and analysis for over thirteen years. Other interests include the synthesis, analysis and occupational exposure to nitrosamines, and synthesis, rearrangements and mechanistic aspects of halogenated ketones. Professor Rappe has authored more than 150 papers in various fields of synthetic and analytical organic chemistry.

Gangadhar Choudhary is a chemist in the Measurement Research Support Branch of the National Institute of Occupational Safety and Health in Cincinnati, Ohio. He has previously worked at the Canadian National Research Council in Ottawa, Canada, Reynolds Metals Company in Sheffield, Alabama and the Center for Disease Control in Atlanta, Georgia, serving as a photochemical kineticist, senior research scientist, GC/MS chemist, organic-analytical group leader, project director and project officer for governmental contracts in special trace organic projects. Dr. Choudhary has been involved with chlorinated dioxins and dibenzofurans for the past seven years and is the author or editor of five books and thirty technical articles involving multidisciplinary chemistry. A native of Uchhati, Bihar State, India, Dr. Choudhary is a naturalized U.S. citizen and received his PhD in physical-organic chemistry from the University of Ottawa, Canada, in 1965.

Lawrence H. Keith is Chemistry Development Coordinator at Radian Corporation in Austin, Texas. Before joining Radian, Dr. Keith was a research chemist with the U.S. Environmental Protection Agency in Athens, Georgia. He received his PhD in chemistry from the University of Georgia in 1966. He is a member of the Executive Committee of the American Chemical Society's Division of Environmental Chemistry, having served as secretary and chairman, and currently as alternate councilor and program chairman. Dr. Keith is an Advisory Board Member of *Environmental Science and Technology* and a delegate to the U.S. National Committee to the International Association for Water Pollution Research. An editor or co-editor of nine books and over fifty publications and author of chapters in various journals and books, Dr. Keith's technical interests center on methods for analysis of organic pollutants in the environment and on the safe handling of hazardous chemicals.

Contents

SECTION I
HUMAN EXPOSURE

SECTION II
INCINERATION EMISSIONS

SECTION III
SOIL CONTAMINATION

SECTION IV
BIOASSAYS

SECTION V
ANALYTICAL

SECTION VI
SYNTHESIS AND DESTRUCTION

SECTION I

Human Exposure

CHAPTER 1

Distribution of Chlorinated Dibenzo-p-Dioxins and Chlorinated Dibenzofurans in Human Tissues from the General Population

John J. Ryan, Arnold Schecter, Wing-F. Sun and Raymonde Lizotte

INTRODUCTION

One method for estimating the exposure of humans to toxic chemicals is to measure levels in tissues. Comparison of the levels found in normal, control or nonexposed tissues to those found in exposed or treated samples can then be used to support or disprove any link between tissue level and adverse effect. Ultimately this information can be used to assess the risks involved due to the presence of a particular chemical. In the last few years the capability to reliably measure picogram (pg) amounts and parts per trillion (ppt; pg/g) levels of chlorinated dibenzo-p-dioxins (PCDDs) and chlorinated dibenzofurans (PCDFs) in biological samples has markedly improved.

Initially, efforts in human tissue were reported on a wet tissue basis and focused on 2,3,7,8-tetrachlorodibenzo-p-dioxin (2,3,7,8-TCDD) in adipose tissue. Ryan et al.[1] reported that 22 of 23 adipose tissues from autopsies collected from older patients who died in Ontario hospitals in 1980 contained an average of about 10 pg/g 2,3,7,8-TCDD, and six out of six of the same adipose tissues contained an average of 17 pg/g 2,3,4,7,8-pentachlorodibenzofuran (PnCDF). This finding of PnCDF was similar to an early report by Miyata et al.[2] on Japanese control tissues. They found PnCDF at a range between 14 and 45 pg/g in four of five fat biopsies and one of four livers from autopsy used as controls in a study of the Yusho rice oil poisoning incident. In this same report no tetrachlorodibenzofurans (TCDFs) nor hexachlorodibenzofurans (HxCDFs) were detected, and PCDDs were not assayed. Gross et al.[3] and Lee and Hobson[4] both reported results on the analysis of 33 adipose biopsy tissues from American military personnel involved in the Vietnam conflict. Although the interpretation of the results differed, the ten control samples in the study averaged about 6 pg/g 2,3,7,8-TCDD on a wet tissue basis. Other reports on adipose tissue by Kutz (cited in Young[5]) from six accident victims from Ohio in 1981 and by Graham et al.[6] on nine autopsy samples from Missouri both showed levels of 2,3,7,8-TCDD between 2 and 15 pg/g. Apart from TCDD, Schecter et al.[7], in a study of patients exposed to a toxic soot from a transformer fire in a New York State office building in 1981, noted that four adipose biopsy samples used as controls appeared to contain 100 to 200 pg/g heptachloro-dibenzo-p-dioxin (HpCDD) and 400–800 pg/g octachlorodibenzo-p-dioxin (OCDD). More recently, Ryan et al.[8] found eight to ten PCDDs and PCDFs, all with 2,3,7,8-chlorine substitution, in a sampling of 46 adipose tissues taken by autopsy in 1976 from accident victims across Canada. The mean values on a wet tissue basis for these samples were 6.2 pg/g for 2,3,7,8-TCDD and 800 pg/g for OCDD, with intermediate values for PnCDF, pentachlorodibenzo-p-dioxin (PnCDD), hexachlorodibenzo-p-dioxin (HxCDD), HxCDF, and HpCDD. No TCDF nor octachlorodibenzo-furan (OCDF) was found. The isomeric pattern of each congener group found in these human tissues with regard to levels and types was consistent, and differed from that found in wildlife and in foods such as poultry and pork. This same pattern but at somewhat lower levels has also been reported by Rappe[9] from a small set of human Swedish control tissues.

Apart from the more readily available adipose tissue, information on the concentrations of PCDDs and PCDFs in other human tissues from the general population is either lacking or available only from tissues taken from abnormal exposures, such as those found in accidents or in occupational situations. Moreover, the questions as to which tissue accumulates PCDDs and PCDFs to the highest degree and whether specific target organs exist in humans have not been answered. Hence we sought information on the distribution and levels of PCDDs and PCDFs in a variety of human tissues collected from the same human subjects. A preliminary account of part of this data has already been reported (Ryan et al.[10]). This more complete study shows that between eight and ten PCDDs and PCDFs with 2,3,7,8-chlorine substitution occur in all American human tissues assayed. The relative amounts and specific isomers are the same in all tissues, and the lipid content is the most important criterion governing the concentration in a particular tissue.

Table 1.   Data on Human Tissue Samples from New York State Donors.

Patient No.	Age	Sex	Cause of Death	Tissues Collected
1	67	F	stroke, pneumonia	fat (subcutaneous, mesenteric), adrenal, bone marrow, liver muscle, spleen, kidney, lung
2	57	F	trauma	fat (subcutaneous, mesenteric, perirenal), adrenal, liver, muscle, spleen, kidney
3	37	M	liver disease, pneumonia	fat (mesenteric), liver, muscle

EXPERIMENTAL

Tissue Samples

Autopsy tissues were collected from three subjects who died in New York State from natural causes. Data on the three subjects are listed in Table 1. The tissue types obtained were: fat (subcutaneous, either abdominal or gluteal, mesenteric abdominal, and perirenal), adrenal, bone marrow, liver, muscle, spleen, kidney and lung. These tissues were chosen on the basis of their known accumulation of lipid-soluble compounds or are tissues and organs of importance in the metabolism of xenobiotics. As far as could be ascertained all subjects had no known abnormal exposure to PCDDs or PCDFs either accidentally or occupationally, although this could not be specifically verified. Tissues were frozen after collection and delivered in the frozen state to the analytical laboratory.

Analytical Methodology

The technique used has been developed to measure all tetra- through octa- PCDDs and PCDFs (49 possible PCDDs and 87 possible PCDFs for a total of 136 analytes) in a single final extract. Crude human tissue as obtained was first homogenized without solvent to a uniform consistency with a tissue homogenizer. An aliquot (2 to 3 g for fat and 5 g for other tissues) was homogenized or shaken with 120 mL of acetone-hexane (2:1 v/v). The extract was then spiked with 10 μL of a toluene solution containing about 10 pg/μL each of seven carbon-13 isotopically labeled PCDDs and five chlorine-37 isotopically labeled PCDFs as listed in Table 2. These internal standards were used to correct for losses of samples during manipulation and chromatography. Each analyte within a specific congener group for either PCDDs or PCDFs was corrected for losses using the internal standard from that group (e.g., all TCDDs were adjusted with ^{13}C-2,3,7,8-TCDD and 2,3,4,7,8-PnCDF was adjusted with ^{37}Cl-2,3,4,7,8-PnCDF). The sample extract was then partitioned against water and the aqueous and solid phases extracted a second time with hexane (Ryan et al.[1]). After

Table 2.   Isotopically Labeled Internal Standards of PCDDs and PCDFs (About 100 pg per Sample Except OCDD) Used for Recovery Correction.

Carbon-13			Chlorine-37			
PCDD	pg/μL	m/z	PCDF	pg/μL		m/z
2378–	7.25	332	2378–	10.0		312
12378–	10.5	366	12378–	10.0		348
			23478–	10.0		348
123478–	2.5	402	123478–	10.0	total	384
123678–	11.5	402	123678–		of both	
123789–	10.0	402	123489–	10.0		384
1234678–	11.0	436				
12346789–	50.0	472				

washing the combined hexane phases with water, a 10% by volume aliquot was taken, evaporated to dryness on a rotary under vacuum and weighed. This weight represented the lipid content of tissue expressed as a percentage. The remaining hexane extracts were then defatted with sulfuric acid and chromatographed on activated Florisil columns to remove PCBs, chlorinated diphenyl ethers, and other nonpolar aromatics (Ryan et al.[11]). Sample extracts were then purified further on activated carbon dispersed on silica gel. The extracts were adsorbed from a hexane solution onto the carbon, the column was washed with cyclohexane-dichloromethane to remove nonplanar aromatics and the planar aromatics such as PCDDs and PCDFs eluted with toluene (U.S. EPA, 1983[12]).

Measurement of the PCDDs and PCDFs in sample extracts was effected by GC-MS. The GC was a Varian 3700 fitted with a 25-m DB-5 (J&W, 0.25 mm ID) bonded-phase fused-silica capillary column. The column was heated in steps rapidly from 80°C to 180°C (ca. 2 min) and then more slowly to 280°C at 5°C per min such that 2,3,7,8-TCDD and OCDD eluted at about 11 and 22 min, respectively. The DB-5 nonpolar column enabled the latest eluting isomer of either PCDDs or PCDFs within a specific congener group to be detected before the elution of the earliest isomer from the next higher congener group. The GC was coupled directly to a VG-Analytical ZAB-2F MS operating at low resolution (2000 at 10% valley). Compounds eluting from the GC were ionized by electron impact and detected by selected ion monitoring of each congener group. Quantitation of analytes in extracts was achieved by comparing the response from sample extracts to that of external standards injected either before or after. Detection limits were about 2 ppt for most analytes, depending on chemical background signals and amount of sample used, except for that of OCDD, which, due to higher reagent blanks from chemical background and lower recoveries, was about 10 ppt. Precision in the measurement was about 15% for most analytes, except near the detection limit, where it was greater, and except for OCDD, which was also greater (20–30%). Method performance was monitored by: the simultaneous processing of laboratory reagent blanks in each set of five samples; the use of an extensive mixture of isotopically labeled internal standards; participation in interlaboratory comparison of values obtained on the same samples; and duplication of complete analysis on selected samples. Confirmation of some samples was carried out by one of the following: (1) further purification of sample extracts on high-performance liquid chromatography (HPLC) (Ryan and Pilon[13]) followed by GC-MS as outlined above; (2) the use of a Sciex TAGA 6000 triple quadrupole MS/MS coupled to a short (12-15m) bonded-phase DB-5 silica column with fast temperature programming (GC cycle time between 12 and 15 min) and monitoring the loss of COCl for each congener; (3) high-resolution (10,000 at 10% valley) MS on the VG instrument coupled to a polar CP Sil 88 (equivalent to an SP-2340) GC capillary column, which has the greatest ability to resolve isomers of PCDDs and PCDFs. With the number of pure standards and their known characteristics on several different types of GC capillary columns available, the method described is isomer specific for all seventeen 2,3,7,8-chlorine-substituted PCDDs and PCDFs except 2,3,7,8-TCDF.

RESULTS

PCDDs and PCDFs were found in all ten tissue types from subjects 1, 2 and 3 (Tables 3, 4 and 5, respectively). As has been noted in a survey of human adipose tissues (Ryan et al.[8]), only PCDDs and PCDFs with 2,3,7,8-chlorine substitution were detected. On a wet tissue basis PCDDs were higher in concentration than PCDFs, with OCDD highest, and values where found decreasing from HpCDD, HxCDD, PnCDD to TCDD. PCDFs were more uniform in their congener distribution with levels of PnCDFs, HxCDFs and HpCDFs being of the same order of magnitude. TCDF and OCDF were not found in these tissues.

Highest concentrations of all analytes were found in adipose tissue. No major differences were seen between abdominal and subcutaneous fat samples or between these two types and perirenal fat when considering the lower lipid content of the latter. Smaller concentrations of PCDDs and PCDFs were measured on a wet weight basis in decreasing order: adrenal, bone marrow, liver, muscle, spleen, kidney and lung. In all fat samples, 2,3,7,8-TCDD was found at levels which would appear to constitute background levels for North Americans (Ryan et al.[1,8], and Schecter et al. [7,14]). Smaller levels of 2,3,7,8-TCDD near the detection limits were also found in bone marrow, both adrenals, two of three livers, one of three muscles and one of three spleens. OCDD was found in all tissues irrespective of type, although in some cases the values found were near the detection limit of about 10 pg/ g. In all cases the relative amounts of the 2,3,7,8-chlorine-substituted PCDDs and PCDFs were approximately similar; e.g., if 1,2,3,6,7,8-

Table 3. 2,3,7,8-Chlorine-Substituted PCDDs and PCDFs in Human Autopsy Number 1 (Nine Tissues); Values in pg/g on a Wet Tissue Basis.

Analyte	Fat		Adrenal	Bone Marrow	Liver	Muscle	Spleen	Kidney	Lung
	Abd.[a]	Subc.[b]							
2378-D[c]	5.7	6.0	3.8	ND[e]	ND	ND	ND	ND	ND
23478-F[d]	17	17	4.9	4.4	ND	1.1	ND	ND	ND
12378-D	7.8	8.2	3.1	12	ND	1.2	11	ND	ND
123478/123678-F	52	22	8.5	9.4	4.2	2.1	ND	2.1	ND
123678-D	64	60	35	30	6.5	7.9	1.9	2.5	1.4
1234678-F	15	12	3.5	2.7	2.1	ND	ND	ND	ND
1234678-D	110	120	55	48	22	14	13	5.2	2.9
12346789-D	680	700	600	540	220	170	46	31	21
% Lipid	75	75	28	26	6.0	9.0	1.8	3.0	2.2

[a] abdominal
[b] subcutaneous
[c] PCDD
[d] PCDF
[e] not detected

Table 4. 2,3,7,8-Chlorine Substituted PCDDs and PCDFs in Human Autopsy Number 2 (Eight Tissues); Values in pg/g on a Wet Tissue Basis.

Analyte	Fat			Adrenal	Liver	Muscle	Spleen	Kidney
	Abd.[a]	Subc.[b]	Peri.[c]					
2378-D[d]	7.4	3.7	1.4	3.7	2.5	ND[f]	1.3	ND
23478-F[e]	14	16	6.5	5.5	5.4	ND	ND	ND
12378-D	10	7.5	5.2	4.8	3.7	1.5	1.1	ND
123478/	29	38	ND	11	17	1.7	ND	3.0
123678-F								
123678-D	61	61	66	39	48	5.8	1.5	4.0
1234678-F	17	24	ND	4.5	7.7	ND	ND	1.7
1234678-D	110	93	53	36	44	5.0	4.7	13
12346789-D	430	590	210	210	350	76	20	39
% Lipid	71	67	46	25	22	7.0	1.7	4.0

[a] abdominal
[b] subcutaneous
[c] perirenal
[d] PCDD
[e] PCDF
[f] not detected

Table 5. 2,3,7,8-Chlorine-Substituted PCDDs and PCDFs in Human Autopsy Number 3 (Three Tissues); Values in pg/g on a Wet Tissue Basis.

Analyte	Fat, Abd[a]	Liver	Muscle
2378-D[b]	8.4	4.6	2.5
23478-F[c]	13	6.9	2.3
12378-D	11	4.5	1.3
123478/123678-F	21	8.5	3.4
123678-D	130	56	21
1234678-F	4.9	ND[d]	ND
1234678-D	84	37	16
12346789-D	560	180	120
Lipid %	69	44[e]	17

[a]abdominal
[b]PCDD
[c]PCDF
[d]not detected
[e]in preservative

HxCDD were high in a particular tissue, then other PCDDs and PCDFs such as 2,3,4,7,8-PnCDF and OCDD were also high, and vice versa. This property, plus the presence of only 2,3,7,8-chlorine-substituted PCDDs and PCDFs (Table 6), resulted in an isomeric pattern which was distinctive for human tissues and differed from that found in wildlife samples (mostly tetra- and penta-isomers) and chicken and pork food samples (mostly hexa-, hepta- and octa-isomers).

The concentrations listed in Tables 3, 4 and 5 for the three subjects are based on a wet weight basis. Since the lipid content of these tissues varied between 2 and 75%, the concentrations of total PCDDs (tetra- to octa-) and total PCDFs (penta-, hexa- and hepta-) can be expressed on a lipid basis as noted in Tables 7, 8 and 9 for subjects 1, 2 and 3, respectively. On a wet tissue basis the ratio between the highest and lowest value of either total

Table 6. 2,3,7,8-Chlorine Substituted PCDDs and PCDFs in Human Adipose Tissues; Number of Positives in Presence Column Indicates Relative Amounts Present.

Congener Group	PCDD		PCDF	
	Isomer	Presence[a]	Isomer	Presence[a]
Tetra	2378	+	2378	
Penta	12378	+	12378	−
Hexa	123478	+	123478	+
	123678	+ +	123678	+
	123789	+	123789	−
			234678	+
Hepta	1234678	+ +	1234678	+
			1234789	+
Octa	12346789	+ + +	12346789	−
Total	7	7	10	6

[a]–usually not present; + usually between 3 and 30 pg/g; + + usually between 30 and 150 pg/g; + + + usually over 150 pg/g.

Table 7. Total PCDD and PCDF Levels (pg/g) in Tissue Samples from Autopsy Number 1.

Tissue	% Lipid	Total PCDDs		Total PCDFs	
		Wet Basis	Lipid Basis	Wet Basis	Lipid Basis
Fat, abdominal	75	870	1160	84	110
Fat, subcutaneous	75	890	1180	51	68
Adrenal	28	690	2440	17	60
Bone marrow	26	630	2440	17	63
Liver	6.0	250	4220	6.3	110
Muscle	9.0	190	2160	3.2	36
Spleen	1.8	72	3980	3.2	180
Kidney	3.0	39	1290	2.1	70
Lung	2.2	25	1150	–	–

PCDDs, total PCDFs or individual isomers in the various tissues is between 20 and 50. However, on a lipid basis this same ratio for the various tissues decreases by about an order of magnitude. On a wet tissue basis, adipose tissue has the highest concentration (total PCDDs, 600 to 900 pg/g) and, on a lipid basis, liver has the highest levels and other tissue levels are more similar.

Table 8. Total PCDD and PCDF Levels (pg/g) in Tissue Samples from Autopsy Number 2.

Tissue	% Lipid	Total PCDDs		Total PCDFs	
		Wet Basis	Lipid Basis	Wet Basis	Lipid Basis
Fat, abdominal	71	620	870	60	85
Fat, subcutaneous	67	750	1120	78	120
Fat, perirenal	46	340	740	6.5	14
Adrenal	25	290	1170	21	83
Liver	22	450	2040	30	140
Muscle	7.0	87	1240	1.7	24
Spleen	1.7	29	1680	–	–
Kidney	4.0	56	1410	4.7	120

Table 9. Total PCDD and PCDF Levels (pg/g) in Tissue Samples from Autopsy Number 3.

Tissue	% Lipid	Total PCDDs		Total PCDFs	
		Wet Basis	Lipid Basis	Wet Basis	Lipid Basis
Fat, abdominal	69	790	1160	38	56
Liver	44[a]	280	640	15	35
Muscle	17	160	950	5.7	34

[a]Presence of formaldehyde preservative may have affected this value.

DISCUSSION

This is the first report to show that about ten 2,3,7,8-chlorine-substituted PCDDs and PCDFs are present not only in adipose tissue from the general or normal North American population but also in all other tissues examined. While highest levels of the analytes on a wet basis were detected in adipose tissue, significant levels, including the toxic 2,3,7,8-TCDD isomer, were also measured in adrenal, bone marrow and some livers. Expression of the concentrations on a lipid basis rather than a wet weight basis caused the variation in the levels among tissues to decrease markedly. Hence the lipid content of the tissue appears to be a major factor in determining the degree of contamination. This property, plus the uniformity of the relative proportions of the individual PCDD and PCDF isomers, would suggest the contaminants are being stored in the lipid and not undergoing rapid metabolism and elimination. For example, in those cases where the liver had a high fat content (autopsies 2 and 3) high levels of PCDDs and PCDFs were also detected. Hence at the level found in the general population, there does not appear to be a single organ which contains most of the contaminants; rather, the PCDDs and PCDFs are generally spread throughout the body lipid.

The distributional data on the general population presented here can be compared to those obtained from accidental or occupational exposure to PCDDs and PCDFs published in the literature, all of which are given on a wet basis. Kuroki and Masuda[15] analyzed liver and fat tissues for five PCDFs from patients who had died in the Yusho poisoning. In the three subjects for which both liver and fat were available, and who died one, four and nine years after consumption of the oil, levels of the five PCDFs were three to ten times higher in fat than in liver, with values greater than ng/g in several instances. In the Seveso accident in Italy in 1976, Facchetti et al.[16] analyzed tissues from a woman who died seven months later, and they detected highest levels of 2,3,7,8-TCDD in the fat and pancreas (1840 and 1040 pg/g, respectively), followed by liver (150 pg/g) and smaller concentrations in kidney, lung, and brain. However, the tendency for fat to have higher values than liver on a wet weight basis in our results and in the Japanese and Italian incidents described above was not found in two other cases. Rappe et al.[17] reported tissue values from a baby who died after exposure to a toxic oil in Taiwan (Yucheng incident). Higher levels of five PCDFs (up to 200 pg/g) were found in liver rather than fat — levels which were generally much lower than the Yusho cases from older subjects cited above.[15] The neonate case from Taiwan was further distinctive in that 1,2,3,7,8-PnCDF was also found at a higher level than 2,3,4,7,8-PnCDF, whereas in the Yusho incident the level of 1,2,3,7,8-PnCDF, although present in the toxic rice oil, was always low in adult tissues. These differences of level and isomer type in the neonate case may reflect the reduced metabolic capacity and increased sensitivity to toxic insult in the young. An additional human case from the Yucheng incident (Chen and Hites[18]) resulted in the analysis of eleven tissues from a single patient of undisclosed age who died probably from the poisoning itself some two years after exposure. Analysis for three PCDFs showed liver concentrations on a wet tissue basis between 3 and 25 ng/g which were two to three times higher than the next tissue, adipose, with significantly smaller amounts in other tissues, such as kidney, spleen and lung. Certainly our results and those from accidental exposure show that adipose and liver are two of the most important tissues as far as body burden is concerned. The question of whether liver levels are higher than fat levels on a wet weight basis when a relatively high dose is given has not yet been resolved and the subject will require more study.

The distributional results reported here for PCDDs and PCDFs in the general population appear to be similar to those found in experimental animals dosed with these compounds. As summarized by Neal et al.[19] for 2,3,7,8-TCDD in mammals, with all values on a wet weight basis, fat and liver are the two most important tissues as far as level is concerned. Generally, fat contains somewhat higher levels than liver, except for a recent exposure or a high dose. To illustrate, Kociba et al.[20] dosed laboratory rats with 2,3,7,8-TCDD at three levels (1, 10 and 100 ng/kg/day) continuously for two years. At the lowest dose, fat and liver levels were similar, and at the two higher doses, liver values were about three times higher than those of fat. In a study of beef cattle, Jensen et al.[21] fed a diet containing 24 ppt 2,3,7,8-TCDD (about 0.8 ng/kg/day) for 28 days and detected a fat level of 80 pg/g and a liver level of about 10 pg/g. However, when the dose was greater, between 0.2 and 22 μg/kg/day of the higher chlorinated dioxins and furans, as reported in the pentachlorophenol feeding study in dairy cattle by Parker et al.[22], residue levels in liver were 2 to 50 times higher than in fat. Studying a species closely related to man, McNulty et al.[23] gave a single oral dose of 1 μg/kg of 2,3,7,8-TCDD to an adult rhesus monkey. After 2 years they found 100 and 15 pg/g in the fat and liver, respectively. There is limited information on the distribution of tissue levels of the other PCDDs and PCDFs in experimental animals.

CONCLUSION

This is the first report to show that several 2,3,7,8-chlorine-substituted PCDDs and PCDFs are present not only in adipose tissues from the general population, but also in all other tissues assayed. The ratios of the PCDDs and PCDFs to each other appear to be approximately the same for each tissue, with overall levels on a wet weight basis decreasing in order: fat, adrenal, bone marrow, liver, muscle, spleen, kidney and lung. The first tissue, fat, is probably a storage tissue, while some of the others could be target organs. If the levels are expressed on a lipid basis rather than a wet weight basis, liver has the highest values and the variation between tissues shows only a two- to fourfold difference. Human tissue data from both the general population and from accidental cases, as well as data from experimental animals, show that levels in adipose and liver tissues contain the highest levels. The question as to whether liver levels increase relative to fat levels on a wet basis in certain instances is not clear.

ACKNOWLEDGMENTS

The authors are grateful to Lawrence Miller, Upstate Medical School, State University New York, Syracuse, for aid in sample collection. Luz Panopio of Health Protection Branch, Ottawa, and Fred Hileman, Monsanto, Dayton, Ohio, are thanked for their contributions in sample purification. The gift of several analytical standards of PCDFs by Yoshito Masuda, Daiichi College, Fukuoka, Japan, is gratefully acknowledged.

REFERENCES

1. Ryan, J. J., D. T. Williams, B. P.-Y. Lau, and T. Sakuma. "Analysis of Human Fat Tissue for 2,3,7,8-Tetrachlorodibenzo-*p*-dioxin and Chlorinated Dibenzofuran Residues," in *Chlorinated Dioxins and Dibenzofurans in the Total Environment II*, L. H. Keith, C. Rappe, and G. Choudhary, Eds. (Stoneham, MA: Butterworth Publishers, 1985), pp. 205–214.
2. Miyata, H., T. Kashimoto, and N. Kunita. "Detection and Determination of Polychlorodibenzofurans in Normal Human Tissues and Kanemi Rice Oils Caused 'Kanemi Yusho,' " *J. Food Hygiene Soc. Japan* 18(3):260–265 (1977).
3. Gross, M. L., J. O. Lay, P. A. Lyon, D. Lippstreu, N. Kangas, R. L. Harless, S. E. Taylor, and A. E. Dupuy. "2,3,7,8-Tetrachlorodibenzo-*p*-dioxin Levels in Adipose Tissue of Vietnam Veterans," *Environmental Res.* 33:261–268 (1984).
4. Lee, L. E., and L. B. Hobson. "2,3,7,8-Tetrachlorodibenzo-*p*-dioxin (TCDD) in Body Fat of Vietnam Veterans and Other Men," in *Chlorinated Dioxins and Dibenzofurans in the Total Environment II*, L. H. Keith, C. Rappe, and G. Choudhary, Eds. (Stoneham, MA: Butterworth Publishers, 1985), pp. 197–204.
5. Young, A. L. "Analysis of Dioxins and Furans in Human Adipose Tissue," in *Public Health Risks of the Dioxins*, W. W. Lowrance, Ed. (Los Altos, W. Kaufmann, 1984), pp. 63–75.
6. Graham, M., F. Hileman, D. Kirk, J. Wendling, and J. Wilson. "Background Human Exposure to 2,3,7,8-TCDD," *Chemosphere* 14(6/7):925–928 (1985).

7. Schecter, A., T. O. Tiernan, M. L. Taylor, G. F. Van Ness, J. H. Garrett, D. J. Wagel, G. Gitlitz, and M. Bogdasarian. "Biological Markers After Exposure to Polychlorinated Dibenzo-*p*-dioxins, Dibenzofurans, Biphenyls, and Biphenylenes – Part I: Findings Using Fat Biopsies to Estimate Exposure," in *Chlorinated Dioxins and Dibenzofurans in the Total Environment II*, L. H. Keith, C. Rappe, and G. Choudhary, Eds. (Stoneham, MA: Butterworth Publishers, 1985), pp. 215–245.

8. Ryan, J. J., R. Lizotte, and B. P.-Y Lau. "Chlorinated Dibenzo-*p*-dioxins and Chlorinated Dibenzofurans in Canadian Human Adipose Tissue," *Chemosphere* 14(6/7):697–706 (1985).

9. Rappe, C. "Chemical Analyses of Adipose Tissues," in *Public Health Risks of the Dioxins*. W. W. Lowrance, Ed. (Los Altos, CA: W. Kaufmann, 1984) pp. 57–61.

10. Ryan, J. J., A. Schecter, R. Lizotte, W.-F. Sun, and L. Miller. "Tissue distribution of Dioxins and Furans in Humans from the General Population," *Chemosphere* 14(6/7):929–932 (1985).

11. Ryan, J. J., R. Lizotte, and W. H. Newsome. "Study of Chlorinated Diphenyl Ethers and Chlorinated 2-Phenoxyphenols as Interferences in the Determination of Chlorinated Dibenzo-*p*-dioxins and Chlorinated Dibenzofurans in Biological Samples," *J. Chromatogr.* 303:351–360 (1984).

12. "Determination of 2,3,7,8-TCDD in Soil and Sediment," U.S. Environmental Protection Agency (EPA), Region VII Laboratory Branch, Kansas City, Kansas, Sept. 1983.

13. Ryan, J. J., and J. C. Pilon. "High-Performance Liquid Chromatography in the Analysis of Chlorinated Dibenzodioxins and Dibenzofurans in Chicken Liver and Wood Shavings Samples," *J. Chromatogr.* 197:171–180 (1980).

14. Schecter, A., J. J. Ryan, R. Lizotte, W.-F. Sun, L. Miller, G. Gitlitz, and M. Bogdasarian. "Chlorinated Dibenzodioxins and Dibenzofurans in Human Adipose Tissue from Exposed and Control New York State Patients," *Chemosphere* 14(6/7):933–938 (1985).

15. Kuroki, H., and Y. Masuda. "Determination of Polychlorinated Dibenzofuran Isomers Retained in Patients with Yusho," *Chemosphere* 7(10): 771–777 (1978).

16. Facchetti, S., A. Fornari, and M. Montagna. "Distribution of 2,3,7,8-Tetrachlorodibenzo-*p*-dioxin in the Tissues of a Person Exposed to the Toxic Cloud at Seveso," *Adv. Mass Spectrometry* 8B:1405–1414 (1980).

17. Rappe, C., M. Nygren, and G. Gustafsson. "Human Exposure to Polychlorinated Dibenzo-*p*-dioxins and Dibenzofurans," in *Chlorinated Dioxins and Furans in the Total Environment*, G. Choudhary, L. Keith, and C. Rappe, Eds. (Stoneham, MA: Butterworth Publishers, 1983), pp. 355–365.

18. Chen, P. H., and R. A. Hites. "Polychlorinated Biphenyls and Dibenzofurans Retained in the Tissues of a Deceased Patient with Yucheng in Taiwan," *Chemosphere* 12 (11):1507–1516 (1983).

19. Neal, R. A., J. R., Olson, T. A. Gasiewicz, and L. E. Geiger. "The

Toxicokinetics of 2,3,7,8-Tetrachlorodibenzo-*p*-dioxin in Mammalian Systems." *Drug Metabolism Rev.* 13(3):355–385 (1982).

20. Kociba, R. J., D. G. Keyes, J. E. Beyer, R. M. Carreon, C. E. Wade, D. A. Dittenber, R. P. Kalnins, L. E. Frauson, C. N. Park, S. D. Barnard, R. A. Hummel, and C. G. Humiston. "Results of a Two Year Chronic Toxicity and Oncogenicity Study of 2,3,7,8-Tetrachlorodibenzo-*p*-Dioxin in Rats," *Toxicol. Applied Pharmacol.* 46:279–303 (1978).

21. Jensen, D. J., R. A. Hummel, N. H. Nahle, C. W. Kocher, and H. S. Higgins. "A Residue Study on Beef Cattle Consuming 2,3,7,8-Tetrachlorodibenzo-*p*-dioxin," *J. Agric, Food Chem.* 29:265–268 (1981).

22. Parker, C. E., W. A. Jones, H. B. Matthews, E. E. McConnell, and J. R. Hass. "The Chronic Toxicity of Technical and Analytical Pentachlorophenol in Cattle. II. Chemical Analyses of Tissues," *Toxicol. Applied Pharmacol.* 55:359–369 (1980).

23. McNulty, W. P., K. A. Nielsen-Smith, J. O. Lay, D. L. Lippstreu, N. L. Kangas, P. A. Lyon, and M. L. Gross. "Persistence of TCDD in Monkey Adipose Tissue," *Food Chem. Toxicol.* 20:985–986 (1982).

CHAPTER 2

Identification of 2,3,7,8-substituted Polychlorinated Dioxins and Dibenzofurans in Environmental and Human Samples

Martin Nygren, Christoffer Rappe, Gunilla Lindström, Marianne Hansson,
Per-Anders Bergqvist, Stellan Marklund, Lennart Domellöf, Lennart Hardell,
and Mats Olsson

INTRODUCTION

Polychlorinated dibenzo-p-dioxins (PCDDs) and dibenzofurans (PCDFs) are two series of tricyclic, almost planar aromatic compounds that exhibit very similar physical, chemical, and biological properties. These compounds have been the subject of much concern in many countries in recent years. The number of chlorine atoms in these compounds can vary between one and eight to produce 75 PCDD and 135 PCDF positional isomers.

Because of the extreme toxicity of some of the PCDD and PCDF isomers (Table 1) as well as the large variation in toxicity between closely related congeners, highly selective, specific, and sensitive analytical techniques are required for the measurements. Detection levels in ecological and human samples should be orders of magnitude below the usual detection levels obtained in pesticide analyses. A detection level of 1 pg (10^{-12} g) or less might be required to find 2,3,7,8-tetra-CDD and the other congeners listed in Table 1 in a l-g sample (1 part per trillion, ppt, or sub-ppt). Analyses at such extremely low levels are complicated by the presence of a multitude of other interfering compounds.

Table I.   The Most Toxic PCDD and PCDF Isomers.

2,3,7,8-tetra-CDD	2,3,7,8-tetra-CDF
1,2,3,7,8-penta-CDD	1,2,3,7,8-penta-CDF
1,2,3,6,7,8-hexa-CDD	2,3,4,7,8-penta-CDF
1,2,3,7,8,9-hexa-CDD	1,2,3,6,7,8-hexa-CDF
1,2,3,4,7,8-hexa-CDD	1,2,3,7,8,9-hexa-CDF
	1,2,3,4,7,8-hexa-CDF
	2,3,4,6,7,8-hexa-CDF

In recent years many analytical methods have been developed for the analysis of trace amounts of PCDDs and PCDFs in environmental samples, especially for the most toxic 2,3,7,8-tetra-CDD. The most specific of these methods are based on mass spectrometry; for recent reviews on the analysis of PCDDs and PCDFs, see references 1–4.

The dioxin problem was first recognized because of teratogenic effects found with the phenoxy herbicide 2,4,5-trichlorophenoxyacetic acid (2,4,5-T). These effects were later found to be caused by 30 ppm of 2,3,7,8-tetra-CDD present in this particular sample. Other products known to be contaminated by PCDDs and/or PCDFs are chlorinated phenols, diphenyl ethers and other chlorophenoxy herbicides, polychlorinated biphenyls (PCBs) and hexachlorophene.

In 1977 Olie et al.[5] reported the occurrence of PCDDs in fly ash from municipal incinerators. Buser et al.[6,7] identified and quantified a number of PCDD and PCDF isomers; in fly ash samples the total amount of PCDDs and PCDFs was found to be about 1 ppm. During the period 1978–84 a number of papers, reports, and reviews were published confirming the original findings.

We have recently analyzed a series of samples of particulate and flue gas condensate and adsorbents from municipal incinerators and hazardous waste incinerators in Sweden, Denmark and Canada.[8,9] In all these samples we found a multitude of PCDD and PCDF congeners at levels corresponding to $\mu g/m^3$–ng/m^3. The 2,3,7,8-tetra-CDD was always found to be a very minor peak, whereas in all samples the 1,2,3,7,8-penta-CDD was found to be a peak of "medium" size. The toxic 2,3,7,8-substituted hexa-CDDs and PCDFs were always medium or major components.

From a toxicological point of view 2,3,7,8-tetra-CDD is the most studied and probably the most potent of all these 210 compounds. This compound is considered to be one of the most toxic chemicals known to man; for the most sensitive species its LD_{50} is approximately 1 $\mu g/kg$ body weight. The spectrum of toxic effects is species dependent but for humans they include chloracne and porphyria cutane tarda and for other animals edema, thymic atrophy, teratogenicity, liver lesions and a slow wasting syndrome followed by death. An identical spectrum of toxicity has been observed for some of the other PCDDs and PCDFs, although they are less potent.[10,11]

The 2,3,7,8-tetra-CDD and a mixture of 1,2,3,6,7,8- and 1,2,3,7,8,9-hexa-CDDs have been tested for carcinogenicity in rats and mice by administering them in the diet and also by gavage. Neoplastic effects were found for extremely low doses (1 ng/kg body weight) of 2,3,7,8-TCDD.[12,13]

Kimbrough et al.[14] have evaluated the human health risk of varying concentrations of 2,3,7,8-tetra-CDD in soil. The virtually safe dose by the linear derived multistage model using an added cancer risk of $1/10^6$ is calculated to be in a dose range of 28 fg/kg body weight per day (0.028 pg) to 1428 fg/kg body weight per day. This calculation is based on data for hepatocellular carcinoma or neoplastic nodules. The increased risk of $1/10^6$ based on data for tissues less sensitive than liver would not be expected to occur until doses as high as 1428 pg/kg body weight per day were administered.

The bioconcentration of 2,3,7,8-tetra-CDD in various aquatic species has been studied under controlled laboratory conditions using static test chambers. Bioconcentration factors as high as 15,000 have been reported by Isensee.[15]

Recent studies by McConnell et al.[16] have shown that guinea pigs and rats given soil contaminated by 2,3,7,8-tetra-CDD by gavage develop characteristic dioxin syndroms. The presence of 2,3,7,8-tetra-CDD in the livers of both species shows that 2,3,7,8-tetra-CDD in soil exhibits high biological availability after ingestion.

Rappe et al.[17] identified a series of 2,3,7,8-substituted tetra- to octa-CDF in fat samples of a snapping turtle from the Hudson River and a gray seal from the Baltic Sea. The total level of PCDFs were 3000 pg/g and 40 pg/g, respectively. Norstrom et al.[18] have analyzed herring gull eggs from the Great Lakes and reported levels of 2,3,7,8-tetra-CDD ranging from 9 to 90 pg/g. Stalling et al.[19] have reported on the occurrence of 2,3,7,8-substituted PCDDs and PCDFs in fish samples from Lake Michigan, Lake Huron and Lake Ontario.

In the present study, we report on the analyses of a series of samples of human adipose tissue and mother's milk as well as a series of environmental samples both aquatic and terrestrial. One objective of the study is to identify background levels of PCDDs and PCDFs in these samples, especially those of human origin. Another objective is to investigate if trace analysis of tissue samples can be used to estimate previous exposure to PCDDs and PCDFs or to products contaminated by these compounds.

MATERIAL

About 10 g of subcutaneous adipose tissue was excised from 13 persons exposed to phenoxy herbicides (11 with neoplastic disease; 5 malignant lymphomas, 3 soft-tissue sarcomas) and 18 nonexposed controls (12 submitted to cholecystectomy and 6 cancer cases). All 31 patients were patients at the Regional Hospital in Umeå, Sweden and they were also carefully interviewed regarding possible exposure. In 1984 the same amount of fat was obtained from a German worker highly exposed to 2,3,7,8-tetra-CDD in the BASF factory (Ludwigshafen, Germany) in November 1953.[20] Since this accident he has suffered from severe chloracne for more than 10 years. Another fat sample was obtained from a chemist, who synthezised more than 80 different PCDF isomers during the last years before the biopsy was taken.

The human milk samples were collected from volunteers in Sweden, Denmark and West Germany. During a visit to Vietnam in January 1983 about 20 mL of milk was collected from each of four mothers living in a province heavily sprayed with Agent Orange during the military operations in the 1960s. These mothers report that their diet primarily consists of fish caught in the local rivers. The four milk samples were pooled before analysis.

Samples of fat and liver from cows were obtained from the province of Närke, Sweden. These samples should represent a background contamination. Other bovine fat samples were obtained from a farm in Scotland, with potential contamination from a hazardous waste and a municipal solid waste incinerator. From the same place we obtained a sample of cow milk. Commercial cow milk was collected in Denmark and commercial cream from the province of Skåne, Sweden.

In the present study we have also chosen to include the analysis of five samples from the Baltic Sea and from the Gulf of Bothnia: (1) an osprey egg from Holmöarna, Sweden (Gulf of Bothnia) collected in 1981, (2) ringed seal fat from Simo, Finland (Gulf of Bothnia) collected in 1978, (3) salmon muscle, from the Ume River, Sweden collected in 1978, (4) a guillemot muscle pooled sample from 10 birds, from Stora Karlsö, Sweden (Baltic Sea) collected in 1971, and (5) a herring muscle pooled sample from 15 fishes, from Utlängan, Sweden (Baltic Sea) collected in 1981.

EXPERIMENTAL

For the analyses of human adipose tissue and the samples of animal fat or milk fat discussed in the present study, we have used the method for sample extraction and containment enrichment discussed in detail by Stalling et al.[19] and Smith et al.[21] This procedure involves three columns in sequence and the key step in this procedure is a column of carbon dispersed on glass fibers. This column adsorbs most planar polychlorinated polynuclear aromatics, the major portion of biological coextractives is, however, not retained. Along with similar chemicals, the PCDDs and PCDFs are then removed from the carbon column by reverse elution with toluene.

Due to differences between Europe and the United States in column material and solvents, some slight modifications of the Stalling methods have been introduced in our laboratory (data to be published). This modified method has recently been verified and validated by analyzing blanks and fish samples fortified at two different levels,[9] and also by analyzing samples of bovine milk fortified at three different levels (data to be published). All the samples analyzed in the present study are fortified at 20–25 ppt with $^{13}C_{12}$-2,3,7,8-tetra-CDD, $^{13}C_{12}$-2,3,7,8-tetra-CDF and $^{13}C_{12}$-octa-CDD. The recovery of each congener was investigated, and normally it was found to be higher than 60%.

In the analysis of the milk and cream samples, the milk fat was extracted prior to the application to the column system. The milk sample (about 200 mL) was mixed with a solution prepared from 2 g of sodium oxalate dissolved in 50 mL of boiling water and added to 250 mL of ethanol. The milk suspension was extracted first with a mixture of 185 mL of diethyl ether and 265 mL of n-hexane and then twice with 100 mL of n-hexane. The combined extracts were washed twice with water and dried by passing through a column of sodium sulfate, evaporated until constant weight and the amount of fat noted.

A 60–m SP 2330 fused-silica column (i.d. 0.26 mm) was used for the isomer-specific separation.[3] The column leads directly into the ion source of the MS instrument. For the determination of 2,3,7,8-tetra-CDD we used a Finnigan 8200 or VG 70–250 (EI) operating at a resolution of 5000–8000. For the higher PCDDs and CL_4-Cl_8 PCDFs we used an updated Finnigan 4500 operating in negative chemical ionization (CH_4) mode. The detection levels are in the range of 0.1–1 pg of each isomer per injection.

Tondeur et al.[22] have reported on a matrix effect with potentially severe consequences during the analysis of trace quantities of 2,3,7,8-tetra-CDD when working with contaminated extracts. The extracts in the present study were found to be very clean; the PCDDs and PCDFs normally constitute the major peaks in the mass fragmentograms. (See Figures 1, 2 and 3, which are the fragmentograms for human adipose tissue mother's milk and guillemot muscle, respectively.)

The criterion for a positive identification of a specific PCDD and PCDF congener is a signal-to-noise ratio greater than 3, the chlorine cluster ± 10% from the theoretical value (normally ± 5% or less) and coelution with $^{13}C_{12}$-labeled compounds or synthetic standards prepared in our laboratory or commercially available. The separation on the GC column was routinely checked by analyzing performance mixtures. Special attention is devoted to known artifacts.[4]

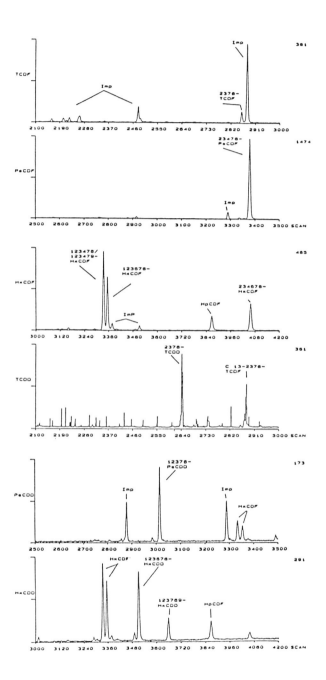

Figure 1. Fragmentograms for tetra-, penta- and hexa-CDFs and CDDs in human adipose tissue.

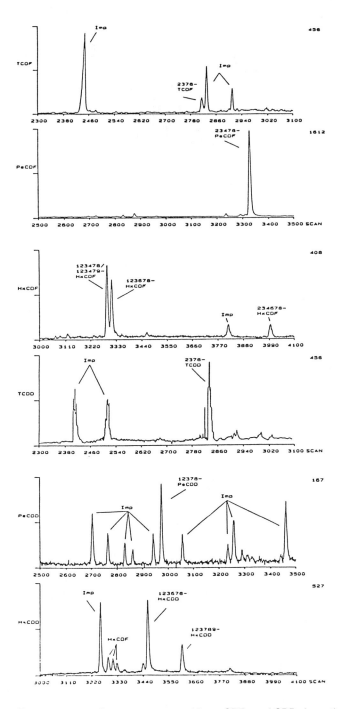

Figure 2.   Fragmentograms for tetra-, penta- and hexa-CDFs and CDDs in mother's milk.

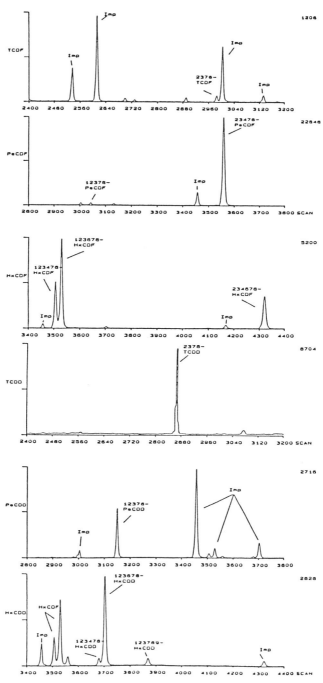

Figure 3. Fragmentograms for tetra-, penta- and hexa-CDFs and CDDs in guillemot muscle.

RESULTS

Adipose Tissue

We have analyzed more than 30 samples of human adipose tissue, and the results of these analyses are given in Table 2. For patient 1 we received sufficient adipose tissue so that three different aliquots were cleaned up and analyzed separately. In the table we report on the results from these triplicate analyses. We also report on the mean values for all cases (n = 31) as well as for subgroups like the cancer patients, noncancer controls, exposed and nonexposed patients. In Table 2 we have also included the results from the German BASF worker and the chemist.

In Figure 1 we have collected representative fragmentograms from the analysis of a human adipose tissue. It is important to point out that all the major peaks are 2,3,7,8-substituted PCDDs and PCDFs. The incorrect isotope clusters revealed that the major peaks in the tetra-CDF channel are artifacts.

In the tetra-CDD channel a response for the ^{13}C-2,3,7,8-tetra-CDF could also be observed ($M^+ + 4 = 320$). This indicates that the mass spectrometer was operating at a resolution of about 6000. A resolution of greater than 8000 could be checked by the absence of the ^{13}C-2,3,7,8-tetra-CDF in this channel:

$$
\begin{array}{lll}
\text{2,3,7,8-tetra-CDD} & (M^+) & = 319.8965 \\
{}^{13}C_{12}\text{-2,3,7,8-tetra-CDF} & (M^+ + 4) & = 319.9360
\end{array}
$$

Mother's Milk

The results from the analyses of mother's milk samples are given in Table 3. It shows the results from individual samples as well as the range found for the Swedish samples and the samples from Germany. The sample from Vietnam is a composite sample from four mothers.

In Figure 2 we have collected the fragmentograms from the analysis of a sample of mother's milk from Sweden. As in Figure 1, all the major peaks are 2,3,7,8-substituted PCDDs and PCDFs. The major peaks in the tetra-CDF channel are artifacts: they were found to have incorrect Cl isotope cluster.

Environmental Samples

The results from the environmental samples are given in Tables 4 and 5. In these tables, we have included aquatic organisms (Table 4) as well as terrestrial animals (Table 5).

In Figure 3 we have collected representative fragmentograms from the

Table 2. Levels of PCDDs and PCDFs Found in Human Adipose Tissue.[a]

Compound	Adipose Pat. 1	Adipose Pat. 1	Adipose Pat. 1	Mean Value n=31	Range	Mean Value Exposed n=13	Range	Mean Value Nonexposed n=18	Range	Mean Value Cancer Pat. n=17	Range	Mean Value Noncancer n=14	Range	German Worker	Chemist
2,3,7,8–TCDD	NA[b]	3	7	3	0–9	2	0–9	3	2–6	3	2–9	3	2–6	100	NA
1,2,3,7,8–PeCDD	NA	24	30	10	3–24	6	3–24	9	4–18	9	4–24	9	3–18	18	5
1,2,3,4,7,8–HxCDD	ND[c]	ND	ND	ND										ND	ND
1,2,3,6,7,8–HxCDD	15	18	21	15	3–55	19	8–55	12	3–18	18	3–55	12	8–18	48	12
1,2,3,7,8,9–HxCDD	5	3	5	4	3–5	5	3–13	4	3–5	4	3–13	4	3–5	12	5
1,2,3,4,6,7,9–HpCDD	ND	ND	ND	ND										ND	ND
1,2,3,4,6,7,8–HpCDD	36	28	32	97	12–380	104	20–380	85	12–176	100	12–380	85	20–168	20	100
OCDD	NA	154	NA	414	90–763	398	90–763	421	98–679	408	90–620	421	182–763	80	374
2,3,7,8–TCDF	4.2	5.1	5.7	3.9	0.3–11	3.7	0–7.2	4.2	0.3–11	3.4	0.3–7.2	4.6	0–11	< 3	7
1,2,3,7,8–PeCDF	ND	ND	ND	ND										ND	ND
2,3,4,7,8–PeCDF	72	67	88	54	9–87	50	15–87	32	9–54	45	9–87	33	11–65	32	26
1,2,3,4,7,8/															
1,2,3,4,7,9–HxCDF	8	4	5	6	1–15	7	2–15	5	1–6	6	1–15	5	2–7	11	12
1,2,3,6,7,8–HxCDF	9	5	6	5	1–13	5	2–13	4	1–5	5	1–13	4	2–7	5	7
2,3,4,6,7,8–HxCDF	2	1	1	2	1–7	2	1–7	2	1–4	2	1–7	2	1–4	2	38
1,2,3,4,6,7,8–HpCDF	7	5	7	11	1–49	14	5–49	10	1–18	13	1–49	10	5–16	37	17
1,2,3,4,6,7,9–HpCDF	ND	ND	ND	ND										ND	ND
1,2,3,4,6,8,9–HpCDF	ND	ND	ND	ND										ND	ND
1,2,3,4,7,8,9–HpCDF	ND	ND	ND	ND										ND	ND
OCDF	NA	NA	NA	4										NA	240

[a] Values given in pg/g on a wet weight basis.
[b] NA, not analyzed.
[c] ND, not detected. (≤1 ppt).

Table 3. Levels of PCDDs and PCDFs Found in Human Milk Samples.[a]

Compound	Sweden n = 4	Range	Germany n = 5	Range	Vietnam n = pooled (4)	Denmark n = 1
2,3,7,8–TCDD	0.6	T[b]–2.3	1.9	1.3–3.3	ND[c]	NA[d]
1,2,3,7,8–PeCDD	6.5	3.5–13.8	12.6	9–18	7.0	31
1,2,3,4,7,8–HxCDD	2.5	0.8–3.6	NI[e]		13.0	26
1,2,3,6,7,8–HxCDD	19	12–23	NI		50	97
1,2,3,7,8,9–HxCDD	6.3	3.9–9.0	NI		24.0	32
Total HxCDD	27.5	17–35	23.4	15–28	87	155
1,2,3,4,6,7,9–HpCDD	ND		ND		ND	14
1,2,3,4,6,7,8–HpCDD	59.5	38–86	72.8	48–92	150	174
OCDD	302	197–484	434	168–623	754	328
2,3,7,8–TCDF	4.2	2.2–8.7	5.4	4.0–8.0	9.4	4.0
1,2,3,7,8–PeCDF	ND < 1		ND < 1		ND < 1	4.5
2,3,4,7,8–PeCDF	21.3	7–53	36.4	24–54	21	31
1,2,3,4,7,8/ 1,2,3,4,7,9–HxCDF	4.7	2.7–8.9	NI		15.0	13
1,2,3,6,7,8–HxCDF	3.4	1.9–6.8	NI		11.0	52
2,3,4,6,7,8–HxCDF	1.4	0.8–2.6	NI		4.2	11
Total HxCDF	7.4	6.0–18.3	26	13–36	30.2	76
1,2,3,4,6,7,8–HpCDF	7.4	4.4–12.0	9.2	4–12	23.0	46
1,2,3,4,6,7,9–HpCDF	ND		ND		ND	ND
1,2,3,4,6,8,9–HpCDF	ND		ND		ND	ND
1,2,3,4,7,8,9–HpCDF	ND		ND		ND	ND
OCDF	3.2	ND–11.0	2.4	1–4	46.0	ND

[a]Values given in pg/g on a fat weight basis.
[b]T, trace.
[c]ND, not detected (< 0.5 pg/g).
[d]NA, not analyzed.
[e]NI, not isomer specific.

Table 4. Levels of PCDDs and PCDFs Found in Aquatic Environmental Samples.

Compound	Osprey[a]	Seal[b]	Salmon[b]	Guillemot[b]	Herring[b]
2,3,7,8–TCDD	NA[c]	28	6	17	0.6
1,2,3,7,8–PeCDD	258	60	22	26	ND[d]
1,2,3,4,7,8–HxCDD	33	2.9	0.8	0.9	ND
1,2,3,6,7,8–HxCDD	141	69	6.6	18	ND
1,2,3,7,8,9–HxCDD	87	3.6	0.2	1.6	ND
1,2,3,4,6,7,9–HpCDD	11	0.4	0.5	ND	ND
1,2,3,4,6,7,8–HpCDD	136	1.2	0.9	0.6	ND
OCDD	275	2.5	4.3	0.2	ND
2,3,7,8–TCDF	63	12	62	0.2	3.9
1,2,3,7,8–PeCDF	27	3.7	24	1.8	1.2
2,3,4,7,8–PeCDF	767	125	82	97	5.6
1,2,3,4,7,8/ 1,2,3,4,7,9–HxCDF	17	1.1	9.1	4.7	0.6
1,2,3,6,7,8–HxCDF	19	2.4	6.3	10	0.4
2,3,4,6,7,8–HxCDF	16	1.5	5.1	4.5	0.7
1,2,3,4,6,7,8–HpCDF	7.5	0.2	41	0.5	0.5
1,2,3,4,6,7,9–HpCDF	ND	ND	0.6	ND	ND
1,2,3,4,6,8,9–HpCDF	ND	ND	2.2	ND	0.2
1,2,3,4,7,8,9–HpCDF	ND	ND	0.9	T[e]	ND
OCDF	5.9	1.8	61	T	ND

[a]pg/egg.
[b]pg/g.
[c]NA, not analyzed.
[d]ND, not detected (< 0.1 pg/g i < 3 pg/egg osprey).
[e]T, trace.

Table 5. Levels of PCDDs and PCDFs Found in Bovine Fat, Bovine Liver, Bovine Milk and Cream Samples.[a]

Compound	Cow 1 Fat, Sweden	Cow 1 Liver, Sweden	Cow 2 Fat, Sweden	Cow 2 Liver, Sweden	Cow 1 Fat, Scotland	Cow 2 Fat, Scotland	Cow 1 Milk, Denmark	Cow 2 Milk, Denmark	Cow 3 Milk, Scotland	Cream, Sweden
2,3,7,8–TCDD	ND[b]	ND	ND	ND	NA[c]	NA	NA	NA	NA	NA
1,2,3,7,8–PeCDD	ND	ND	ND	ND	2	ND	T[d]	6	5	ND
1,2,3,4,7,8–HxCDD	ND	ND	ND	ND	1	2	ND	ND	3	ND
1,2,3,6,7,8–HxCDD	ND	ND	ND	ND	2	2	6	2	6	18
1,2,3,7,8,9–HxCDD	ND	ND	ND	ND	ND	ND	6	4	2	ND
1,2,3,4,6,7,8–HpCDD	3	4	3	3	3	4	9	7	2	42
OCDD	9	10	4	39	3	4	< 86	< 45	3	16
2,3,7,8–TCDF	ND	ND	ND	ND	ND	ND	ND	ND	ND	2
2,3,4,7,8–PeCDF	ND	ND	ND	ND	5	4	3	3	8	4
1,2,3,4,7,8/ 1,2,3,4,7,9–HxCDF	ND	ND	ND	ND	1	1	ND	ND	3	3
1,2,3,6,7,8–HxCDF	ND	ND	ND	ND	1	1	8	7	2	T
2,3,4,6,7,8–HxCDF	ND	ND	ND	ND	1	1	4	2	2	3
1,2,3,4,6,7,8–HpCDF	ND	ND	ND	ND	2	3	< 7	< 5	3	6
OCDF	ND	ND	ND	ND	ND	ND	< 7	< 7	< 15	< 20

aValues given in pg/g on a wet weight basis.
bND, not detected (< 1.0 ppt).
cNA, not analyzed.
dT, trace.

analysis of the guillemot muscle. Here we also find that all the major peaks are the toxic 2,3,7,8-substituted PCDDs and PCDFs and the major TCDF peaks are artifacts (wrong chlorine cluster).

CONCLUSIONS

In this study we identified a series of PCDDs and PCDFs in all the samples of human adipose tissue and mother's milk we have analyzed. This is a strong indication that we have a background of these compounds in the general population in Europe. We have also identified a series of PCDDs and PCDFs in all aquatic samples from the Baltic Sea and the Gulf of Bothnia. The levels in the human and aquatic samples are normally in the low ppt range; the mean value for 2,3,7,8-tetra-CDD is 2.5 ppt. The highest values are reported for octa-CDD in adipose tissue (mean value 414 ppt). The levels found in aquatic animals are similar to those reported earlier.[9]

Discussing individual isomers, we have observed that there is a great similarity between the specific PCDDs and PCDFs found in human adipose, mother's milk and environmental samples (see Figures 1, 2 and 3). All the isomers identified represent the toxic 2,3,7,8-substituted toxic congeners listed in Table 1. However, of the 12 congeners listed here, 1,2,3,7,8-penta-CDF and the 1,2,3,7,8,9-hexa-CDF are missing in all samples.

Human exposure to tetra-CDD during the use of phenoxy herbicides in Vietnam has been the subject of much concern. Young et al.[23] discuss a study in which very low levels of 2,3,7,8-tetra-CDD were detected in adipose tissue from some Vietnam veterans. However, the levels do not correlate well with known exposure data. To the contrary, Gross et al.[24] reported that Vietnam veterans heavily exposed to phenoxy herbicides have elevated levels of 2,3,7,8-tetra-CDD.

Ryan et al.[25] reported on the levels of 2,3,7,8-tetra-CDD in 23 autopsy samples from the Great Lakes area in Canada. They found 22 of the samples to be positive, with levels ranging from 4.1 to 130 ppt. Excluding an outlying high sample, the average value was 10.7 ± 5.4 ppt.

In the study of human adipose tissue we have a group of cancer patients (soft tissue sarcomas, lymphomas) and a group of controls. We cannot see any difference in the pattern of congeners between these two groups (see Table 2).

However, a slight difference was observed between the levels found in exposed patients and those of nonexposed controls (see Table 2). The largest difference was found for 2,3,4,7,8-penta-CDF, a compound which, however, cannot be associated with specific exposure to PCDDs.

The analysis of the adipose tissue of the heavily exposed BASF worker shows a value for 2,3,7,8-tetra-CDD of 100 ppt, which is about 25–30 times higher than the mean value found in the Swedish samples. Therefore, it is evident that this analytical method can be used to monitor earlier exposure. The values of the sample from the chemist are also in agreement with this statement. Very specific congeners related to his earlier exposure could be found in this particular sample.

A study of 103 samples of breast milk from mothers living in areas in the United States sprayed with 2,4,5-T revealed no 2,3,7,8-tetra-CDD at a detection level of 1 to 4 pg/g.[26] Langhurst and Shadoff[27] report that 6 of 9 samples showed 2,3,7,8-tetra-CDD at levels slightly higher than the detection level (0.2 to 0.7 pg/g). The authors consider these results unconfirmed because of the lack of validation and studies on the accuracy and precision of the data.

In the present study we report that, in addition to 2,3,7,8-tetra-CDD, a series of toxic 2,3,7,8-substituted PCDDs and PCDFs can be identified in all the samples of mother's milk we have analyzed so far. This is the first report on the occurrence of these compounds in mother's milk.

The failure to identify 2,3,7,8-tetra-CDD in the composite mother's milk sample from Vietnam is somewhat unexpected. However, it can be explained by the age of the mothers and a rapid sedimentation rate in the local rivers.

The levels found in mother's milk might be of special interest from a toxicological point of view. A 5-kg baby consuming 850 mL of milk a day will receive a dose of 2,3,7,8-tetra-CDD of about 5 pg/kg body weight per day, which is much higher than the virtually safe dose discussed by Kimbrough et al.[14] Contrary to the scenario discussed by Kimbrough et al. (Times Beach, Mo.), mother's milk is also found to be contaminated by several other toxic PCDDs and PCDFs. In this chapter various methods are discussed to report levels of PCDDs other than 2,3,7,8-tetra-CDD and PCDFs as "toxic" or 2,3,7,8-tetra-CDD equivalents.[28-30] Using this approach, the dose for the nursing babies discussed above is increased by a factor of 10–20 or even higher.

A WHO (World Health Organization) consultation was held in Bilthoven, Holland in January, 1985 to discuss this specific problem. The following conclusion was reached at this meeting.[31]

> For PCDDs and PCDFs, available data from animal and human exposure strongly suggest that man belongs most probably to a species less sensitive to their effects. However, animal experiments suggest that newborns may have a higher sensitivity.
>
> The toxicological evaluations temporarily adopted in several countries have been based on the determination of either a no-effect level in animal studies or on extrapolation from long-term cancer studies in animals. This has resulted in temporary tolerable daily intakes of 2,3,7,8-TCDD in the range of 1–5 pg/kg body weight calculated on a lifetime basis.
>
> To estimate the tolerable daily intake for the sum of PCDDs and PCDFs, an approach using calculation of so-called "TCDD (toxicity) equivalents" has

been used. These calculations have been made by comparing data from various short-term *in vitro* and *in vivo* studies on various congeners. Although such an approach imposes a high degree of uncertainty, it has value and no better approach appears to be available at present.

The relatively few data available for human milk show levels of PCDDs and PCDFs that were probably lower than those at which adverse effects may be expected, but which may nevertheless approach or exceed the presently proposed calculated temporary tolerable daily intakes.

2,3,7,8-tetra-CDD is reported to be slowly excreted from the bodies of all animals tested. For small rodents the half-life in the body is 10–43 days.[32–34] However, McNulty et al.[35] reported a half-life of 2,3,7,8-tetra-CDD in fat from a female rhesus macaque of about one year. The data from the BASF worker indicate a much longer half-life in humans than in small rodents and monkeys.

Discussing the levels and the pattern of the different congeners or groups of congeners in the human samples, we find great similarities between adipose tissue and mother's milk (see Figure 4). In this figure we have also included an aquatic mammal, the seal. Here we can see some interesting

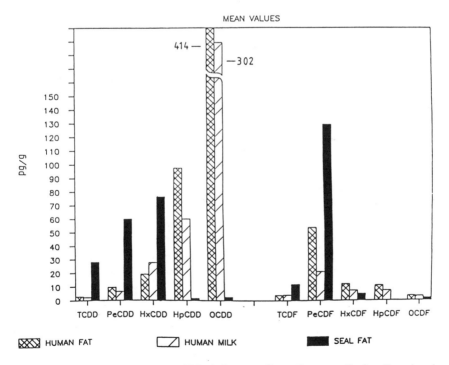

Figure 4. Levels of PCDDs and PCDDs in human adipose tissue, mother's milk and seal fat.

differences. Normally, the seal fat is more contaminated than the human samples; however, the levels of hepta- and octa-CDD are much lower. These congeners have been reported as impurities in commercial pentachlorophenol formulations; consequently, these data could indicate a direct human exposure to products containing pentachlorophenol, although differences in solubility and/or metabolism could also account for these observations.

Using GC (packed column) and high-and low-resolution MS, Mahle et al.[36] failed to identify and quantify 2,3,7,8-tetra-CDD in bovine milk from farms where 2,4,5-T had been used. Lamparski et al.[37] have analyzed samples of quarantine milk from Michigan, USA. They found no hepta-CDD or octa-CDD at detection levels of 25 pg/g and 50 pg/g, respectively.

We have also analyzed bovine samples: fat, liver and milk. In these samples the levels of PCDDs and PCDFs are lower — normally, close to the detection limit. This indicates that the background levels are higher in the aquatic environment than in the terrestrial environment, which is in agreement with earlier experience for substances, such as DDT and PCB. The human exposure to PCDDs and PCDFs by milk consumption is not negligible.

It is evident that it is of particular importance to identify the ultimate source or sources of the toxic 2,3,7,8-substituted PCDDs and PCDFs found as background constituents in the environmental and human samples, especially the samples of mother's milk, which are of toxicological interest. Analysis of historical samples and samples from various locations could reveal a time trend and a geographic trend in these levels. It has been pointed out above that the levels found for octa- and hepta-CDDs might reflect a direct exposure to pentachlorophenol, but the source for the lower chlorinated congeners, which are the most biologically active, are possibly found elsewhere.

Of special importance in this discussion is the observation of 1,2,3,7,8-penta-CDD, normally present at levels of about 10 ppt, or higher even than levels of 2,3,7,8-tetra-CDD. However, this particular congener has never been reported as a contaminant in any commercial product, but it has always been found in samples from municipal and hazardous wastes incinerators.[3] At the present time it cannot be excluded that incineration of various wastes or chlorinated products (PVC, VCM-tax, hexachloroethane and chlorinated solvents) might contribute to the background levels of PCDDs and PCDFs found in the environment and in the general population. A one-year moratorium for new municipal incinerators was recently issued in Sweden.[38] Other sources of PCDDs and PCDFs recently identified are copper melters using recycled copper (PVC coated wires) and electrical furnaces in the steel industry.

REFERENCES

1. Tiernan, T. In *Chlorinated Dioxins and Dibenzofurans in the Total Environment*, G. Choudhary, L. Keith and C. Rappe, Eds. (Stoneham, MA: Butterworth Publishers, 1983) p. 211.
2. Crummett, W. B. *Chemosphere* 12:429 (1983).
3. Rappe, C. *Environ. Sci. Technol.* 18:78 A (1984).
4. Rappe, C., M. Nygren, and H. R. Buser. In *Application of New Mass Spectrometry Technique in Pesticide Chemistry.*, J. Rosen, Ed. (New York, NY: John Wiley & Sons, Inc. 1985).
5. Olie K., P. L. Vermeulen, and O. Hutzinger. *Chemosphere 1977, 6*, 455
6. Buser, H. B., H.-P. Bosshardt, and C. Rappe. *Chemosphere*, 7:109 (1978).
7. Buser, H. B., H.-P. Bosshardt, and C. Rappe. *Chemosphere* 7:419 (1978).
8. Rappe, C., S. Marklund, L.-O. Kjeller, P.-A. Bergqvist, M. Nygren, and M. Hansson. 1st Conference on Toxic Substances, Montreal, Canada (1984).
9. Rappe, C., P.-A. Bergqvist, and S. Marklund. In *Chlorinated Dioxins and Dibenzofurans in the Total Environment II.* L. H. Keith, C. Rappe, and G. Choudhary, Eds., (Stoneham, MA: Butterworth Publishers, 1985) p. 125.
10. McConnell, E. E. In *Halogenated Biphenyls, Terphenyls, Naphthalenes, Dibenzodioxins and Related Compounds*, R. D. Kimbrough, Ed. (Amsterdam: Elsevier–North Holland, 1980) p. 109.
11. Goldstein, J. A. In *Halogenated Biphenyls, Terphenyls, Naphtalenes, Dibenzodioxins and Related Compounds*, R. D. Kimbrough, Ed. (Amsterdam: Elsevier–North Holland, 1980).
12. Kociba, R. J. D. G. Keyes, J. E. Beyer, R. M. Carreon, et al. *Toxicol. Appl. Pharm.* 46:279 (1978).
13. National Toxicology Program. DHHD Publication No. 80-1754; National Institute of Health, Washington, DC (1980).
14. Kimbrough, R. D., H. Falk, P. Stehr, and G. Fries. *J. Toxicol. Environ. Health.* In press.
15. Isensee, A. R. In "Chlorinated Phenoxy Acids and Their Dioxins," *Ecol. Bull.*, 27:255 (1978).
16. McConnell, E. E., G. W. Lucier, R. C. Rumbaugh, P. W. Albro, D. J. Harvan, J. R. Hass, and M. W. Harris. *Science* 223:1077 (1984).
17. Rappe, C., H. R. Buser, D. L. Stalling, L. M. Smith, and R. C. Dougherty. *Nature* 292 (1981).
18. Norstrom, R. J., D. J. Hallett, M. Simon, and M. J. Mulvihill. In *Chlorinated Dioxins and Related Compounds*, O. Hutzinger, R. W. Frei, E. Merian, and F. Pocchiari, Eds. (Oxford, UK: Pergamon Press, 1982) p. 173.
19. Stalling, D. L., L. M. Smith, J. D. Petty, J. W. Hogan, J. L. Johnson, C. Rappe, and H. R. Buser. In *Human and Environmental Risks of Chlorinated Dioxins and Related Compounds*, R. E. Tucker, A. L.

Young, and A. P. Gray, Eds. (New York, NY: Plenum Press, 1983) p. 221.

20. Thiess, A. M., R. Frentzel-Beyme, and R. Link. *Am. J. Ind. Med.* 2:179 (1982).
21. Smith, L. H., D. L. Stalling, J. L. Johnson. *Anal. Chem.* 56:1830 (1984).
22. Tondeur, Y., P. W. Albro, J. R. Hass, D. J. Harvan, and J. L. Schroeder. *Anal. Chem.* 56:1344 (1984).
23. Young, A. L. H. K. Kang, and B. M. Shepard. *Environ. Sci. Technol.* 17:530 A (1983).
24. M. L. Gross, J. O. Lay, P. A. Lyon, D. Lippstreu, N. Kangas, R. L. Harless, S. E. Taylor, and A. E. Dupuy. *Environ. Res.* 33:261 (1984).
25. Ryan, J. J., D. T. Williams, B. P.-Y. Lau, and T. Sakuma. In *Chlorinated Dioxins and Dibenzofurans in the Total Environment II.* L. H. Keith, C. Rappe, and G. Choudhary. Eds. (Stoneham, MA: Butterworth Publishers 1985) p. 205.
26. Esposito, M. P., T. O. Tiernan, and F. E. Dryden. *Dioxins*, EPA–600/ 2–80–197, Cincinnati, OH (1980).
27. Langhurst, M. L., and L. A. Shadoff. *Anal. Chem.* 53:2037 (1980).
28. Kim, N. K., and J. Hawley. "Risk Assessment of Binghamton State Office Building." Division of Health Control, New York State Department of Health, Albany, NY (1983).
29. Grawitz, N., A. Fan, and R. R. Neutra. "Interim Guidelines for Acceptable Exposure Levels in Office Settings Contaminated with PCB and PCB Combustion Products." California Department of Health Services, Berkeley, CA (1983).
30. Milby, T. H., T. H. Miller, and T. L. Forrester. *J. Occup. Med.* 27:351 (1985).
31. WHO, "Organohalogen Compounds in Human Milk and Related Hazards," Report on a WHO Consultation. WHO Regional Office for Europe, ICP/CEH 501/m 05 (1985).
32. Rose, J. Q., J. C. Ramsey, T. H. Wentzler, R. A. Hummel, and P. J. Gehring. *Toxicol. Appl. Pharmacol.* 36:209 (1976).
33. Nolan, R. J., F. A. Smith, and J. G. Hefner. *Toxicol. Appl. Pharmacol.* 48:A162 (1979).
34. Olson, J. R., T. A. Gasiewicz, and R. A. Neal. *Toxicol. Appl. Pharmacol.* 56:78 (1980).
35. McNulty, W. P. K. A. Nielsen-Smith, and J. O. Lay Jr. *Food Cosmet. Toxicol.*, 20:985–987 (1982).
36. Mahle, N. E., H. S. Higgins, and M. E. Getzendaner, *Bull. Environ. Contamin. Toxicol.* 18:123 (1977).
37. Lamparski, L. L., N. H. Mahle, and L. A. Shadoff. *J. Agric. Food Chem.* 26:1113 (1978).
38. SNV. Statens Naturvårdsverk, Stockholm, Sweden. Press release, Feb. 13, 1985.

CHAPTER 3

Chlorinated Dioxins and Dibenzofurans in Human Tissues from Vietnam, 1983–84

Arnold J. Schecter, John. J. Ryan, Michael Gross, N. C. A. Weerasinghe, and John D. Constable

INTRODUCTION

2,3,7,8-tetrachlorinated dibenzo-p-dioxin (2,3,7,8-TCDD) is one of the extremely toxic chemicals produced as an unwanted contaminant of certain chemical manufacturing processes.[1] It is characteristically found in 2,4,5-T, one of two herbicides used in Agent Orange, which was a 50/50 mixture of 2,4,5-T and 2,4-D. Agent Orange was used as a defoliant in the south of Vietnam during the Vietnam War, also known as the Second Indochina War, between 1964 and 1971. In 1983 and in 1984, two of the authors (JDC and AJS) brought back for dioxins content analysis breast milk and adipose tissue from Vietnamese, living in Vietnam, who were believed to have been exposed to dioxin from Agent Orange and who were therefore potentially at risk for any sequelae that might occur from such exposure. In 1984 we also analyzed adipose tissue from nonexposed individuals as controls.

Figure I shows a map of Vietnam, including Hanoi, in the north, where adipose tissue specimens from patients not exposed to Agent Orange were obtained, and in the south, Ho Chi Minh City (formerly Saigon), where specimens from presumably Agent Orange-exposed individuals were obtained.

2,3,7,8-TCDD has been found as recently as 1980 in soil from Vietnam.[2]

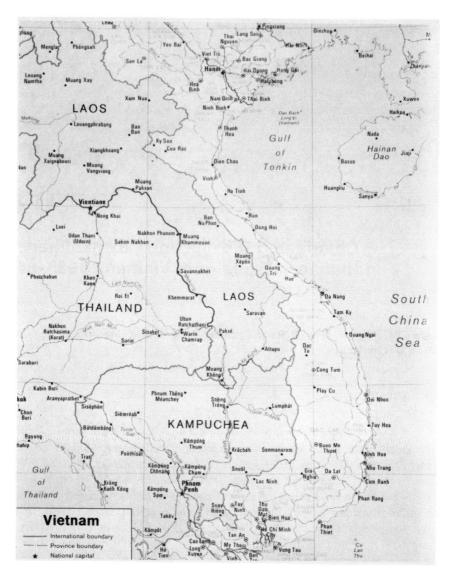

Figure 1.   Specimens were collected from Hanoi, in the north of Vietnam and in Ho Chi Minh City, far to the south. Agent Orange was sprayed in the south of Vietnam only.

In previous work by Baughman and Meselson it was shown that 2,3,7,8-TCDD could be found in human breast milk and also in fish obtained in 1970 and 1973 from the south of Vietnam.[3,4] In previous studies, elevated levels of 2,3,7,8-TCDD in the adipose tissue of Vietnam veterans exposed to Agent Orange were reported approximately ten years after the last known exposures.[5] Previous studies also demonstrated elevated levels of dioxin and furans in human adipose tissue years after exposure in a different incident.[6-8]

The usefulness of chemical analysis of adipose tissue biopsies many years after the last exposure for the estimation of human exposure to chlorinated dioxins, dibenzofurans and PCBs has been demonstrated.[9-11] While PCBs are found in human blood at levels high enough to be used to estimate exposure, dioxins and dibenzofurans are found in parts per trillion (ppt) in adipose tissue in the general population, and because they are fat soluble, they are best measured in human adipose tissue.

The exposure of humans to 2,3,7,8-TCDD is unique in Vietnam, because over 400 pounds of pure 2,3,7,8-TCDD[2] were sprayed over discrete areas in the form of Agent Orange. Questions remain as to the possible health consequences to exposed humans in Vietnam and elsewhere. Until recently, chemical analysis of small samples of tissue was not sufficiently sensitive or specific to permit accurate estimation of 2,3,7,8-TCDD exposure of humans.

On the basis of animal laboratory studies, there is no doubt as to the extreme toxicity of 2,3,7,8-TCDD and other 2,3,7,8 halogenated dioxins and dibenzofurans. Small doses cause death after a wasting syndrome. Cancers may be produced in various organs of the body. Increased numbers of congenital malformations and spontaneous abortions occur in female animals dosed with dioxin. Immune deficiency occurs, as does liver damage and central and peripheral nervous system pathology. Skin lesions are seen in some cases. Gastrointestinal epithelial cell hyperplasia and pathology of the urinary and hematological systems occur. Monkeys and guinea pigs are especially sensitive, followed by certain strains of rats and mice. Hamsters are more resistant to certain acute effects, such as LD_{50}, but not to others, such as ED_{50} for liver enzyme induction. There is also transplacental and breast milk transfer of the lipid-soluble dioxins and related chemicals, such as polychlorinated biphenyls.[1,13]

Vietnam is now a country of 60 million persons. Presumably, a large population exists in the south which has been exposed to 2,3,7,8-TCDD. The exposure could have occurred by direct contact with the spray, or by secondary contamination of food or water, or by a combination of these two modalities. Because dioxins are extremely stable chemicals, they do not break down easily, and they persist in the environment. In addition to the large population of unknown size in the south exposed to 2,3,7,8-TCDD, there is a population in the north of Vietnam presumably unexposed, since herbicides were not used in the north. In addition, agricultural and industrial chemicals have not been used as extensively in the north of Vietnam because of economic constraints.

Pentachlorophenol (PCP), an antifungal agent used extensively as a wood and paper preservative by the Americans (and possibly the French) in the south of Vietnam, is also not known to have been used in the north. PCP typically contains decreasing amounts of octa- through pentachlorinated dioxins and dibenzofurans. Therefore, the introduction into the food chain of octa- through hexa- and possibly also the pentachlorinated dioxins, and to a lesser extent, dibenzofurans, found as contaminants of technical-grade PCP might have occurred in the south, but is less likely to have occurred in the north of Vietnam.

However, there might be a population in the north with some exposure to dioxins. Former soldiers who spent several years in dioxin-contaminated jungles in the south during the war, then returned to the north of Vietnam, would be expected to have an intermediate exposure as compared to the other two groups. They would be more comparable to American Vietnam veterans exposed during the war, with the exception that most U.S. soldiers were rotated into Vietnam for one year, while Vietnamese soldiers served for long periods. Unlike the U.S. soldiers serving in the south of Vietnam, these northern troops, along with their southern allies, would have had the potential for significant dietary exposure to dioxins, since their food was provided from Agent Orange-sprayed areas in some instances.

In recent years, we have found, as have others, that all adults in the United States and Canada, like those in other industrial countries such as Sweden, have detectable levels of approximately 14 isomers of 2,3,7,8-chlorine-substituted tetra- through octa- dioxins and dibenzofurans in adipose tissue, and to a lesser extent in other organs or tissues. The average wet weight levels are approximately 1000 + ppt in adipose tissue for total chlorinated dioxins and furans and approximately 6–10 ppt for 2,3,7,8-TCDD.[6-8,12,14,15] Therefore, epidemiologic or clinical studies of Americans exposed to 2,3,7,8-TCDD from Agent Orange as a result of military service in Vietnam, or exposed to other dioxins as a result of industrial accidents, suffer because both the potentially exposed group and the control group have baseline levels of dioxins and furans. Without exact measurement of levels of dioxin and furan isomers in body tissues, estimation of exposure is difficult if not impossible.

Therefore, the possibility of finding a large population within a given country with a 2,3,7,8-TCDD exposure and also another genetically similar population within the same country with no known exposure to dioxins would be of scientific and public health significance. Vietnam would seem potentially to offer a wealth of valuable scientific information concerning the extent of dioxin pathology in exposed and slightly exposed humans. By utilizing the developing tool of adipose tissue biopsy followed by sensitive isomer-specific quantification of the dioxins and furans in both cases and controls, accurate estimates of exposure are now possible, and future clinical and epidemiological studies assessing the extent of the toxicity of furans and dioxins toward humans may be based on a more scientific basis than previously possible.

METHODS

While attending a conference in Ho Chi Minh City (which is the former city of Saigon and surrounding urban areas) in January, 1983, one of the authors (JDC) brought back breast milk samples from presumed Agent Orange-exposed patients for dioxin analysis. Relatively small amounts of milk, less than 30 mL per sample, were obtained from each mother. To preserve the specimen each was diluted by 50% with a 1:1 mixture of ethanol in water and delivered to the University of Nebraska at Lincoln for analysis.

During a 1984 visit to lecture at Hanoi and Ho Chi Minh City medical schools and public health departments, another of the authors (AJS) was asked to bring back adipose tissue specimens from hospitalized patients undergoing surgery. The adipose tissue was removed by Vietnamese surgeons at Hanoi and Ho Chi Minh City hospitals. The results of analysis of twenty of these specimens, along with five of the earlier milk samples, are described in this chapter.

The adipose tissue samples were frozen in Vietnam and carried back to the United States and Canada. The Ottawa laboratory served as the primary laboratory, analyzing all specimens, quantifying all tetra- through octachlorinated dioxin and dibenzofuran isomers down to a level of 2 or 3 ppt for 2,3,7,8-TCDD. The University of Nebraska laboratory served as confirmation laboratory for 2,3,7,8-TCDD by analyzing homogenates of 17 specimens for which there was sufficient tissue. Both laboratories were unaware of the geographical source of the specimens and were given only a coded number randomly assigned to the sample.

These tissues were analyzed by the methods reported in the related chapter by Ryan et al. in this volume. The secondary laboratory used methods described previously.[5] Summaries of both methods and others used for analyzing dioxins in human tissues are also discussed elsewhere.[16,17]

RESULTS

Human Adipose Tissue

The values of 2,3,7,8-TCDD found in 7 human adipose tissue specimens from Hanoi hospitals and 13 specimens from Ho Chi Minh City (Saigon) hospitals , as extracted and analyzed in Ottawa, are presented in Table 1. Of special interest is the finding of nondetectable levels of 2,3,7,8-TCDD in 7 patients from the north (hospitalized at Viet Duc or the Gynecological and Obstetrical Hospitals in Hanoi) who were thought by their physicians to have never been in the south and, therefore, not to have been exposed to 2,3,7,8-TCDD from Agent Orange. The finding of 10 of 13 patients in the

Table 1.   2,3,7,8–TCDD Levels in Human Adipose Tissue from the North and South of Vietnam – 1984 – Values in Parts per Trillion Wet Weight Basis.

North – 7 Samples	South – 13 Samples	
ND^a(2)^b	6.7	
ND(2)	ND(3)	
ND(2)	8.3	10 Positive of 13
ND(2)	79.4	
ND(3)	4.2	Mean of Positives 22.1
ND(3)	ND(2)	
ND(2)	3.6	Range 3.6 – 79.4
	ND(2)	
	56.7	
	16.9	
	7.6	
	15.4	
	22.4	

^{a}ND, not detected.
^{b}( ) refers to detection level in parts per trillion.

south (from Cho Ray and Tu Du hospitals) to be positive for 2,3,7,8-TCDD is also noteworthy in that they were believed to have been exposed to Agent Orange. Prior to the visit, the Vietnamese physicians had little advance notification concerning the possibility of saving for analysis adipose tissue specimens, which would otherwise have been disposed of. Patients were selected on short notice during AJS's one week visit in Hanoi, followed by one week in Ho Chi Minh City (Saigon) in November and December of 1984.

The mean of the 10 positive samples, 22.1 ppt, and the range of 3.6–79.4 ppt for 2,3,7,8-TCDD on a wet weight basis, are higher than would be expected on the basis of results obtained for North American patients, for whom a mean approximation of 8 ppt has been obtained. Certainly, there is a striking difference between the two groups in this small sample. Markedly higher levels of 2,3,7,8-TCDD, presumably from Agent Orange, were found for the southern Vietnamese as compared to northern Vietnamese patients with no known exposure to Agent Orange.

A more detailed presentation of data is given in Table 2. The results are calculated as concentrations on a wet weight and extractable lipid basis. This is done because the lipid content of adipose tissue specimens varies. We have generally found lower lipid content, as low as 14%, in intraabdominal as compared to subcutaneous adipose tissue specimens, which, in addition to fat, may contain blood, serum, collagen, blood vessels, nerves and other diluting materials. Subcutaneous adipose tissue generally has a higher lipid content, in the vicinity of 70–90%. For this study, our ranges for 2,3,7,8-TCDD are 3.6–79.4 and 9.2–103 ppt, and the means are 22.1 and 34.2 ppt on a wet weight as compared to a lipid basis, respectively. The mean values are summarized in Table 3.

In Table 4, results of an isomer-specific analysis for detectable tetra-

Table 2. 2,3,7,8–TCDD Levels in Human Adipose Tissue from the North and South of Vietnam – 1984 – Values in ppt.

North – 7 Samples		South – 13 Samples	
Wet Weight	Lipid Basis	Wet Weight	Lipid Basis
ND[a](2)[b]	ND(3.5)	6.7	15.2
ND(2)	ND(3.1)	ND(3)	ND(4.1)
ND(2)	ND(11.8)	8.3	11.4
ND(2)	ND(11.8)	79.4	103
ND(3)	ND(3.8)	4.2	9.7
ND(3)	ND(3.8)	ND(2)	ND(3.6)
ND(2)	ND(3.6)	3.6	7.8
		ND(2)	ND(4.4)
		56.7	67.5
		16.9	56.3
		7.6	9.2
		15.4	34.2
		22.4	27.9
Mean of Positives		22.1	34.2
Range		3.6 – 79.4	9.2 – 103

[a]ND, not detected.
[b]( ) refers to detection level in parts per trillion.

through octa- dioxins and dibenzofurans for the north and south Vietnamese samples are presented. Of interest is the finding that 9 (5 dioxins and 4 furans) 2,3,7,8-PCDD/PCDF isomers which we are accustomed to finding in previously described North American human adipose samples were detected above the current detection limits. The pattern seen in the samples from persons in the south of Vietnam is similar to that seen in the U.S. and Canada, with the octachlorodibenzodioxin predominating, followed by the hepta-, the hexa-, and then the pentachlorodibenzodioxin. However, for North American specimens the lowest dioxin is usually the tetra-, whereas in the samples from the south of Vietnam the levels of 2,3,7,8-tetrachlorinated isomer are higher than the pentachlorodibenzodioxin. The octa- through hexa-, and possibly pentachlorodibenzodioxin levels in persons from industrial countries are similar to those seen here. The chlorinated dibenzofuran levels here, as usual, are lower than dioxin levels in human tissues.

Again, as is the case to date for persons from industrial countries, only

Table 3. 2,3,7,8–TCDD Mean Levels in Human Adipose from the North and the South of Vietnam – 1984 – Values in ppt on Wet Weight and Lipid Basis.

North – 7 Samples		South – 13 Samples	
Wet Weight	Lipid Basis	Wet Weight	Lipid Basis
ND[a](2.3)[b]	ND(5.9)	22.1	34.2

[a]ND, not detected.
[b]( ) refers to detection level in parts per trillion.

Table 4. Dioxin and Furan Mean Levels in ppt in Vietnamese Adipose Tissue Samples — 1984 — Wet Weight Basis.

	North (7 Samples)		South (13 Samples)	
Analyte	Mean	Positive Samples	Mean	Positive Samples
2378–D[a]	ND[c](2)[d]	0	22.1	10
12378–D	ND(2)	0	9.9	11
123678–D	4.6	4	46.7	13
1234678–D	19.0	3	105	13
12346789–D	36.1	6	514	13
23478–F[b]	9.7	4	13.0	13
123478/123678–F	9.3	4	31.7	13
1234678–F	4.2	3	17.0	13
Lipid(%)	51.0	–	60.0	–

[a]D, dibenzo-p-dioxin isomer.
[b]F, dibenzofuran isomer.
[c]ND, not detected.
[d]( ) refers to detection level in parts per trillion.

2,3,7,8-substituted PCDDs or PCDFs are found at the current detection limits. The 2,3,7,8 lateral chlorine substitution pattern is the dioxin pattern characteristic of the most toxic isomers.

The finding of an average of 51% lipid in the adipose tissue from persons in the north of Vietnam and 60% in the tissue of those from the south of Vietnam is similar to findings reported elsewhere for intraabdominal and subcutaneous adipose tissue; there was no striking difference observed in adipose tissue lipid content between samples from the north and samples from the south of Vietnam.

In order to compare levels found for tissue from persons in Vietnam with values being reported elsewhere by us and others, Table 5 contains not only

Table 5. 2,3,7,8–TCDD Mean Levels (ppt) in Human Adipose Tissue Among Countries — Wet Weight Basis, 1976–1985.[15]

Country	Sample Origin	Mean Level (ppt)
Vietnam (North)	7 – mostly biopsy	ND[a](2)[b]
Vietnam (South)	13 – mostly biopsy	22.1
		(10 positives)
Canada	46 – accident	6.2
	1976 – average age 36	(21 positives)
Canada – Ontario	22 hospital	10.7
	1980 – average age > 60	
U.S.A.	8 – biopsy/autopsy	7.2
	1982-83	
U.S.A.	10 military personnel	5.6
U.S.A. – Missouri	9 – accident	5.4
	1983	

[a]ND, not detected.
[b]( ) refers to detection level in parts per trillion.

Table 6. Dioxin and Furan Mean Levels (ppt) in Adipose Among Countries — Wet Weight Basis.

ANALYTE	N. Vietnam 1984 7 Samples	S. Vietnam 1984 13 Samples	Canada 1976 46 Samples	Canada 1980 10 Samples	New York 1982–3 8 Samples
2378–D[a]	ND[c](2)[d]	22.1	6.2	10.0	7.2
12378–D	ND(2)	9.9	10.4	13.2	11.1
123678–D	4.6	46.7	79.6	90.5	95.9
1234678–D	19.0	105	137	116	164
12346789–D	36.1	514	796	611	707
Total Dioxins	59.7	697.7	1029.2	840.7	985.2
23478–F[b]	9.7	13.0	16.8	18.4	14.3
12378/ 123678–F	9.3	31.7	17.3	17.3	31.3
1234678–F	4.2	17.0	32.7	39.4	16.5
Total Furans	23.2	61.7	66.8	75.1	62.1
Total Dioxins and Furans	82.9	759.4	1096	915.8	1047.3
Lipid %	51	60	90	–	71

[a]D, dibenzo-p-dioxin isomer.
[b]F, dibenzofuran isomer.
[c]ND, not detected.
[d]( ) refers to detection level in parts per trillion.

these new Vietnamese data, but also recent findings in Canada and the United States. With the exception of the last data,[15] these values are from our previously cited United States and Canadian adipose tissue analyses of specimens obtained between 1976 and 1984. This table emphasizes the 5 to 11 ppt levels for 2,3,7,8-TCDD found in tissue from persons residing in industrial countries, and the higher level, 22.1 ppt, reported here in tissue of patients thought to have been exposed to Agent Orange. In order to visualize the levels of those isomers which can be currently identified using state-of-the-art chemical methods, analytical data for persons from Canada and the U.S. are compared with those from Vietnam in Table 6. The data suggest that the lowest human dioxin contamination seen to date is in the north of Vietnam. There is nearly as much body burden of the penta-through octachlorodibenzodioxins and dibenzofurans in human adipose tissue from the south of Vietnam as in samples from the U.S. and Canada. However, more 2,3,7,8-TCDD is seen in tissue from patients from the south (with presumed Agent Orange exposure) than the average for persons from the U.S. and Canada.

Table 7 shows several calculations relating current mean of 22 ppt with possible original adipose tissue levels based on 1- and 5-yr estimated human half-life values.

A comparison of the results of the lead laboratory in Ottawa and the findings of the confirmatory laboratory in Nebraska are listed in Table 8. In general, the values obtained using different extraction, cleanup, and identification techniques correspond quite well. The results for sample V–8 differ

Table 7.   Vietnam Herbicide Spraying Chronology and Possible Half-Life Data.[a]

1965 – 1970	Agent Orange plus other herbicides, southern provinces
1984	adipose samples collected (14 yr later) Average 22 ppt (wet weight)
assuming 1-yr half-life (monkey)	
1970	14 half-lives
	360 ppb or 360,000 ppt in 1970
1965	19 half-lives
	11.5 ppm pr 11,500,000 ppt
assuming 5-yr half-life in humans[19]	
1970	2.8 half-lives
	153 ppt
1965	3.8 half-lives
	306 ppt

[a]Half-life calculations assume no further intake of 2,3,7,8-TCDD after initial exposure.

in the two laboratories, and a minor difference is seen for samples V–11 and V–20. The recovery of the internal standard was low for the first analysis of sample V–39 at Nebraska, and the result does agree with that obtained in Ottawa. The results for a repeat analysis show good agreement. In our view, it is not unexpected that differences sometimes existed in state-of-the-art laboratories when approaching low ppt levels when small adipose tissue samples are available.

Table 8.   Comparison Table: Primary Laboratory and Confirming Laboratory Analysis of Human Fat from Vietnam.

ID	UN-L Weight (g)	Concentration in ppt		UN-L % Recovery
		UN-L[a]	Ottawa	
V–1	3.29	ND[c](13)[d]	6.7	55
V–3	4.84	ND(3.1)	ND	50
V–4	15.42	ND(1.0)	ND	40
V–5	6.93	ND(1.5)	ND	70
V–7	5.60	ND(10)	ND	45
V–8	5.56	45(20)	8.3	45
V–11	5.30	50(30)	79.4	65
V–13	8.68	ND(2.0)	ND	45
V–15	5.52	ND(5.0)	4	30
V–16	8.97	ND(2.0)	ND	40
V–19	4.82	ND(30)	ND	20
V–20	6.16	40(20)	56.7	40
V–22	3.80	17(17)	16.9	80
V–34	5.01	ND(5.0)	7.6	50
V–36	4.58	14(10)	15.4	70
V–39	16.03	ND(3.3)	22	30
V–39 (re-run)	4.4	25(20)	22	70

[a]UN-L: University of Nebraska at Lincoln, M. Gross & N. C. A. Weerasinghe.
[b]Ottawa: Health & Welfare Canada, Ottawa, J. J. Ryan and B. P.-Y. Lau.
[c]ND, not detected.
[d]( ) refers to detection level in parts per trillion.

Table 9. Analysis of Human Milk from the South of Vietnam — 1983 — Wet Weight Basis (ppt) University of Nebraska Analysis.

Sample Identification	Weight[a] (g)	Concentration (d1)[a] (ppt)	% Recovery
Vietnam 1–4	49.07	ND[b](0.5)[c]	55
Vietnam 2–2	33.35	ND(0.85)	40
Vietnam 1–3	51.81	ND(0.5)	55
Vietnam 2–4	28.98	ND(0.8)	60
Vietnam 2–8	32.34	ND(0.32)	70

[a]Total weight with alcohol and water. 50% ethanol (water) mixture added to equal volumes of the human breast milk above.
[b]ND, not detected.
[c]( ) refers to detection level in parts per trillion.

Human Milk

Dioxin levels from five samples of human milk analyzed by the Nebraska laboratory are shown in Table 9. No 2,3,7,8-TCDD could be detected in any of the 5 specimens at the detection limits ranging from 0.32 to 0.85 ppt on a total weight basis. These levels are calculated for whole milk diluted to 50% of a 50:50 ethanol:water solution. These findings are similar to those presented by Rappe (Chapter 2 of this volume) for milk specimens provided by JDC from the same Vietnam trip; i.e., higher chlorinated dioxins were detected but no 2,3,7,8-TCDD was found.

For mothers in the United States, the amount of milk that would be expected to be expressed would average 80–100 mL per breast per feeding. The first milk expressed would be low in fat content (less than 2%) and the last part of the feeding would be higher (about 5–6%). Likewise, milk from very early feeding, immediately after birth, would have lower lipid content, and hence lower dioxin content, than milk taken later in the nursing.[13] Unfortunately, the lipid content of these milk samples was not determined. The 15 to 26 mL of milk obtained from each woman represents less than the nursing output of one breast if Vietnamese women in the south of Vietnam are similar in this respect to mothers in the United States. Information about these Vietnamese women is sparse; their physicians believed they had been exposed to dioxin from Agent Orange between 1964–1971.

DISCUSSION

The discovery of dioxins in adipose tissue samples from patients in Ho Chi Minh City hospitals suggests that 2,3,7,8-TCDD persisted in 1984 at elevated levels as a result of exposure to Agent Orange. Patients from the south were similar to patients from industrialized countries in their levels of specific isomers of penta- through octachlorodibenzodioxins and dibenzo-furans. The results differ from our previous findings in that the levels of 2,3,7,8-TCDD are about 3 times higher in tissue from persons in Ho Chi

Minh City (Saigon plus surrounding areas) than in tissue from persons in North America and Canada. Even more striking is the difference in levels of 2,3,7,8-TCDD in patients from southern Vietnamese hospitals, compared with those in the north where no 2,3,7,8-TCDD could be detected at a 2 or 3 ppt detection limit.

We tentatively conclude that elevated body burden of 2,3,7,8-TCDD, probably from Agent Orange either directly or secondarily through the diet, persists in the bodies of some Vietnamese living in the south of Vietnam despite the long time period (about 13 years) since the last application of Agent Orange in Vietnam. The other dioxin and dibenzofuran isomers are found in tissue from persons from the south at levels similar to those in tissue of persons in industrial countries, but the levels are much lower in specimens from the north. This may be because of the alleged extensive use of technical grade pentachlorophenol (PCP), a wood preservative and anti-fungal agent, during the years 1954–1973, when the U.S. was extensively involved in the south of Vietnam. The finding of low levels of all dioxins and furans in Hanoi patients is consistent with what might be expected in a country with sparse financial resources where chemicals are not used to the extent that they are in industrialized countries. Incineration may also con-tribute to contamination of human tissue in southern Vietnam, but the extent is not known.

With respect to the finding of nondetectable levels of 2,3,7,8-TCDD in the fat content of the milk, one can only speculate as to the amount expected when one extrapolates from adipose tissue data. Fat-soluble chlo-rinated chemicals, such as dioxins, PCBs, and dibenzofurans, are found passively distributed in fat stores in the body. The levels in adipose tissue can often be related to those in breast milk by simple calculations, as has been shown with PCBs. Because the age, exposure, nursing history, decrease in body burden of 2,3,7,8-TCDD to be expected from nursing, and fat content of these women are not known, further speculation is not justi-fied. It may be that these women were young, lived far from sprayed areas, ate insufficient fish or other contaminated food, or expressed the milk sample at times when the fat content was low.

CONCLUSION

The findings of high levels of the chemically and toxicologically similar chlorinated dioxins and dibenzofurans in tissues of individuals in industrial-ized countries suggests a difficulty in performing clinical or epidemiological studies of the human health effects of dioxins in those countries. With the development of analytical methods for fat biopsies as a sensitive and spe-cific marker to estimate exposure, it may now be possible to identify more accurately levels of exposure in a quantitative fashion, and hence improve clinical and epidemiological studies. Such research in Vietnam may present an opportunity to determine more rapidly the extent of toxicity of dioxins to humans than has been the case in past and some present clinical and epidemiological studies.

Fat biopsies, which are quite simple, easily tolerated, and relatively painless, especially when done by suction aspiration, or during the course of surgery for other conditions, in combination with chemical extraction, separation, and analysis, constitute a powerful approach for estimating body burden of specific isomers even though the metabolic fate and time course of the various dioxins and furans of interest and concern in man are not known. Present studies suggest a 1- to 5-yr half-life for certain dioxin isomers in humans.[18,19] With the recent identification of the dioxin receptor in humans, the need to investigate the presence or absence of the dioxin receptor in study patients may also be of critical importance in human health effect studies.

Vietnamese epidemiologists believe that liver cancer, hydatidiform mole, spontaneous abortions, and neural tube defects are caused by the dioxin in Agent Orange in their country.[20] Using adipose tissue biopsy and dioxin analysis as tools in epidemiological research, it may be possible to prove or disprove these hypotheses.

It should also be considered that there are over 500,000 Vietnamese in the United States at this time, some of whom may be at increased health risk because of previous exposure to 2,3,7,8-TCDD from Agent Orange applied during 1964–1971, when most of them lived in Vietnam. In addition, there are over 3,000,000 United States personnel who served in Vietnam, some of whom may have been significantly exposed to Agent Orange. The detection of 2,3,7,8-TCDD from Agent Orange exposure after at least ten years in some U.S. servicemen[5] who served in Vietnam is consistent with exposure.

At the present time there is no generally recognized way to remove dioxin from the human body. Sucrose polyesters, mineral oil, charcoal, and severe weight loss have been considered but no clinical trials have demonstrated efficacy, nor are any trials under way at present in the U.S. to the best of our knowledge.[21,22] This is an important area for future medical research.

SUMMARY

In contrast to findings in previous studies of breast milk obtained from south Vietnam in 1970 and 1973 no 2,3,7,8-TCDD was detected in the 1983 breast milk samples. However, higher mean levels of 2,3,7,8-TCDD, the dioxin characteristic of 2,4,5-T (a component of Agent Orange), were found in some of the adipose tissues from the south of Vietnam than were previously found in most U.S. studies. In samples from Hanoi, where there was no exposure to Agent Orange, the level was nondetectable at a 2 to 3 part per trillion detection level. This population had tissue levels of 2,3,7,8-TCDD lower than any other known population studied to date. Hanoi patients also had lower penta- through octachlorinated dioxin levels than have been detected in other populations from industrial countries. Ho Chi Minh City patients had levels of penta- through octachlorinated dioxins and furans similar to those found in the U.S., Canada and Europe, suggesting technical grade pentachlorophenol, food and incineration sources for these other non-Agent Orange associated dioxins. A potentially large population of heavily dioxin-exposed persons may exist in the south of Vietnam, while an equally large group with a low level of dioxin and furan exposure can be presumed to exist in the north. We suggest that this provides a valuable setting for clinical and epidemiological studies of the possible human health effects of dioxin.

ACKNOWLEDGMENTS

The authors wish to acknowledge generous financial support from grants from the Christopher Reynolds Foundation, Inc., and the Samuel Rubin Foundation to the Research Foundation of the State University of New York. In addition, we wish to thank the VVA Foundation, Inc., for financing and arranging the 1984 trip to Vietnam. The research at the University of Nebraska was supported by a regional instrumentation facility grant from the U.S. National Science Foundation (No. CHE-8211164). In Vietnam, we especially wish to thank our colleagues working on dioxin research, including, Dr. Hoang Dinh Cau, Dr. Ton Duc Lang, Dr. Nguyen Can, Dr. Duong Thi Cuong, Dr. Le Cao Dai, and Dr. Nguyen Tien Thinh, of Hanoi, Ministry of Health, the Hanoi Medical School, Viet Duc Hospital, the Institute (Hospital) for Gynecology and Obstetrics (or Hospital for Mothers and Newborn Infants) and Dr. Nguyen Thi Ngoc Phuong, Dr. Pham Hoang Phiet, Cho Ray Hospital, and the Ho Chi Minh City Medical School, as well as many other colleagues who worked with us in Vietnam and also in the United States.

REFERENCES

1. *Halogenated Biphenyls, Terphenyls, Naphthalenes, Dibenzodioxins and Related Products, Vol. 4*, R. D. Kimbrough, Ed. (New York: Elsevier/North Holland Biomedical Press, 1980).
2. *Herbicides in War: The Long-Term Ecological and Human Consequences* A. H. Westing, Ed. (London: Taylor & Francis, 1984).
3. Baughman, R. W. "Tetrachlorodibenzo-p-Dioxins in the Environment: High Resolution Mass Spectrometry at the Picogram Level," PhD Thesis, Harvard University, Cambridge, MA (1974).
4. Baughman, R., and M. Meselson. "An Analytical Method for Detect-

ing TCDD (Dioxin): Levels of TCDD in Samples from Vietnam," in *Environ. Health Pers.* (September, 1973), pp. 27–35.

5. Gross, M. L., J. O. Lay, Jr., P. A. Lyon, D. Lippstreu, N. Kangas, R. L. Harless, S. E. Taylor, and S. E. Dupuy, Jr. "2,3,7,8-Tetrachloro-dibenzo-p-Dioxin Levels in Adipose Tissue of Vietnam Veterans," in *Environ. Res.* 33:261–268 (1984).

6. Schecter, A., T. Tiernan, F. Schaffner, M. Taylor, G. Gitlitz, G. F. Van Ness, J. H. Garrett, and D. J. Wagel. "Patient Fat Biopsies for Chemical Analysis and Liver Biopsies for Ultrastructural Characterization After Exposure to Polychlorinated Dioxins, Furans and PCBs," in *Environ. Health Pers.* 60:241–254 (1985).

7. Schecter, A., T. Tiernan, M. Taylor, G. Van Ness, J. Garrett, D. Wagel, G. Gitlitz, and M. Bogdasarian. "Biological Markers After Exposure to Polychlorinated Dibenzo-p-Dioxins, Dibenzofurans, Biphenyls and Biphenylenes. Part I: Findings Using Fat Biopsies to Estimate Exposure," in *Chlorinated Dioxins and Dibenzofurans in the Total Environment II*, L. Keith, C. Rappe, and G. Choudhary, Eds. (Stoneham, MA: Butterworth Publishers, 1985), pp. 215–245.

8. Schecter, A., J. J. Ryan, R. Lizotte, W. F. Sun, and L. Miller. "Chlorinated Dibenzo-Dioxin and Dibenzofuran Levels in Human Adipose Tissues in Exposed and Control Patients," in *Chemosphere* 14(6/7):933–938 (1985).

9. Wolff, M., J. Thornton, A. Fischbein, R. Lilis, and I. Selikoff. "Disposition of Polychlorinated Biphenyl Congeners in Occupationally Exposed Persons," in *Toxocol. Appl. Pharmacol.* 62:294–306 (1982).

10. Wolff, M., A. Fischbein, J. Thornton, C. Rice, R. Lilis, and I. Selikoff. "Body Burden of Polychlorinated Biphenyls Among Persons Employed in Capacitor Manufacturing," in *Int. Arch. Occup. Environ. Health* 49:199–208 (1982).

11. Masuda, Y., R. Kagawa, and M. Kuratsune. "Polychlorinated Biphenyls in Yusho Patients and Ordinary Persons," in *Fukuoka Acta Medica* 65(1) (January, 1974).

12. Ryan, J. J., R. Lizotte, and B. P. -Y. Lau. "Chlorinated Dibenzo-p-Dioxins and Chlorinated Dibenzo-p-Furans in Canadian Human Adipose Tissue," in *Chemosphere* 14(6/7):697–706 (1985).

13. Neville, M. C., and M. R. Neifert, Eds. *Lactation: Physiology, Nutrition, and Breast-Feeding.* (New York: Plenum Press, 1983), 74–75, 381–391.

14. Rappe, C., M. Nygren, and G. Gustafsson. "Human Exposure to Polychlorinated Dibenzo-p-Dioxins and Dibenzofurans," in *Chlorinated Dioxins and Dibenzofurans in the Total Environment*, G. Choudhary, L. H. Keith, and C. Rappe, Eds., (Stoneham, MA: Butterworth Publishers, 1983), pp. 355–365.

15. Graham, M., F. Hileman, D. Kirk, J. Wendling, and J. Wilson. "Background Human Exposure to 2,3,7,8-TCDD," in *Chemosphere* 14:925–928 (1985).

16. Stanley, J. S. "Analysis of PCDDs in Human Adipose Tissue," Special

Report—*Meeting Summary for E. P. A.*, Midwest Research Institute, Kansas City, MO, (June, 1983), pp. 1–23.

17. Albro, P. W., W. B. Crummett, A. E. Dupuy, Jr., M. L. Gross, M. Hanson, R. L. Harless, F. Hileman, D. Hilker, C. Jason, J. L. Johnson, L. L. Lamparski, B. P. -Y. Lau, D. D. McDaniel, J. L. Meehan, T. J. Nestrick, M. Nygren, P. O. O'Keefe, T. L. Peters, C. Rappe, J. J. Ryan, L. M. Smith, D. L. Stalling, N. C. A. Weerasinghe, and J. M. Wending. "Methods for the Quantitative Determination of Multiple, Specific Polychlorinated Dibenzo-p-Dioxin and Dibenzofuran Isomers in Human Adipose Tissue in the Parts-per Trillion Range. An Interlaboratory Study," in *Anal. Chem.* 57:2717–2725 (1985).

18. Gorski, T., L. Komopka, and M. Brodzki. "Persistence of Some Polychlorinated Dibenzo-p-Dioxins and Polychlorinated Dibenzofurans of Pentachlorophenol in Human Adipose Tissue," in *Roczn. Phz.* 35:297–301 (1984).

19. Poiger, H., and C. Schlatter. "Pharmacokinetics of 2,3,7,8-TCDD in Man," *Chemosphere* (1986) in press.

20. Proceedings, Conference on the Effects of the Use of Herbicides in War, 1983, Vietnam Health Ministry Publications, 1983.

21. Manari, L., P. Coccia, and T. Croci. "Prevention of TCDD Toxicity in Laboratory Rodents by addition of Charcoal or Cholic Acids to Chow," in *Fd. Chem. Toxic.* 22(10):815–818 (1984).

22. Manari, L., P. Coccia, and T. Croci. "Persistent Tissue Levels of TCDD in the Mouse and Their Reduction of Toxicity," in *Drug Metab. Rev.* 13(3):423–466 (1982).

CHAPTER **4**

Chlorinated Dioxin and Dibenzofuran Levels in Human Adipose Tissues from Exposed and Control Populations

Arnold Schecter, John J. Ryan, and George Gitlitz

INTRODUCTION

We have previously presented findings of elevated dibenzofuran levels in human adipose tissue following the Binghamton State Office Building PCB transformer fire of February 5, 1981, as well as findings of elevated blood levels of PCBs. The PCB mixture found in the transformer was similar to the pattern of PCB and furan isomers found in the adipose tissue, the soot, and an Arochlor–1254 sample.[1-7] With improved chemical techniques, and improvements in obtaining fat samples from patients in a less invasive fashion, isomer-specific characterization of the levels of furans and dioxins in human adipose tissue is now possible. It can be used to characterize exposure to an extent not previously possible for many of the dioxins and furans, down to the level seen in the general adult population in industrial countries.[8-10]

Without such markers of ingestion (by dermal, gastrointestinal or respiratory routes) it is not possible to quantitate the dose actually taken into the body. Without some quantitative or semiquantitative means of estimating dose in a given patient, it is difficult in many instances, such as the PCB transformer accident in Binghamton, to differentiate persons with or without intake of these chemicals. It has become routine to measure blood levels

51

of PCBs after such incidents and then again some time afterwards if the levels appear possibly elevated for a given patient, to determine if the level has decreased over time, as would be expected with intake after an acute incident. Until chemical techniques for extraction, separation and isomer-specific identification of furans and dioxins measured down to background levels in adipose tissue where these fat-soluble compounds accumulate became available, it had not been possible to perform the sort of laboratory determinations for the furans and dioxins that clinicians are accustomed to employing in estimating patient PCB (or other chemical) exposure. This has presented major problems in evaluating clinical effects of these chemicals on humans.

This chapter extends and refines the development of the technique of fat biopsy followed by isomer-specific identification and quantification of dioxins and furans. Also, a larger number of general population patients from the North American continent (U.S. and Canada) is now available for purposes of comparison. It is now evident that the general U.S. and Canadian populations are not nonexposed to dioxins and furans; rather, there is chronic exposure, probably from food and air, from municipal and industrial incineration products, to a considerable amount of dioxins and furans. These are found at a total level of about 1000 parts per trillion (ppt) in adipose tissue, as well as at lower levels in other tissues, in adults in industrial countries studied to date. Further, these dioxins and furans found in human tissue all appear to be 2,3,7,8 laterally substituted, and thus are more toxic than other dioxins and furan isomers. We have usually found 9 to 14 of the more than 200 theoretically possible isomers of dioxins and furans in adult human tissues at current detection limits.[8-10]

METHODS

Adipose tissue biopsy can be accomplished in several ways. Originally, a small, (ca. 2 in.) incision was employed after local anesthetic was injected in the skin and subcutaneous tissue of the buttock or abdomen; this is an outpatient procedure and can be performed in a physician's office. A portion of fat the size of a large marble or table tennis ball, 5 to 20 g in weight, is removed and placed in a chemically clean container, after which the specimen is frozen. The incision is closed with sutures and the patient can then return to work.

Recently, we have used suction to remove adipose tissue. With this procedure, pioneered by plastic or reconstructive surgeons, local anesthesia is injected in the usual fashion. A small opening is made in the skin with a scalpel and a suction device is placed in the small incision. The adipose tissue specimen is aspirated in several minutes. A suture or two may be used to close the small opening in the skin, if needed. A pressure dressing is applied, kept in place for one day, and then discarded. The patient can return to work immediately after the biopsy. The content of the sample is sometimes more diluted with blood and other nonlipid tissue components when this technique is employed.

The chemical techniques used are described elsewhere in this volume and will not be detailed here. Homogenization of tissue is followed by digestion, chromatographic cleanup on Florsil and carbon columns, gas chromatographic separation, and mass spectroscopic identification using appropriate standards, with correction of values to reflect recovery percentages. Recently, we have begun to report dioxin and furan levels for adipose tissue specimens, and sometimes for liver tissue as well, in ppt on a lipid basis, rather than on the common wet weight basis.

FINDINGS

Figure 1 is a schematized representation of typical findings from the North American adult control population. It represents a mass spectroscopic pattern for the tetra- and pentafurans and dioxins, with specific isomers labeled. Figure 2 illustrates the typical findings for the hexa; hepta- and octachlorinated furans and dioxins.

Table 1 shows the dioxin and furan isomers detected in a soot sample from the Binghamton State Office Building after the 1981 PCB transformer fire.[11] Other soot samples showed slightly different values, as might be expected due to soot or laboratory variability, but this typical sample illustrates significant levels of furan isomers presumably formed from heating of the PCBs, and also the lower levels of dioxins presumably formed from heating the tri- and tetrachlorinated benzenes which constituted 35% of the original transformer mixture. Noteworthy are the many 2,3,7,8 laterally substituted furans, and to a lesser extent, dioxins. Also of interest is the finding of high levels of hexa-; followed by penta- and then heptafurans. The isomers found in the Yusho patients, 2,3,4,7,8-PCDF, 1,2,3,4,7,8-HxCDF, and 1,2,3,6,7,8-HxCDF, chemicals with documented human toxicity, are present in the soot mixture.[12-15]

Table 2 reviews our findings of the isomer-specific adipose tissue levels of dioxins and furans from the general adult North American population. Dioxins are present at higher levels than furans and, in general, the highest levels are found of the more highly chlorinated PCDFs and PCDDs than of lesser chlorinated ones. In general, there are the highest levels of octa-, followed by hepta-, then hexa-, penta- and finally the least amount of tetrachlorinated dioxin. Total dioxin isomers found in human adipose tissue from adults at this time in the United States and Canada, on a wet weight basis, are approximately 1000 ppt (here 1088 ppt), and the total furan isomers are less (here 60.8 ppt). Total dioxin plus furan isomers here are just

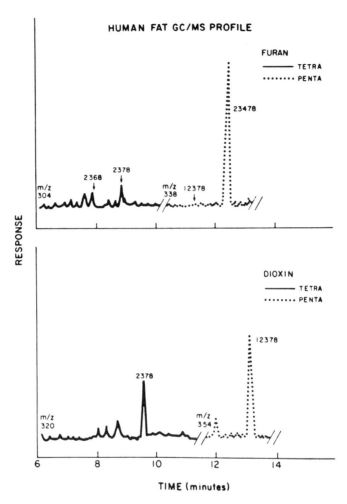

Figure 1. This schematic shows typical mass spectrometric pattern for tetra- and pentachlorinated dioxins and furans seen in the samples from industrialized countries.[9]

over 1000 ppt (here 1149 ppt). The mean level of 2,3,7,8-TCDD here is found to be 6.4 ppt, with a range of 2.0–12.7 ppt, excluding nondetectable specimens.

Table 3 shows dioxin and furan levels from four general population adult upstate New York patients, labeled 1, 2, 3, and 4, with adipose tissue either intraabdominal in origin, e.g., omentum or mesentery (Abd), or subcutaneous in origin (Subcu), and a Binghamton State Office Building (BSOB) exposed worker, number 5. Adipose tissue was obtained from Patients 3 and 4 during the course of autopsies, thus permitting adipose tissue from

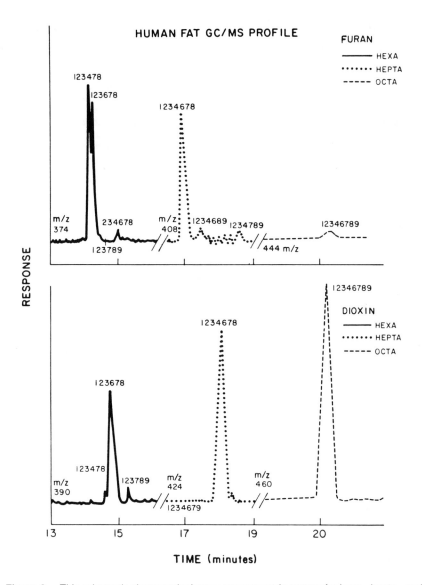

Figure 2.   This schematic shows typical mass spectrometric pattern for hexa-, hepta-, and octachlorinated dioxins and furans seen in industrialized countries.[9]

two frequently sampled adipose tissue anatomical sites to be easily obtained in sufficient quantity to compare levels of these chemicals in the two tissue sites, presumably in equilibrium, of the same patient. Little or no difference is apparent in dioxin or furan levels from the same patient from these two adipose tissue sites. Of particular interest is the higher level of dioxins and furans found in Patient 5 (1418 ppt) as compared with the under 1000 ppt

Table 1.   Dioxin and Furan Levels (ppm) in Binghamton, New York Soot Sample from
PCB Transformer Fire[11]

Dioxins	Level	Furans	Level
2,3,7,8-Tetra-CDD	0.6	2,3,8,9-Tetra-CDF	12.0
Others (4)[a]	0.6	1,3,7,9-Tetra-CDF	1.0
		Others	15.0
1,2,3,7,8-Penta-CDD	2.5		
Others (7)	2.5	1,2,3,7,8-Penta-CDF	310.0
		2,3,4,7,8-Penta-CDF	48.0
1,2,3,4,7,8-Hexa-CDD	0.7	1,2,4,7,8-Penta-CDF	25.0
1,2,3,6,7,8-Hexa-CDD	0.6	1,2,4,7,9-Penta-CDF	22.0
1,2,3,7,8,9-Hexa-CDD	0.4	1,3,4,7,8-Penta-CDF	65.0
1,2,3,4,6,8/1,2,4,6,7,9/1,2,4,6,8,9-Hexa-CDD	1.2	1,2,6,7,8-Penta-CDF	25.0
1,2,3,6,8,9/1,2,3,6,7,9-Hexa-CDD	1.3	1,2,3,6,7-Penta-CDF	60.0
1,2,3,4,6,7-Hexa-CDD	0.5	2,3,4,6,7-Penta-CDF	12.0
		Others (12)	110.0
1,2,3,4,6,7,9-Hepta-CDD	4.0		
1,2,3,4,6,7,8-Hepta-CDD	3.0	1,2,3,4,7,8-Hexa-CDF	310.0
		1,2,3,6,7,8-Hexa-CDF	150.0
Octa-CDD	2.0	2,3,4,6,7,8-Hexa-CDF	10.0
		1,2,3,4,6,8-Hexa-CDF	30.0
Total	19.9	1,2,3,6,8,9-Hexa-CDF	38.0
		1,2,4,6,7,8-Hexa-CDF	50.0
		1,3,4,6,7,8-Hexa-CDF	125.0
		Others (8)	250.0
		1,2,3,4,6,7,8-Hepta-CDF	230.0
		1,2,3,4,6,7,9-Hepta-CDF	120.0
		1,2,3,4,6,8,9-Hepta-CDF	55.0
		1,2,3,4,7,8,9-Hepta-CDF	55.0
		Octa-CDF	40.0
		Total	2068

[a]( )refers to the number of isomers.

levels seen in the control specimens from the same geographical region
(mean 835 ppt and range 704–971 ppt). Of interest is a mean furan level in
the controls of 70.4 ppt, compared with 402 ppt total furans in the adipose
tissue of the exposed patient, an almost sixfold increase in the furans which
predominated in the soot. The total dioxin mean of 835 ppt in the controls is
only slightly below the 1015 ppt total dioxin content of the exposed patient,
presumably reflecting the smaller amounts of dioxins in the soot as compared to the furans and the higher background level of dioxins in human
adipose tissue. Striking, also, in analyzing Patient 5, is the finding of elevated levels of six of the isomers found in the original soot mixture to which
the patient had been exposed during several months of work in the building
two years previously. The three isomers found in Yusho patients, 2,3,4,7,8-
PCDF, 1,2,3,4,7,8-HxCDF and 1,2,3,6,7,8-HxCDF, found in all the
patients, are seen to be especially elevated in this patient, at 74.7, 149 and
112, three years after his exposure to them in the soot and air. Also of

Table 2. Dioxin and Furan Levels in Human Adipose Tissues from the General Canadian Adult Population on a Whole or Wet Weight Basis; Values in Parts per Trillion for 46 Samples Collected in 1976.

Analyte	Mean of Positives	Number Positive	Range
2,3,7,8-D[a]	6.4	25	ND,[c]2.0 – 13
1,2,3,7,8-D	10	46	3.4 – 34
1,2,3,6,7,8-D	87	46	19 – 356
1,2,3,4,6,7,8-D	135	46	1.6 – 580
1,2,3,4,6,7,8,9-D	850	46	202 – 2961
Total Dioxins	1088		
2,3,4,7,8-F[b]	15	46	4.2 – 36
1,2,3,4,7,8/			
1,2,3,6,7,8-F	16	34	ND,6.2 – 54
1,2,3,4,6,7,8-F	30	44	5.4 – 76
Total Furans	61		
Total D & F	1149		

[a]D, dibenzo-p-dioxin isomer.
[b]F, dibenzofuran isomer.
[c]ND, not detected.

Table 3. Adipose Tissue Levels of Dioxins and Furans (in ppt) from One Exposed and Six General Population[a] Specimens from Binghamton, New York, on a Wet Weight Basis, 1983-84.

Patient	1(Subcu)[b]	2(Subcu)	3(Subcu)	3(Abd)[c]	4(Subcu)	4(Abd)	5(Subcu)
2,3,7,8-D[d]	8.3	7.2	6.0	5.7	3.7	7.4	11.6
1,2,3,7,8-D	13.8	10.3	8.2	7.8	7.5	10.5	15.0[i]
1,2,3,6,7,8-D	46.2	54.5	60.3	64.2	60.4	60.9	72.6
1,2,3,7,8,9-D	7.4	7.5	7.4	9.4	6.8	10.6	7.3
1,2,3,4,6,7,9-D	ND[g]	ND	2.7	2.6	5.3	3.0	9.6
1,2,3,4,6,7,8-D	95.8	39.4	119.0	114.0	93.1	110.0	209.0
OCDD	534	493	675	675	586	428	690
2,3,7,8-F[e]	ND	4.1	ND(2)	ND	ND	ND	ND
2,3,4,7,8-F[f]	12.5	10.9	17.0	16.7	16.5	14.5	74.7[i]
1,2,3,4,7,8-F[f]	11.4	9.3	13.0	26.4	22.9	16.3	149.0[i]
1,2,3,6,7,8-F[f]	5.6	5.8	8.8	25.5	15.4	11.9	112.0[i]
1,2,3,4,6,7,8-F	16.3	13.7	12.5	15.0	23.8	16.8	39.3[i]
1,2,3,4,7,8,9-F	ND	ND	19.6	ND	20.6	13.7	25.9[i]
OCDF	ND(20)[h]	ND(20)	1.2	2.5	1.5	0.9	1.6
Total D	706	712	899	879	762	630	1015
Total F	45.8	43.8	72.1	86.1	100.7	74.1	402
Total D & F	751	755	971	965	863	704	1418

[a]Mean of general population is 835 ppt.
[b]Subcu, subcutaneous abdominal or gluteal area adipose tissue.
[c]Abd, abdominal cavity adipose tissue.
[d]D, dibenzo-p-dioxin isomer.
[e]F, dibenzofuran isomer.
[f]Toxic isomers thought to have accounted for the Yusho rice oil toxicity.
[g]ND, not detected.
[h]( ) indicates detection limit.
[i]Isomers found in the exposed patient which were reported by Rappe et al.[15] as being prominent in Binghamton State Office Building soot.

Table 4.   Control and Exposed Patients' Adipose Tissue Levels of Dioxins and Furans (ppt).

	Wet Weight Basis[a]						
	1-S[b]	2-S	3-S	3-A[c]	4-S	4-A	5-S
Total Dioxins	706	712	899	879	763	630	1015
Total Furans	45.8	43.8	72.1	86.1	101	73.6	403
Total D & F	752	756	971	965	864	704	1418
Lipid %	88.9	83.7	75	75.2	66.6	70.9	85.2

	Extractable Lipid Basis						
	1-S	2-S	3-S	3-A	4-S	4-A	5-S
Total Dioxins	794	851	1200	1170	1150	890	1190
Total Furans	51.5	52.3	96.1	114	151	104	472
Total D & F	846	903	1296	1284	1301	994	1662

[a]1, 2, 3 and 4 are general population control patients; 5 represents adipose tissue from a Binghamton State Office Building exposed patient.
[b]S, Subcutaneous adipose tissue.
[c]A, Abdominal mesenteric adipose tissue.

interest, but difficult to explain, is the elevated hepta isomer 1,2,3,4,6,7,8-HpCDD, which, at a level of 209, is higher than the general population, and may reflect absorption and retention from the Binghamton building's soot or air. This patient has no other known source of the elevated chemicals which could be found on repeated, thorough, medical evaluations. Noteworthy also is the usual finding of similar levels of dioxins and furans in the adipose tissue of the same individuals regardless of the anatomical site from which it is derived. We interpret the small differences occasionally noted to reflect laboratory variability rather than anatomical or physiological variation. It is also noteworthy that some isomers found in high quantity in the soot are not elevated in the patient, but whether this reflects lesser intake or more rapid elimination is not clear.

Table 4 displays the total dioxin and furan values on a wet weight basis as well as an extractable lipid basis for these same seven samples. Either presentation of the data may be useful, depending on which organ is being analyzed or considered, and whether or not one wishes to attempt a conversion of values to a standard reference base. Adipose tissue samples have yielded considerable variation in lipid content, quite possibly due to the specimen containing blood, serum, and tissue components other than pure fatty tissue, when a surgical sample is presented for chemical analysis. In these seven samples, the lipid content varies only from 66% to 88.9%, but in other samples we have analyzed, lipid values as low as 14% to 17% have been measured. If one wishes to make calculations estimating total body burden or milk content of these chemicals from adipose tissue data, extractable lipid basis may well be preferable. When the values are presented on an extractable lipid basis, the concentration appears higher than when a wet weight or whole tissue weight basis is employed. Thus, our mean rises here

Table 5. Dioxin and Furan Adipose Tissue Levels (ppt) in 8 Control and 4 Exposed Samples from Binghamton, New York

Analyte	Control Mean	Control Range	Exposed A	Exposed B	Exposed C	Exposed D	Mean Exposed
2,3,7,8-D[a]	7.2	1.4–17.7	13.3	28.3	16.2	11.6	17.4
1,2,3,7,8-D	11.1	5.2–25.2	2.2	11.1	5.7	15.0	8.5
1,2,3,6,7,8-D	95.9	46.2–355	66.5	181	193	73	128
1,2,3,4,6,7,8-D	164	53–691	72	531	325	209	284
1,2,3,4,6,7,8,9-D	707	214–1931	166	946	948	690	688
2,3,4,7,8-F[b]	14.3	3.1–19.7	ND(3)[d]	24.3	45.6	74.7	48.2(3)[e]
1,2,3,4,7,8/1,2,3,6,7,8-F	31.3	ND[c],15.1–46.9	ND(3)[d]	13.5	97	261	124(3)[e]
1,2,3,4,6,7,8-F	16.5	ND,12.5–23.8	8.6	14.5	17.7	39.3	20
Lipid %	70.6	46–88	69	65	83	85	76

[a]D = dibenzo-p-dioxin isomers.
[b]F = dibenzofuran isomers.
[c]ND, not detected.
[d]Detection limit in parentheses.
[e]Mean number of samples in parentheses.

from below 1000 ppt to above that level. In comparing values in the literature, such different presentations must be clearly specified and kept in mind. Liver also has varied considerably in lipid content; increased intracellular lipid is frequently seen in response to toxic chemical injury in liver cells.

Table 5 expands the Binghamton State Office Building exposed cohort to four patients and a geographical adult comparison group to eight patients. As might be expected, a variation exists in the levels of isomers such that there is an overlap between the general population and the exposed group of patients. The mean levels are higher, on an isomer-specific basis, for those exposed to furans, and to a lesser extent, dioxins, in the Binghamton building. The data suggest varying levels of ingestion and/or retention of the chemicals while workers were in the building. The finding of 48.2 ppt vs 14.3 ppt mean values among positives for 2,3,4,7,8-PCDF, of 124 ppt vs 31.3 ppt for 1,2,3,4,7,8-HxCDD plus 1,2,3,6,7,8-HxCDF and of 284 ppt vs 164 ppt for 1,2,3,4,6,7,8-HpCDD are consistent with ingestion and retention of chemicals from the Binghamton soot. More striking and convincing is the finding of a specific patient with both elevation of total furan isomers and elevation of specific isomers to which the patient was exposed, designated as Patient D in Table 5, whose dioxin and furan levels were also shown as Patient 5 in Table 3.

Table 6 presents summarized data on the total dioxin and furan values from the exposed and control patients from the previous table, with an outlier (Control 8) presented separately. The furan and dioxin values calculated on a wet weight and lipid basis are both shown. With expression of the data on a lipid basis the concentrations are higher, and the fixed base, lipid, may be more useful or meaningful in characterizing content of these chemi-

Table 6. Total Dioxins and Furans (ppt) in Adipose from Control and Binghamton, New York Exposed Subjects.

		Control			
		Total Dioxins		Total Furans	
Number	Lipid %	Wet	Lipid	Wet	Lipid
1	88.9	698	785	45.8	51.5
2	83.7	704	842	39.7	47.4
3	75.0	889	1184	51.3	68.4
4	75.2	867	1152	83.6	111
5	66.6	751	1127	78.6	118
6	70.9	617	870	78.6	111
7	46.0	340	738	6.5	14
8	59.6	3020	5067	83.9	141
Mean	70.7	695(7)[a]	956(7)	54.7(7)	74.5(7)
		986(8)[b]	1471(8)	58.5(8)	82.8(8)
		Exposed			
1	69	320	464	8.6	12.5
2	65	1697	2611	52.3	80.4
3	83	1488	1793	160	193
4	83.2	998	1172	375	440
Mean	75.6	1126	1510	149	181

[a]Mean values calculated from seven control samples, and excluding one outlier.
[b]Mean values calculated from all eight control samples.

cals. The presence of one control patient with a high level of total dioxins tends to obscure somewhat the difference between the two groups. However, the difference between the two groups is still quite clear when the total furans, the major chemical here to which patients were exposed, are considered. The furan value of 181 ppt on a lipid basis is notably higher than either 74.5 ppt or 82.8 ppt, with or without outlier Control 8. The dioxin values for the exposed patients on a lipid basis, 1510, is somewhat higher than either the 1471(8) or 956(7) ppt of the controls. The same mean elevation of dioxin levels is shown here in exposed patients (1126 ppt) as compared to 986(8) or 695(7) on a wet weight basis for the control patients. After PCB transformer incidents or other PCB, PCDD or PCDF potential exposures, persons who wish to be placed on a medical surveillance in order to determine whether or not they have ingested chemicals, and experienced pathology from such ingestion, may in some cases be found to have ingested little if any of these chemicals. This is apparent in Table 6, where it can be seen that exposed Patients 3 and 4 appear to have evidence of exposure, with levels of furans of 193 and 440 ppt, respectively (on a lipid basis), as compared with potentially exposed Patients 1 and 2, where the finding of 12.5 and 80.4 suggests markedly less ingestion of furans from the Binghamton incident or any other source of furans; the time after the 1981 exposure is between two and three years for the patients in this table. Previous values are not known, however, and in some cases serial biopsies, or in the case of

Table 7.   Tissue Levels (ppt) in Five Yusho Victims After the 1968 Rice Oil PCB and Furan Incident in Japan.[14]

Analyte	Subject 1 Liver 1969	Subject 2 1969 Fat	Liver	Subject 3 1972 Fat	Liver	Subject 4 Fat 1975	Subject 5 1977 Fat	Liver
2,3,6,8-F[a]	700	600	80	80	30	400	ND(5)	ND(5)
2,3,7,8-F	300	300	20	ND[b](5)[c]	7	ND(5)	ND(5)	ND(5)
1,2,4,7,8-F	7100	1000	400	200	90	800	200	20
2,3,4,7,8-F	6900	5700	1200	800	300	100	500	100
1,2,3,4,7,8/ 1,2,3,6,7,8-F	2600	1700	300	200	30	500	ND(5)	40

[a]F, dibenzofuran isomers.
[b]ND, not detected.
[c]( ) indicates detection limit.

PCBs, serial blood samples, must be taken over time to rule out ingestion of these chemicals from this incident, since preexisting levels are usually not known for a given patient or group. Control Patient 8 stands out because of quite high total dioxin levels, with the source of exposure not known. It is not, however, the Binghamton Building. This conclusion is based on the patient's history and also on his low levels of furans, the predominating chemical contaminating that building.

Table 7 presents findings from an incident where some of the same furan isomers were ingested in rice oil in 1968, along with PCBs, in Japan, and where tissue levels were measured at various times afterward. These values are noted frequently to be higher than those found in our patients, e.g., the 5700 or 1700 ppt, respectively of 2,3,4,7,8-PCDF and a mixture of 1,2,3,4,7,8- and 1,2,3,6,7,8-PCDF in adipose tissue of Patient 2, in 1969. It is also striking that quite low or nondetectable levels of these furans were found in Patient 5 in 1977, 9 years after the incident. It is not clear whether the amount of material ingested or absorbed was similar for different patients. Nevertheless, the ranges found in Japan, where human illness was attributed to ingestion of certain penta- and hexachlorinated furans, over-lap with levels seen in Binghamton patients two or three years after the incident.

SUMMARY AND DISCUSSION

A group of PCB-, furan- and dioxin-exposed patients had adipose tissue biopsies performed two years or more after exposure to these chemicals in the course of their work in a building contaminated with these compounds. Elevated levels of furans, especially those isomers found in the building, were found in exposed workers, and were especially apparent in certain individual patients. To a lesser extent, dioxins, to which the patients were exposed to a lesser degree than the furan isomers, were also elevated in exposed patients. There was an overlap of some values with those seen in control patients from the same geographical area, reflecting, we believe, variations in individual doses of furans and dioxins. Some persons had evidence of high absorption and/or retention of these chemicals, while this was not obvious in other potentially exposed patients. In our clinical experience using serial blood PCB measurements to monitor worker exposure after PCB incidents, the majority of those in most medical surveillance programs show little or no evidence of intake of PCBs and can thus be reassured.

Levels of certain penta- and hexachlorinated furan isomers were found to be similar, although lower, than those seen in patients exposed several years previously to toxic levels of these chemicals in the Yusho or rice oil PCB and furan poisoning incident of 1968, which occurred in Fukuoka, Japan. The Binghamton patient with the highest adipose tissue level of furans (and also of serum PCBs) also had elevated levels of serum triglycerides and cholesterol; in addition, he was found to have hypertension. The levels of his serum PCBs dropped between 1981 and 1985 from approximately 60 ppb to approximately 27 ppb; serial furan and dioxin levels will be reported later. Weight loss following rigid dietary control also occurred in this time period, as did a lowering of blood pressure, serum cholesterol and triglyceride level. Whether this was due to diet or to the excretion of these chemicals over time is not apparent. No other illness was observed in this patient, although liver abnormalities and skin cancer, including melanoma, were found and described in other patients.[5,6,16] Excess suicides were also reported to have occurred in this BSOB-exposed group.[17]

The clinical usefulness of adipose tissue biopsies in identifying patients with evidence of elevated body burden of furan or dioxin isomers to which they were exposed is demonstrated here. The use of suction or aspiration to remove subcutaneous adipose tissue from a patient simplifies the process and reduces costs, bypassing, when desired, the hospital and surgical suite. At the next stage in the development of this technique we anticipate that a needle biopsy of adipose tissue, which currently yields 0.25 g of fat tissue, should suffice, and be a trivial office procedure. Eventually, chemical techniques may be sufficiently sensitive and specific to permit blood samples to be used to determine whether patients have elevated body burdens of dioxins and furans.

The finding of surprisingly high levels of total 2,3,7,8-chlorine-substituted (and hence more toxic) dioxins and furans in the general population suggests an additional reason why previous epidemiological studies have yielded less evidence of pathology of dioxins to humans than was expected based on animal toxicological data. Our work demonstrates that patients in control populations in industrial countries do not have an absence of exposure to chlorinated dioxins and dibenzofurans. In addition, not all persons who had an opportunity for ingestion of these chemicals through the respiratory, dermal or gastrointestinal route actually have evidence of absorption and/or retention. Without clearly characterizing levels of exposure in controls, as well as *dose* taken into the body and/or found in a target organ in cases, meaningful clinical and epidemiological studies are less probable, and scientifically valid conclusions cannot be reached. The relatively new tool of adipose tissue or fat biopsy followed by isomer-specific dioxin or furan analysis provides a basis for quantitative or semiquantitative estimation of intake or body burden of these toxic chemicals. Still needed are more studies establishing the human metabolism and kinetics of these compounds.[15,18,19] Some such studies suggest half-lives in humans of about 5 years for 2,3,7,8-TCDD in a small number of exposed patients monitored over time. Incidents such as those at Binghamton and Fukuoka provide opportunities to obtain further data on human metabolism and toxicity of these compounds.

The finding of dioxin receptors in humans adds to the previous environmental health occupational medicine literature suggesting that humans are susceptible to the toxic effects of the dioxins and of dioxin-like compounds. The finding that not all humans can be demonstrated to have dioxin receptors (if this finding holds up and is not merely due to a technical problem) suggests genetic heterogeneity of humans with respect to dioxin.[20] If this is the case, then characterizing the extent of human pathology caused by dioxins, furans, PCBs, PBBs and related chemicals, including chlorinated naphthalenes and biphenylenes, may require measurement of dioxin and other chemical tissue levels, as well as characterization of patients as "dioxin receptor positive" or "dioxin receptor negative."

The question of the human health effects of these chemicals on the patients described in this chapter is not addressed here; rather, this chapter is devoted to characterizing and illustrating the development of a tool which will allow correlation of body levels of these chemicals with illness. If, as with asbestos and other carcinogens, a latency period of decades occurs before disease becomes manifest, then patients with elevated body levels should be followed for decades after such exposure is detected, for the sake of the patient potentially at increased risk, as well as to expand medical knowledge regarding these compounds. Likewise, patients with no elevation of levels of these chemicals can be reassured and dropped from medical surveillance.

Whether humans will ultimately be shown to be as sensitive to dioxins as guinea pigs or monkeys, or somewhat less sensitive (as hamsters with respect to LD_{50}'s, although not necessarily hepatic enzyme induction), the direct measurement of these chemicals in humans, along with correlation of these levels with illness, should ultimately hasten the day when the human health effects of dioxins and related compounds will be more adequately and fully characterized. Without precise quantitative tools to estimate the body and tissue burden of exposed and control patients, epidemiological research suffers from great uncertainty. With the current improvements in chemical methodology, dioxin research need no longer be conducted without such quantitative markers.

REFERENCES

1. Schecter, A., G. Haughie, and R. Rothenberg. "PCB Transformer Fire—Binghamton, New York," *Morbidity and Morality Weekly Report* 30(16):187–193 (1981).
2. Schecter, A. "Contamination of an Office Building in Binghamton, New York by PCB's, Dioxins, Furans and Biphenylenes after an Electrical Panel and Electrical Transformer Incident," *Chemosphere* 12(4/5):669–680. (Elmsford, NY: Pergamon Press, Inc., 1983).
3. Tiernan, T., M. Taylor, G. Van Ness, J. Garrett, S. Bultman, C. Everson, D. Hinders, and A. Schecter. "Analysis of Human Tissues for Chlorinated Dibenzoparadioxins and Chlorinated Dibenzofurans—The Current State of the Art," in *Public Health Risks of the Dioxins. Proceedings of a Rockefeller University Symposium*, W. W. Lowrance, Ed. (Los Altos: Kaufman Publishing Co., 1984), pp. 31–56.
4. Tiernan, T., M. Taylor, J. Garrett, G. Van Ness, J. Solch, D. Wagel, G. Ferguson, and A. Schecter. "Sources and Fate of Polychlorinated Dibenzodioxins, Dibenzofurans and Related Compounds in Human Environments," *Environ. Health Pers.* 59:145–148 (1985).
5. Schecter, A., T. Tiernan, F. Schaffner, M. Taylor, G. Gitlitz, G. Van Ness, J. Garrett, and D. Wagel. "Patient Fat Biopsies for Chemical Analysis and Liver Biopsies for Ultrastructural Characterization After Exposure to Polychlorinated Dioxins, Furans and PCBs," *Environ. Health Pers.* 60:241–254 (1985).
6. Schecter, A., and T. Tiernan. "Occupational Exposure to Polychlorinated Dioxins, Polychlorinated Furans, Polychlorinated Biphenyls and Byphenylenes after an Electrical Panel and Transformer Accident in an Office Building in Binghamton, New York," *Environ. Health Pers.* 60:305–313 (1985).
7. Schecter, A., T. Tiernan, M. Taylor, G. Van Ness, J. Garrett, D. Wagel, G. Gitlitz, and M. Bogdasarian. "Biological Markers after Exposure to Polychlorinated Dibenzo-p-Dioxins, Dibenzofurans, Biphenyls, and Biphenylenes. Part I: Findings Using Fat Biopsies to Estimate Exposure," in *Chlorinated Dioxins and Dibenzofurans in the Total Environment II*, L. Keith, C. Rappe, and G. Chourdhary, Eds. (Stoneham, MA: Butterworth Publishers, 1985), pp. 215–145.
8. Schecter, A., J. J. Ryan, R. Lizotte, W.-F. Sun, and L. Miller. "Chlorinated Dibenzo-Dioxin and Dibenzofuran Levels in Human Adipose Tissues in Exposed and Control Patients," *Chemosphere* 14(6/7):933–938 (1985).
9. Ryan, J. J., R. Lizotte, and B. P.-Y. Lau. "Chlorinated Dibenzo-p-Dioxins and Chlorinated Dibenzofurans in Canadian Human Adipose Tissue," *Chemosphere* 14(6/7):697–706 (1985).

10. Ryan, J. J., A. Schecter, R. Lizotte, W.-F. Sun, and L. Miller. "Distribution of Dioxins and Furans from Human Autopsy Tissue from the General Population," *Chemosphere* 14(6/7):929–932 (1985).
11. Buser, H. R. "Formation, Occurrence and Analysis of Polychlorinated Dibenzofurans, Dioxins and Related Compounds," *Environ. Health Pers.* 60:259–267 (1985).
12. Masuda, Y., R. Kagawa, and M. Kuratsune. "Polychlorinated biphenyls in Yusho Patients and Ordinary Persons," *Fukuoka Acta Medica* 65(1):January, (1984).
13. Kuroki, H., and Y. Masuda. "Determination of Polychlorinated Dibenzofuran Isomers Retained in Patients with Yusho," *Chemosphere* 7(10):771–777 (1978).
14. Masuda, Y., H. Kuroki, T. Yamaryo, K. Haraguchi, M. Kuratsune, and S. T. Hsu. "Comparison of Causal Agents in Taiwan and Fukuoka PCB Poisoning," *Chemosphere* 11(2):199–206 (1982).
15. Rappe, C., M. Nygren, and G. Gustafsson. "Human Exposure to Polychlorinated Dibenzo-p-Dioxins and Dibenzofurans," in *Chlorinated Dioxins and Dibenzofurans in the Total Environment*, G. Choudhary, L. H. Keith, and C. Rappe, Eds. (Stoneham, MA: Butterworth Publishers, 1983), pp. 355–365.
16. Schecter, A., F. Schaffner, T. Tiernan, and M. Taylor. "Ultrastructural Alterations of Liver Mitochondria in Response to Dioxins, Furans, PCBs and Biphenylenes," *Banbury Report* 18:177–190 (1984).
17. Fitzgerald, E., S. S. Standfast, J. M. Melius, et al. "Binghamton State Office Building Medical Surveillance Report," New York State Department of Health (1985).
18. Gorski, T., L. Konopka, and M. Brodzki. "Persistence of Some Polychlorinated Dibenzo-p-Dioxins and Polychlorinated Dibenzofurans of Pentachlorophenol in Human Adipose Tissue," *Roczn. pzh* 35 (1984), pp. 297–301.
19. Poiger, H., and C. Schlatter. "Pharmacokinetics of 2,3,7,8-TCDD in Man," *Chemosphere* (1986), in press.
20. Roberts, E. A., N. H. Shear, A. B. Okey, and D. K. Manchester. "The Ah Receptor and Dioxin Toxicity: From Rodent to Human Tissues," *Chemosphere* 14(6/7):661–674 (1985).

CHAPTER 5

Investigation of Health Effects Due to 2,3,7,8-Tetrachlorodibenzo-p-Dioxin— Missouri, 1983

Paul A. Stehr, Gary Stein, Karen Webb, Wayne Schramm, Henry Falk, and Eric Sampson

BACKGROUND AND CHARACTERIZATION OF THE ENVIRONMENTAL CONTAMINATIONS

In 1971, approximately 29 kg of 2,3,7,8-tetrachlorodibenzo-p-dioxin (TCDD)-contaminated sludge wastes, which originated as a by-product of hexachlorophene production in a southwest Missouri plant, were mixed with waste oils and sprayed for dust control throughout the state. Almost 250 residential, work, and recreational areas (including several horse arenas) were thought to be contaminated, including the town of Times Beach. As of April, 1985, 42 sites had been confirmed as having at least 1 part per billion (ppb) of TCDD in soil; investigations were still under way at that time to evaluate the remaining sites. In 1974, levels as high as 35,000 ppb were measured in soil at 1 of these 42 sites; in 1985, isolated levels over 2,000 ppb were found at these contaminated areas, but most detectable levels in soil samples range from less than 1 ppb to several hundred ppb.

About one-third of the 42 confirmed sites were contaminated with peak levels in excess of 100 ppb; one-half of these were in residential areas. These sites varied widely in their potential for leading to human exposure due to the lack of uniformity in geography, topography, geology, and characteristic land use. This presented difficulties in the public health decision-making process. Sites at which the levels of contamination were high and which were in areas of frequent and regular access constituted the greatest public health risk; however, at other sites, dioxin contamination was in clearly circumscribed areas, at subsurface depths exceeding 15 ft, under paved areas, or in areas with limited land use. All of these considerations were taken into account in assessing the risk of exposure for an estimated 5,000 individuals from these contaminated areas from 1971–1983.

The earlier phases of this investigation focused on several sites in eastern Missouri, but subsequent activities include all 42 contaminated sites. The Chronic Diseases and Clinical Chemistry Divisions in the Center for Environmental Health of the Centers for Disease Control (CDC) had previously worked with the Missouri Division of Health (MDH) in 1971 at the time the initial contaminations occurred after receiving a report of an exposed child with hemorrhagic cystitis; in 1974, this work culminated in the laboratory identification of TCDD in the waste oil. With further discoveries of widespread contaminations in mid-1982, MDH and CDC in consultation reinitiated public health activities on the basis of new information and additional environmental data.

RISK TO HUMAN HEALTH

The case of chlorinated dioxin illustrates many of the difficulties encountered in assessing health risks following long-term, low-dose exposure to environmental chemical contaminations. At the time of our initial investigations, there was no widely available method for directly measuring chlorinated dioxin levels in human tissue. The lack of any direct measure of body burden or exposure substantially hindered attempts to assess the degree of exposure to and concomitant health risk posed by environmental chlorinated dioxins.

Exposure Assessment

We estimated the long-term risk of exposure in the areas contaminated with these dioxins by considering the excess risks of developing specific adverse health effects as a result of an estimated total cumulative dose. This dose is a function of several factors:

1) concentration of environmental contamination
2) location of and access to contaminated areas
3) types of activities conducted in contaminated areas
4) duration of exposure

Our risk assessments were concerned primarily with health risks in regard to contamination of soils in residential areas.[1]

Risk Evaluation

To estimate exposure, we made assumptions regarding the bioavailability and absorption of TCDD from soil as well as other metabolic parameters; principal routes of uptake were thought to be through dermal absorption, ingestion, and inhalation of contaminated dirt/dust particles.

Animal studies have shown great species variability in both acute and chronic responses to TCDD exposures; where humans fit on this response scale is not clear.[2] However, common findings from both animal toxicological work and limited data on cases of high-dose, accidental exposures of humans have included prominent effects on several organ systems: liver changes such as diminished function, hepatocellular necrosis, tumor induction (in animals), and microsomal enzyme induction; other effects include chloracne, depressed cell-mediated immunity, reproductive abnormalities (in animals), and peripheral neuropathy[3-10]; some studies have suggested that occupational exposures to TCDD may induce an excess risk of developing soft tissue sarcomas[11,12] but the only adequate dose-response data available for use in the risk assessment calculations were from animal carcinogenicity studies.[8] A linear, nonthreshold dose-response model was used to calculate increased lifetime cancer risk, and the calculation methods incorporated guidelines that a group of outside consultants recommended to CDC.[1]

Risk Management

Based on the above calculations, we concluded that residential soil TCDD levels of $\geq$ 1 ppb pose a level of concern for delayed health risks. In highly contaminated areas (e.g., areas with soil contamination levels > 100 ppb) with a high degree of access and concomitant exposure, the estimated incremental lifetime cancer risk may accumulate rapidly and be orders of magnitude higher than 1 per million. Therefore, MDH and CDC issued advisories which stated that the continued, long-term exposure to persons living in specified residential areas with 1 ppb or more TCDD contamination in the soil posed an unacceptable health risk.

These public health advisories and the U.S. Environmental Protection Agency's (EPA) consideration of the available remedial options were the basis on which site-specific decisions to eliminate or mitigate these exposures were made. The time frame for such decisions was dependent on the degree of contamination and on the degree to which continued exposures could be prevented while temporary or permanent remedial actions were considered and/or executed. In most cases, the EPA opted for temporary environmental cleanup, stabilization, or restriction of access to contaminated areas because of limited, well defined areas of contamination, relatively low TCDD soil levels, or relative inaccessibility of contaminated areas. However, in several noteworthy situations (such as the case of Times Beach), it was decided that permanent relocation of residents was the most prudent action.

PUBLIC HEALTH ACTIVITIES

In addition to ongoing review and assessment of EPA environmental sampling data, in January, 1983, MDH and CDC initiated four distinct public health actions:[13]

1) *Providing health education for both the medical and public health community and the general public about current understandings of the health effects of TCDD exposures.* To this end, a summary of the medical/ epidemiological literature was prepared and sent to physicians in eastern Missouri. On January 18, 1983, experts from government, academic institutions, and industry were brought together to give a seminar for the local medical community. Individual consultations and toll-free hotlines were established to answer questions from and address concerns of the general public.

2) *Providing a dermatologic screening clinic to the general public.* This clinic was intended to screen for cases of chloracne as an indication of possible TCDD exposure. In February, 1983, on consecutive weekends, all residents of eastern Missouri who had reason to suspect that they had been exposed and who had current skin problems were invited to be seen at these screening clinics.

3) *Creating and maintaining a central listing of potentially exposed individuals.* This listing will enable public health agencies to keep in touch with and locate potentially exposed individuals for educational purposes or possible epidemiologic and/or clinical followup. Specifically, when a reliable screening method for TCDD in serum becomes available, we will be better able to assess their exposure status and concomitant health risks. Baseline and identifying information was collected in the form of a Health Effects Survey questionnaire designed to elicit information on possible routes of exposure, lifestyle habits, residential and occupational histories, and medical history. It was also intended to serve (a) as a screening tool for identifying a "highest risk" cohort on whom intensive medical evaluations were focused and (b) in compiling a community-based data set from which epidemiologic inferences might be drawn.

4) *Designing and implementing a pilot medical study of a "highest risk" cohort.* This research was conceived as a pilot study of a group of persons presumed to be at highest risk of exposure to environmental TCDD and was intended to provide preliminary information on possible health effects from these exposures so as to enable investigators to develop more refined and specific epidemiologic protocols to be used in further investigations. A complete report of our methods and results has been presented elsewhere.[14]

MATERIALS AND METHODS

In this study, we assessed potential health effects related to chlorinated dioxin exposures by three means. First, as previously mentioned, we developed a Health Effects Survey questionnaire to elicit information on each person's exposure risk, medical history, and potentially confounding influences. We sought data for individuals believed to be at risk of exposure because they lived near, worked at, or frequently participated in activities near a contaminated site. Second, we sponsored the dermatology screening clinic mentioned above.

Third, we reviewed approximately 800 completed questionnaires and selected 122 persons for inclusion in a pilot medical study. We selected a high-risk group which comprised 82 individuals who reported—

  a) living or working in TCDD-contaminated areas *OR*
  b) participating more than once per week, on the average, in activities that involved close contact with the soil (such as gardening, field/court sports, horseback riding, or playing in soil), in contaminated areas

with TCDD levels of between 20 and 100 ppb for at least 2 years or levels greater than 100 ppb for at least 6 months. We also selected a low-risk comparison group of 40 persons who reportedly had had no access to or regular high-soil-contact activities in any known contaminated areas.

Of the 122 persons in the study group, 17.1% of the high-risk group and 10.0% of the low-risk group either refused to participate or failed to appear for the examinations (this difference was not significant at the 0.05 level), yielding a study population of 104 (68 at high risk and 36 at low risk of exposure).

In addition to being compared according to their responses on the Health Effects Survey questionnaire, these 104 persons were assessed under a clinical protocol that included the following elements:

  1) physical examination
  2) neurologic examination
  3) dermatologic examination
  4) laboratory analyses
  5) immune response tests
  6) collection of serum for use in future TCDD analyses

RESULTS

A summary of our results is presented in Table 1. The high- and low-risk groups were comparable in terms of age, race, sex, education of head of household, and interview respondent distributions. The two groups did not differ significantly in reporting other potential sources of exposure or the use of prescription medicines. The only significant difference in lifestyle

Table 1. Summary of Findings.[a]

Category	Statistically Significant Individual Findings	Evidence of Patterns or Trends
Demographics	Regular exercise HR > LR	None
General Disorders	None	Diminished peripheral pulses HR > LR
		Minor musculoskeletal abnormalities HR > LR
Routine Diagnostics	Platelet count HR > LR	None
Cardiopulmonary	None	Minor pulmonary abnormalities HR > LR
Reproductive Health	None	None
Dermatological	None	None
Neurological	None	None
Immunological	Palpable axillary nodes LR > HR	Palpable nodes LR > HR
Hepatic	"Other liver diseases" LR > HR	
	Urinary heptacarboxylporphyrin LR > HR	None
Renal/Urinary Tract	None	All previously diagnosed problems HR > LR
		Abnormal urinalyses results HR > LR

[a]HR, high-risk group; LR, low-risk group; >, greater prevalence or mean value in former risk group.

habits was that the high-risk group reported exercising more regularly (p < 0.01).

We found no differences or consistent trends regarding the prevalence of generalized disorders as reported in the questionnaires, the results of the general physical examinations, or the routine hematology tests (except for a higher mean platelet count and a nonsignificant trend of diminished peripheral pulses in the high-risk group).

No consistent overall trends or statistically significant individual diagnostic differences were detected for reproductive health outcomes from the questionnaire material. No birth defects were reported among children born to women in the high-risk group after the time at which exposures could have occurred.

In the dermatologic screening, no cases of chloracne were seen in the 140 persons examined from the general community or for the 104 persons in the study groups. In addition, no significant differences in all other dermatological findings were demonstrated by either medical histories or physical examination for the study population.

Results of the neurological examinations showed no significant differences or patterns between the two groups for the self-reported neurological conditions or from the neurological examinations, although a non-statistically significant diminution of vibratory sensation at 256 Hz was noted in the high-risk group.

As reported in the medical histories, there were no differences in prevalence of immune disorders. On physical examination, the only significant difference was a suggestion of a greater prevalence of palpable nodes in the low-risk group. Laboratory analyses showed no statistically significant differences between the two groups in regard to total induration in response to the antigenic skin tests, the *in vitro* lymphocyte proliferative responses, or in comparisons of parameters from T cell subset assays.

In regard to the hepatic system, no trends or significant specific problems were reported in the medical histories. On physical examination, there was a greater prevalence of hepatomegaly in the high-risk group, but this finding also was not statistically significant. There were no statistically significant differences between the two groups on tests of hepatic function except for higher mean urinary heptacarboxylporphyrin in the low-risk group. However, the two groups showed no difference in urinary porphyrin patterns, and no cases of overt porphyria cutanea tarda (PCT) or any precursor conditions (latent PCT or Type B porphyria) were detected.

There appeared to be a trend of increased urinary tract problems among the high-risk cohort on the basis of the medical history section of the questionnaire, although no statistically significant differences were demonstrated. Urinalyses also suggested a consistent pattern of abnormal findings, with a non-statistically significant higher prevalence of pyuria (defined as > 5 WBC/hpf) and microscopic hematuria (> 3 RBC/hpf) in the high-risk group.

DISCUSSION

The potential health effects considered in this study were based primarily on the animal toxicology of chlorinated dioxin and results from studies of long-term industrial and accidental acute human exposures. Animal toxicity studies are commonly used to predict health effects in humans (although, as discussed above, the existence of species-specific and even organ-specific effects of TCDD make extrapolations tenuous); the organ systems most prominently affected are the liver (acute toxicity and hepatocarcinogenesis), the immune system (thymic atrophy and decreased cell-mediated immunity), the skin (chloracne-like changes), and reproductive effects.[3-10]

Most of our direct knowledge of human health effects has been obtained from workers who were exposed to chlorinated dioxin during the production or subsequent handling of 2,4,5-trichlorophenol (2,4,5-TCP, or simply 2,4,5-T). In some plants, chloracne, but no systemic illnesses, developed in exposed workers.[15] Others have reported weight loss, easy fatigability, myalgias, insomnia, irritability, and decreased libido; the liver has been shown to become tender and enlarged, and sensory changes, particularly in the lower extremities, have occurred; total serum lipids may be increased, and the prothrombin times may be prolonged.[16,17] PCT, an acquired form of porphyria characterized by chronic skin lesions and other symptoms, has also been observed.[16,18,19] The most specific of these findings are chloracne (which can also be caused by other structurally related compounds such as polychlorinated biphenyls [PCBs] and chlorinated naphthalenes) and PCT (which also has a variety of potential causes). As mentioned above, a number of studies addressing the association of TCDD exposures to soft tissue sarcoma have been conducted in the industrial setting, including studies which reported a sixfold increase in risk of soft tissue sarcomas among persons exposed to chlorphenols and phenoxy herbicides.[11,12,20]

Information on health effects involving nonoccupational environmental exposure is sparse. After an explosion in 1976 at the ICMESA plant in Seveso, Italy, chloracne developed in exposed children, some elevated liver function test results were detected in the exposed population,[21] and the incidence of abnormal nerve conduction tests was statistically significantly elevated in subjects with chloracne.[22] A child in Missouri who played in dirt in a riding arena contaminated with up to 33 ppm TCDD contracted hemorrhagic cystitis.[23]

These analyses did not produce any firm indications of increased disease prevalence directly related to the putative exposures. These results did, however, offer some insights and leads for further study. Of interest is the trend indicative of urinary tract abnormalities in the high-risk group (especially in light of the previously reported finding of hemorrhagic cystitis in an exposed person). The findings of no significant differences in liver function are important; however, hepatic function is being examined in subsequent studies because of other animal and human toxicologic data suggesting hepatotoxic effects of TCDD. Although none of the findings from the immune function tests and assays demonstrated statistically significant differences, several results were noteworthy, such as slight relative anergy and an increased prevalence of helper:suppressor T-cell ratios < 1.0 in the high-risk group, although the functional tests of the immune system revealed no overall abnormalities; followup and/or further investigation of these effects in exposed cohorts will be conducted before conclusions can be drawn.

Several factors could explain at least part of the overall negative findings of this pilot study:

1. restricted statistical power to detect significant differences due to the relatively small sample size
2. introduction of potential biases, since a large percentage of the pool of persons from which the study and comparison groups were chosen was self-selected
3. absence of an objective direct measure of exposure status, leading to misclassification errors
4. inability to detect effects with long latency periods or subtle health effects for which our tests were not sensitive

5. uptake of TCDD from contaminated soils may have been generally less than estimated
6. chronic exposures to low-level environmental TCDD may actually induce little or no adverse health effects.

CONCLUSION

These actions represent the first phase in the investigation of chlorinated dioxin contaminations in Missouri. The public health agencies involved continue to review environmental sampling data on new suspected sites and to develop public health advisories. No overall definitive conclusion should be based just on the initial pilot study, although the results appear to be largely negative. Results are pending from a more refined epidemiologic study which was planned to test and confirm the results of this pilot study. Also, research into standardizing laboratory methods and establishing reference ranges for TCDD body burden are currently being pursued.

Finally, public health policy in situations such as this environmental contamination with TCDD must continue to be focused on the prevention of any potential health effects, even if such effects were not demonstrated in a small pilot study. For this reason, all appropriate efforts need to be made to prevent human exposure.

REFERENCES

1. Kimbrough, R. D., H. Falk, P. A. Stehr, and G. Fries. "Health Implications of 2,3,7,8-Tetrachlorodibenzodioxin (TCDD) Contamination of Residential Soil," *J. Toxicol. Environ. Health.* 14:47–93 (1984).
2. Kociba, R. J., and B. A. Schwetz. "Toxicity of 2,3,7,8-Tetrachlorodibenzo-p-dioxin (TCDD), *"Drug Metab. Rev.* 13:387–406 (1982).
3. Allen, J. R., D. A. Barsotti, L, K. Lambreckt, and J. P. Van Miller. "Reproductive Effects of Halogenated Aromatic Hydrocarbons on Nonhuman Primates," *Ann. N. Y. Acad. Aci.* 320:419–25 (1979).
4. Gupta, B. N., J. G. Vos, J. A. Moore, J. G. Zinkl, and B. C. Bullock. "Pathologic Effects of 2,3,7,8-TCDD in Laboratory Animals," *Environ. Health Pers.* 5:125–40 (1973).
5. Hook, G. E. R., J. K. Haseman, and G. W. Lucier. "Induction and Suppression of Hepatic and Extrahepatic Microsomal Foreign-Compound Metabolizing Enzyme Systems by 2,3,7,8-TCDD," *Chem. Biol. Interact.* 10:199–214 (1975).
6. Hook, J. B., K. M. McCormack, and W. M. Kluwe. "Renal Effects of 2,3,7,8-TCDD," *Environ. Sci. Res.* 12:381–88 (1978).
7. Kimmig, J., and K. H. Schulz. "Occupational Acne (Chloracne) Caused by Chlorinated Aromatic Cyclic Ethers," *Dermatologica* 115:540–46 (1957).
8. Kociba, R. J., D. G. Keyes, J. E. Beyer, et al. "Results of a Two-Year

Chronic Toxicity and Oncogenicity Study of 2,3,7,8-TCDD in Rats," *Toxicol. Appl. Pharmacol.* 46:279–303 (1978).

9. Thigpen, J. E., R. E. Faith, E. E. McConnell, and J. A. Moore. "Increased susceptibility to Bacterial Infection as a Sequela of Exposure to 2,3,7,8-TCDD," *Infect. Immunol.* 12:1319–24 (1975).

10. Van Miller, J. P., J. J. Lalich, and J. R. Allen. "Increased Incidence of Neoplasms in Rats Exposed to Low levels of 2,3,7,8-TCDD," *Chemosphere* 10:625–32 (1977).

11. Eriksson, M., L. Hardell, N. Berg, T. Moller, and O. Axelson. "Soft Tissue Sarcomas and Exposure to Chemical Substances: A Case-Referent Study," *Brit. J. Ind. Med.* 38:27–33 (1981).

12. Hardell, L., and A. Sandstrom. "Case-Control Study: Soft Tissue Sarcomas and Exposure to Phenoxyacetic Acids and Chlorophenols," *Brit. J. Cancer* 39:711–17 (1979).

13. Stehr, P. A., D. L. Forney, G. F. Stein, et al. "The Public Health Response to 2,3,7,8-TCDD Environmental Contaminations in Missouri," *Public Health Reports* 100(3):289–93 (1985).

14. Stehr, P. A., G. F. Stein, K. Webb, et al. "A Pilot Epidemiologic Study of Health Effects due to 2,3,7,8-Tetrachlorodibenzodioxin Contaminations in Missouri," *Arch. Environ. Health* (in press 1986).

15. May, G. "Chloracne from the Accidental Production of Tetrachlorodibenzodioxin," *Brit. J. Ind. Med.* 30:276–83 (1973).

16. Bleiberg, J., M. Wallen, R. Brodkin, and I. L. Applebaum. "Industrially Acquired Porphyria," *Arch. Dermatol.* 89:793–97 (1964).

17. Bauer, H., K. H. Schulz, and U. Spiegelberg. "Berufliche Vergiftungen bei der Herstellung von Chlorphenol-Verbindungen," *Arch. Gwerbepathol. Gwer. Behyg.* 18:538–55 (1961).

18. Jirasek, L., J. Kalensky, K. Kubec, J. Pazderova, and E. Lukas. "Chlorakne, Porphyria Cutanea Tarda und andere Intoxikationen durch Herbizide," *Hautarzt* 27:328–33 (1976).

19. Poland, A. P., D. Smith, G. Metter, and P. Possick. "A Health Survey of Workers in a 2,4-D and 2,4,5-T Plant," *Arch. Environ. Health* 22:316–327 (1971).

20. "Chlorinated Dibenzodioxins," *Monographs on the Evaluation of the Carcinogenic Risk of Chemicals to Man* 15 (Lyon, France: IARC, 1977), 41–102.

21. Reggiani, G. "Acute Human Exposure to TCDD in Seveso, Italy," *J. Toxicol. Environ. Health* 6:27–43 (1980).

22. Fillipini, G. B. Bordo, P. Crenna, N. Massetto, G. Musicco, and R. Boeri. "Relationship between Clinical and Electrophysiological Findings and Indicators of Heavy Exposure to 2,3,7,8-TCDD," *Scand. J. Work. Environ. Health* 7:257–62 (1981).

23. Carter, C. D., R. D. Kimbrough, J. A. Liddle, et al. "Tetrachlorodibenzodioxin: An Accidental Poisoning Episode in Horse Arenas," *Science* 188:738–40 (1975).

SECTION II

Incineration Emissions

CHAPTER **6**

Determination of PCDDs and PCDFs in Incineration Samples and Pyrolytic Products

S. Marklund, L.-O Kjeller, M. Hansson, M. Tysklind, C. Rappe, C. Ryan,
H. Collazo, and R. Dougherty

INTRODUCTION

On February 13, 1985 the Swedish EPA issued a moratorium on the construction of new municipal solid waste (MSW) incinerators. The main reason for this moratorium was the dioxin issue, especially the potential connection between incineration and background levels of PCDDs and PCDFs found in human adipose tissue and mother's milk. The 1,2,3,7,8-penta-CDD, which has never been reported as a contaminant in technical products but always found in incineration samples, can serve as a marker.[1-3]

During 1985 a research program was launched in Sweden in order to investigate the emission from various types of incinerators and also to correlate these data with potential health risks. In this chapter we will discuss some of the data collected so far for various types of incinerators or incineration models.

Emissions from incinerators constitute a multitude of different PCDD and PCDF isomers; a few of these have been found to have much higher toxicity than others.[1,3] Various models have been used to convert a multitude of levels of more or less toxic PCDDs and PCDFs into a more simple expression like "TCDD equivalents" or "toxic equivalents." (For a discussion of different approaches see Milby et al.[4] and Ahlborg.[5]) In Sweden and in this report the approach discussed by Eadon et al.[6] has been used.

EXPERIMENTAL

Sampling and Cleanup

The sampling from the MSW incinerators studied here was performed by Lars Lunde'n and Curt-Åke Boström, IVL-IPK, Gothenburg, Sweden. In the sampling train the following samples were collected:

- Particulate (glassfiber filter)
- Condensate
- Adsorbent tube (XAD–2)

In order to validate the sampling procedure, 13–C-labeled surrogates were added to the filter and to the XAD–2 adsorbent before the collection. The sample volume varied between 1 and 3 Nm^3 dry gas (dg).

For the extraction and sample cleanup we mainly used conventional methods (Figures 1–3). However, the final step, using a Carbopack C cleanup system, was slightly modified and a validation of this column is presented here.

Validation of Carbopack C Column

We mixed 0.5 g Carbopack C and 2.3 g Celite 545 and activated this mixture at 130 °C for 6 hr. The mixture was packed in a disposable pipet fitted with a small plug of glass wool (Figure 3). The column was preeluted with 20 mL of toluene followed by 9 mL of a mixture of methylene chloride / c-hexane (1:1) and 9 mL of n-hexane. The extract or the combined extracts were dissolved in 3×1 mL of n-hexane and added onto the column, followed by 3mL of methylene chloride / c-hexane (1:1). The PCDD and PCDF fraction was collected by elution with 20 mL of toluene. To the extract 10 μL of tetradecane was added and the volume was reduced to 10 μL.

In the validation study, three levels with 12 different PCDF congeners (Table 1) were eluted through the Carbopack C column. The levels were 250, 2500 and 12,500 pg/congener. Every level was studied in triplicate. The results are shown in Table 1 and the mean values in Figure 4. No significant trend or difference could be found between isomers of the same chlorination level.

High-Resolution Gas Chromatography/Mass Spectrometry

We were using a 60–m Supelco SP2330 or a SP2331 fused-silica capillary column for the isomer-specific analysis of PCDDs and PCDFs. This

EXTRACTION

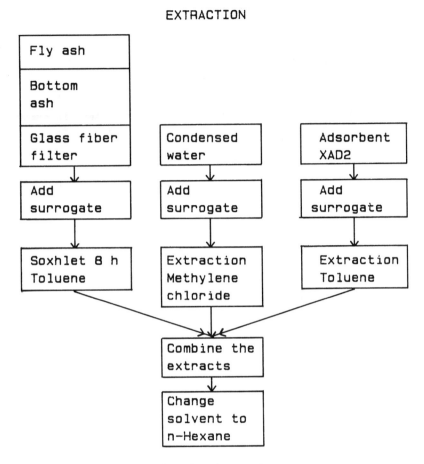

Figure 1. Extraction procedure for incineration samples.

column separates all toxic congeners from the less toxic congeners except 1,2,3,7,8–PeCDF from 1,2,3,4,8–PeCDF and 1,2,3,4,7,8–HxCDF from 1,2,3,4,7,9-HxCDF. Separation of 1,2,3,7,8-PeCDF from 1,2,3,4,8-PeCDF can be done on a 50-m OV–1701 column.[1]

All samples were analyzed on an updated Finnigan 4021 mass spectrometer in negative ion chemical ionization (NCI) mode using methane as reagent gas. However, due to the poor response of TCDD using NCI mode, electron impact (EI) mode was used to quantify 2,3,7,8–TCDD. This resulted in somewhat less sensitivity for TCDD than for the other homologs.

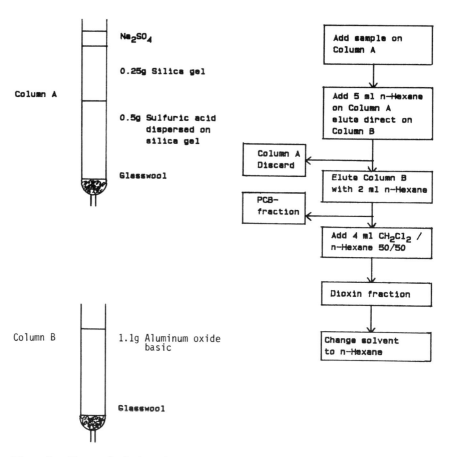

Figure 2.   Cleanup for incineration samples; Part I.

MATERIAL

MSW Incinerator, Umeå

During the fall of 1984, an extensive investigation took place at the MSW Incinerator in Umeå, Sweden. This incinerator is of the cross-grate type, built in 1970 and equipped with a boiler and an electrostatic precipitator (ESP). The incinerator is charged with raw refuse at a rate of 6 metric tons/hr and the effect is 10 MW. We performed 15 experiments during the fall of 1984, and three additional measurements were done during the spring of 1985 (Table 2). Bottom ash and fly ash from the ESP were also analyzed from every experiment. The results are not reported here. These experiments indicate that, of the total, about 20% of the PCDDs and PCDFs are

SAMPLE PURIFICATION II   (EPRI 613)

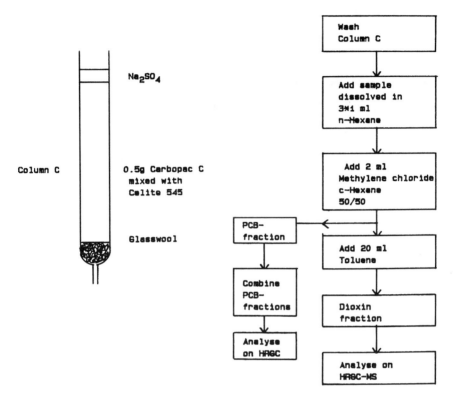

Figure 3.   Cleanup for incineration samples; Part II.

Table 1.   Recovery of All Added PCDF Congeners in Every Experiment (%).

Experiment	250 pg			2500 pg			12,500 pg		
	1	2	3	1	2	3	1	2	3
1,3,6,8–	108	112	112	116	102	105	109	102	92
2,3,7,8–	105	121	113	120	105	106	105	102	106
1,3,4,6,8–	106	124	114	101	127	101	118	108	119
1,2,3,7,8–	105	107	104	101	107	101	99	105	98
2,3,4,8,9–	115	77	107	105	126	95	107	113	104
2,3,4,6,7–	124	113	117	98	117	111	117	105	113
1,2,3,4,6,8–	110	90	94	95	98	104	97	90	99
1,2,3,4,7,8–	119	112	105	102	104	103	102	100	105
1,2,3,7,8,9–	66	67	51	90	87	94	103	89	101
2,3,4,6,7,8–	116	93	94	99	88	103	97	91	101
1,2,3,4,7,8,9–	118	122	104	99	109	98	101	103	95
Octa–	77	74	60	57	48	50	49	60	66

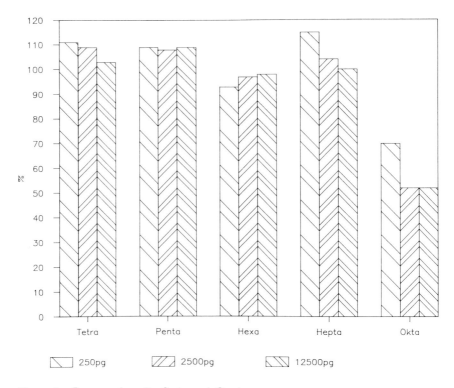

Figure 4. Recovery from the Carbopack C column.

found in the bottom ash, another 20% occur in the fly ash, and the rest are in the true emissions.

Peat Incineration

This incinerator, also in Umeå, was built in 1981, and it is a cross-grate type equipped with a boiler and an ESP. The capacity of this incinerator

Table 2.   Experiments at MSW Incinerator in Umeå.

Experiments	Number of Experiments	Temperature, °C Mean	Range
Fall 1984			
Normal conditions	3	803	736–846
Normal conditions with wooden chips	2	764	737–790
Normal conditions with oil burner	2	827	811–842
Low temperature	3	539	484–580
Low temperature with oil burner	3	625	602–658
Startup procedure	1		20–790
Startup procedure with oil burner	1		20–816
Spring 1985			
Normal conditions	3	784	700–850

using peat, wood chips or oil as fuel is 6 metric tons/hr and the maximum effect is 25 MW. In our experiment peat was the only fuel used.

Copper Smelter

Scrap copper containing PVC plastic and PVC-coated wires and cords was burned and melted in a converter. The temperature in the converter was about 1500 °C, but during the feeding process the temperature was much lower. Two samples of gas exhaust were collected during the feeding process and analyzed.

Steel Mill

A dust sample was collected from the baghouse of a Swedish steel mill. In this steel mill a high portion of the metal (stainless steel) was recycled. This recycled material can be contaminated by PVC or by other organochlorine additives, like polychlorinated paraffins used in cutting oils and similar products for metal treatment.

Smoke Generator

A military smoke torch containing hexachloroethane and zinc was burned in a hood. Two gas samples were collected on an adsorbent (XAD-2) and analyzed.

Laboratory Pyrolysis

We have studied the pyrolysis of a mixture of octachlorodibenzofuran and octachlorodibenzo-p-dioxin in the following manner. In a glass ampule 2 mg of OCDF and 1.4 mg of OCDD were combined. The glass ampule was sealed and heated for 90 sec in an oven preheated to 600°C. After cooling, the ampule was opened and the content extracted with methylene chloride. The extract was cleaned as described above for the samples from the MSW incinerator.

PVC and saran pyrolysis were preformed by H. Collazo at the Florida State University, Tallahassee, Florida, USA.

RESULTS

MSW Incinerator, Umeå

The levels of 2,3,7,8-substituted isomers, as well as the total amount of each group of congener and the levels of TCDD equivalents, are given in Table 3. The isomeric pattern of the PeCDFs in a typical sample is shown in Figure 5 (upper curve).

Before the fall measurements we spiked each filter and adsorbent with 1 ng of ^{13}C-2,3,7,8-TCDD, ^{13}C-2,3,7,8-TCDF and ^{13}C- OCDD. We ana-

Table 3.   Levels of PCDD and PCDF from MSW Incineration, Umeå. (ng/Nm³ dg 10% CO₂)

Experiment	Normal	Normal Chips	Normal Oil	Low Temperature	Low Temperature Oil	Start	Start Oil	Normal
Season	Fall	Fall	Fall	Fall	Fall	Fall	Fall	Spring
Number of Experiments	3	2	2	3	3	1	1	3
2378–TCDF	2.5	2.3	2.4	2.6	2.1	9.5	2.3	0.85
Total TCDF	86	75	68	87	75	260	80	19
2378–TCDD	0.5	0.6	0.7	0.4	0.3	1.3	0.7	<0.1
Total TCDD	43	45	52	54	47	100	49	<10
12378/								
12348–PeCDF	9.0	8.3	9.8	8.3	7.1	52	9.0	2.5
23478–PeCDF	6.1	7.3	7.8	7.4	6.5	40	9.0	3.9
Total PeCDF	97	100	120	110	87	520	120	43
12378–PeCDD	2.5	3.6	3.6	3.2	3.6	14	3.9	2.4
Total PeCDD	53	70	76	80	70	280	90	49
123478/								
1234789–HxCDF	3.6	4.6	5.6	5.2	3.6	48	5.7	4.5
123678–HxCDF	3.7	4.6	5.5	5.0	3.4	40	5.7	4.6
123789–HxCDF	0.8	0.8	1.2	1.2	1.4	52	2.3	3.6
234678–HxCDF	2.6	4.4	4.3	5.1	4.2	36	5.4	4.3
Total HxCDF	33	46	51	50	37	380	54	43
123478–HxCDD	1.6	2.9	3.5	5.1	3.1	31	4.4	2.3
123678–HxCDD	3.7	5.6	6.5	9.5	6.6	56	7.9	5.8
123789–HxCDD	1.3	2.4	3.0	3.8	2.6	20	3.5	2.0
Total HxCDD	32	53	72	82	57	400	70	55
Total HpCDF	34	73	94	67	40	380	51	49
Total HpCDD	18	29	37	54	36	380	38	56
OCDF	10	52	50	23	25	180	41	33
OCDD	12	19	31	14	20	130	27	53
TCDD Equivalents	9.1	10	11	11	10	53	12	5.6

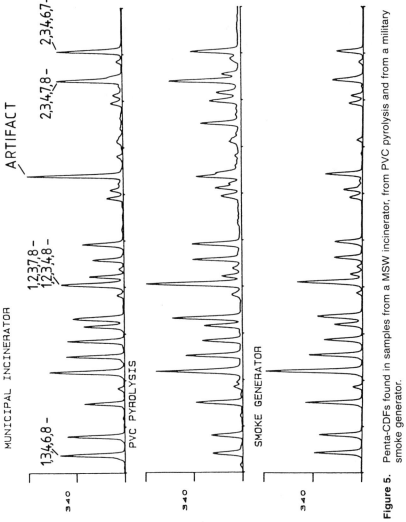

Figure 5.   Penta-CDFs found in samples from a MSW incinerator, from PVC pyrolysis and from a military smoke generator.

Table 4. Distribution of Spike in Sample Train (ng).

	^{13}C-TCDF	^{13}C-OCDD
Filter	0.1	0.5
Condensate	0.45	0.5
Adsorbent	1.45	1.0

lyzed the filter, the condensate and the adsorbent separately and found the distribution given in Table 4.

During the spring measurements we added a second adsorbent tube after the first one. We found that less than 5% of the total amount of TCDD-equivalents were collected on the second XAD-2 adsorbent tube.

Other MSW Incinerators and Industrial Incineration

In Table 5 we have summarized the results from various Swedish incinerators. Umeå 1 is the average of the fall measurements and Umeå 2 is the spring value. Avesta 1 is before and Avesta 2 is after the oven was modified.

Peat Incineration and Laboratory Pyrolysis

Table 6 shows the value of 2,3,7,8–substituted isomers and total congeners from peat incineration, a smoke generator and laboratory pyrolysis. Due to the low amount, it was not possible or practical to calculate the TCDD-equivalents. The pattern of individual penta-CDF isomers from PVC pyrolysis and the smoke generators is shown in Figure 5.

In the OCDD and OCDF pyrolysis it was not possible to detect higher chlorinated isomers than penta, due to overload from unpyrolyzed octa-CDD and octa-CDF on the hexa and hepta channels. The pattern of penta-CDF cogeners is shown in Figure 6.

CONCLUSIONS

The formation of PCDDs and PCDFs in MSW incinerators is now well documented and not controversial. However, this seems to be the first report where the stack sampling and laboratory cleanup operations have

Table 5. Results of Total Emissions of TCDD Equivalents from Various Incineration Processes (ng/Nm^3 dg 10% CO_2).

Place	Type	Mean Value
Umeå 1	MSW cross-grate	10
Umeå 2	MSW cross-grate	5.6
Avesta 1	MSW cross-grate	80
Avesta 2	MSW cross-grate	2.0
Borås	MSW cross-grate	38
Mid Sweden	MSW fluidized bed	1.8
Rönnskär	Indust. Copper Smelter	11
Mid Sweden	Indust. Steel mill	0.8 ng/g dust

Table 6. Results from Peat Incineration and Laboratory Pyrolysis.

Experiment	Peat Incinerator (pg/Nm³)	Smoke Generator (ng/g)	Saran Pyrolysis (ng/g)	OCDD & F Pyrolysis (μg/g)
2,3,7,8–TCDF	0.4	0.075	0.06	16
Total TCDFs	13	1.5	0.46	240
2,3,7,8–TCDD	< 2	NAa	NA	NA
Total TCDDs	< 20	NA	NA	NA
1,2,3,4,8–/ 1,2,3,7,8–PnCDF	< 0.2	0.4	0.1	110
2,3,4,7,8–PnCDF	1	0.2	0.2	4.6
Total PnCDFs	10	4.4	2.2	240
1,2,3,7,8–PnCDD	< 0.6	< 0.04	< 0.06	< 3
Total PnCDDs	< 6	< 0.4	< 0.6	22
1,2,3,4,7,9–/ 1,2,3,4,7,8–HxCDF	0.4	1.1	0.7	NDb
1,2,3,6,7,8–HxCDF	0.4	1.0	0.6	ND
1,2,3,7,8,9–HxCDF	0.2	0.4	0.6	ND
2,3,4,6,7,8–HxCDF	0.4	0.3	0.5	ND
Total HxCDFs	10	8.0	6.6	ND
1,2,3,4,7,8–HxCDD	< 1	< 0.04	< 0.06	ND
1,2,3,6,7,8–HxCDD	< 1	< 0.04	< 0.06	ND
1,2,3,7,8,9–HxCDD	< 1	< 0.04	< 0.06	ND
Total HxCDDs	< 6	< 0.3	< 0.5	ND
Total HpCDFs	20	8.6	46	ND
Total HpCDDs	20	< 0.6	0.5	ND
OCDF	< 20	6.0	24	ND
OCDD	70	< 1	1	ND

aNA, not analyzed.
bND, not detected due to overload of unreacted material.

been controlled and validated by the use of ^{13}C–labeled standard compounds. The low recovery of the ^{13}C–2,3,7,8-tetra-CDF in the filter is worth noting: sampling of particulates at elevated temperatures without appropriate backup equipment yields erroneous results.

The levels of PCDDs and PCDFs in the MSW incinerator in Umeå were found to vary quite little over the period of time the samples were taken (10 days), even in the case where the burning conditions were quite different (Table 3). However, in all cases the burning conditions were fully acceptable.

We have analyzed samples from a series of MSW incinerators in Sweden. The levels are given in Table 5, where we also show data from other types of industrial incinerators.

In another Swedish MSW incinerator (Avesta), the emissions were originally quite high. Later, when the turbulence within the incinerator was optimized and the air flow leaking into the oven was minimized, the levels were reduced by a factor of about 50.

We have made another interesting observation. The emissions were found to be higher in the fall, when the amount of wet leaves in the MSW incinerator was quite high compared to winter and spring.

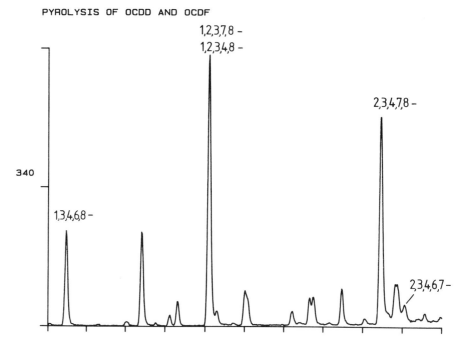

Figure 6.   Penta-CDFs from pyrolysis of octa-CDF.

It is interesting to compare the PCDD and the PCDF emissions from MSW incinerators with other energy sources. Table 6 presents the results from incineration of peat. The emission here was much lower than in the MSW incinerator — close to a factor of 10^3. This observation is in agreement with earlier observations from coal-fired power plants where Junk and Richard[7] and Kimble and Gross[8] were unable to identify 2,3,7,8–tetra–CDDs in the fly ash samples they analyzed.

The levels found in the emissions of industrial incinerators seem to be quite similar to those found in properly operating MSW incinerators. Both the copper smelter and the steel mill produced PCDDs and PCDFs. This strongly indicates that the total emissions from industrial incinerators could be of the same magnitude or even higher than the emissions from MSW incinerators.

The laboratory pyrolysis of PVC and Saran clearly shows that PVC and other organochlorine polymers can be precursors to the PCDDs and PCDFs found in various incinerators. This is a very important observation because the New York City Department of Sanitation[9] recently claimed "PVC has never been shown to be a precursor of PCDF/PCDD." This statement is based on investigations by Karasek et al.[10] and Olie et al.[11] A recent German pamphlet arrives at the same erroneous conclusion.[12]

Table 7.   Exposure of TCDD Equivalents for a 55-kg Person or 5-kg Baby.

Exposure		pg/kg body weight/day	Reference
Inhalation	(20 m³/day)	0.02	this study
Milk	(1 L/day)	0.05 – 5	16
Salmon	(100 g/week)	20	17
Breast Milk	(850 mL/day) 5 kg baby	20 – 200	5

In a recent study in Baltimore County, MD, and Brooklyn, NY, USA, it was found that plastics and paper are the two major sources of water-insoluble chlorine content of MSW incinerators. The total level of chlorine was found to be about 1%.[13]

In addition to the levels of PCDDs and PCDFs found in the emissions of various incinerators, it can be of interest to study the pattern of individual PCDD and PCDF congeners found in these samples. In Figure 5 we have made a comparison of the penta-CDF patterns found in a typical sample from a MSW incinerator, from PVC pyrolysis and from the pyrolysis of hexachloroethane (smoke generator). The patterns of individual isomers in these samples are very much the same, in spite of the fact that the chlorine content was found to vary between 1 and 90%.

However, the laboratory pyrolysis of OCDD and OCDF shows a completely different isomeric pattern (Figure 6). Here the 2,3,7,8–substituted isomers were found to dominate, indicating a preferential loss of chlorine atoms in the peri positions.

The identification of lower chlorinated PCDFs like penta-CDFs in pyrolysis of hexachloroethane indicates dechlorination to be an important pathway for the formation of PCDDs and PCDFs. This is in agreement with our earlier observations of tetra- and penta-CDDs from the pyrolysis of very pure pentachlorophenol.[14]

The environmental and human health impact of the emissions of PCDDs and PCDFs from MSW incinerators have been discussed in Umeå as well as at other places in Sweden and in other countries. Using the observed stack gas levels in Umeå, the Swedish Meteorological Institute has calculated the average air concentration in different parts of the surrounding area.[15] The highest levels were calculated for two hills about 3 km north of the MSW incinerator. One of these is a residential area. The annual mean level here was calculated to be 0.055 pg/m³. The daily exposure by inhalation (20 m³/day for a 55–kg person) is 0.02 pg/kg body weight. This value should be compared to the ADI value discussed,[5] which is 1–5 pg/kg body weight/day, and also to some other calculated values of exposure via food (Table 7).

From this table it is clear that the inhalation exposure is marginal compared to the exposure via food, especially fish and other foodstuffs from the aquatic food web. However, a correlation between incinerators of various kinds and the environmental levels of PCDDs and PCDFs cannot be excluded; consequently, all such emissions should be controlled and minimized.

REFERENCES

1. Rappe, C., S. Marklund, L.-O. Kjeller, P.-A. Bergqvist, and M. Hansson, in *Chlorinated Dioxins and Dibenzofurans in the Total Environment II*, L. H. Keith, C. Rappe, G. Choudhary, Eds. (Stoneham, MA: Butterworth Publishers, 1984), p. 401.
2. Rappe, C., P.-A. Bergqvist, M. Hansson, L.-O. Kjeller, G. Lindström, S. Marklund, and M. Nygren. In: *Banbery Report 18, Biological Mechanisms of Dioxin Action*, A. Poland and R. D. Kimbrough, Eds. (Cold Spring Harbor Laboratory 1984) p. 17.
3. Nygren, M., C. Rappe, G. Lindström, M. Hansson, P.-A. Bergqvist, S. Marklund, L. Domellöf, L. Hardell, and M. Olsson. Chapter 2, this volume.
4. Milby, T. H., T. H. Miller, and T. L. Forrester. *J. Occup. Health* 27:351–355 (1985).
5. Ahlborg, U. G. In: *Organo Halogen Compounds in Human Milk and Related Hazards*, Report on a WHO Consultation. WHO Regional Office for Europe IPC/CHE 501/m 05, August 1985.
6. Eadon, G., K. Aldous, D. Hilker, P. O'Keefe, and R. Smith, "Chemical data on air samples from the Binghamton State Office building," Memo from Center for Lab Research, New York State Department of Health, Albany NY 12201, 7/7/83 (1983).
7. Junk, G. A., and J. Richard, *Chemosphere* 10:1237–1241 (1981).
8. Kimble, B. J., and M. L. Gross, *Science* 207:59–61 (1981).
9. "Dioxin Study for Proposed Resource Recovery Plant" New York City Department of Sanitation, New York (1984).
10. Karasek, F. W., A. C. Viau, G. Guiochon, and M. F. Gonnord. *J. Chromatog.* 270:227–234 (1983).
11. Olie, K., M. Van den Berg, and O. Hutzinger. *Chemosphere* 12:627–636 (1983).
12. "PVC-Ursache fur Dioxin-Bildung?" Verband Konststofferzeugende Industrie e. V. (VKE). Frankfurt (June 1985).
13. Churney, K. L., A. E. Ledford, Jr., S. S. Bruce, and E. S. Domalski. In: U. S. Department of Commerce Report NBSIR 85-3213 (1985).
14. Rappe, C., S. Marklund, H. R. Buser, and H.-P. Bosshart. *Chemosphere* 7:269–272 (1978).
15. Kindell, S. "Spridningsberäkningar för en sopförbränningsanläggning i Umeå. "SMHI rapport 1985(5) Norrköping, Sweden (1985).
16. Bulletin de l'Office federal de la santé publique No. 8., 28. 2. Bern, Switzerland (1985).
17. Rappe, C., P.-A. Bergqvist, and S. Marklund. "Analysis of Polychlorinated Dioxins and Dibenzofurans in Aquatic Samples," Presented at Pori, Finland, August 1983. Publications of the Water Research Institute, Helsinki, Finland.

CHAPTER 7

Sampling, Analytical Method and Results for Chlorinated Dibenzo-p-Dioxins and Chlorinated Dibenzofurans from Incinerator Stack Effluent and Contaminated Building Indoor Air Samples

R. M. Smith, P. W. O'Keefe, D. R. Hilker, K. M. Aldous, L. Wilson,
R. Donnelly and R. Kerr, A. Columbus

INTRODUCTION

Municipal incinerators[1-5] and overheated PCB-filled transformers[6] are examples of different sources of potentially large amounts of chlorinated dibenzofurans (CDFs) and/or chlorinated dibenzo-p-dioxins (CDDs) produced by unknown combustion mechanisms. There is currently no standard method for sampling or analysis of the complex combustion products.

Recent analytical efforts have been directed to the analysis of these compounds in stack emissions from a municipal incinerator and air from a contaminated building. A comparison of the concentrations of CDFs and CDDs found in each study is of interest for the following reasons: 1) A combustion process produced the CDFs and CDDs in both cases, although the CDF/CDD precursors may have been different, and 2) Gaseous and particulate-bound CDFs and CDDs were collected as separate fractions. Although the processes of adsorption and desorption of CDFs/CDDs from particulates may occur during sampling, we decided to analyze the particulate filter and adsorbent separately, aware that only the total may be quantitatively meaningful.

Municipal Incinerator

The energy recovery incinerator tested was designed in 1974 and is located at Sheridan Avenue, Albany, New York. It combusts approximately 600 tons/day of uncharacterized shredded municipal wastes from which metals have been removed. Samples of stack effluent were collected on three consecutive days under isokinetic conditions from a sampling port halfway up the stack. Samples were taken with and without supplementary natural gas fuel. The incinerator's particulate removal system consisted of an electrostatic precipitator. A complete description of the overall series of tests is available and will not be discussed here.[7]

Contaminated Office Building

The Binghamton, New York State Office Building was contaminated with PCBs and combustion products in 1981 when a fire caused a Pyranol-filled transformer to rupture, spreading soot throughout the 18 floors.[8,9] Analyses of the soot showed the presence of CDFs, presumably formed from the PCBs, along with smaller amounts of CDDs, including 2,3,7,8-TCDD.[10,11] After the building contents were removed and the remaining walls and ductwork were extensively cleaned, air samples were collected for analysis.

Sampling Equipment

The ambient air sampling device used at Binghamton is shown in Figure 1.[12] A cartridge design, it consists essentially of two stages: a glass fiber particulate filter (Figure 1,1), and an extraction thimble containing silica gel adsorbent (Figure 1,2). Each may be analyzed separately. This sampler has been validated for ambient temperature collection of 2,3,7,8-TCDD.

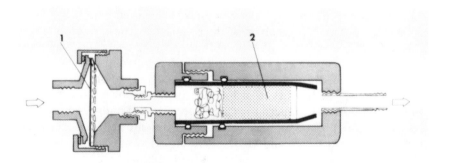

Figure 1. Ambient air sampler to collect 100 m^3.

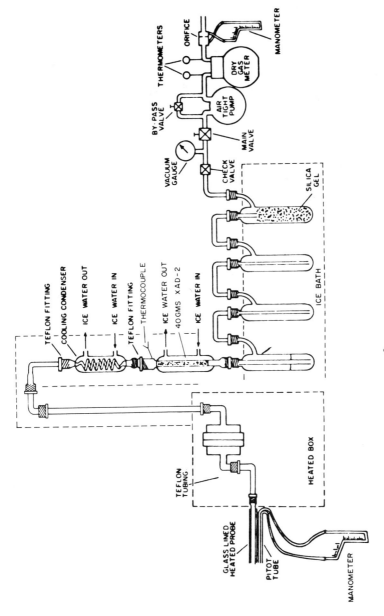

Figure 2. Incinerator sampling train to collect 3 m³.

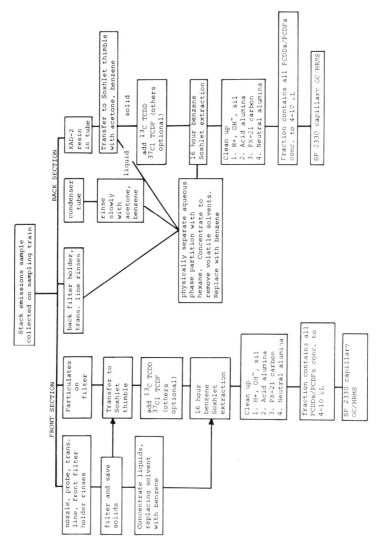

Figure 3. Workup of incinerator samples.

ANALYTICAL METHOD

- add ^{13}C 1234-TCDD internal standard to adsorbent prior to sampling

- take air or stack effluent sample

- separate quantification: particulate-adsorbed

- add internal standard(s) for quantification, cleanup recovery

- benzene Soxhlet extraction 16 h

- 1 fraction sample cleanup: 1 H_2SO_4 silica/K silicate/silica gel

  2 acid alumina

  3 adsorptive carbon

  4 neutral alumina

- concentrate to 4-10 µL

- SP 2330 GC/HRMS, ±10 mmu, 2 ions

Figure 4. Summary of the analytical method.

The sampling train used for collection of hot, acid stack effluents is shown in Figure 2. An all-glass/Teflon® modified method 5 Train, meeting the guidelines of ASME,[13] it consists essentially of three main stages: a glass fiber particulate filter, a cooling condenser and a temperature-controlled XAD-2-containing tube held in a vertical position. The component parts of the train are combined (as shown in Figure 3) into two samples: 1) the probe residue and filter particulates, and 2) the condenser rinses and XAD-adsorbed compounds. These were analyzed separately.

Analytical Method

Both ambient and stack samples were analyzed using the same analytical method (Figure 4). A single fraction containing all Cl_4-Cl_8 CDFs and CDDs is prepared by Soxhlet extraction followed by semiautomated three-column cleanup of the extract.[14] A portion of the cleanup extract is then injected onto a SP2330 GC Column which provides isomer specificity; detection is by multiple-ion monitoring high-resolution mass spectrometry (HRMS). (Note: because of space limitations, only one ion is shown in the following figures).

Isotope-labeled internal standards were used to provide a measure of both the potential for CDF/CDD breakthrough during each sampling event and the efficiency of the sample cleanup procedure. The chromatogram in Figure 5 shows the recovery of 10 ng ^{13}C-1,2,3,4-TCDD from exposed XAD is typically 50–70%. For building air only one standard was used (^{13}C-

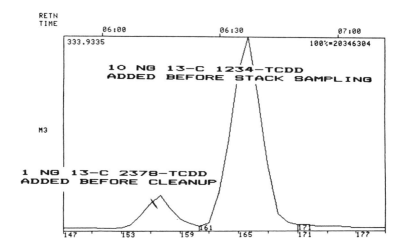

Figure 5. Recovery of ^{13}C-1,2,3,4-TCDD from spent XAD-2.

2,3,7,8-TCDD. This standard was added to the silica gel cartridge prior to sampling and measures both sample breakthrough and cleanup efficiency.

RESULTS AND DISCUSSION

TCDDs

Incinerator TCDD ion chromatograms (one ion shown) are given in Figure 6. The sampling, analytical and day-to-day combustion processes appear to be reproducible, with a total particulate and adsorbed concentration of TCDDs of 14 ± 3.6 ng/m^3 (supplementary gas off testing, n = 3). Isomer distributions for the three test days were found to be essentially the same. Significant quantities of TCDDs were found both on the particulate and XAD sections of the sampling train. The most abundant isomers were the 1,3,6,8-, 1,3,7,9-, with 2,3,7,8-TCDD constituting only 2%. This distribution of isomers was nearly identical to that reported by others, e.g., fly ash,[5] as shown in Figure 6.

A contaminated building air sample TCDD ion chromatogram (one ion shown) and mass profile are given in Figure 7. Very low concentrations of TCDDs were detected in several building air samples, (e.g., 2,3,7,8-TCDD = 0.4 pg/m^3). Several TCDD peaks were observed, including a peak at the retention time of 2,3,7,8-TCDD, which appears to be a major isomer. Pentachlorobiphenylene at m/e 322 was also observed, but was adequately resolved from the TCDDs using HRMS.

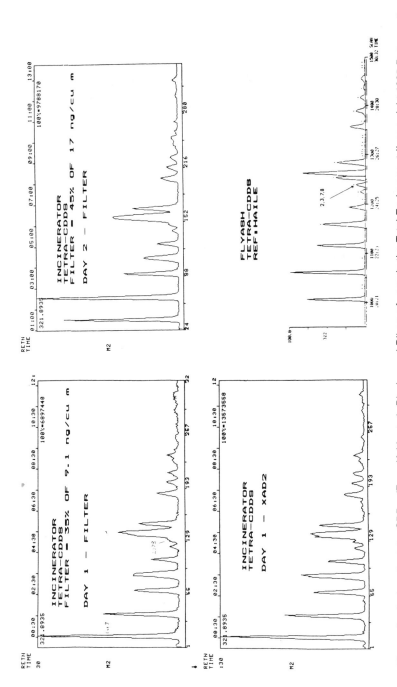

Figure 6. Incinerator tetra-CDDs. (From Keith et al. *Dioxins and Dibenzofurans in the Total Environment II*, copyright 1985 Butterworth Publishers. Used with permission.)

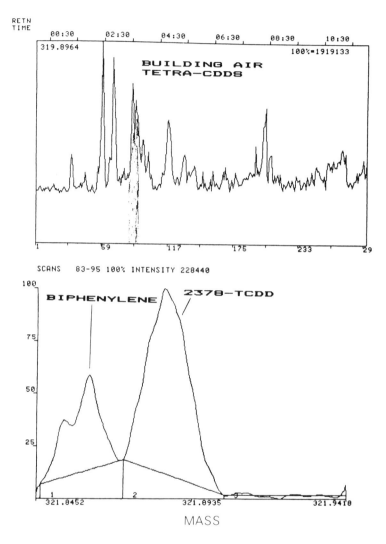

Figure 7. Building air tetra-CDDs.

TCDFs

Typical tetra-CDF ion chromatograms for the incinerator effluent and building air samples are shown in Figure 8. Although many isomers were found in both cases, the relative proportions of isomers were significantly different. The distribution of incinerator effluent isomers found was not unlike that reported for an incinerator in another study[6] (Figure 8). As expected, the major TCDF isomers reported as present in the contaminated

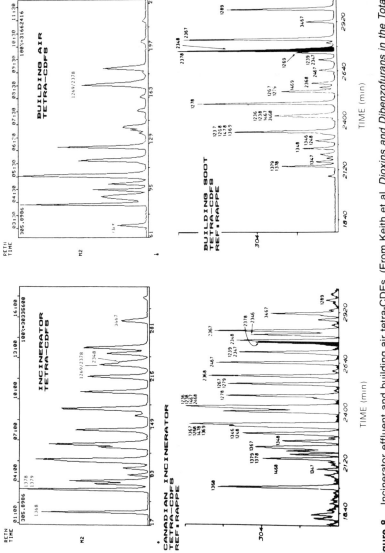

Figure 8. Incinerator effluent and building air tetra-CDFs. (From Keith et al. *Dioxins and Dibenzofurans in the Total Environment II,* copyright 1985 Butterworth Publishers. Used with permission.)

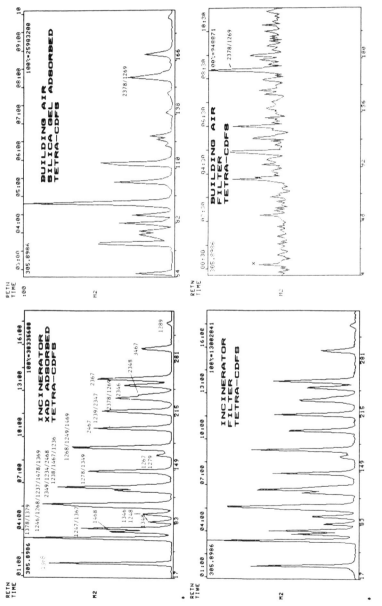

Figure 9. Adsorbed and particulate tetra-CDFs.

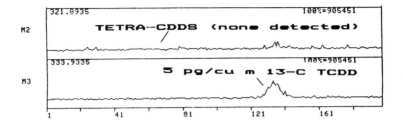

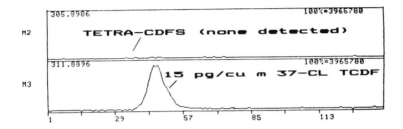

Figure 10. 80-m^3 air samples taken downwind of the incinerator.

building soot[6] were also found in the air samples, although the relative concentrations are slightly different.

A significant difference was observed between incinerator effluent and building air with respect to the particulate/vapor phase distribution of CDFs, as shown in Figure 9. Similar to the TCDDs, incinerator effluent TCDFs collected on both the particulate filter and XAD-2 adsorbent; identical isomer patterns of TCDFs were also found. However, the contaminated building TCDFs, collected at ambient temperature, were found predominantly on the silica-gel adsorbent and only trace amounts on the particle filter. While it is possible that differences in temperature and flow rate influenced the observed distributions, the data suggest that the CDFs found in the building were in the gas phase.

RELATED STUDIES AND HIGHER CHLORINATED CONGENERS

As a separate phase of the incinerator testing, an attempt was made to detect tetra-CDFs and CDDs in ambient air directly downwind of the incinerator. The 2,3,7,8-substituted internal standards may be seen at concentrations of 5 and 15 pg/m^3, although no tetra-CDDs (1,2,6,8-/7,9- were the major effluent isomers) or tetra-CDFs (more abundant in the effluent than the TCDDs) were observed (Figure 10) at detection limits of approximately 1 pg/m^3. In contrast, the tetra-CDFs in air from inside the contaminated

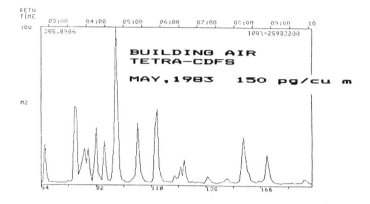

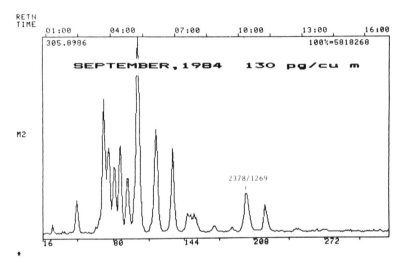

Figure 11. Building air tetra-CDFs—1983 to 1984.

building were easily detected. These compounds showed little change either in concentration or isomer distribution of tetra-CDFs over a time period of 1 year, as shown in Figure 11. The concentration of total tetra-CDFs for a group of samples ranged from 121–195 pg/m^3 in April, 1983, to 130–260 pg/m^3 in September, 1984. Penta-CDFs were also detected in the building air. As shown in Figure 12, a mixture of isomers was found at concentrations ranging from 29 to 185 pg/m^3.

To confirm our findings of pg/m^3 quantities of 2,3,7,8-substituted and other CDFs in the contaminated building air, three samples were collected, extracted with benzene, and split to be analyzed by another laboratory using isomer-specific techniques. The results given in Table I show good agree-

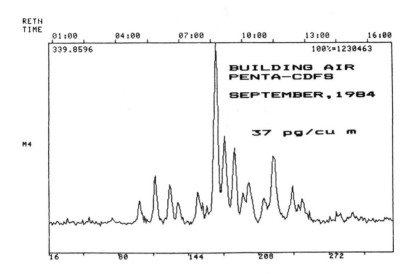

Figure 12. Building air penta-CDFs.

ment between the two laboratories and confirm our previous findings of tetra-, penta- and hexa-CDFs.

Finally, the overall distribution of Cl_4 to Cl_8 congeners is shown in Figure 13 for an incinerator stack emission composite (supplementary gas on, three-sample composite) and for the contaminated building air, soot, and vinyl wall wipes (averages). Data on the higher chlorinated congeners was obtained using either selected ion monitoring (SIM) low-resolution MS, full-scan HRMS or multiple-injection SIM HRMS. As shown in the figure, the incinerator emissions contained predominantly tetra- and penta-CDF congeners and penta-, hexa- and hepta-CDDs. The building soot contained predominantly tetra-, penta- and hexa-CDFs; however, the concentration of the lighter tetra congeners was greatest in the air samples; thus it appears that these compounds may have volatilized from the soot into the building air.

In conclusion, the above comparison of recently acquired data is presented to summarize and highlight results which constitute part of more extensive studies. Detailed descriptions of the studies and their findings will be published as they are completed.

Table 1. Interlaboratory Comparison of Air Results (split Soxhlet extracts, pg/m^3).[a]

Sample		Tetra-CDFs		Penta-CDFs			Hexa-CDFs		
		2,3,7,8	Total	1,2,3,7,8 1,2,3,4,8	2,3,4,7,8	Total	1,2,3,4,7,8 1,2,3,4,7,9	1,2,3,6,7,8	Total
1 Building Air	Lab A	5.9	117	10	1.4	71	2.2	1.1	8.8
	NYS Health Lab	6.4	102	9	1.0	48	1.8	NA[b]	7.0
2 Building Air	Lab A	2.3	46	3.9	0.5	27	0.8	0.4	3.2
	NYS Health Lab	1.8	52	4.8	0.2	37	0.8	NA	5.8
3 Outside Air	Lab A	0.1	0.9			0.6			
	NYS Health Lab	ND[c](0.4)	ND(0.6)			ND(4.6)			

[a]Uncorrected for internal standard recovery.
[b]NA, not analyzed.
[c]ND, not detected.

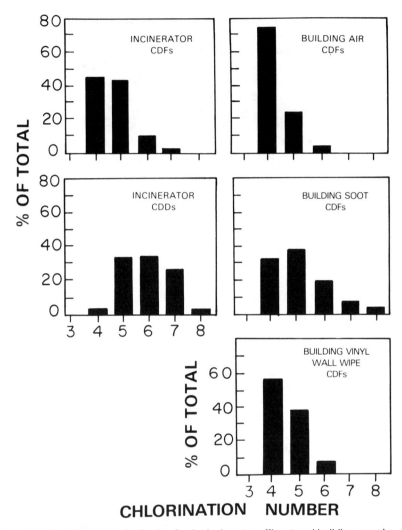

Figure 13.   Congener distribution for the incinerator effluent and building samples.

REFERENCES

1. Lustenhouwer, J. W. A., K. Olie, and O. Hutzinger. "Chlorinated Dibenzo-p-dioxins and Related Compounds in Incinerator Effluents: A Review of Measurements and Mechanisms of Formation," *Chemosphere* 9 (1980).
2. Bumb, R. R., et al. "Trace Chemistries of Fire: A Source of Chlorinated Dioxins," *Science* 210:385–390 (1980).
3. Tiernan, T. O., et al. "Chlorodibenzodioxins, Chlorodibenzofurans and Related Compounds in the Effluents from Combustion Processes," *Chemosphere* 12(4/5): 595–606 (1983).

4. Ozvacic. V., et al. "Emissions of Chlorinated Organics from Two Municipal Incinerators in Ontario," paper presented at the 77th Annual Meeting, San Francisco, California (1984).

5. Haile, C. L., et al. "Emissions of Polychlorinated Dibenzo-p-dioxins and Polychlorinated Dibenzofurans from a Resource Recovery Municipal Incinerator," in *Chlorinated Dioxins and Dibenzofurans in the Total Environment II*, Keith, L. H., C. Rappe and G. Choudhary, Eds. (Stoneham, MA: Butterworth Publishers, 1985), p. 439.

6. Rappe, C., et al. "Composition of Polychlorinated Dibenzofurans (PCDF) Formed in PCB Fires," in *Chlorinated Dioxins and Dibenzofurans in the Total Environment II*, Keith, L. H., C. Rappe and G. Choudhary, Eds. (Stoneham, MA: Butterworth Publishers, 1985), p. 401.

7. Kerr, R., and A. Columbus. "Sheridan Avenue Municipal Incinerator Stack Emissions Report," New York State Department of Environmental Conservation Report (1985).

8. Haughie, G. F., A. J. Schecter and R. Rothenberg. "PCB Transformer Fire—Binghamton, New York," *MMWR* 30:187–188, 193 (1981).

9. Schecter, A. J. "Contamination of an Office Building in Binghamton, New York by PCBs, Dioxins, Furans and Biphenylenes after an Electrical Panel and Electrical Transformer Incident," *Chemosphere* 12:669–680 (1983).

10. O'Keefe, P. W., et al. "Chemical and Biological Investigations of a Transformer Accident at Binghamton, N. Y.," in *Environmental Health Perspectives*, Lucier G. W. and G. Hook, Eds. (U. S. Government Printing Office NIH Publication NIH 85–218, 1985).

11. Smith, R. M., P. W. O'Keefe, D. R. Hilker, B. L. Jelus-Tyror, and K. M. Aldous. "Analysis for 2,3,7,8-Tetrachlorodibenzofuran and 2,3,7,8-Tetrachlorodibenzo-p-dioxin in a Soot Sample from a Transformer Explosion in Binghamton, New York," *Chemosphere* 11(8):715–720 (1982).

12. Smith, R. M., P. W. O'Keefe, D. R. Hilker, and K. M. Aldous. "Determination of pg/m^3 Concentrations of Chlorinated Dibenzofurans and Dibenzo-p-dioxins in Air Samples from a Contaminated Building by High Resolution Gas Chromatography/High Resolution Mass Spectrometry," Unpublished results submitted to *Anal. Chem.* (1985).

13. "Environmental Standards Workshop on Developing Draft Standards for Sampling and Measuring Trace Chlorinated Emissions from Waste-to-Energy Combustion Facilities," Draft protocols ASME, Tysons Corners, Virginia, January 23–26, 1984.

14. O'Keefe, P. W., et al. "A Semiautomated Cleanup Method for Polychlorinated Dibenzo-p-dioxins and Polychlorinated Dibenzofurans in Environmental Samples," in *Chlorinated Dioxins and Dibenzofurans in the Total Environment II*, Keith, L. H., C. Rappe and G. Choudhary, Eds. (Stoneham, MA: Butterworth Publishers, 1985), p. 111.

CHAPTER 8

Thermal Combustion
of Octachlorodibenzofuran
to Form Lower PCDFs

Steven E. Swanson, Mitchell D. Erickson, Leslie Moody, and Daniel T. Heggem

INTRODUCTION

Several studies have examined the distribution of polychlorinated dibenzo-p-dioxin (PCDD) and polychlorinated dibenzofuran (PCDF) congeners in combustion effluents and soot.[1-10] Generally, similar distributions of PCDDs and PCDFs have been found. An example of a typical PCDD and PCDF distribution in flue gas condensate is shown in Figure 1. Similar distributions have been observed in the products of thermal degradation of both PCB-containing and non-PCB-containing dielectric fluids (Figures 2 and 3).

The chemistry of PCDD and PCDF formation probably involves several intermediates, and may involve sequential chlorination or formation of higher chlorinated homologs (hepta or octa) followed by sequential dechlorination.[9,14] To examine this latter route, octachlorodibenzofuran (OCDF)-spiked mineral oil was fed into a thermal combustion system operated at 675°C and the products examined.

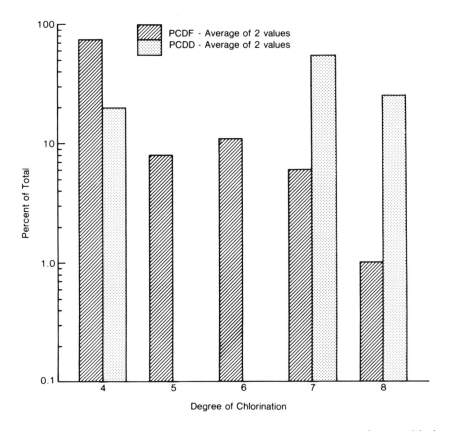

Figure 1. Typical PCDD and PCDF distributions in flue gas condensate from municipal refuse incinerators.[4]

EXPERIMENTAL

Feed Material

A 500 $\mu g/mL$ solution of OCDF in mineral oil was used as the feed material. The OCDF (99%) was purchased from Analabs, North Haven, CT (Part No. RCS 021). The mineral oil is Exxon type HPLX 355077.

Apparatus and Procedures

The thermal combustion system and its operation, the chemical analysis procedures, and other experimental details are described in detail by Erickson et al.[11] and Swanson et al.[12]

Samples were fed into a thermal combustion system at 14 $\mu L/min$ and combusted under controlled temperature (675°C), residence time (0.8 sec),

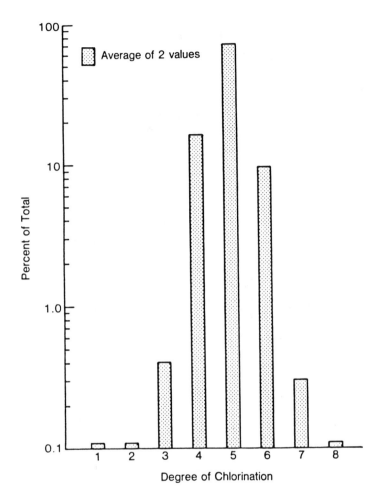

Figure 2.   PCDF distribution in products of thermal degradation of PCB-containing dielectric fluid (70% Aroclor 1260/30% trichlorobenzene).[11]

and effluent oxygen content (8%). These conditions were previously shown
to give the optimum yield of PCDFs from PCBs.[11] The total gas (N^2, O^2,
CO^2, CO) flow rate through the system was held at about 1.2 L/min. The
precise conditions and effluent gas compositions are listed in Table 1. The
combustion efficiency (C.E.), also presented in Table 1, is calculated by:

$$C. E. = \frac{[CO_2]}{[CO_2] \text{ and } [CO]} \times 100$$

where the CO_2 and CO concentrations are expressed in the same units.

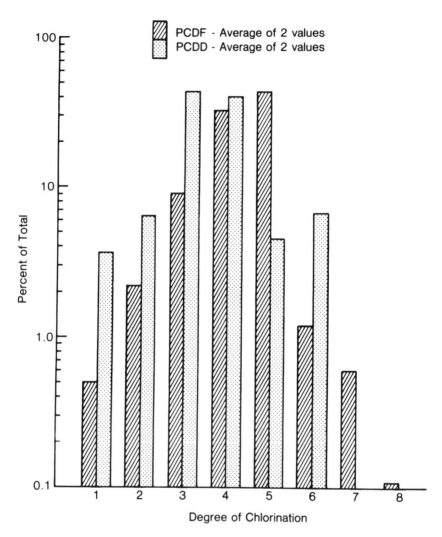

Figure 3. PCDD and PCDF distribution in products of thermal degradation of non-PCB-containing dielectric fluid (trichlorobenzene).[12]

Organic Analysis

The combustion effluent gas was collected on a 15-g XAD-2 cartridge. The effluent gas was also monitored for CO, CO_2, and O_2 content. The XAD-2 resin was spiked with the surrogate compounds ^{13}C-2,3,7,8-TCDF and ^{13}C-OCDD and then Soxhlet extracted with hexane. This extract was concentrated to 2 mL, then split. Half of the extract was cleaned up using the acidified silica and acidified alumina column technique.[13] The cleaned extracts were analyzed by high-resolution gas chromatography/electron

Table 1. Combustion System Operating Conditions.

Feed Material	Combustion Temperature (°C)	Run Time (min)	Excess Oxygen (%)	Residence Time (sec)	CO_2 (%)	CO (%)	Combustion Efficiency (%)	Pyrolysis Temperature (°C)
OCDF in Mineral Oil–1	678	37	7.9	0.81	0.14	0.33	29	397
OCDF in Mineral Oil–2	679	35	7.9	0.78	0.15	0.33	30	398
Blank (No Feed)	677	55	7.9	0.77	0.06	0.0003	99	398

Table 2.   Results of Analysis of Polychlorinated Dibenzofurans in OCDF-Spiked Mineral Oil Feed.

Feed	Tetra-CDF (total ng)	Penta-CDF (total ng)	Hexa-CDF (total ng)	Hepta-CDF (total ng)	Octa-CDF (total ng)	Total PCDF (total ng)
OCDF–1	3,600	6,200	29,000	41,000	14,000	94,000
OCDF–2	3,700	4,300	42,000	50,000	20,000	120,000
Blank	170	120	190	260	150	890

impact ionization mass spectrometry with selected ion monitoring (HRGC/EIMS-SIM) for PCDDs and PCDFs. The other half of each extract was analyzed by HRGC/EIMS with full scan mode for other chlorinated organics.

RESULTS AND DISCUSSION

Two runs followed by a blank run (no feed material) were conducted with OCDF in mineral oil at 500 μg/mL as the feed. As noted in Table 2 and in Figure 4, substantial amounts of the lower PCDF homologs were formed. Only about 8% of the OCDF feed passed through the reaction system unchanged. No PCDDs were detected. Nonquantitative analyses indicated that mono- through tri-CDFs were also formed in amounts representing less than 1% of the total PCDF.

The distribution of PCDF homologs obtained from combustion of OCDF (Figure 4) is quite different from that observed in flue gas condensate (Figure 1) and in thermal degradation products from combusting dielectric fluids (Figures 2 and 3). In Figures 1–3 the tetra-CDF or penta-CDF represents the major homolog concentration, whereas the hepta-CDF homolog represents the greatest concentration in the OCDF combustion effluent. It is also noteworthy that the isomer distribution observed in the OCDF combustion effluents is different from that in the combustion of PCBs and trichlorobenzene.[11,12] Figures 5 through 7 present extracted ion current profiles (EICPs) of the tetra- through octa-CDFs which show the isomer distribution for each homolog. The difference in isomer distribution is particularly apparent for the tetra-CDFs, as can be seen by a comparison of Figures 5 and 7.

The full-scan HRGC/EIMS analysis results of the OCDF combustion effluents showed that only a few chlorinated organics were present in these samples above the 1- to 10-μg detection limits, as shown in Table 3. Many chlorinated combustion products have been detected in incineration and fire samples;[1,3,4,6-8,13] however, the lack of major chlorinated products in the OCDF samples may be due to the single compound tested.

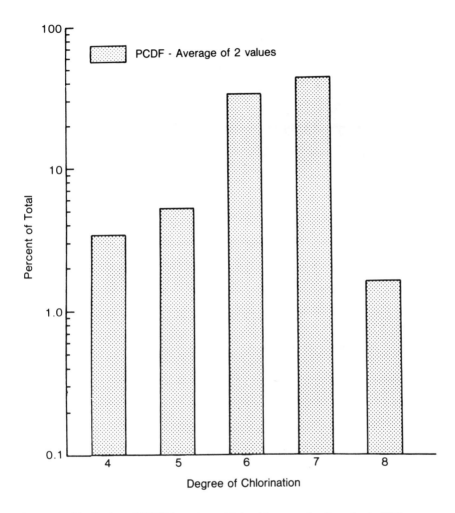

Figure 4.   Distribution of PCDF homologs obtained from combustion of octa-CDF.

SUMMARY

A broad distribution of PCDFs is formed during the thermal degradation of PCB and other organochlorine dielectric fluids.[11,12] The mechanism for this formation may involve sequential chlorination of dibenzofuran, or formation of higher chlorinated dibenzofurans (such as heptachloro or octachloro) followed by dechlorination. To test the latter route, OCDF-spiked mineral oil was fed into a thermal combustion system operated at 675°C. The results showed that only 8% of the OCDF feed passed through the reaction system unchanged. Substantial amounts of lower PCDF homologs were formed, with greatest concentration being hexa- and hepta-CDFs.

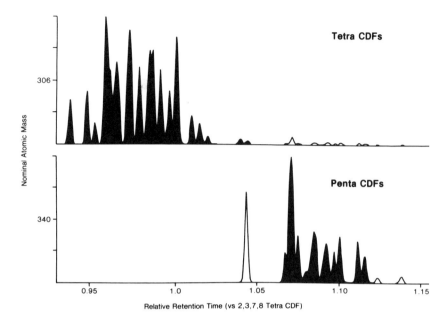

Figure 5. Extracted ion current profiles of tetra- and penta-CDFs from combustion of OCDF, showing isomer distribution for each homolog.

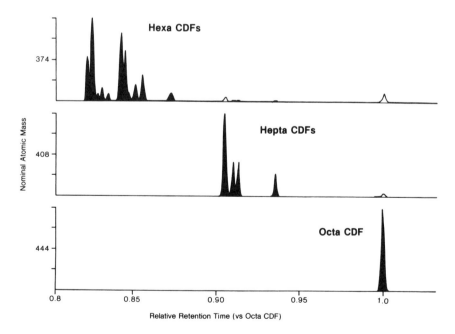

Figure 6. Extracted ion current profiles of hexa-, hepta-, and octa-CDFs from combustion of OCDF, showing isomer distribution for each homolog.

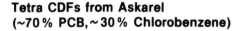

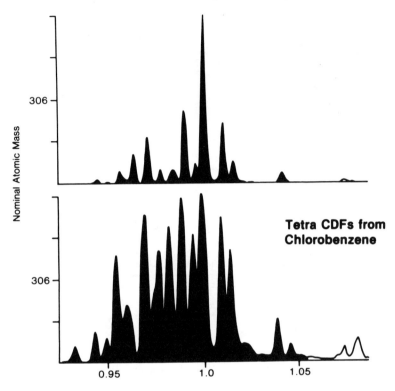

Figure 7. Extracted ion current profiles of tetra-CDFs from combustion of PCBs and trichlorobenzene, showing isomer distribution for each homolog.

The total tetra- through octa-CDFs measured in the products accounted for 35 to 50% of the OCDF fed into the heated zone. These data indicate that dechlorination of higher chlorinated CDFs is a facile reaction at the temperature tested (675°C), which can be found in the cooler zones of combustion sources. A similar dechlorination of PCDDs may also be assumed to occur under similar conditions. These dechlorination reactions may account for at least part of the similarity of the PCDF and PCDD profiles observed from various feed materials combusted under different conditions.

Table 3.   Results of Full-Scan GC/EIMS Analysis of Combustion Effluents of OCDF in Mineral Oil.

Analyte	Feed		
	OCDF–1 (total μg)	OCDF–2 (total μg)	Blank (total μg)
Trichlorobenzene	< 0.5	16	< 20[a]
Tetrachlorobenzene	< 0.6	22	< 30[a]
Pentachlorobenzene	< 0.8	4	< 0.8
Hexachlorobenzene	< 1	< 1	< 1
Monochlorobiphenyl	< 0.4	< 0.2	< 0.4
Dichlorobiphenyl	< 0.3	< 0.6	< 0.3
Trichlorobiphenyl	< 1	< 0.6	< 0.1
Tetrachlorobiphenyl	< 1	< 1	< 1
Pentachlorobiphenyl	< 2	< 2	< 2
Hexachlorobiphenyl	< 2	16	< 2
Heptachlorobiphenyl	< 3	4	< 3
Octachlorobiphenyl	< 3	< 3	< 3
Nonachlorobiphenyl	< 3	< 3	< 3
Decachlorobiphenyl	< 4	< 4	< 4
Chlorophenol (total)	< 5	< 5	< 5
Chloronaphthalene (total)	< 5	< 5	< 5
Chloropyrene (total)	< 5	< 5	< 5
Chlorochrysene (total)	< 5	< 5	< 5
Chlorobiphenylene (total)	< 10	< 10	< 10
Chloroxanthene (total)	< 10	< 10	< 10

[a]High background from previous GC/MS run.

ACKNOWLEDGMENTS

The work upon which this chapter is based was performed pursuant to Contract No. 68–02–3938 with the U. S. Environmental Protection Agency.

REFERENCES

1. Czuczwa, J. M., and R. A. Hites. "Environmental Fate of Combustion-Generated Polychlorinated Dioxins and Furans," *Environ. Sci. Technol.* 18:444–450 (1984).

2. Kooke, R. M. M., J. W. A. Lustenhouwer, K. Olie, and O. Hutzinger. "Extraction Efficiencies of Polychlorinated Dibenzo-p-dioxins and Polychlorinated Dibenzofurans from Fly Ash," *Anal. Chem.* 53:461–463 (1981).

3. Lustenhouwer, J. W. A., K. Olie, and O. Hutzinger. "Chlorinated Dibenzo-p-dioxins and Related Compounds in Incinerator Effluents: A Review of Measurements and Mechanisms of Formation," *Chemosphere* 9:501–522 (1980).

4. Rappe, C., S. Marklund, P. A. Bergqvist, and M. Hansson. "Polychlorinated Dibenzo-p-dioxins, Dibenzofurans and Other Polynuclear Aromatics Formed During Incineration and Polychlorinated Biphenyl Fires," in *Chlorinated Dioxins and Dibenzofurans in the Total Environment*, G. Choudhary, L. H. Keith, and C. Rappe, Eds. (Stoneham, MA; Butterworth Publishers, 1983), p. 99–124.

5. Rappe, C., S. Marklund, L.-O. Kjeller, P. A. Bergqvist, and M. Hansson. "Composition of Polychlorinated Dibenzofurans (PCDF) Formed in PCB Fires," in *Chlorinated Dioxins and Dibenzofurans in the Total Environment II*, L. H. Keith, C. Rappe, G. Choudhary, Eds. (Stoneham, MA: Butterworth Publishers, 1985), p. 401–424.

6. Redford, D. P., C. L. Haile, and R. M. Lucas. "Emissions of PCDDs and PCDFs from Combustion Sources," in *Human and Environmental Risks of Chlorinated Dioxin and Related Compounds*, R. E. Tucker, D. L. Young, and A. P. Bray, Eds. (New York: Plenum Press, 1983), p. 143–152.

7. Haile, C. L., R. B. Blair, J. S. Stanley, D. P. Redford, D. Heggem, and R. M. Lucas. "Emissions of Polychlorinated Dibenzo-p-dioxins and Polychlorinated Dibenzofurans from a Resource Recovery Municipal Incinerator," in *Chlorinated Dioxins and Dibenzofurans in the Total Environment II*, L. H. Keith, C. Rappe, G. Choudhary, Eds. (Stoneham, MA: Butterworth Publishers, 1985).

8. Taylor, M. L., T. O. Tiernan, J. H. Garrett, G. F. Van Ness, and J. G. Solch. "Assessment of Incineration Processes as Sources of Supertoxic Chlorinated Hydrocarbons: Concentrations of Polychlorinated Dibenzo-p-dioxins/Dibenzofurans and Possible Precursor Compounds in Incinerator Effluents," in *Chlorinated Dioxins and Dibenzofurans in the Total Environment*, G. Choudhary, L. H. Keith, and C. Rappe, Eds. (Stoneham, MA: Butterworth Publishers, 1983), p. 125–164.

9. Marklund, S., L.-O. Kjeller, M. Hansson, M. Tysklind, C. Ryan, J. de Kanel, and R. C. Dougherty. "Determination of PCDDs and PCDFs in Incineration Samples and Pyrolytic Products," Chapter 6, this volume.

10. Thompson, H. C., Jr., D. L. Kendall, W. A. Korfmacher, J. R. Kominsky, L. G. Rushing, K. L. Rowland, L. M. Smith, and D. L. Stalling. "Polychlorinated Dibenzo-p-dioxins, Polychlorinated Dibenzofurans and PCBs at a Contaminated Multibuilding Facility," Chapter 6, this volume.

11. Erickson, M. D., C. J. Cole, J. D. Flora, P. G. Gorman, C. L. Haile, G. D. Hinshaw, F. C. Hopkins, and S. E. Swanson. "Thermal Degradation Products from Dielectric Fluids," for U. S. Environmental Protection Agency, Office of Toxic Substances, Washington, DC, Report No. EPA 560/5-84-009 (1985), p. 98.

12. Swanson, S. E., M. D. Erickson, and L. Moody. "Products of Thermal Degradation of Dielectric Fluids," for U. S. Environmental Protection Agency, Office of Toxic Substances, Washington, DC, Report No. EPA 560/5-85-022 (1985), p. 30.

13. "Determination of 2,3,7,8-TCDD in Soil and Sediment," U. S. Environmental Protection Agency, Region VII Laboratory, Kansas City, Kansas (1983).

14. Choudhry, G. G., and O. Hutzinger. *Mechanistic Aspects of Thermal Formation of Halogenated Organic Compounds Including Polychlorinated Dibenzo-p-Dioxins* (New York: Gordon and Breach Science Publishers, 1983), p. 194.

CHAPTER 9

Polychlorinated Dibenzo-p-Dioxins, Polychlorinated Dibenzofurans and Polychlorinated Biphenyls at a Contaminated Multibuilding Facility

H. C. Thompson, Jr., D. C. Kendall, W. A. Korfmacher, J. R. Kominsky, L. G. Rushing, K. L. Rowland, L. M. Smith, and D. L. Stalling

INTRODUCTION

The pyrolysis of polychlorinated biphenyls (PCBs) and electrical equipment containing PCB askarel has been shown to generate polychlorinated dibenzofurans (PCDFs), polychlorinated dibenzo-p-dioxins (PCDDs) and polychlorinated biphenylenes.[1-5] Typically, PCDFs are found at a concentration greater than the PCDDs in the transformer fires and capacitor explosions that have been investigated.[2,3,5,6] PCDFs and PCDDs have also been identified in fly ash from incinerators.[7-9]

In this report, a unique site of PCB, PCDD and PCDF contamination is described. The contamination is believed to be due to the improper incineration of PCB askarel at the site over 12 years ago. Contamination of the multibuilding facility by PCBs, PCDFs and PCDDs has been found to be widespread. Based on the observed levels of contamination in the buildings and the distribution throughout the facility, it is postulated that products of incomplete combustion of PCB askarel were airborne, and entered the buildings via air intake vents.

EXPERIMENTAL

The cyclohexane and hexane used for sampling and cleanup of samples was pesticide residue grade obtained from Burdick and Jackson (Muskegon, MI). The sulfuric acid used was Reagent A. C. S. grade obtained from Fisher.

Sampling Procedure

Two groups of surface wipe samples were collected. The first group of samples (I) was collected by wiping an area of 0.25 m^2 with toluene using two or three absorbent tissues (12.5 cm × 21.5 cm Kimwipes). The tissues were then placed into a 38mm × 200mm culture tube, which was sealed with a Teflon®-lined screw cap.

The second group of samples (II) was collected by wiping an area of 100 cm^2 with a 5-cm by 5-cm cyclohexane-extracted cotton gauze swatch wetted with cyclohexane. The cotton gauze swatch was then placed in a 17-mm × 125-mm culture tube, which was sealed with a Teflon®-lined screw cap.

Areas designated for sampling which contained excessive amounts of dust and debris were sampled by collecting the residue in a 17-mm × 125-mm culture tube, which was sealed with a Teflon®-lined screw cap.

Dust samples (0.1-1g) were weighed into Soxhlet thimbles and extracted with 100 mL of hexane for 2 hr. For comparison, a second portion of some samples was Soxhlet extracted with benzene for 6 hr, as noted in the text. The extracts were then taken to dryness on a rotary evaporator and redissolved in 5–7 mL of hexane for cleanup and gas chromatography/electron-capture detection (GC/ECD) analysis.

For an interlaboratory methods comparison, four dust samples were collected from horizontal surfaces of four buildings and each was made into a composite sample containing 30–40 g of dust. Each composite sample was homogenized by rotating it in a bottle containing stainless steel ball bearings. Ten-gram portions of the four dust samples were analyzed by both the National Center for Toxicological Research (NCTR) GC/ECD method[10] and by the Columbia National Fisheries Research Laboratory (CNFRL) GC/mass spectrometric (MS) method.[11]

Cleanup of Extracts for PCB Analysis

Group I samples were prepared for analysis by adding 50 mL of hexane to the culture tube, extracting for 1 hr on a mechanical shaker, transferring 5 mL of the extract to a 17-mm × 125-mm culture tube containing 5 mL of concentrated sulfuric acid with subsequent shaking for 30 sec. After allowing phase separation, the hexane phase (top layer) was ready for PCB analysis. The remainder (45 mL) of the sample extract was reserved for heptachlorodibenzo-p-dioxin (HpCDD) and octachlorodibenzo-p-dioxin (OCDD) analysis. Group II samples were prepared for analysis by adding 10 mL of hexane to the culture tube, extracting for 2 hr on a mechanical shaker, transferring 5 mL of the extract to a 17-mm x 125-mm culture tube containing 5 mL of concentrated sulfuric acid, followed by shaking for 30 sec. After allowing phase separation, the hexane phase (top level) was ready for analysis. The remainder (5 mL) of the sample extract was reserved for OCDD/HpCDDs analysis.

Cleanup of Extracts for OCDD/HpCDDs Analysis

Each sample extract was transferred into a 15-mL conical-tipped centrifuge tube equipped with a Teflon®-lined screw cap and diluted with hexane to 9–10 mL. Five milliliters of $1M$ KOH solution was added to each tube and the tubes were then rotated by mechanical rotation (Fisher Roto-Rack, Model 343) at medium speed for 15 min. The samples were next centrifuged at 2000 rpm (Dupont Sorvall centrifuge, model GLC-2B) for 5 min, and then the aqueous phase was removed using a Pasteur pipet. Five milliliters of 1% H_2SO_4 was added to the tubes, and they were again rotated for 15 min. After centrifuging the samples at 2000 rpm for 5 min, the aqueous phase was removed using a Pasteur pipet. Then 5 mL of deionized water was added to the tubes and they were rotated for 15 min. The samples were centrifuged as before and the aqueous phase was discarded. Each sample was then percolated through a 3-cm × 2-cm plug of anhydrous sodium sulfate followed by four rinses of the sodium sulfate using 4-mL portions of hexane. The sodium sulfate bed was on top of glass wool in a glass funnel. (The assembly was prewashed with hexane before the sample was applied.) The sample was collected in a 50-mL round bottom flask and was taken to dryness under a water pump vacuum in a 50°C water bath.

The high-pressure liquid chromatographic (HPLC) cleanup procedure has been described previously.[10] The HPLC column was prepared by mixing 5% (by weight) Amoco active carbon PX-21 (sample # C540, from the Amoco Research Corp., Chicago, IL) and 20-μm silica (S20W from Spherisorb, Queensperry, Clwy, UK). This packing was stirred and sieved through a 40-μm screen to remove any clumps. The material that passed through the screen was used to dry pack the 2-mm × 70-mm stainless steel column.

All solvents were pesticide grade, suitable for gas chromatography. Each sample was dissolved in 500 μL of 50/50 (v/v) cyclohexane/dichloromethane and injected onto the column via the Rheodyne (Cotati, CA) injector. Samples that were heavily colored or of an oily nature were diluted to 5 mL before removing 500 μL for injection to prevent overloading of the cleanup column. Then 30 mL of 50/50 cyclohexane/dichloromethane solvent was pumped at a constant rate of 2 mL/min through the column; 10 mL of the second solvent (75/20/5 [v/v/v] dichloromethane/methanol/benzene) followed. Using a second valve, the column was then backflushed with 30 mL of toluene, which eluted the OCDD and HpCDDs, which were collected in a 100-mL round bottom flask. The column was cleaned by pumping an additional 30 mL of solvent four (methanol), followed by 24 mL of toluene. The second valve was returned to its original position (i.e., forward flow through the column) and 26 mL of solvent one was pumped through the column to reequilibrate it in preparation for the next sample. The cleaned up samples (each 30 mL) were evaporated in a 75°C water bath using a water pump vacuum and then redissolved in 1 mL of hexane and transferred to an 8-mL culture tube which was sealed with a Teflon®-lined screw cap. A 100-μL aliquot of each sample was transferred to a second 8-mL tube and evaporated with a stream of air. The samples were reconstituted in 1 mL of n-hexadecane (Burdick and Jackson, Muskegon, MI) for subsequent analysis by capillary GC/ECD.

GC/ECD Analysis of PCBs

An aliquot of the cleaned up hexane extract was transferred to a sample vial (1 mL, Wheaton #223682) and sealed with a Teflon®-lined aluminum seal (11 mm, Wheaton #224211). High-concentration samples were diluted with hexane such that the resultant concentration was within the approximate range of 500–2000 ng/mL of the Aroclor. Low-concentration samples were concentrated using a stream of nitrogen such that the resultant concentration was within the approximate range of 100–2000 ng/mL of the total Aroclor.

The gas chromatograph used for the analysis of PCBs was a Hewlett-Packard Model 5880 GC equipped with an electron-capture detector and automatic sampler (Hewlett-Packard Model 7627). The column was a 2-m × 2-mm i.d. glass column packed with 3% OV–101 on Supelcoport®, 80/100 mesh. The gas chromatographic conditions were 240°C, 220°C, and 300°C for the injection port, oven, and detector, respectively. The carrier gas was 95% argon, 5% methane and flowed at a flow rate of approximately 35 mL/min.

Figure 1 shows the chromatographic patterns for Aroclors 1254 and 1260. Peaks used for quantitation of Aroclor 1254 are also present in Aroclor 1260, making accurate quantitation of Aroclor 1254 difficult in samples which contained both Aroclors. The peaks used for quantitation of Aroclor 1260 were relatively specific for Aroclor 1260, and are not present in significant quantity in Aroclor 1254.

In samples which contained both Aroclor 1254 and Aroclor 1260, to approximate the quantity of Aroclor 1254 present, the contribution from Aroclor 1260 to the total peak height was calculated. A ratio between the peaks used for quantitation of Aroclor 1254 and Aroclor 1260 was determined based on an Aroclor 1260 standard. Using this ratio and the calcu-

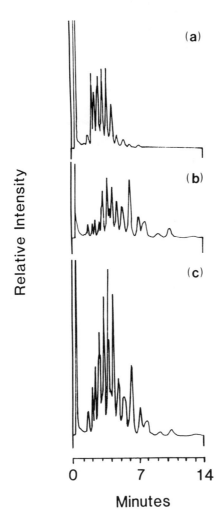

Figure 1. Typical GC/ECD results for (a) Aroclor 1254 standard, (b) Aroclor 1260 standard, and (c) wipe sample containing PCBs.

lated amount of Aroclor 1260 present, the contribution to Aroclor 1254 was subtracted; then, based on the Aroclor 1254 standard, the amount of Aroclor 1254 present in the sample was calculated. For any sample in which the calculated concentration of one Aroclor was greater than five times the concentration of the other Aroclor, the Aroclor with the lowest concentration was reported as none detected. The Aroclor levels were then summed and are reported as "total PCBs."

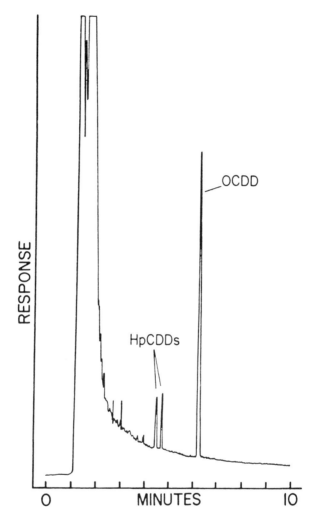

Figure 2. GC/ECD results for a wipe sample. The surface was found to contain 50 μg/m^2 OCDD and 12 μg/m^2 HpCDDS.

GC/ECD of OCDD/HpCDDs.

The cleaned up samples were assayed by on-column injection of 1 μL of the sample in n-hexadecane (b.p 287°C) into a Varian (Walnut Creek, CA) Model 3700 gas chromatograph equipped with a Varian on-column injector (purge outlet open) and a Varian (constant-current, pulse-modulated) electron-capture detector with an 8-mCi ^{63}Ni foil. The capillary fused-silica column was a J&W (Cordova, CA) 25-m × 0.25-mm DB-1 (0.25-μm bonded film) column. The injector and column were operated isothermally at 290°C and the electron-capture detector was set at 320°C. The helium carrier gas flowed at 1 mL/min, and 30 mL/min of N_2 was used as makeup gas for the detector. The chromatograms were recorded on a Spectra Physics (San Jose, CA) computing integrator model SP4100. Quantitation was performed using standards at 5,16.8, and 168 ng/mL in n-hexadecane prepared from an OCDD (catalog #RPE-17, RFR Corp., Hope, RI) standard. Because no quantitative HpCDD standards were available, the relative response for the HpCDDs was assumed to be the same as for OCDD.

NCTR GC/MS Analysis

Selected samples were analyzed by GC/MS to confirm the GC/ECD results. The GC/MS method used was the method of Korfmacher et al.[10] and is based on a 30-m DB-5 fused-silica column connected directly to a Finnigan 4023 GC/MS system.

PCDD/PCDF Analysis

Selected samples were analyzed by the CNFRL using the method of Smith et al.[11] Dust samples were Soxhlet extracted for 24 hr with benzene as the solvent and the extract was treated as described by Smith et al.[11] For these analyses, the marker compounds [UL-^{13}C]-2,3,7,8-TCDD, [UL-^{37}Cl]-OCDD, and a mixture of six [UL-^{37}Cl]-TCDFs including [UL-^{37}Cl]-1,2,7,8- and [UL-^{37}Cl]-2,3,7,8-TCDF as the major components were spiked into each sample at the beginning of the enrichment procedure.[11] Using these marker compounds, quantitation was typically performed by the internal standard technique. For those samples which contained PCDDs or PCDFs that were outside of the linear calibration range, the external standard technique was used for quantitation.

RESULTS AND DISCUSSION

Preliminary samples were analyzed for PCDD and PCDF content in order to define the nature of the problem at the facility. The results of the analyses of dust samples are shown in Table 1, and of surface wipe samples in Table 2. Additional preliminary samples were analyzed by GC/MS and found to contain PCBs at levels of 100–1000 times that of the PCDFs and PCDDs. These initial results demonstrated that a significant contamination problem existed at the facility.

Due to the large number of samples needed to determine the extent of the contamination, a multitiered approach was applied to the sample analysis. After the preliminary samples were analyzed by GC/MS using the method of Smith et al.[11] to determine the level and ratios of the tetra- to octa-CDFs and the tetra- to octa-CDDs, these same samples were also analyzed for

Table 1. PCDD and PCDF Analysis of Dust Samples.

Sample	Homologue Type	2,3,7,8– Tetra (ng/g)	Sum of Isomers (ng/g)[a]				
			Tetra	Penta	Hexa	Hepta	Octa
D1	PCDDs	ND(0.5)	ND(0.5)	ND	62	1160	1170
D1	PCDFs	6	6	35	180	220	110
D2	PCDDs	3	3	ND	200	2800	18000
D2	PCDFs	350	700	1400	600	440	540
D3	PCDDs	ND(0.5)	ND(0.5)	ND	96	600	2430
D3	PCDFs	11	11	44	182	190	160
D4	PCDDs	ND(0.5)	ND(0.5)	ND	55	475	1200
D4	PCDFs	10	10	9	17	36	30
D5	PCDDs	ND(0.5)	ND(0.5)	ND	ND	130	810
D5	PCDFs	4	4	4	20	38	32
D6	PCDDs	ND(0.5)	ND(0.5)	ND	3	24	16
D6	PCDFs	ND(0.5)	ND(0.5)	ND	1	19	0.6

[a]ND, none detected. The number in parentheses is the detection limit per isomer.

PCB levels. A larger set of samples was analyzed by GC/ECD for OCDD and HpCDDs using the method of Korfmacher et al.[10] All of the samples were analyzed for PCBs. By this approach, a detailed picture of the extent of contamination was obtained.

Figure 1(c) shows the typical results of a surface wipe sample found to contain PCBs. This particular sample was determined to have a level of 4.8 mg PCBs/m^2.

The GC/ECD method for the analysis of surface wipe and dust samples for OCDD and HpCDDs was found to work well with these samples. Figure 2 shows the GC/ECD results of a surface wipe sample found to have 50 μg

Table 2. PCDD and PCDF Analysis of Group I Surface Wipe Samples.

Sample	Homologue Type	2,3,7,8– Tetra (ng/m^2)	Sum of Isomers (ng/m^2)[a]				
			Tetra	Penta	Hexa	Hepta	Octa
S1	PCDD	ND(0.8)	ND(0.8)	68	270	1520	840
S1	PCDF	20	36	64	72	44	ND(2.0)
S2	PCDD	ND(0.8)	ND(0.8)	ND	7	39	320
S2	PCDF	1	1	ND	10	18	36
S3	PCDD	ND(0.8)	ND(0.8)	3	38	2200	3440
S3	PCDF	3	10	ND	16	440	480
S4	PCDD	ND(0.8)	ND(0.8)	ND	28	192	1080
S4	PCDF	6	13	12	8	10	17
S5	PCDD	ND(0.8)	ND(0.8)	I[b]	1980	4160	4520
S5	PCDF	16	16	60	220	360	48
S6	PCDD	ND(0.8)	ND(0.8)	12	67	1304	2200
S6	PCDF	7	16	24	18	52	32
S7	PCDD	ND(0.2)	ND(0.2)	ND(0.2)	23	420	470
S7	PCDF	8	16	17	17	89	21
S8	PCDD	ND(0.2)	ND(0.2)	ND(0.2)	24	260	450
S8	PCDF	5	10	7	10	20	10

[a]ND, none detected. The number in parentheses is the detection limit per isomer.
[b]I, interferences prevented detection.

OCDD/m^2 and 12 μg HpCDDs/m^2. Figure 3 shows the results of a dust sample analyzed by (a) the GC/ECD method and (b) the NCTR GC/MS method. Figure 3 demonstrates that the peaks observed in the GC/ECD analysis were due to HpCDDs and OCDD.

Table 3 lists the results of the interlaboratory comparison of the NCTR GC/ECD method[10] to the CNFRL GC/MS method[11] for the analysis of the same four dust samples. The results in Table 3 indicate favorable agreement between these two methods of analysis.

Figure 4 shows the relative ratios of the tetra- to octa-CDDs and tetra- to octa-CDFs for samples analyzed from the facility. This figure shows that OCDD and the HpCDDs were the major components of the tetra- to octa-CDDs and tetra- to octa-CDFs, which is why the GC/ECD method[10] was developed specifically for determination of these compounds.

Table 4 lists the levels of PCBs found in one of the buildings at the site. Samples were taken on each of the seven floors in this building. As shown in this table, the level of PCBs found on the floors and the elevated horizontal surfaces (EHS) were significantly higher than the levels found on the (interior) walls of the building. These results suggest that the source of the PCBs was airborne and that it settled out onto the floors and horizontal surfaces of the building. This data set also shows that the level of PCBs was fairly uniform throughout this building. The range of the floor samples varied from 5.1 mg/m^2 to 11 mg/m^2, with the average level for the floor concentration being 8.9 mg/m^2.

Table 5 lists the results for surface wipe samples analyzed for PCBs, OCDD and HpCDDs. Table 6 lists the results for dust samples analyzed for PCBs, OCDD and HpCDDs. Together these tables show that PCBs, OCDD and HpCDDs were found in the same samples. These data also show that the ratio of the total PCBs to OCDD was in the range of 400–1000:1. Detection limits are sample dependent.[10]

Table 7 lists the average level of PCBs found in 32 buildings at the site. This table lists averages of surface wipe samples from floors and elevated horizontal surfaces. As this table shows, the average level of PCBs varied from none detected to 121 mg/m^2. Building 25, with the highest level of PCBs, is the building that was the site of a known PCB spill; therefore the majority of the PCBs in this building are assumed to be due to this spill and not the airborne contamination that is thought to be the cause of the PCB contamination at other buildings. The next highest level is building 27, which has an average level of 40 mg/m^2. Building 27 is located near the incinerator, as is building 21, which has an average PCB level of 9.1 mg/m^2.

Figure 5 shows a schematic diagram of the multibuilding facility indicating the average levels of the PCBs found on the surfaces of various buildings at the facility. It can be seen that the PCB levels were higher on one side of the incinerator than on the other side. It is suspected that the contamination is due to the incomplete incineration of several kilograms of PCBs in

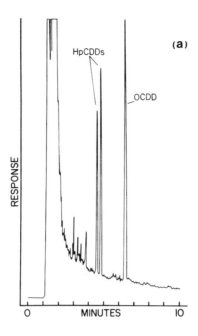

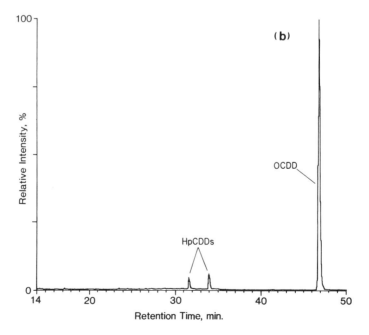

Figure 3. Results of the analysis of a dust sample by (a) GC/ECD method, and (b) NCTR GC/MS method.

Table 3. Interlaboratory Comparison of OCDD and HpCDD Concentrations in Four Dust Samples.

	NCTR GC/ECD Method[a]		CNFRL GC/MS Method[11]	
Sample	OCDD (μg/g)	HpCDDs (μg/g)	OCDD (μg/g)	HpCDDs (μg/g)
1	2.2[b]	0.7[b]	2.4[b]	1.4[b]
	1.7[c]	0.4[c]		
2	1.6[b]	0.4[b]	1.2[b]	0.5[b]
	1.1[c]	0.3[c]		
3	0.5[c]	0.05[c]	0.8[b]	0.1[b]
4	0.1[c]	< 0.04[c]	0.02[b]	0.02[b]

[a]Results are not corrected for recovery.
[b]Sample extracted by benzene Soxhlet extraction.
[c]Sample extracted by hexane Soxhlet extraction.

the incinerator. It is also unknown what effect, if any, the long period of time between the incident and the sampling of the facility had on the results obtained.

It was learned that in the 1960s large diffusion pumps were used in building 25 which were charged with 100 gal of Aroclor 1254. The pumps were occasionally recharged, and the waste PCBs were incinerated. It was estimated that in the late 1960s or early 1970s (pre-1974) 200 gal of the Aroclor 1254 was burned in the incinerator, probably on two separate occasions of 100 gal per burn. Unfortunately, the incinerator was not designed to destroy PCBs. It was typically operated at approximately 650°C with a 2-sec residence time. Current research has shown that to destroy PCBs, an incinerator should operate at 1200°C and have a 2-sec residence time.[12] Therefore, it appears that the incineration of the Aroclor 1254 under the

Table 4. Level of PCBs Found in Building 21.

	Sample Type			
			EHS[b]	
Floor #	Floor[a] Wipe, mg/m²	Wall[a] Wipe, mg/m²	Wipe[c], mg/m²	Dust[d], μg/g
1	10.	1.5	14.	–
2	5.1	1.1	–	590.
3	7.3	0.8	7.4	635.
4	9.4	1.0	4.1	210.
5	11.	5.1	8.5	620.
6	11.	1.0	20.	505.
7	8.4	2.5	4.8	160.
Average	8.9	1.9	9.8	453
s.d.[e]	2.1	1.5	6.1	213.

[a]Average of three samples.
[b]Elevated horizontal surfaces.
[c]0–3 samples per floor.
[d]0–3 samples per floor.
[e]Standard deviation.

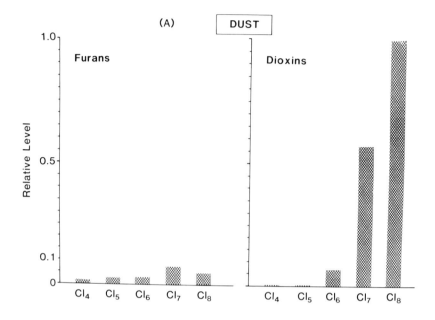

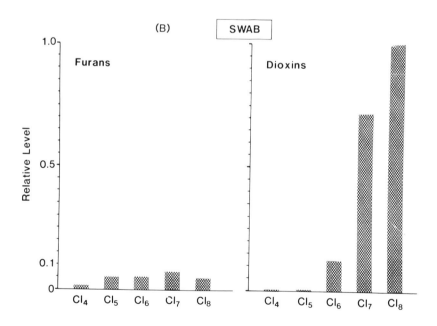

Figure 4.   Relative ratios of the tetra to octa-CDFs and tetra to octa-CDDs for dust samples (a) and surface wipe samples (b).

Table 5. Results Summary for Wipe Samples Analyzed for PCBs, OCDD and HpCDDs.[a]

Building	Total PCBs (mg/m²)	OCDD (μg/m²)	HpCDDs (μg/m²)
1	12.	23.	6.
12	0.01	4.	< 4.
12	0.005	< 4.	< 4.
22	2.1	2.	1.
23	0.065	< 2.	< 2.
23	11.	14.	5.
23	11.	25.	10.
25	16.	56.	27.
25	42.	5.	< 2.
26	1.7	4.	< 2.
26	10.	< 2.	< 2.
27	150.	54.	15.
27	530.	92.	27.
27	42.	50.	13.
27	18.	22.	19.
29	3.7	16.	6.
31	58.	5.	< 2.

[a]OCDD and HpCDD levels obtained by GC/ECD method.

described conditions resulted not only in the incomplete destruction of the PCBs, but also in the production of PCDDs and PCDFs.

These results demonstrate that the multibuilding facility was contaminated by PCBs, PCDDs and PCDFs, and that the PCB source was most likely due to the improper incineration of PCB askarel at an incinerator at

Table 6. Results Summary for Dust Samples Analyzed for PCBs, OCDD and HpCDDs.[a]

Building	Total PCBs, (μg/g)	OCDD (ng/g)	HpCDDs (ng/g)
5	11.	83.	< 34
5	47.	138.	< 38.
10	25.	136.	< 71.
12	16.	215.	< 108.
22	4.6	35.	15.
23	51.	585.	224.
23	14.	33.	< 7.
23	590.	1680.	443.
25	54.	1000.	295
25	32.	256.	66.
26	930.	540.	45.
27	1400.	282	116.
27	300.	120.	< 120.
28	4.3	84.	< 16.
29	75	158.	82.
29	160.	170.	26.
29	39.	1110.	260.
31	9.8	9.	< 8.

[a]OCDD and HpCDD levels obtained by GC/ECD method.

Table 7.   Average Level of PCBs in Buildings.[a]

Building	Number of Samples Analyzed	Average PCB level[b] (mg/m^2)
1	3	4.0
2	6	< 0.01
3	5	0.10
4	10	0.16
5	10	0.11
6	5	< 0.01
7	3	0.04
8	2	0.05
9	2	0.37
10	1	0.07
11	5	0.34
12	5	0.016
13	2	0.15
14	2	0.019
15	2	< 0.015
16	3	0.028
17	2	0.093
18	2	0.50
19	2	0.041
20	3	0.097
21	30	9.1
22	5	0.53
23	54	4.8
24	17	0.81
25	26	121.[c]
26	15	1.7
27	19	40.
28	4	0.089
29	69	5.9
30	8	0.18
31	6	11.
32	2	4.3

[a]Includes floor wipe samples and wipe samples from elevated horizontal surfaces.
[b]Sum of Aroclor 1254 and Aroclor 1260.
[c]Building includes a known PCB spill.

the site. The pattern of PCDDs and PCDFs, however, does not match that predicted from the incineration of PCBs. The PCDD, PCDF pattern does indicate pentachlorophenol contamination, but no evidence of penta-chlorophenol incineration or contamination was found at the facility. It is also unknown what effect, if any, the long period of time between the incident and the sampling of the facility had on the results obtained.

Efforts have been made to clean up this facility and the results of the initial cleanup efforts are reported elsewhere.[13]

ACKNOWLEDGMENT

The authors would like to thank Lyle Davis and Patricia Bulloch for typing this manuscript.

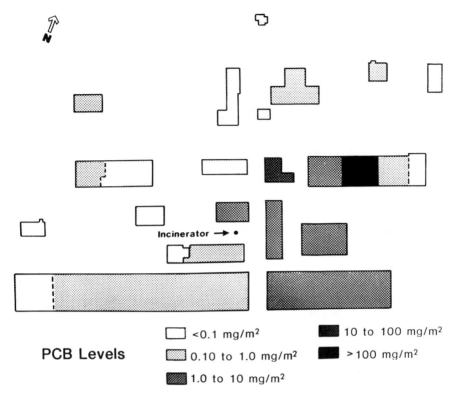

PCB Levels

☐ <0.1 mg/m²   ■ 10 to 100 mg/m²

▨ 0.10 to 1.0 mg/m²   ■ >100 mg/m²

▨ 1.0 to 10 mg/m²

Figure 5. Schematic diagram of the facility showing the average level of PCBs in the various buildings.

REFERENCES

1. Buser, H. R., H. P. Bosshardt, and C. Rappe. "Formation of Polychlorinated Dibenzofurans (PCDFs) from the Pyrolysis of PCBs," *Chemosphere* 7:109–119 (1978).
2. Rappe, C., S. Marklund, P. A. Bergqvist, L.-O. Kjeller, and M. Hansson. "Composition of PCDFs formed in PCB Fires," in *Chlorinated Dioxins and Dibenzofurans in the Total Environment II*, L. H. Keith, C. Rappe, and G. Choudhary, Eds. (Stoneham, MA: Butterworth Publishers, 1984).
3. Williams, C. H., Jr., C. L. Prescott, P. B. Stewart, and G. Choudhary. "Formation of Polychlorinated Dibenzofurans and Other Potentially Toxic Chlorinated Pyrolysis Products in PCB Fires," in *Chlorinated Dioxins and Dibenzofurans in the Total Environment II*, L. H. Keith, C. Rappe, and G. Choudhary, Eds. (Stoneham, MA: Butterworth Publishers, 1984).
4. Buser, H. R., and C. Rappe. "Isomer-Specific Separation of 2,3,7,8-Substituted Polychlorinated Dibenzo-p-dioxins by High Resolution

Gas Chromatography/Mass Spectrometry," *Anal. Chem.* 56:442–448 (1984).

5. Rappe, C., S. Marklund, P. Bergqvist, and H. Hansson. "Polychlorinated Dioxins (PCDDs), Dibenzofurans (PCDFs) and Other Polynuclear Aromatics (PCPNAs) Formed During PCB Fires," *Chemica Scripta.* 20:56–61 (1982).

6. Schecter, A. "Contamination of an Office Building in Binghamton, New York, by PCBs, Dioxins, Furans and Biphenylenes after an Electrical Panel and Electrical Transformer Incident," *Chemosphere* 12:669–690 (1983).

7. Buser, H. R., H. P. Bosshardt, and C. Rappe. "Identification of Polychlorinated Dibenzo-p-dioxin Isomers Found in Fly Ash," *Chemosphere* 2:165–172 (1978).

8. Olie, K., P. L. Mermeulen, and O. Hutzinger. "Chlorodibenzo-p-dioxins and Chlorodibenzofurans are Trace Components of Fly Ash and Flue Gas of some Municipal Incinerators in the Netherlands," *Chemosphere* 6:455–459 (1977).

9. Eiceman, G. A., A. C. Vlan, and F. W. Karasek. "Ultrasonic Extraction of Polychlorinated Dibenzo-p-dioxins and Other Organic Compounds from Fly Ash from Municipal Incinerators," *Anal. Chem.* 52:1492–1496 (1980).

10. Korfmacher, W. A., L. G. Rushing, D. M. Nestorick, H. C. Thompson, Jr., R. K. Mitchum, and J. R. Kominsky. "Analysis of Dust and Surface Swab Samples for Octachlordibenzo-p-dioxin and Heptachlorodibenzo-p-dioxins by Fused Silica Capillary GC with EC Detection," *J. High Resolut. Chromatogr. Chromatogr. Commun.* 8:12–19 (1984).

11. Smith, L. M., D. L. Stalling, and J. C. Johnson. "Determination of Part-per-Trillion Levels of Polychlorinated Dibenzofurans and Dioxins in Environmental Samples," *Anal. Chem.* 56:1830–1842 (1984).

12. Ackerman, D. G., L. L. Scinto, P. S. Bakshi, R. G. Delumyea, R. J. Johnson, G. Richard, and A. M. Takata. "Guidelines for the Disposal of PCBs and PCB Items by Thermal Destruction," prepared by TRW, Inc. for EPA-600/2-81-022, February, 1981.

13. Kominsky, J. R., D. L. Kendall, H. C. Thompson, K. L. Rowland, W. A. Korfmacher, and L. G. Rushing. Unpublished results.

SECTION III

Soil Contamination

CHAPTER 10

A Sampling Strategy for Remedial Action at Hazardous Waste Sites: Cleanup of Soil Contaminated by Tetrachlorodibenzo-p-Dioxin

Jurgen H. Exner, William D. Keffer, Richard O. Gilbert, and Robert R. Kinnison

INTRODUCTION

Cleanup of hazardous waste sites presents an impressive social and technical challenge. Reduction of these hazards requires a balance between technical, economic, legal, and social considerations. Sample acquisition and analysis are a considerable cost in defining the problem and developing solutions. In general, the initial site surveys involve obtaining samples on a regular pattern, such as a grid, and scrutinizing the data using judgment or sophisticated geostatistical techniques.[1-3] Similar approaches are used at the conclusion of cleanup to verify that a site has been cleaned. However, if we wish to be very confident that a cleanup level has been achieved because, for example, the pollutant is acutely toxic, and if the attendant analysis is expensive and time-consuming, the normal sampling approaches become unattractive.

This chapter presents a sampling strategy for verifying to a desired confidence that a previously contaminated area meets criteria of cleanliness. The approach is illustrated for the proposed excavation of soil contaminated by 2,3,7,8-tetrachlorodibenzo-p-dioxin (dioxin). The general strategy, however, can be useful in any cleanup of materials that are toxic and difficult to analyze.

Dioxin Contamination in Missouri

Spraying dioxin-contaminated waste oil for dust control led to contamination of more than 36 areas in Missouri. At these sites, contaminated soil can range from 500 to 200,000 yd^3, with dioxin concentrations ranging from less than 1 up to 1800 ppb at depths ranging from a few inches to several feet. Survey sampling, which used judgment and extensive compositing techniques in addition to sampling on grids, defined the areal contamination reasonably well. However, the vertical contamination remains less well defined because of the nonhomogeneity of the contamination and the difficulty of obtaining representative samples at depth.[4]

Health advisories by the Center for Disease Control (CDC)[5] suggesting limits on exposure of inhabitants to dioxin, and attendant social and political concern, have led to far-ranging investigations of potential emergency measures, remedial options,[6] and terminal disposal methods.[7] Excavation of contaminated soil and subsequent restoration are activities common to most proposed remedial actions. In order to excavate large areas and volumes of contaminated soil in a safe, efficient manner, it was proposed to excavate contaminated areas and to test for cleanliness. The following constraints were placed on the sampling and analytical techniques.

- About 500–1000 yd^3 can be excavated per day.
- Results must be available within 24 hr to reduce waiting time of idle equipment and labor, and to minimize exposure of potentially contaminated soil to erosion.
- The quantity of uncontaminated soil should be minimized, since the ultimate storage or disposal could cost $100–400/yd^3.
- We should be greater than 95% certain that an area declared clean is indeed clean.
- Available analytical resources of about 40–80 analyses per day and anticipated sampling and analysis costs of about $700 per sample should be considered.

IMPORTANT CLEANUP CONSIDERATIONS

Cleanup of a contaminated area requires definitions of: (1) what is being measured, (2) what criterion is used to make cleanup decisions, (3) various statistical quantities that define a decision rule for when to remove soil, (4) a field sampling plan for obtaining representative dioxin concentration data, and (5) action guides.

Concerning item 1, in the present case 2,3,7,8-tetrachlorodibenzo-p-dioxin is the major toxicant of concern. However, since this dioxin isomer is 98 to 100% of the total dioxin concentration at Missouri sites[8], the cleanup criterion can be set equally well for total tetrachlorinated dibenzodioxins. The use of this definition can result in a slightly faster analysis than for the specific isomer.

Item 2 requires definition of a cleanup unit (area) and an acceptable average dioxin concentration (decision criterion). Selection of a cleanup unit size depends on site characteristics, exposure estimates, and practical concerns. The sampling strategy developed below defines the decision criterion, D, to be that true mean concentration in the top 2 in. of soil in the entire cleanup unit that does not require the removal of soil. Selection of a specific value for D is beyond the scope of this chapter, but such a selection must be based on a risk assessment of human and environmental exposure, environmental fate, and on legal, social and political factors. For illustration purposes we use D = 1 ppb in this chapter. We also assume the cleanup unit is 20 × 250 ft.

Item 3 concerns the definition of a decision rule that makes use of D and data from the cleanup unit in question to decide whether soil removal is needed. The rule suggested here is to compute an upper confidence limit on the true concentration for the unit and to remove soil if that limit exceeds D. The computation of the confidence limit requires the specification of α, the prespecified small risk (probability) of not removing soil when in fact the true average concentration for the unit exceeds D. We must also assume that the composite sample means are normally (Gaussian) distributed.

Item 4 concerns the definition of the number and location of soil samples removed from the unit, whether compositing of samples is done, and the number of dioxin analyses conducted. To reduce analytical costs and satisfy the assumption of normally distributed composite means mentioned above, the use of composite sampling is suggested. However, it must be understood that the compositing approach is not ideal if the primary goal is to find small hot spots, since compositing dilutes (averages out) hot spots. Furthermore, compositing requires a procedure for thoroughly mixing and homogenizing individual soil samples. If the mixed composite sample is nonhomogeneous, then the standard deviation of the composite means, s, will be too large, and the decision to remove soil will be made more frequently. Hence, to avoid unnecessary removal of soil, a good mixing procedure is needed.

Item 5 (action guides) refers to developing clear responses to the following questions.

1. If the decision rule indicates soil removal is required, must the top layer of soil over the entire cleanup unit be removed?
2. If points of contamination (hot spots) are found, must the whole top layer of soil or just the hot spot be removed?

The answer to the first question would appear to be "yes" if the sampling strategy described below is used, i.e., if composites are formed by mixing small soil samples collected from all parts of the unit. Concerning the second question, if a hot spot is found and only that spot removed, individual or composite samples must be collected to provide assurance that the remainder of the unit meets the decision criterion. In practice it may be simpler to always remove the top layer of soil from the entire unit, unless the unit is very large, generating large amounts of soil to transport and store.

A SAMPLING STRATEGY

Main Features

The sampling strategy developed here has five main features.

1. Soil removal decisions are made for entire cleanup units.
2. Soil removal with depth occurs in stages.
3. Each stage involves collecting composite samples from the exposed soil surface. Randomly chosen aliquots from each composite are analyzed for dioxin.
4. Soil removal decisions are made individually for each cleanup unit by comparing a computed upper confidence limit against the decision criterion D.
5. Lateral soil removal occurs sequentially by sampling and applying the decision criterion to cleanup units adjacent to units where soil removal has occurred.

Establishing Cleanup Units

The assumption is made here that prior sampling for dioxin has identified areas where soil removal is clearly required. Based on this information, cleanup units are placed in a manner that minimizes wide ranges in concentration. Surface soil in these areas will be removed to a depth deemed appropriate on the basis of past data. This soil will be either temporarily stored at the site or loaded immediately on trucks for transport to a suitable disposal area. The area where soil removal has occurred is then divided into cleanup units. Decisions concerning future soil removal are made for individual cleanup units, so that any additional soil removal proceeds unit by unit.

Next to each outermost unit in the area where soil has been initially removed (which includes areas where the original soil surface has been substantially disturbed or where soil from the soil removal operation may have been inadvertently deposited), an adjacent unit is established (Figure 1). These adjacent units are subjected to the same sampling and compositing scheme and the same decision criterion and decision rule as the original units. Figure 1 shows four cleanup units, U_1, U_2, U_3, and U_4, along a road where initial soil removal has occurred. Also shown are adjacent units that will be sampled and evaluated for possible soil removal. If soil removal is

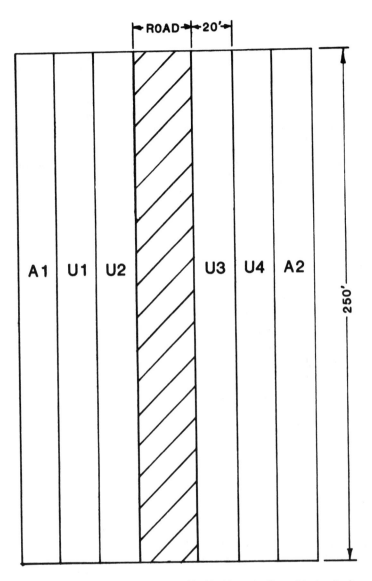

Figure 1. Illustration of excavation units U_1, U_2, U_3, U_4 and adjacent test units A_1 and A_2 for cleanup of contaminated road.

necessary in any adjacent unit, then another unit adjacent to it is established, and the same sampling strategy and decision criterion is applied.

For each cleanup unit, soil removal occurs in stages with depth. Soil samples are collected from the top 2 in. of exposed soil and an additional layer of soil removed if use of the decision criterion so indicates.

Using the above approach, excavation is continued until no soil removal

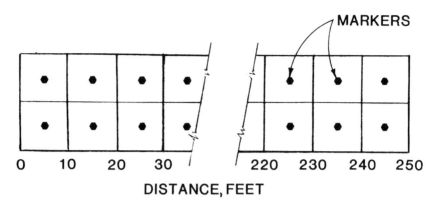

Figure 2. A 20 × 250 ft cleanup unit divided into 50 blocks of equal size.

is required in any unit at any depth. Note that this sequential approach assumes that an absence of dioxin at one depth implies an absence of dioxin at greater depths. This assumption may be reasonable, based on a knowledge of how dioxin was originally applied and its movement through soil, or on information from the samples initially taken to define the original soil removal area. If reasonable doubt remains, then some proportion of the cleanup units should be sampled at depth, using trenching techniques.

In a few locations, it will not be reasonable to exactly follow the sampling protocol specified above because of such problems as steep terrain, obstruction, etc. With adequate planning, these situations can be identified in advance of the field operations, and an alternative and equivalent cleanup area may be chosen through consultation between the scientific and field personnel. Any such alterations must be thoroughly documented in order to not invalidate the data analysis.

Sampling and Compositing

As indicated above, we assume that each cleanup unit is 20 × 250 ft. If other sizes are used, the general sampling and compositing approach described here can be adapted easily.

Each cleanup unit is divided into 50 equal blocks of 10 × 10 ft by setting up two lines parallel to the long axis of the unit, 10 ft apart and 5 ft from each side of the unit. Markers are then placed every 10 ft along these lines, starting 5 ft from one end. Each marker is at the center of a 10 × 10 ft block (Figure 2).

A minimum of three composite samples should be obtained from each cleanup unit according to the systematic pattern shown in Figure 3. Referring to Figure 3, Composite 1 consists of 50 soil samples pooled together, where a single sample is collected within each of the 50 one-square-foot areas labeled with the number 1 that lie around the periphery of the cleanup

50 SOIL SAMPLES COMPOSITED

TO FORM COMPOSITE 2

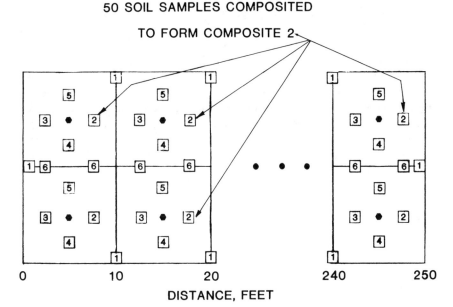

Figure 3. Systematic sampling design for forming the first six composite samples.

unit. Similarly, Composite 2 consists of 50 samples pooled together, where each sample is taken 3 ft north of a stake, and so on for the remaining composites. The "sample" within each one-square-foot area consists of four cores or spoonfuls of soil of approximately equal weight taken from the top 2 in. of soil. Hence, a composite sample consists of 200 aliquots of soil collected in a container that will allow homogenization by ball-milling, blending, or some other mechanical procedure. The use of spoons for obtaining each "sample" will allow for rapid collection of the 50 samples needed for each composite. However, a preferred method is to use a small soil corer of constant size and depth at each of the 50 locations. This would provide a consistent soil volume and depth.

If four, five, or six composites are collected, they should be taken at the locations indicated in Figure 3. If more than six composite samples are required, each additional composite should be obtained by choosing at random a location within a 10 × 10-ft block and collecting a sample (four spoonfuls) at the same position in all 50 blocks, and pooling the samples.

Following thorough mixing and homogenization of each composite, one or more (m) aliquots from each composite are chosen at random and analyzed for dioxin. If n composites are collected, then a total of nm data are available for computing the upper confidence limit for making the soil removal decision as described below.

The sampling and compositing plan given above has two important advantages over analyzing single grab samples for dioxin. First, by pooling many small samples across the entire unit, each dioxin datum is an estimate of the average for the entire unit, not just for a small local area. This compositing is important, since the decision criterion D is defined to be the acceptable average concentration for the entire unit. Second, the compositing process is a mechanical way of averaging out variabilities in concentrations from place to place over the unit. Hence, the resulting dioxin concentrations should tend to be more normally (Gaussian) distributed than individual grab samples. This is important, since normality is required when computing the upper confidence limit. However, these two advantages will be lost unless the 50 samples going into each composite are thoroughly mixed and homogenized.

Making Cleanup Decisions

The decision whether to remove the surface soil that has been sampled in a particular unit is made using the following decision rule: Remove soil if and only if

$$\overline{x} + t_{\alpha,n-1} s/\sqrt{n} \geq D \tag{1}$$

where $\overline{X} + t_{\alpha, n-1} s/\sqrt{n}$ is the estimated upper $100 (1 - \alpha)\%$ confidence limit
on the true mean for the unit
D is the preset decision criterion discussed above
$t_{\alpha,n-1}$ is obtained from a table of Student's t-distribution

This decision rule is a one-tailed test of the null hypothesis

$$H_0 : \text{True dioxin mean} \geq D$$

versus the alternative hypothesis

$$H_A : \text{True dioxin mean} < D$$

We reject H_0, and hence do not remove soil, if Equation 1 is satisfied, i.e., if

$$\overline{x} + t_{\alpha,n-1} s/\sqrt{n} < D$$

Clearly, to use this decision rule we must compute $\overline{X}$ and s, where

$$\overline{x} = (mn)^{-1} \sum_{i=1}^{n} \sum_{j=1}^{m} x_{ij} \tag{2}$$

equals the arithmetic mean of the nm dioxin concentrations x_{ij},

$$s = \left[(n-1)^{-1} \sum_{i=1}^{n} (\overline{x}_i - \overline{x})^2 \right]^{1/2} \tag{3}$$

equals the standard deviation of the n composite means $\overline{x}_i$,

$$\overline{x}_i = m^{-1} \sum_{j=1}^{m} x_{ij} \qquad (4)$$

equals the arithmetic mean of the m aliquot concentrations from the ith composite.

We also need $t_{\alpha,n-1}$, which is the value that cuts off $100\alpha\%$ of the upper tail of the t distribution with n-1 degrees of freedom, α is the prespecified small risk (probability) of not cleaning a dirty area, when in fact the true mean for the unit (in top 2 in. of soil) equals or exceeds D. Hence, the decision procedure is to choose a value for D and for α (e.g., $\alpha = 0.01$ or 0.05), find $t_{\alpha,n-1}$ in the t tables and see whether the upper confidence limit equals or exceeds D. If it does, then the rule requires the removal of soil. If not, the rule requires no removal of soil.

The tabled value $t_{\alpha,n-1}$ changes depending on n for a given α. For example, if $\alpha = 0.05$, then $t_{0.05,n-1}$ varies from 2.92 for n = 3 to 2.01 for n = 6, to 1.80 for n = 12. If we set $\alpha = 0.01$, then $t_{0.01,n-1}$ varies from 6.96 to 3.36 to 2.72 for n = 3,6, and 12, respectively. The t tables from which values of $t_{\alpha,n-1}$ are obtained and found in most statistics books (such as *Statistical Methods* [9]).

Note that if Equation 1 is solved for $\overline{x}$, we obtain

$$\overline{x} \geq D - t_{\alpha,n-1} \, s/\sqrt{n} \qquad (5)$$

Hence, for specified values of D, α, s and n, Equation 5 gives the value of $\overline{x}$ below which the decision rule in Equation 1 indicates that no soil removal is required.

Rather than specify s, we may choose to specify the relative standard deviation of the composite means, $C = s/\overline{x}$, in which case we replace s in Equation 1 with $C\overline{x}$. (In general, we expect C to be more constant than s from one cleanup unit to the next. Hence, C is usually preferred for planning purposes.) Suppose for illustration that D = 1 ppb. Solving Equation 1 for $\overline{x}$ gives

$$\overline{x} \geq 1/[1 + t_{\alpha,n-1} \, C/\sqrt{n}]. \qquad (6)$$

For illustrative purposes, Table 1 gives values of $\overline{x}$ obtained using Equation 6 for selected values of C and n for $\alpha = 0.05$, 0.01 and D = 1. For example, if $\alpha = 0.01$, n = 3 and $C = s/\overline{x} = 0.25$, then soil must be removed if $\overline{x} > 0.50$ ppb. But if the standard deviation s is larger, so that, e.g., C. = 0.50, then soil removal is required if $\overline{x} \geq 0.33$ ppb.

Table 1. Observed Average Dioxin Concentrations x̄ (ppb) Below Which no Soil Removal is Required When the Decision Criterion D is 1 ppb and when the Relative Standard Deviation of the Composite Means, C, Equals 0.50, 0.25, or 0.10.

Number of Composites, n	C[a] = 0.50		0.25		0.10	
	α[b] = 0.01	0.05	0.01	0.05	0.01	0.05
2	0.08	0.31	0.15	0.47	0.31	0.69
3	0.33	0.49	0.50	0.66	0.71	0.86
4	0.47	0.63	0.64	0.77	0.81	0.89
5	0.54	0.68	0.70	0.81	0.86	0.91
6	0.59	0.71	0.74	0.83	0.88	0.92
12	0.72	0.79	0.84	0.89	0.93	0.95
30	0.82	0.87	0.90	0.93	0.96	0.97

[a]C = s/ x̄.
[b]α = prespecified probability we are willing to accept of not removing soil when in fact the true mean for the unit equals or exceeds D.

Choosing the Number of Composites

We suggested above that a minimum of three composite samples be obtained from each unit and the first (up to six) composites be collected according to the pattern in Figure 3. If six composites are taken, this pattern gives good coverage of the entire unit.

In this section we give a method (using Equation 7 below) for choosing n that is based on controlling the chances of making cleanup decision errors to acceptably low levels. This approach may indicate an n greater than 6. In that case, we suggest each additional composite sample also be composed of 50 small samples collected over the 50 blocks, as explained above. The relative location where each small sample is taken for a given composite should be the same in each block, that location being chosen at random. If the approach for n given below should result in an n less than 6, we suggest the composite samples be chosen in the order of their number in Figure 3. For example, if n = 4, then composites numbered 1, 2, 3 and 4 in Figure 3 are collected. However, if fewer than six composites are taken, the advantage of good coverage of the entire unit is not realized. This may be a good reason to require n > 6.

The method for determining n given below requires an estimate of the variance σ^2, of all possible composite means that could conceivably be obtained from the unit. In practice, σ^2 is estimated by collecting several composites in a preliminary study in one or more cleanup units. Also, as cleanup units are sampled during the cleanup process, the estimate of σ^2 can be updated using the additional data. We will see below that if σ^2 is large, more composites are required.

The choice of n using the method given below also depends implicitly on budget constraints, turnaround time of the dioxin analytical procedure and other practical constraints. It also depends explicitly on the value of D relative to a smaller mean value μ_0, (explained below), and on the risks (probabilities) we are willing to assume of making the two types of cleanup decision errors. These errors are called Type I and Type II errors and are defined as follows:

Type I: Error of not removing soil when the true mean concentration equals or exceeds D, i.e., of not cleaning a dirty area.

Type II: Error of removing soil when the true mean concentration equals μ_0 and where $\mu_0 < $ D, i.e., of cleaning a clean area.

The probability of a Type I error is denoted by α, the same quantity used in Equations 1, 5, and 6 above. The probability of a Type II error is denoted by β. Ideally, we would like both α and β to be very near zero, but this may require collecting many composites. In practice there is a trade off between what the budget and other practical concerns will allow, and the complete assurance ($\alpha = \beta = 0$) we would ideally like to achieve that no decision errors are made.

The method suggested for choosing n or for evaluating the costs and benefits of choosing various values for α, β, D and μ_0 is to compute (see Burr[10] for derivation)

$$n = (Z_\alpha + Z_\beta)^2 \, [\sigma/(D - \mu_0)]^2 \qquad (7)$$

where D is the chosen decision criterion

Z_α is the value that cuts off 100 $\alpha\%$ of the upper tail of a standard normal (Gaussian) distribution

Z_β is the value that cuts off 100 $\beta\%$ of the upper tail of a standard normal distribution

σ is the standard deviation of all possible composite means that could conceivably be obtained from the cleanup unit

μ_0 is a mean concentration less than D, such that, if actually present, the probability of removing soil from the unit is β

Values of Z_α and Z_β are tabled in most statistics books. Values of Z_α for $\alpha = 0.05$ and 0.01 are 1.654 and 2.33, respectively.

Equation 7 gives the number of composites that must be collected to assure that the probability is not greater than α of failing to remove soil when $\mu \geq$ D, and the probability is no greater than β of incorrectly removing soil when $\mu \leq \mu_0$. The relationship between the chosen values of α, β, D and μ_0 is shown in Figure 4. In practice, β might be chosen to be larger than α, since it is more important to limit undue exposure to higher than allowed mean levels of dioxin than to prevent unnecessary removal of soil. The validity of Equation 7 depends on the composite means being normally distributed, and on an advance estimate of σ for the unit. An advance estimate of σ may be obtained by conducting preliminary sampling studies as indicated above. The normality assumption may not be unreasonable, since each composite sample is the sum of 50 smaller soil samples.

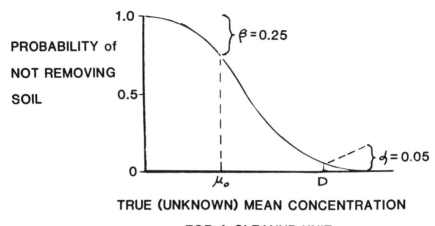

PROBABILITY of

NOT REMOVING

SOIL

TRUE (UNKNOWN) MEAN CONCENTRATION

FOR A CLEANUP UNIT

Figure 4. Probability of not removing a layer of soil from the cleanup unit for a range of possible values of the true mean dioxin concentration.

Hence, assuming the mixing process thoroughly homogenizes and mixes the small samples, the Central Limit Theorem[11] should apply. This theorem states that the average of several data values is closer to normality than the data values themselves. In the case of composite samples, the mixing process is a mechanical way of averaging the 50 small samples. The normality assumption should be evaluated statistically on the basis of preliminary data and data obtained during the cleanup operation.

Table 2 gives values of n computed using Equation 7 for the case where D = 1 ppb and for various choices of α, β, μ_0, and σ. Table 3 gives values of $(Z_\alpha + Z_\beta)^2$ that may be used in Equation 2. Our understanding of Figure 4 and the results in Table 2 may be aided by considering μ_0 and D as defining "good" and "bad" units, in the sense that we have a strong preference for not removing soil when the true mean concentration is less than μ_0, and we have a strong preference for removing soil when the true mean equals or exceeds D. If the true mean is greater than D or between zero and μ_0, we are willing to tolerate only small probabilities of making wrong decisions. If the true mean is between μ_0 and D, we are less concerned whether or not soil is removed. Once the pairs (α, D) and (β, μ_0) are chosen, and if a good estimate of σ is available, Equation 4 gives the number of composites needed to achieve this specification.

A potential problem with the use of Equation 7 is that the value of σ is likely to depend on the true mean concentration level, μ, present in the unit. For example, if $\mu = D$, a different value for σ should be used than if $\mu = \mu_0$. In practice, one could use an upper and then a lower limit for σ and see how n changes. Data obtained during the cleanup of initial units should help define the extent of this problem.

Table 2. Number of Composites, n, Obtained Using Equation 7 When D = 1 ppb.

			σ		
α	β	μ_0	**0.20**	**0.40**	**0.60**
0.01	0.25	0.20	3	5	8
		0.50	4	8	15
		0.70	6	18	38
		0.80	8	38	83
		0.85	18	66	146
0.01	0.45	0.20	3	4	6
		0.50	3	6	11
		0.70	5	13	26
		0.80	8	26	57
		0.85	13	45	99
0.05	0.25	0.20	3	4	6
		0.50	3	6	10
		0.70	5	12	24
		0.80	8	24	51
		0.85	12	41	89
0.05	0.45	0.20	3	3	4
		0.50	3	4	7
		0.70	4	8	15
		0.80	6	15	31
		0.85	8	25	53

Dealing with Hot Spots

Thus far in this chapter we have assumed that the average soil concentration (to some specified depth) over the entire cleanup unit (e.g., 20 × 250 ft) is the preferred criterion for deciding whether or not to remove additional soil from the unit. However, suppose the unit is "clean" except for one or more small hot spots. Then there is a finite probability that the individual samples collected over the unit (those that are composited) will not be taken at hot spot locations. In that case the unit will not be cleaned. But indeed even if the hot spots are sufficiently large to have a high probability of being sampled, compositing 50 individual samples, only one or two of which have high concentrations, may result in the composite average being so low that the decision rule (Equation 1) will still indicate cleanup is not required.

To illustrate this latter point, suppose six composite samples are formed, where each composite is obtained by pooling 50 individual samples collected over the cleanup unit as illustrated in Figure 3. Suppose 299 of the 300 individual samples contain no dioxin, but one sample has a concentration of 99.5 ppb. Then, five of the composite means will be zero and one composite mean will be 99.5/50 = 1.99 pbb (assuming perfect mixing of the 50 individual samples). Is cleanup required in this case? What does the use of Equation 1 indicate? Suppose we choose $\alpha = 0.05$; then $t_{0.05,5} = 2.015$

Table 3.   Values of $(Z_\alpha + Z_\beta)^2$ for Use in Equation 7 to Estimate n When the Normality Assumption is Tenable. α and β are Probabilities of Cleaning a Dirty Area and of Cleaning a Clean Area, Respectively.

β/α	0.0001	0.001	0.01	0.05	0.10	0.15	0.20	0.25	0.30	0.35	0.40	0.45
0.0001	55.32	46.37	36.55	28.77	25.01	22.61	20.80	19.30	18.01	16.85	15.78	14.78
0.001	46.37	38.20	29.34	22.42	19.11	1703	15.46	14.17	13.07	12.08	11.18	10.34
0.01	36.55	29.34	21.65	15.77	12.02	11.31	10.04	9.005	8.13	7.353	6.654	6.012
0.05	28.77	22.42	15.77	10.82	8.564	7.189	6.183	5.380	4.706	4.122	3.603	3.135
0.10	25.01	19.11	13.02	8.564	6.570	5.373	4.508	3.826	3.262	2.779	2.356	1.980
0.15	22.61	17.03	11.31	7.189	5.373	4.296	3.527	2.927	2.436	2.021	1.633	1.350
0.20	20.80	15.45	10.04	6.183	4.508	3.527	2.833	2.299	1.866	1.505	1.119	0.936
0.25	19.30	14.17	9.005	5.380	3.826	2.927	2.299	1.820	1.437	1.123	0.861	0.640
0.30	18.01	13.07	8.13	4.706	3.262	2.436	1.866	1.437	1.100	0.828	0.605	0.423
0.35	16.85	12.08	7.353	4.122	2.779	2.021	1.505	1.100	0.828	0.5938	0.408	0.261
0.40	15.78	11.18	6.654	3.603	2.356	1.663	1.119	0.861	0.605	0.408	0.2566	0.144
0.45	14.78	10.34	6.012	3.135	1.980	1.350	0.936	0.640	0.423	0.261	0.144	0.0632
0.50	13.83	9.55	5.410	2.706	1.643	1.074	0.708	0.455	0.275	0.148	0.064	0.0158

(from the t tables). Also, the reader may verify that for this scenario, the value of s is calculated to be 0.812414. Therefore, Equation 1 is

$$\bar{x} + t_{0.05,5} \, s/\sqrt{n} = \frac{99.5}{300} + 2.015 \, (0.812414)/\sqrt{6} = 1 \text{ ppb}$$

Hence, if D = 1 is used, the entire unit would be cleaned. However, if the one hot spot concentration had been less than 99.5 ppb, say 99.2 ppb, then $\bar{x} + t_{0.05,5} \, s\sqrt{6}$ would be less than 1 ppb. Then the unit would not be cleaned and the hot spot would remain. For the above scenario, the concentration of the single hot spot could be as high as 99.4 ppb, and Equation 1 would still indicate no additional cleanup required. Clearly, the possibility of leaving a hot spot (or several hot spots) is a disadvantage of the compositing method and the use of Equation 1 as discussed in this chapter.

As another example, suppose one circular hot spot of 100 ft^2 (diameter 11.28 ft) and concentration 50 ppb is present within the cleanup unit. Suppose it is located so that one of the individual samples in each of the six composites hits the spot, e.g., the hot spot might cover the upper left 10 × 10 ft square in Figure 3. Then each composite mean will have a concentration of 50 ppb/50 samples = 1 ppb (assuming perfect mixing) and the average of the six composite means will also be 1. Since all composite means are identical, the standard deviation, s, of the composite means is zero. Then Equation 1 gives $\bar{x} + 0 = 1$ ppb, which indicates cleanup is required if D has been set at 1 ppb.

Another scenario is where the contamination is uniform and slightly greater than 1 ppb over most of the cleanup unit, but a few local areas have zero concentrations. Hence, most of the unit should be cleaned if the true situation were known. However, if the zero concentration areas happen to be sampled, compositing may result in $\bar{x} + t_{\alpha,n-1} \, s\sqrt{n}$ being less than D = 1. In that case no cleanup would be done.

There are many alternatives to the compositing design developed in this chapter. For example, the size of the cleanup unit could be reduced and the number of composite samples increased. This would tend to reduce the dilution effect and increase the chances of cleaning units that contain hot spots. Or, the use of compositing could be abandoned and cleanup decisions made entirely on the basis of whether concentrations of individual (rather than composite) samples exceed D. However, if very small hot spots are important to find and remove, many individual samples would be required to have a high probability of finding them all. (These probabilities can be found using the techniques of Gilbert[12] and Zirschky and Gilbert.[13]) The dioxin analysis costs could be excessive in this case.

In practice, there must be a balance between compositing and "looking for hot spots." People will differ in their assessments of what the optimum balance should be, especially since there is at present no definitive statistical guidance on optimum sampling strategies for cleanup situations. The approach in this chapter puts more emphasis on compositing than on finding small hot spots. If the detection of hot spots is of overriding concern, then it becomes very important to define the size of hot spot that must be found and an acceptable risk of not finding it given that a specified grid spacing is used.[12,13]

As an approximation to the methodology of Gilbert[12] and Zirschky and Gilbert,[13] we may state that in order to have a reasonable chance (greater than 90%) of finding hot spots, the sampling grid must be approximately the same size as the diameter of the hot spots. Thus, for any practical sampling protocol it must be accepted that hot spots smaller than the design criteria will be missed.

HEALTH RISK ESTIMATES AND HOT SPOTS

The Center for Disease Control (CDC) recently constructed a health risk assessment on exposure of humans to 2,3,7,8-tetrachlorodibenzo-p-dioxin.[5] The assessment estimated that a daily human intake of 28 to 1428 fg/kg body weight/day poses a risk of one excess lifetime cancer per million persons exposed. Similarly, 276 fg to 14.3 pg/kg body weight/day poses a risk of one excess lifetime cancer per 100,000 persons exposed. By assuming absorption of dioxin from soil via dermal, oral, or respiratory routes, and considering exposure to children in residential areas, CDC declared 1 ppb in soil as the level for concern. The assessment considers the average daily dose that could be received if 100, 10, or 1% of dioxin-contaminated areas were available to exposure at various concentrations, and estimates the range of 10^{-5} and 10^{-6} cancer risk for people over a 70-yr lifetime.

In considering cleanup, the risk assessment provides additional support for the concept of using an average concentration as the criterion for decision, and relieves concerns about potential hot spots. If we assume that 1 ppb is the decision level, and if 2% of the area were at 50 ppb, the daily dose would still fall within the 10^{-6} excess lifetime cancer risk range. It is important to emphasize that sampling and analytical procedures are much more precise, within error of 10 to 50%, than the assumptions of the risk assessment, which may cover several orders of magnitude. In summary, health risk assessments are based on an average potential exposure to the population, and include in their estimation small variations in the concentration of dioxin.

ACKNOWLEDGMENTS

We thank G. Flatman and P. Richitt for reviewing this work and S. Wojinski for discussions on suitable analysis methods. Portions of this work were carried out under contract to the U. S. Environmental Protection Agency. The contents do not necessarily reflect the views and the policies of the U. S. EPA, nor does mention of trade names or commercial products constitute endorsement or recommendations for use.

REFERENCES

1. U. S. Environmental Protection Agency. "Test Methods for Evaluating Solid Waste," SW–846 (1982).
2. Thomas, V. W., and R. R. Kinnison. "Recommended Sampling Strategies for Spatial Evaluation of Windblown Contamination Around Uranium Tailings Piles," NUREG/CR–3479 (1983).
3. Parkhurst, D. F. *Environ. Sci. Technol.* 18:521 (1984).
4. Harris, D. J. U. S. EPA Region VII, Draft Report on TCDD Sampling Methods (December 1983).
5. Kimbrough, R. D., H. Falk, P. Stehr, and G. Fries. "Health Implications of 2,3,7,8 Tetrachlorodibenzodioxin (TCDD) Contamination of Residential Soil," *J. Tox. Env. Health* 14:47 (1984).
6. Exner, J. H., D. G. Erikson, R. Cibulskis, and W. Keffer. "Disposal, Treatment, and Mitigation Options for Missouri Dioxin Sites," *1984 Hazardous Materials Spill Conference Proceedings*, J. Ludwigson, Ed. (Rockville, MD: Government Institutes, Inc., 1984).
7. Exner, J. H., E. S. Alperin, A. Groen, C. E. Morren, V. Kalcevic, J. J. Cudahy, and D. M. Pitts. *Chlorinated Dioxins and Dibenzofurans in the Total Environment II*, L. H. Keith, C. Rappe, and G. Choudhary, Eds. (Stoneham, MA: Butterworth Publishers, 1984), pp. 47–56.
8. Kleopfer, R. U. S. EPA Region VII, Kansas City, KS, private communication, February 1984.
9. Snedecor, G. W., and W. G. Cochran. *Statistical Methods*, 6th ed. (Ames, IA: Iowa State University Press, 1967).
10. Burr, I. W. "Statistical Quality Control Methods," (New York: Marcel Dekker, 1976), pp. 325–328.
11. Hole, P. J., S. C. Port, and C. J. Stone. "Introduction to Probability Theory," (Boston: Houghton Mifflin, 1971).
12. Gilbert, R. O. *Tran-Stat, 19*, "Some Statistical Aspects of Finding Hot Spots and Buried Radioactivity," Battelle Pacific Northwest Laboratory, Richland, WA, PNL-SA-10274 (March 1982).
13. Zirschky, J., and R. O. Gilbert. "Detecting Hot Spots at Hazardous Waste Sites," *Chem. Eng.* (July 1984).

CHAPTER 11

Desorption of 2,3,7,8-TCDD from Soils into Water/Methanol and Methanol Liquid Phases

Richard W. Walters, Annette Guiseppi-Elie, M. M. Rao, and Jay C. Means

INTRODUCTION

The chemical 2,3,7,8-tetrachlorodibenzo-p-dioxin (abbreviated simply as TCDD in this chapter) is one of the most toxic compounds known to man. Experimental studies of the toxicity of TCDD using guinea pigs have shown that the LD_{50} can range as low as 0.6 to 2.0 μg/kg of body weight. Environmental sources of this chemical include the uncontrolled burning of halogenated organic chemicals and possibly the burning of organic materials in the presence of chlorinated compounds. This chemical is also present as a by-product in effluents from processes which produce chlorophenols and related compounds, and has perhaps received most widespread national attention because it was present in waste oils used for dust control in Missouri.

Soil samples collected from a number of sites in the U. S. have been found to contain significant levels of TCDD. For example, an average TCDD soil concentration of 1.3 parts per billion (ppb) has been determined for soil samples collected at a disposal site near Jacksonville, Arkansas.[1] Soil samples in the Times Beach area have been found to contain TCDD at levels as high as 1600 ppb.[2] The extent to which this compound is transported as a result of leaching from soil into groundwater has not been reported in the literature. However, previous studies of sorptive transport with other water-insoluble, or hydrophobic, compounds (e.g., polycyclic aromatic hydrocarbons [PAHs] and polychlorinated biphenyls [PCBs]) have been performed.[3-6] The results of these studies may be quite similar to what is expected for TCDD, because these solutes also have very low water solubilities.[7] The equilibrium aspect of the transport of these latter solutes in soil and sediment systems can be expressed in terms of the water-soil partition coefficient (K_D). These partition coefficients have been found to be related to the octanol-water partition coefficient (K_{ow}) and aqueous solubility (S) of the solute and the fraction organic carbon content f_{oc} of the soil or sediment. An equation describing the relationship for chlorinated hydrocarbons and water has been proposed by Means et al.[4]

$$\log(K_{oc}) = -0.686 \log(S) + 4.273 \tag{1}$$

where S is water solubility in $\mu g/mL$.

K_{oc} is the partition coefficient normalized on organic carbon content ($K_{oc} = K_D/f_{oc}$).

A number of additional factors have been found to influence sorptive transport of these compounds, including pH, ionic strength, and temperature. The presence in water of solvents — miscible and/or immiscible — may enhance liquid phase solubility and hence improve transport through soils.[9] Kinetically, the sorption and desorption processes appear to involve two steps; the initial step is typically rapid (i.e., on the order of minutes), over which most of the sorption/desorption occurs, followed by a slower second step, which may occur over periods of several hours or longer.[10]

The literature regarding the reversibility of sorption of various classes of organic contaminants from soils and sediments is limited and has led to varied interpretations. Leenheer and Ahlrichs[11] have reported complete reversibility of sorption of carbaryl and parathion by humified soil organic matter. In contrast, DiToro and Horzempa[12] and Horzempa and DiToro[13] report that for studies of desorption of PCBs on lake sediments, a significant portion of sorbed solute appeared to be persistently sorbed. Isaacson and Frink[14] studied the reversibility of sorption of phenol, 2-chlorophenol and 2,4-dichlorophenol on lake sediments, and reported that up to 90% of the solute was irreversibly sorbed.

The extent to which these observations apply to TCDD sorption/desorption has been a major focus of this study. The sorptive and desorptive behavior of TCDD has been investigated using soils of various organic carbon contents and a range of equilibration periods from 1 to 90 days. These studies have employed water, methanol, toluene, and mixtures of water/methanol and water/toluene as the liquid phase. These solvents have been chosen as solvents which are associated with chlorinated hydrocarbon manufacturing processes or as solvents which may be disposed in landfills.

Results of sorption isotherm testing are reported by Walters et al.[15] This chapter presents results of desorption studies involving water/methanol (50/50 by volume) and methanol.

METHODS

Details regarding the procedures used in this experimental work are provided by Walters et al.[15] A summary of these procedures is provided below.

Materials

Solvents used in these studies were pesticide grade from Fisher. Radiolabeled (C–14) TCDD was obtained from Cambridge Radioisotope Laboratories with a specific activity of 33.24 mCi/mmol. Insta-gel liquid scintillation cocktail was obtained from United Technologies. These materials were used as received, without additional purification. GC/MS analyses confirmed no significant levels of contamination of the TCDD standard.

Soils

Two soils were used in these studies. Soils were air dried and sieved through 0.3-mm standard sieves. Sieved soils were characterized for organic matter content, mineralogy, and texture using techniques described previously.[16] Results of soil characterization are reported by Walters et al.[15] Desorption experiments reported here involved a low organic carbon content soil (soil 91, $f_{oc} = 0.0062$) and a high organic carbon content soil (soil 96, $f_{oc} = 0.0765$).

Desorption Isotherms

Consecutive desorption isotherms were generated for each solvent system with soils 91 and 96, in which the soils had been previously equilibrated with TCDD for 1 and 90 days. These sorption equilibration periods were chosen to investigate whether aging had an effect on the rate and/or degree of desorption of TCDD. Tubes previously equilibrated for one day in sorption represented freshly contaminated soils, and were expected to exhibit some extent of desorption. Tubes previously equilibrated for ninety days in sorption represented aged contaminated soils, and were expected to exhibit similar or less desorption.

Isotherm points were determined in triplicate by batch shake-testing. Tests were performed using 15-mL glass centrifuge tubes fitted with Teflon®-lined screw caps. The inside wall of the tubes was precoated with 0.4 μg TCDD by adding TCDD in methanol followed by evaporation of the methanol into nitrogen. Soil (50 mg for water/methanol and 300 mg for methanol) and 12 mL of the liquid phase were added to the centrifuge tubes, and the tubes were sealed and equilibrated for 1 or 90 days. For equilibration, tubes were placed horizontally on a shaking table, and mixing was provided at low speed for 2 min at 30–min intervals over the initial day of each isotherm. After the first day, tubes for long-term isotherms (90 days) were agitated daily by manual shaking. This practice was employed to minimize soil-particle breakup, which was found to occur under continuous, rigorous agitation. Following equilibration, tubes were centrifuged for 6 min at 3000 rpm ($800 \times$ g), and three 1-mL aliquots of centrifuged liquid phase were withdrawn for liquid scintillation counting (LSC). Liquid phase TCDD concentrations were determined directly from the LSC results. Soil concentrations of TCDD were determined by difference from total initial TCDD minus total TCDD in the liquid phase.

Consecutive desorption experiments were performed by replacing the three 1-mL aliquots of liquid phase removed for LSC counting with 3 mL of fresh liquid. This strategy was used as an alternative to complete replacement of liquid phase to conserve total TCDD in the system and to minimize extraction of organic matter from the soil. With as much as 80% of the TCDD in the shake test tubes associated with the liquid phase, reduction of TCDD in the system to the point where equilibrium concentrations below detection limits and/or significantly below the range of corresponding sorption isotherm data was anticipated after only a few desorption tests. Both 1-day and 90-day sorption tubes were sampled after 1, 3, 10, 30 and 90 days of desorption.

The centrifugation procedure employed here achieved complete separation of particles of diameter $\geq$ 0.5 μm from the liquid phase based on Stoke's Law settling velocities. This extent of centrifugation was found to be sufficient to minimize potential errors in sorption and desorption experiments associated with incomplete removal of small particles (colloidal, microparticulate, or macromolecular) as discussed by Gschwend and Wu.[17] Further, these errors are particularly important when the total mass of partitioning compound in the liquid phase of the experimental shake test tube is negligible compared to the total mass of compound in the solid phase. For the data reported here, the equilibrated liquid phase typically contained 50–80% of the total TCDD in the system (liquid plus solid phase).

Radiochemical Analyses

Liquid samples of 1-mL volume were added to 10 mL of counting cocktail and counted using a Packard Model 4000 LSC instrument.

Isotherm Data Analysis

Isotherm data were evaluated by linear regression to determine best fit parameters for two equations. The first equation is the linear (Henry's Law) relationship

$$q_e = K_D C_e \qquad (2)$$

Table 1. Consecutive Desorption Data for Soil 96 and Water/Methanol With Prior Sorption for 1 Day.[a]

	Initial Total System TCDD, μg	C_e (μg/mL)	q_e (μg/g)	Apparent K_D
S1	0.3377	0.0109	4.19	384
	0.3377	0.0108	4.19	388
	0.3377	0.00892	4.59	515
D1	0.3049	0.00838	4.15	495
	0.3053	0.00916	3.93	429
	0.3109	0.00587	4.79	816
D3	0.2798	0.00633	4.13	653
	0.2778	0.00671	3.97	591
	0.2933	0.00497	4.65	936
D10	0.2608	0.00432	4.24	980
	0.2577	0.00469	4.05	863
	0.2784	0.00340	4.73	1390
D30	0.2478	0.00254	4.41	1740
	0.2436	0.00269	4.25	1580
	0.2682	0.00236	4.78	2030
D90	0.2402	0.00143	4.52	3160
	0.2355	0.00141	4.39	3120
	0.2611	0.00161	4.81	2980

[a]The desorption series employed 50 mg soil in 12 mL liquid. S1 represents the initial 1-day sorption. D1 through D90 represent the consecutive desorption measurements following 1 through 90 days of desorption.

where q_e = equilibrium soil concentration of TCDD (μg/g)
C_e = equilibrium liquid concentration of TCDD (μg/mL)
K_D = distribution coefficient (mL/g)

Some isotherms were observed to be nonlinear over the range of concentration studied. Hence, a second equation was used to define this relationship:

$$\log(q_e) = 1/n \log(C_e) + \log(K_f \tag{3}$$

This equation plots C_e versus q_e as a straight line on a log-log plot with a slope $1/n$ and with $K_f = q_e$ at a value of $C_e = 1$ μg/mL. This equation is the logarithmic transformation of the Freundlich equation

$$q_e = K_f C_e \tag{4}$$

RESULTS AND DISCUSSION

Raw desorption isotherm data for the solvent/soil systems investigated are presented by Guiseppi-Elie et al.[18] Results of a typical desorption series are shown in Table I, which shows results from experiments with the water/methanol liquid phase and soil 96 using a 1–day sorption equilibration period.

Log-log plots of the desorption data are presented in Figures 1–4. In these plots, data points for only the initial sorption and the desorption points are plotted. Desorption points from soils equilibrated in sorption for 1-day (S1)

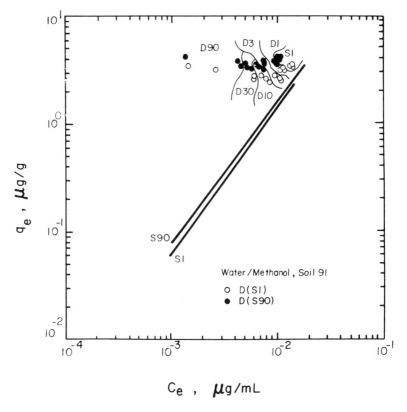

Figure 1. Desorption points for TCDD from Soil 91 into water/methanol.

are indicated with open circles, and desorption points from soils equilibrated in sorption for 90 days (S90) are indicated with closed circles. Sorption isotherm lines appearing in these figures are from Walters et al.[15] and
are the best-fit lines determined by regression using Equation 3 plotted over
the range of C_e for which data were generated.

Before reviewing the desorption data, it is worthwhile to consider potential interpretations of desorption points relative to their corresponding sorption isotherms. An illustrative sorption isotherm line, and three potential
trends in desorption data, are shown in Figure 5. These paths are potential
results for a simple system in which liquid and solid are distinctly defined
and noninteractive. In this context, simple systems are typified by water/
soil, where equilibrium partitioning of solute occurs between soil organic
matter and water, and sorption kinetics are primarily related to surface or
interparticle diffusion of solute into soil. Desorption trends shown in Figure
5 are based on consecutive desorption experiments in which either a portion

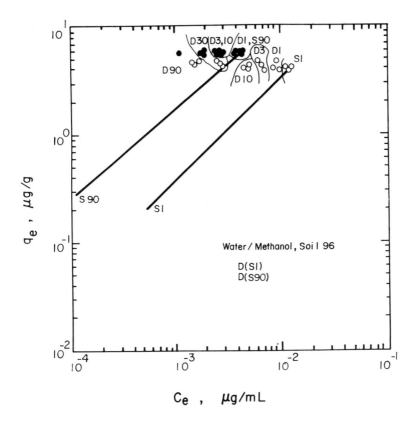

Figure 2.   Desorption points for TCDD from Soil 96 into water/methanol.

of the centrifuged liquid (e.g., 25%) or all is removed and replaced with fresh liquid to set up each subsequent desorption test.

Path A in Figure 5 would result for a system in which sorption/desorption were completely reversible and kinetics were relatively rapid (e.g., equilibrium achieved within one day). Desorption points would follow the sorption isotherm downward as new equilibrium points are established following the reduction of total system solute (by removing liquid phase containing solute). Path B in Figure 5 would result for a system in which desorption was completely irreversible, regardless of kinetics. In this system, q_e would remain constant through the consecutive desorption series, with C_e being reduced in proportion to the fraction of liquid phase withdrawn. For systems which are reversible but subject to slow desorption kinetics (slow is relative to the removal rate of liquid phase in the consecutive desorption series), or for systems with reversibly and nonreversibly sorbed solute components, pathways between those of A and B might be expected.

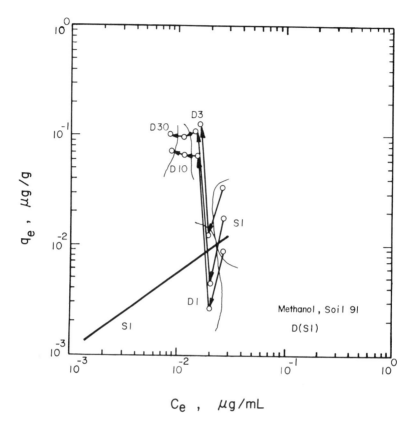

Figure 3. Desorption points for TCDD from Soil 91 into methanol.

A third possibility occurs for systems in which sorption is very slow. In particular, if sorption equilibrium is not achieved prior to the setup of subsequent consecutive desorption points, these points might follow a path similar to that shown for path C. Path C could result if a system is tending to sorb solute and approach equilibrium despite consecutive partial removals of solute by liquid phase withdrawal. The exact path depends upon the relative rates of sorption and removal of total system solute. If solute removal is slower than sorption, path C will result. If solute removal is much more rapid than the rate of sorption, the slope of the desorption path will be reduced in the direction of path B. In fact, if sufficient solute is removed from the liquid phase, the driving force for sorption could be reversed, and desorption may occur. Hence, in this event the desorption pathway may actually fall below path B.

The conclusion from these considerations is that, even for a relatively simple water/soil system, a number of possible desorption pathways can

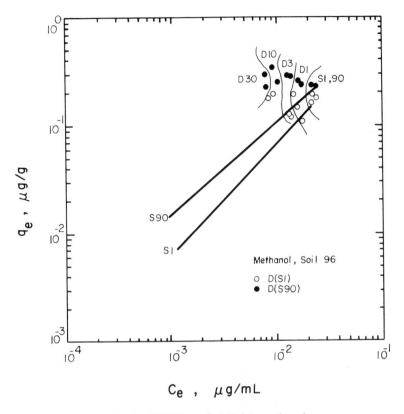

Figure 4.   Desorption points for TCDD from Soil 96 into methanol.

result. These pathways depend not only upon the extent of reversibility of sorption/desorption but also upon the rates of sorption and desorption.

Figure 1 shows the S1 and S90 isotherms and desorption points for soil 91 with water/methanol (W/M). The two sorption isotherm lines indicate that sorption is relatively rapid, with no significant change in soil loading between 1 and 90 days. The desorption points move left away from the initial sorption point in a manner which suggests that soil loading (q_e) remains constant and that desorption does not occur—hence, an irreversible system. However, there is a slight though discernable trend in both D(S1) and D(S90) data in that soil loading decreases through the D1 and D3 points (suggesting desorption) remain constant through D10 and D30, and increase slightly at D90.

The desorption points for soil 96 in W/M are shown with S1 and S90 isotherms in Figure 2. The sorption isotherms suggest that sorption is not rapid, with a significant increase in soil loading from S1 to S90. However, data presented by Walters et al.[15] suggest little change between S30 and S90

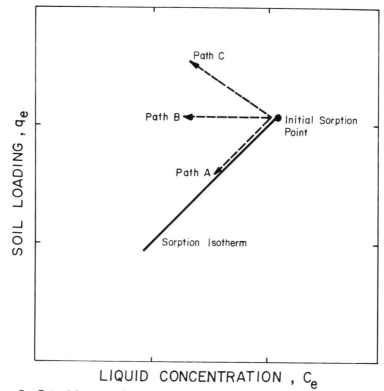

Figure 5. Potential consecutive desorption point pathways.

isotherms, suggesting attainment of equilibrium at S90. Desorption points for the S1 system show a slight increase in soil loading from D1 through D90 (displaced to the left of the sorption isotherm). These data are roughly consistent with the system moving towards sorption equilibrium through the time frame of the desorption series (i.e., path C in Figure 5). Hence, it is not possible to make a conclusion regarding the reversibility of sorption for this system. For the S90 system, the desorption points indicate an essentially constant soil loading over the entire time series studied. Because the S90 sorption points have been found to correspond to sorption equilibrium, the constant loading observed during desorption suggests irreversible sorption.

Similar desorption points for Soils 91 and 96 using methanol as a liquid phase are shown in Figures 3 and 4. The S1 desorption points for both soils show a trend similar to that exhibited for W/M Soil 91 desorption points; q_e initially decreases from the sorption point and then increases at the second or third consecutive desorption test. For Soil 91 and methanol data shown in Figure 3, the increase between D1 and D3 points is particularly dramatic, and is followed by q_e values which are roughly constant from D3 through

D30. The desorption data for the S1 system using Soil 96 shown in Figure 4 show a similar though less dramatic decrease in q_e between the initial sorption point and D3, though both S1 and S90 systems show ultimate increases in q_e through the desorption series with roughly constant q_e values at D10 and D30.

The basic conclusions from these data are that (1) sorption of TCDD from soil in the presence of methanol appears to be highly irreversible and (2) for systems in which some desorption is apparent, desorption points follow an unusual sequence in which, over consecutive points, q_e initially decreases and later increases. Irreversibility has also been observed in subsequent experiments involving complete liquid phase removal. These observations have important implications on modeling TCDD transport through soils in the presence of solvents. The advection/dispersion model of Roberts et al.[19] may not be applicable to this system because this model assumes completely reversible sorption. This behavior is not readily understood, though these observations are believed to be a result of the presence of the cosolvent (methanol) in the liquid phase, and soil-liquid interactions involving the partitioning of the organic matter phase of the soil.

Sluffing of organic material from the soil during equilibration is a possible explanation for at least part of the observed behavior. Organic materials which sluff may be truly dissolved materials (e.g., simple phenolics and/or carboxylic acids with low molecular weight) or suspended materials (e.g., colloidal, macromolecular and/or microparticulate). Centrifuging may not completely separate the liquid phase and sluffed material, resulting in an increase in TCDD in the liquid phase as a result of either TCDD being associated (sorbed) with suspended material and/or enhanced solubility of TCDD in the liquid phase. Differences in the types of organic materials associated with each soil may account for the differences observed. For soil 91, for example, organics may be comprised of more simple compounds which themselves equilibrate rapidly. On the other hand, soil 96 may be comprised of a more complex array of organics which partition more slowly. Desorption data would consequently be sensitive to the rates at which these materials are washed from the experimental system. Qualitatively, the washing of organic materials from the system may explain the data reported here. Specific evaluation of this effect requires careful evaluation of the relative rates of partitioning of organics from the soil. Based on a model proposed previously,[20] the following reactions involving TCDD, soil as a separable solid, nonseparable suspended materials (SS), and dissolved organic material (DOC) can be proposed.

$$\text{TCDD} + \text{Soil} = = = = \text{TCDD-Soil} \quad K_1 \qquad (5)$$

$$\text{TCDD} + \text{SS} = = = = \text{TCDD-SS} \quad K_2 \qquad (6)$$

$$\text{SS} + \text{Soil} = = = = \text{SS-Soil} \quad K_3 \qquad (7)$$

$$DOC + Soil = = = = DOC\text{-}Soil \quad K_4 \qquad (8)$$

Investigation of these reactions, including determination of rate and equilibrium constants, is a subject of future study.

ACKNOWLEDGMENTS

This research was supported by the U. S. Environmental Protection Agency under cooperative agreement CR–811743–01–0 with the Robert S. Kerr Environmental Research Laboratory, Ada, Oklahoma. Carl G. Enfield and Marvin D. Piwoni were project officers for this work, and commented on the manuscript. The authors are grateful to Zohreh Yousefi, Stan Ostazeski and Yu-Ping He for their contributions to the experimental and data manipulation support. This chapter is Contribution Number 1684 of the Center for Environmental and Estuarine Studies.

REFERENCES

1. Josephson, J. "Chlorinated Dioxins and Furans in The Environment," *Environ. Sci. Technol.* 17(3):124A–28A (1983).
2. Exner, J. H., et al. "Detoxication of Chlorinated Dioxins," in *Chlorinated Dioxins and Dibenzofurans in the Total Environment II*, L. H. Keith, C. Rappe, and G. Choudhary, Eds. (Stoneham, MA: Butterworth Publishers, 1985).
3. Walters, R. W., and R. G. Luthy. "Liquid/Suspended Solid Phase Partitioning of Polycyclic Aromatic Hydrocarbons in Coal Coking Wastewaters," *Water Res.* 18(7):795–809 (1984).
4. Means, J. C., S. G. Wood, J. J. Hassett, and W. L. Banwart. "Sorption of Polynuclear Aromatic Hydrocarbons by Sediments and Soils," *Environ. Sci. Technol.* 14(12):1524–8 (1980).
5. Karickhoff, S. W., D. S. Brown, and T. A. Scott. "Sorption of Hydrophobic Pollutants on Natural Sediments," *Water Res.* 13:241–8 (1979).
6. Weber, W. J., Jr., T. C. Voice, M. Pirbazari, G. E. Hunt, and D. M. Ulanoff. "Sorption of Hydrophobic Compounds by Sediments, Soils and Suspended Solids-II. Sorbent Evaluation Studies," *Water Res.* 17(10):1443–52 (1983).
7. Hay, A. *The Chemical Scythe: Lessons of 2,4,5-T and Dioxin,* (New York: Plenum Press, 1982).
8. Chiou, G. T., L. J. Peters, and V. H. Freed. *Science* 206:831 (1979).
9. Rao, P. S. C., A. G. Hornsby, and P. Nkedi-Kizza. "Influence of Solvent Mixtures on Sorption and Transport of Toxic Organic Compounds in the Subsurface Environment," paper presented before the Division of Environmental Chemistry, ACS, Washington, DC (1983).
10. Karickhoff, S. W. "Sorption Kinetics of Hydrophobic Pollutants in Natural Sediments," paper presented before the Environmental Chemistry Division of the ACS, Washington DC (August, 1983).

11. Leenheer, J. A., and J. L. Ahlrichs. *Soil Sci. Soc. Am. Pro.* 35:700–5.
12. DiToro, D. M. and L. M. Horzempa. "Reversible and Resistant Components of PCB Adsorption-Desorption: Isotherms," *Environ. Sci. Technol.* 16(9):594–602 (1982).
13. Horzempa, L. M. and D. M. DiToro. "The Extent of Reversibility of Polychlorinated Biphenyl Adsorption," *Water Res.* 17(8):851–9 (1983).
14. Isaacson, P. J. and C. R. Frink. "Nonreversible Sorption of Phenolic Compounds by Sediment Fractions: The Role of Sediment Organic Matter," *Environ. Sci. Technol.* 18(1):43–8 (1984).
15. Walters, R. W., A. Guiseppi-Elie, M. M. Rao, and J. C. Means. "Sorption of 2,3,7,8-TCDD on Soils from Water/Methanol Liquid Phases," submitted for publication.
16. Means, J. C., J. J. Hassett, S. G. Wood, and W. L. Banwart. "Sorption Properties of Energy-Related Pollutants and Sediments," in *Polynuclear Aromatic Hydrocarbons*, P. W. Jones and P. Leber, Eds. (Stoneham, MA: Butterworth Publishers, 1979).
17. Gschwend, P. M. and S. Wu. "On the Constancy of Sediment Water Partition Coefficients of Hydrophobic Organic Pollutants," *Environ. Sci. Technol.* 19(1):90–6 (1985).
18. Guiseppi-Elie, A., R. W. Walters, M. M. Rao, and J. C. Means. "Evaluation of Cosolvent Effects on the Sorptive Transport of Dioxin in Soils," Interim Report to the U. S. Enviromental Protection Agency, (1986).
19. Roberts, P. V., M. Reinhard, and A. J. Valocchi. "Movement of Organic Contaminants in Groundwater: Implications for Water Supply," *J. AWWA* 74(8):408–13 (1982).
20. Voice, T. C., et al. Paper presented before the Environmental Division of the ACS, Washington, DC (1983).

CHAPTER 12

Environmental Mobility
of 2,3,7,8-TCDD and Companion Chemicals
in a Roadway Soil Matrix

Raymond A. Freeman, Jerry M. Schroy, Fred D. Hileman, and Roy W. Noble

INTRODUCTION

During the early 1970s, Northeastern Pharmaceutical Company and Chemical Company (NEPACCO) produced hexachlorophene at a plant located in Verona, Missouri. An unwanted by-product, 2,3,7,8-tetrachloro-dibenzo-p-dioxin (TCDD), was produced. A mechanism for the formation of TCDD has been proposed by Esposito et al.[1] and is shown in Figure 1. The TCDD was primarily contained in a still bottoms waste. NEPACCO contracted with a waste-oil dealer for disposal of this still bottoms waste. The waste-oil dealer used the still bottoms waste and other oils for dust control at many horse arenas and at Times Beach, Missouri. This chapter presents data on the levels of a cocontaminant, 1,2,4,5,7,8-hexachloroxanthene (HCX), identified in the still bottoms waste and in the soil at Times Beach, Missouri.

FORMATION OF HEXACHLOROXANTHENE

The formation of HCX from hexachlorophene has been described by Gothe and Wachtmeister.[2] The reaction involves the removal of a water molecule from a hexachlorophene molecule to form an oxygen bridge. The

171

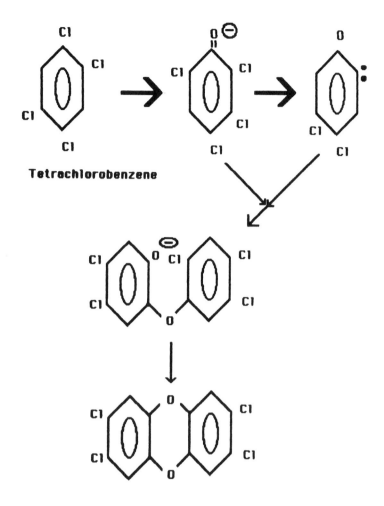

Tetrachlorobenzene

2,3,7,8-Tetrachlorodibenzo-p-dioxin

Figure 1. Formation of 2,3,7,8–TCDD.

overall reaction is given in the last step of Figure 2. Gothe and Wachtmeister found that very pure HCX could be formed by heating hexachlorophene and sodium hydroxide in a solvent at 140°C with p-toluenesulphonyl chloride added to improve the ring closure reaction. Thus, HCX can be expected to be found in still bottom residues remaining after hexachlorophene is purified.

The very limited data available on the properties of HCX are summarized in Table 1. The vapor pressure reported by Gothe and Wachtmeister was determined during product purification. The value was obtained from a

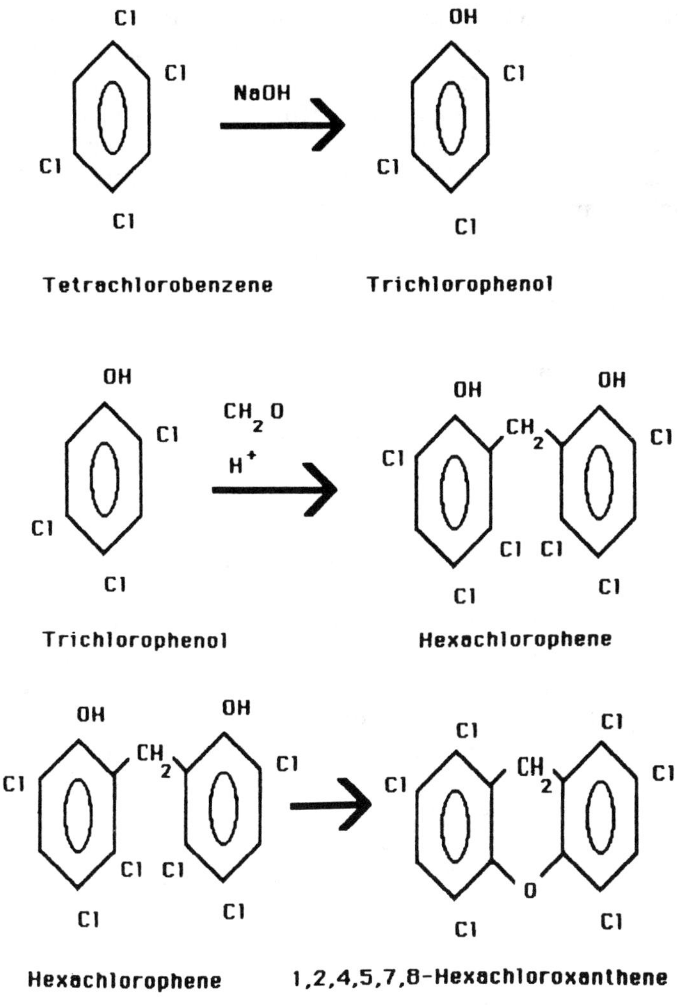

Figure 2.   Formation of Hexachloroxanthene.

Table 1.   Physical Properties of HCX.

Name	1,2,4,5,7,8-Hexachloroxanthene	
CAS	38178–99–3	
Formula	$C_{13}H_4Cl_6O$	Gothe and Wachtmeister[2]
Molecular Weight	388.892	Calculated
Melting Point	261 – 262°C	Gothe and Wachtmeister[2]
Solid Density	1.85 g/cm³	Soderholm et al.[3]
Vapor Pressure	0.05 mm Hg[a]	Gothe and Wachtmeister[2]

[a]At 140° C.

sublimation point determination and was not a measurement of the vapor pressure. This method may overestimate the actual vapor pressure. Figure 3 presents the vapor pressure point of Gothe and Wachtmeister and the vapor pressure of TCDD as determined at Monsanto.[4] The vapor pressure of HCX, measured by Gothe and Wachtmeister, is higher than TCDD at the same temperature (140°C). The water solubility of HCX constitutes important data that are missing.

ANALYTICAL METHODS

Soil Sample Preparation

Soil and clay samples were dried, weighed and spiked with 24.0 ng of ^{13}C-labeled 2,3,7,8-TCDD. Ten grams of washed anhydrous sodium sulfate were added and the samples were stirred, allowed to stand and stirred again. Samples were mixed again immediately before the addition of extraction solvents. The extraction process involved the addition of 15 mL of methanol (all solvents are Burdick and Jackson [Muskegon, MI] "distilled in glass" grade), stirring, and the subsequent addition of 150 mL of hexane. The samples were then shaken for 1 hr on a wrist-action shaker and the extraction solvent decanted from the soil. The solvent was then successively extracted with 50 mL of water, concentrated sulfuric acid (until clear), water, 20% potassium hydroxide and water. The hexane was then dried with 10 g of anhydrous sodium sulfate and passed through a mixed-phase column. The column was prepared from a 25-mL disposable pipet containing (top to bottom) 2 g of silica, 4 g of silica plus 40% sulfuric acid, 1 g of silica, 2 g of silica plus 36% potassium hydroxide and 1 g of silica. The column was eluted with an additional 50 mL of hexane, and the total eluant was concentrated to 2 mL under a stream of purified nitrogen. This concentrate was placed onto a column prepared from a 10-mL disposable pipet containing 2.5 of Woelm basic alumina. The alumina column was first eluted with 10 mL of 2% methylene chloride/hexane (volume/volume) and this fraction was discarded. An additional elution was made with 15 mL of a 50% methylene chloride/hexane (volume/volume) solution to elute the TCDD and HCX off the column. To this fraction was added 50 μL of dodecane keeper. The sample was then concentrated to 50 μL under a gentle stream of nitrogen and submitted for GC/MS analysis.

Analysis

The GC/MS analyses were conducted on a Hewlett Packard 5987 GC/MS instrument. The gas chromatographic separations were carried out with a 30-m DB-5 capillary column. The mass spectrometer was operated in the

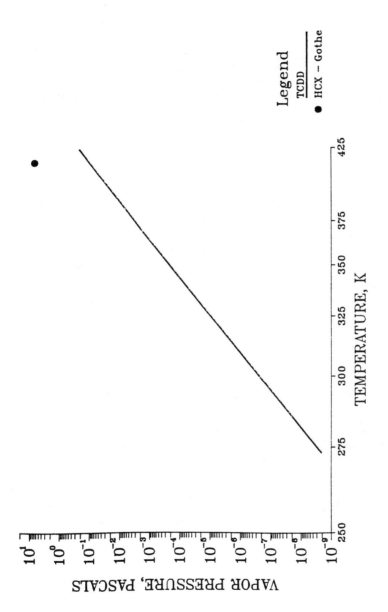

Figure 3. Vapor Pressures of 2,3,7,8–TCDD and 1,2,4,5,7,8–HCX.

Table 2.   Analysis of Chemical Residue from NEPACCO Plant at Verona, Missouri.

Replicate Number[a]	TCDD[b] (ppm)	HCX[b] (ppm)
1	120.	178.
2	112.	188.
3	104.	171.
4	120.	212.

μ_{TCDD} = 114 ppm s.d. = 7.66 ppm
μ_{HCX} = 187.3 ppm s.d. = 17.9 ppm

[a]Replicate numbers 1 and 2 done by analyst A. Replicate numbers 3 and 4 done by analyst B.
[b]Detection limits on control blanks were 1 ppm for TCDD and 2 ppm for HCX.
[c]s.d., standard deviation.

selected ion monitoring mode, analyzing for ions characteristic of the native (unlabeled) 2,3,7,8-TCDD (m/z 320, 322),[13]C-labeled 2,3,7,8-TCDD (m/z 322, 334) and the 1,2,4,5,7,8-HCX (m/z 388, 390). Quantitation was achieved by ratioing the areas of either the native 2,3,7,8-TCDD or the 1,2,4,5,7,8-HCX to the area of the ^{13}C-labeled 2,3,7,8-TCDD. The ratio was then compared to that obtained for a set of analytical standards, and the nanograms of 2,3,7,8-TCDD or 1,2,4,5,7,8-HCX were determined. The levels present in the original soil were then determined by dividing the weight of the soil into the amount of 2,3,7,8-TCDD or 1,2,4,5,7,8-HCX found.

FIELD SAMPLES

Because HCX can be expected to be found in the wastes from a hexachlorophene plant, a sample previously collected at the NEPACCO Verona, Missouri plant by Dr. Patrick Phillips (Missouri Department of Health), was analyzed for both TCDD and HCX. Dr. Phillips collected several samples from the residue tank at the NEPACCO plant during August of 1974. Some of these samples were analyzed by the Centers for Disease Control and have previously been reported as containing 343 ppm of TCDD. The last sample collected by Dr. Phillips was collected in a glass jar and packed inside a sealed steel can. This sample was stored in a laboratory until being opened for analysis during 1985. The results of these analyses are presented in Table 2. The mean TCDD concentration was found to be 114 ppm and the mean HCX concentration was 187 ppm.

Core samples previously taken from a roadside at Times Beach were reanalyzed for TCDD and HCX. The core samples were segmented into 2.5-cm (1-in.) slices for analysis. The concentrations of TCDD and HCX as a function of depth are given in Tables 3 and 4, and are shown in Figure 4 in graphic form. The concentration of TCDD was found to increase to a

Table 3. Analysis of Times Beach Core 2 for TCDD and HCX.[a]

Depth (cm)	TCDD (ppb)	HCX (ppb)
0.00 – 2.54	52	18
2.54 – 5.08	141, 97.4	54, 19
5.08 – 7.62	196	72
7.62 – 10.16[b]	154, 274	52, 120
10.16 – 12.70	119, 101	49, 82
12.70 – 15.24	60	27
15.24 – 17.78	13	5.2
17.78 – 20.32	5.2	2.6
20.32 – 22.86	2.7	ND[c][1ppb]

[a]Analysis of core during March 1985.
[b]Small sample (approximately 1 g).
[c]ND, none detected; detection limit in [ ].

maximum at a depth of approximately 8 cm and then declined at deeper depths.

Table 5 presents the measured ratio of the HCX to TCDD concentration as a function of depth. The concentration ratio for Core 2 is very much different then that for Core 3. This suggests a difference in the initial oil loading of the soil surface at the two points where the cores were taken. The two cores were taken within 1 m of each other. This implies that a large variability in the oil application pattern existed at the time of road spraying. The difference may be because of one core being sprayed by only one nozzle and the second core being sprayed by two or more nozzles.

Table 4. Analysis of Times Beach Core 3 for TCDD and HCX.[a]

Depth (cm)	TCDD[b] (ppb)	HCX[b] (ppb)
0.00 – 2.54	41	54
2.54 – 5.08	105	101
5.08 – 7.62	92, 112	71, 80
7.62 – 10.16	14	10
10.16 – 12.70	0.8	ND[c] [< 3 ppb]
12.70 – 15.24	3	8
15.24 – 17.78	2	3
17.78 – 20.32	0.9	ND [< 3 ppb]
20.32 – 22.86	7.7	9

[a]Analysis of core during March 1985.
[b]Small sample (approximately 1 g).
[c]ND, none detected; detection limit in [ ].

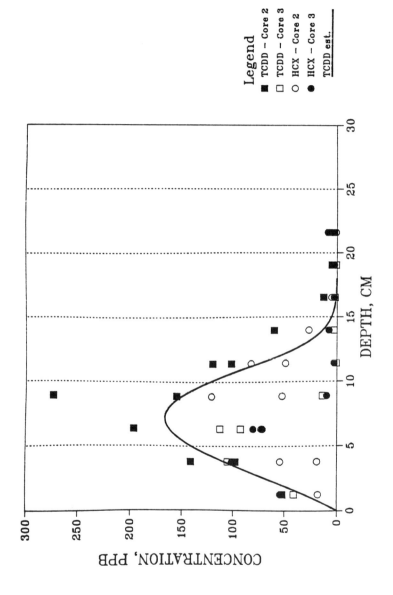

Figure 4.   Hexachloroxanthene and TCDD Profile at Times Beach, Missouri.

Table 5.   Relative Ratio of HCX and TCDD at Times Beach.

Depth (cm)	Core 2 Ratio[a]	Core 3 Ratio[a]
0.00 – 2.54	0.346	1.317
2.54 – 5.08	0.383, 0.195	0.962
5.08 – 7.62	0.367	0.772, 0.714
7.62 – 10.16	0.338, 0.438	0.714
10.16 – 12.70	0.412, 0.812	–
12.70 – 15.2	0.450	2.667
15.24 – 17.7	0.400	1.500
17.78 – 20.32	0.500	–
20.32 – 22.86	–	1.168

[a]Ratio = concentration HCX / concentration TCDD.

MODELING THE SOIL MOVEMENT OF CHEMICALS*

Freeman and Schroy[5,6] have constructed a model for TCDD transport based on a solid and a vapor phase being in contact with each other. The model uses the following material balance equations.

$$\frac{\partial C_d}{\partial t} = \frac{2 \, \rho_{molar} \, D_{ab}}{\Phi} \left[\frac{C_a}{M_w \, \rho_{molar}} - \frac{K \, C_d \, P^0}{P_t} \right] \frac{M_w \, a}{\rho_{soil}} = R \tag{1}$$

$$\frac{\partial C_a}{\partial t} = \frac{h}{\partial Z} \frac{\partial}{\partial Z} \left(D_{ab} \frac{\partial C_a}{\partial Z} \right) - \frac{R}{\epsilon} \tag{2}$$

These two nonlinear partial differential equations must be solved simultaneously. Studies on the material balance equations have allowed some simplifications to be made.

The vapor phase equation changes five orders of magnitude faster than the soil phase equation. Thus, the vapor phase is always in equilibrium with the soil phase. Therefore, the material balance equations may be rewritten as

$$\frac{\partial C_a}{\partial t} = \frac{(\epsilon^2/\tau)}{\epsilon + P_t/(K \, P^0 \, M_w \, \rho_{molar})} \frac{\partial}{\partial Z} \left(D_{ab} \frac{\partial C_a}{\partial C} \right) \tag{3}$$

Since the vapor pressure, diffusivity, and molar density are dependent on the temperature, a method of predicting the soil temperature profile is needed. The method used is to solve the transient energy balance on a soil column for the temperature as a function of time. The energy balance

---

*Abbreviations used are listed at the end of the chapter.

equation is the same as previously presented and is described below. The one-dimensional transient energy balance equation may be written as

$$\rho_{soil}\, C_P\, \frac{\partial T}{\partial t} = k\, \frac{\partial^2 T}{\partial Z^2} \tag{4}$$

Solving this equation for the temperature waves that pass through a soil column requires two boundary conditions.

Boundary Condition 1—Surface Energy Flux

$$at\ Z = 0;\ k\frac{\partial T}{\partial Z} = q_r + q_c + q_b \quad for\ all\ t \geq 0 \tag{5}$$

Boundary Condition 2—Constant Temperature
at Some Depth in Ground

$$at\ Z = L;\ T = T_g \quad for\ all\ t \geq 0 \tag{6}$$

The term q_r represents the radiative solar input into the soil column. The term q_c is convective heat transfer between the soil and the air. The term q_b represents the black body radiative loss of energy from the soil surface. These soil surface energy fluxes are complex functions of soil temperature, weather, and site location. Schroy and Weiss[7] has previously presented methods for the computation of q_r, q_c and q_b.

The model, using the simplified material and energy balance equations, has been used to predict the soil TCDD profiles measured at Eglin Air Force Base in Florida.[6] The model has been applied to the data for TCDD presented in Tables 3 and 4. The curve labeled "TCDD (Est.)" in Figure 4 is the model fit to the TCDD concentration profile. The model fits the data reasonably well for Core 2. An attempt to simulate the HCX concentration profile was completed using a vapor pressure equation estimated from the one data point of Gothe and Wachtmeister[2] and the vapor pressures of various chlorodioxins as reported by Rordorf.[8] The model predicted a deeper HCX penetration than observed for either Core 2 or Core 3. A study of the model assumptions showed that the vapor pressure of HCX was the most uncertain parameter.

Because of the importance of vapor pressure data in the prediction of environmental fate of HCX, a collaborative research program with Dr. Frank Rordorf (Ciba Geigy – Basel, Switzerland) has been undertaken to measure the vapor pressure of HCX. The results of this program are planned for presentation at the 5th International Symposium on Chlorinated Dioxins and Related Compounds to be held in Bayreuth, Germany during September 1985.

DISCUSSION

The variability of chemical loading in the tested soil column cores was significant. The variation in both specific component loading and the relative loading of cocontaminants can not be attributed to any specific cause. However, it is clear that no single soil sample can represent all Missouri experiences.

Before findings on the toxicity of a given soil can be attributed to TCDD or any specific component, all other contaminants must be identified and factored out of the results. The acute toxicity of TCDD is believed to be related to the flat structure and the location of the chlorine atoms in the TCDD molecule. Finding a chemical with a structure similar to TCDD in the same soil as TCDD, such as HCX, should concern anyone doing toxicological studies on the soil. To allow for proper interpretation of the resulting data, the soil should be analyzed for other chemicals besides TCDD. The fate of all these chemicals in thes test organisms should be determined.

The conclusions derived from this work are:

1. HCX is a marker for chemical wastes from hexachlorophene production.
2. The variability of chemical loading from core to core at Times Beach is significant and is due to a number of physical factors involved in the spraying of the waste material.
3. The concentration profile for TCDD at Times Beach is consistent with an average initial concentration of 18 ppm in the top 1 cm of soil.
4. The physical properties of HCX should be measured, specifically the vapor pressure and the water solubility.
5. The biodegradation of HCX should be studied. If HCX is slowly degraded, its usefulness as a marker compound would be reduced.

ACKNOWLEDGMENTS

The authors wish to thank the Missouri Department of Health and the Missouri Department of Natural Resources for their assistance and for allowing the remaining sample of NEPACCO residue to be analyzed. The authors also thank Dr. Frank Rordorf for his assistance in studying the vapor pressure of HCX.

ABBREVIATIONS

a	air-soil interfacial area per unit soil volume
C_a	concentration of TCDD in air
C_d	concentration of TCDD in soil
C_p	heat capacity of soil
D_{ab}	diffusivity of TCDD in air
h	hinderance factor, $h = \epsilon/\tau$
K	empirical equilibrium partitioning coefficient between soil and air
k	thermal conductivity of soil
L	soil depth where temperature does not change during a year
M_w	molecular weight of TCDD
P_o	vapor pressure of TCDD
P_t	total barometric pressure
q_b	black body radiation loss to the sky
q_c	convective energy exchange between soil and atmosphere
q_r	radiative energy received by soil from the sun
R	volumetric rate of volatilization of TCDD into air voids
T_g	soil temperature at a depth L
t	time
z	depth into the ground

GREEK SYMBOLS

ϵ	soil void fraction
ρ_{molar}	molar density of air in soil void space
ρ_{soil}	density of soil
τ	tortuosity factor ($\tau = 2$ for an average soil)
φ	average diameter of soil particle

BIBLIOGRAPHY

1. Esposito, M. P., T. O. Tiernan, and F. E. Dryden. *Dioxins*, EPA report number EPA-600/2-80-197, NTIS report number PB82-136847, pg. 74 (November 1980).
2. Gothe, R. and C. A. Wachtmeister. "Synthesis of 1,2,4,5,7,8-Hexa-chloroxanthene," *Acta Chemica Scandinavica* 26:(6) (1972).
3. Soderholm, M. U. Sonnerstam, R. Norrestam, and T. B. Palm. "Structural Studies of Polychlorinated Hydrocarbons II. Hexachloroxanthene and Hexachloroxanthone," *Acta Cryst.* B32:3013-3018 (1976).
4. J. M. Schroy, F. E. Hileman, and S. C. Cheng. "Physical and Chemical Properties of 2,3,7,8 TCDD: The Key to Transport and Fate Characterization," paper presented at the 8th ASTM Aquatic Toxicology Symposium held in Fort Mitchell, Kentucky, April 15-17, 1984.

5. Freeman, R. A., and J. M. Schroy. "Environmental Mobility of Dioxins," paper presented at the 8th ASTM Aquatic Toxicology Symposium held in Fort Mitchell, Kentucky, April 15–17, 1984.
6. Freeman, R. A., and J. M. Schroy. "Modeling the Transport of 2,3,7,8-TCDD and Other Low Volatility Chemicals in Soils," paper presented at the AICHE Summer National Meeting, Philadelphia, Pennsylvania, August 19–22, 1984.
7. Schroy, J. M., and J. S. Weiss. "Prediction of Wastewater Basin Temperatures: A Design and Operating Concern," paper presented at the AICHE Symposium, Washington, DC, November 4, 1983.
8. Rordorf, B. F. "Thermal Destruction of Polychlorinated Compounds: The Vapor Pressures and Enthalpies of Sublimation of Ten Dibenzo-para-Dioxins," *Thermochemica Acta* 85: 435–439, (1985).

CHAPTER **13**

Solubility of 2,3,7,8-TCDD
in Contaminated Soils

D. R. Jackson, M. H. Roulier, H. M. Grotta, S. W. Rust, and J. S. Warner

INTRODUCTION

2,3,7,8-Tetrachlorodibenzo-p-dioxin (2,3,7,8-TCDD) has been associated with the manufacture of 2,4,5-trichlorophenoxyacetic acid (2,4,5-T), hexachlorophene, and other pesticides having 2,4,5-trichlorophenol as a precursor.[1] The persistence and toxicity of 2,3,7,8-TCDD have created concern on a national scale for the cleanup and safe disposal of contaminated soil.[2] The potential mobility of 2,3,7,8-TCDD in soil water is a factor in selecting remedial action alternatives for contaminated sites. Previous investigators[3-6] have documented that 2,3,7,8-TCDD is extremely immobile in most soils due to its low solubility in water (200 ppt).[7] For example, mobility of 2,3,7,8-TCDD in soils was limited to the upper 8 cm one year after an accident which contaminated a large area in Seveso, Italy.[8]

The octanol-water partition coefficient (K_{ow}) for hydrophobic compounds such as 2,3,7,8-TCDD is considered one of the most important physicochemical characteristics related to sorption on soils and sediments.[9] Hydrophobic compounds generally have log K_{ow}'s greater than 10[10] and are highly partitioned onto soils.[11] Dragun[12] has calculated the log K_{ow} for 2,3,7,8-TCDD to be equal to 7.16 using methods from Leo[13] and Hansch and Leo.[14] This value is in general agreement with the range of log K_{ow} values of 6.15–7.28 for 2,3,7,8-TCDD reported by the U. S. EPA.[15]

Sorption of many hydrophobic compounds on soils and sediments has been related to the organic matter content of the sorbent and the K_{ow} of the hydrophobic compound.[10,16-20] The method generally employed in these studies is to determine a partition coefficient (K_p) over a broad range of aqueous concentrations. K_p is defined by,

$$K_p = \frac{X}{C} \tag{1}$$

where X denotes the concentration of the solute on the solid phase in ng/g;
C is the equilibrium solute concentration in ng/mL

Karickhoff et al.[19] found that individual K_p's determined for pyrene and methoxychlor on particle size isolates of three sediments were highly correlated with the organic carbon content of each particle size fraction. A single partition coefficient (K_{oc}) was determined to describe partitioning for all the soil particle size fractions by normalizing the K_p of each fraction for its organic matter content. The following formula describes this relationship.

$$K_{oc} = K_p/f_{oc} \tag{2}$$

where K_{oc} is the organic carbon partition coefficient
f_{oc} is the fractional mass of organic carbon in the sediments

Karickhoff et al.[19] extended this relationship using ten hydrophobic compounds to relate K_{oc} to K_{ow} using the linear regression equation

$$\log K_{oc} = \log K_{ow} - 0.21 \tag{3}$$

Hassett et al.[10] confirmed these sorption relationships using dibenzothiophene (an unchlorinated dioxin-like compound) and 14 soil and sediment samples.

Previous investigations of 2,3,7,8-TCDD sorption have been conducted under idealized conditions; 2,3,7,8-TCDD has been spiked as a single compound on uncontaminated soil, the soil mixed with water, and a partition coefficient calculated from measurements of the amount of 2,3,7,8-TCDD in solution. These experimental conditions may not address the chemical and physical reactions of 2,3,7,8-TCDD that occur over an extended time period in the soil environment, particularly when 2,3,7,8-TCDD is a trace contaminant in a complex mixture of other organic chemicals that have been applied to the soil.

This report examines the sorption behavior of ten soils contaminated with 2,3,7,8-TCDD more than ten years ago. The objectives of this investigation were to determine water/soil partition coefficients (K_p) for 2,3,7,8-TCDD and analyze relationships between K_p values and soil physicochemical parameters.

SITE HISTORY

Samples of contaminated soils were collected from Missouri and New Jersey. The Missouri samples were collected from two horse arenas, two residential sites, and four sites adjacent to roadways. These sites, listed in Table 1, were contaminated with 2,3,7,8-TCDD as a result of being sprayed with various mixtures of waste oil. While no official records are available on the chemical composition of these oily mixtures, investigations by the St. Louis Post-Dispatch[21] indicate that waste crankcase oil was often sprayed in addition to chemical still bottoms containing 2,3,7,8-TCDD and waste transformer oil which contained PCBs. These materials were applied on soils to control dust.

The soil sample from the slough area at Shenandoah Stables contained the highest level of 2,3,7,8-TCDD (2,400 ng/g) in soil collected from Missouri. Soil from this site was collected from a deposit of contaminated soil which had been scraped from the adjacent indoor arena. A second horse arena was sampled at Bubbling Springs Stables. This outdoor arena also had been scraped to remove visibly contaminated soil prior to any knowledge of the presence of 2,3,7,8-TCDD on the site. According to the St. Louis Post-Dispatch[21] investigation, material from this arena was deposited at the Minker residence as fill material to prevent erosion. Two samples from the Minker residence were collected (Table 1). These samples were higher in 2,3,7,8-TCDD concentration than the one collected from the Bubbling Springs Stables. This observation is representative of the fact that soil deposited at the Minker site originated from the more highly contaminated surface layers of the arena at Bubbling Springs Stables.

Two soil samples were collected adjacent to Piazza Road (Table 1). These samples, although located within one-half mile of each other, varied in 2,3,7,8-TCDD concentration by approximately fivefold. Samples from Sontag Road and Times Beach, both adjacent to residential streets, represent the lowest 2,3,7,8-TCDD concentrations from locations in Missouri known to have been sprayed with contaminated oil. Due to the inherent variability expected in soil 2,3,7,8-TCDD concentrations as a result of oil spraying, samples collected for this investigation may not be representative of the overall contamination level at each of the sites.

Samples were collected from two different locations at the Minker residence and at the Piazza Road site because enough data were available from previous work at these sites to identify areas with high and low levels of 2,3,7,8-TCDD contamination. The two samples from each location differed considerably in 2,3,7,8-TCDD concentration (Table 1); this facilitated the statistical analysis of the relations between soil properties and partition coefficients (K_p).

The soil sample containing the highest concentration of 2,3,7,8-TCDD (26,000 ng/g) was collected from a defunct chemical manufacturing plant in

Table 1. Physical and Chemical Properties of Soils.

No.	Location	Site	Total 2,3,7,8-TCDD Concentration	SEOM[a]	Peak Area[b]	Organic Carbon	CEC[c]	pH	EC[d]	Sand	Silt	Clay	Texture
1	New Jersey	Chemical Plant	26,000	22	270	8.0	13	7.5	65	81	18	1	Loamy Sand
2	Missouri	Shenandoah Stables[e]	2,400	7.1	15	6.0	39	7.3	195	42	52	6	Silty Loam
3	Missouri	Piazza Road-1	1,100	5.2	2	2.7	35	6.7	95	57	30	12	Sandy Loam
4	New Jersey	Scrap Yard	1,100	19	21	3.7	18	6.7	99	72	23	5	Sandy Loam
5	Missouri	Minker-2	760	5.3	4	2.5	24	7.7	139	34	59	7	Silty Loam
6	Missouri	Minker-1	610	7.0	4	2.2	19	8.0	119	56	37	7	Sandy Loam
7	Missouri	Bubbling Springs Stables	280	3.4	2	1.5	13	7.6	101	61	37	2	Sandy Loam
8	Missouri	Piazza Road-2	230	0.7	0.6	7.6	54	7.5	153	60	30	9	Sandy Loam
9	Missouri	Sontag Road	63	5.2	0.7	6.0	37	8.4	350	38	58	3	Silty Loam
10	Missouri	Times Beach	8	0.3	0.4	1.7	23	7.7	107	52	43	4	Sandy Loam

[a]Solvent-extractable organic matter.
[b]Total peak area of solvent-extractable chlorinated organic compounds detected by GC-ECD analysis.
[c]Cation exchange capacity.
[d]Electrical conductivity.
[e]From slough area.

New Jersey. A second sample was collected from a nearby metal scrap yard where used reactor vessels from the chemical plant were collected and disassembled. A soil concentration of 1100 ng/g was found at this site. Contamination of 2,3,7,8-TCDD at these two sites, which occurred between 1968 and 1971, represents spillage of 2,4,5-trichlorophenol (2,4,5-TCP) still bottoms from the manufacture of certain chlorinated pesticides and 2,4,5-T.

MATERIALS AND METHODS

One background soil sample was collected outside each of the contaminated areas at Shenandoah Stables, Piazza Road, Bubbling Springs Stables, Sontag Road, and Times Beach. These samples were analyzed for 2,3,7,8-TCDD to confirm that they were uncontaminated (less than 1.0 ng/g). These soils were analyzed for solvent-extractable organic matter (SEOM) as a basis for estimating the amounts of organic contaminants (other than 2,3,7,8-TCDD) in SEOM from the contaminated samples. These background samples were not used in any other phase of this study.

Bulk soil samples were collected by excavating soil to a depth of approximately 15 cm. The soil was mixed and placed in plastic-lined 5-gal steel containers. These samples comprised a variety of physicochemical characteristics (Table 1) and total 2,3,7,8-TCDD concentrations (Figure 1). Soil pH was determined using a soil-to-water ratio of 1:1.[22] Electrical conductivity (EC) was measured using a suspension of soil and water mixed in a ratio of 1:2.[23] Soil organic carbon content was determined by oxidation with $K_2Cr_2O_7$ and colorimetry.[23] Cation exchange capacity (CEC) was based on soil saturation with NH_4^+ followed by displacement with K^+; analysis of displaced NH_4^+ was conducted using an ammonium ion-specific electrode.[24] Particle size analysis (sand, silt, and clay content) was conducted by the hydrometer method.[25]

Soil Extracts

The analytical method for determining 2,3,7,8-TCDD in soil extracts required a sample volume of 1 L. Therefore, a procedure for generating soil extracts was selected to accommodate this requirement. A batch extraction procedure was used to determine the effect of sequential extraction on the partitioning of 2,3,7,8-TCDD between soil and water.[26] This method is similar to the Standard Leaching Test (SLT) developed by Ham et al.[27] For this procedure, two 100-g aliquots from the contaminated soil sample for each location were sequentially extracted twice with 1 L of distilled water by tumbling for 18 hr at 30 rpm using an electric-powered rotary extractor. The 18-hr extraction time was used to facilitate efficient laboratory operations and to minimize biological activity which a longer equilibration time

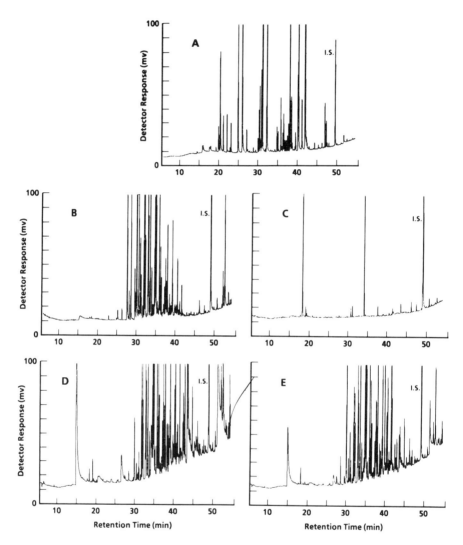

Figure 1. Gas chromatogram obtained in analysis of solvent extracts of five contaminated soils; (A) Chemical plant, (B) Shenandoah Stables, (C) Piazza Road-2, (D) Minker-2, and (E) Bubbling Springs Stables. I.S. refers to the internal standard (octachlorobiphenyl).

would encourage. Due to this time constraint, the partition coefficients resulting from this method are not necessarily representative of equilibrium conditions in the soils evaluated. Kinetics of 2,3,7,8-TCDD desorption from soil were not within the scope of this investigation. The suspensions were placed in pressurized filtration vessels and filtered through Millipore hydrophilic cellulose acetate and nitrate membranes with a pore size of 0.45

μm. The filtrates were analyzed for 2,3,7,8-TCDD content. Each 100-g aliquot of contaminated soil was extracted a second time by repeating the above procedure. This procedure results in leachates representing two sequential extractions of each of two 100-g aliquots of each soil sample. Each extract was analyzed for 2,3,7,8-TCDD using a modified version of the U. S. EPA Region VII protocol.[28]

Solvent-Extractable Organic Matter

Solvent-extractable organic matter of soils was determined by extracting a suspension of 30 g of soil in 30 mL of an aqueous solution containing 30% NaCl and 2% KH_2PO_4 with 30mL of methyl tert-butyl ether (MTBE).[29] Extraction was facilitated by mixing the soil-liquid suspension for 18 hr using a tumbling apparatus. The MTBE phase was separated from the aqueous phase using a separatory funnel. The MTBE layer was dried with 2 g of anhydrous $MgSO_4$. The dried MTBE extract was filtered through a 0.5-μm PTFE membrane filter. A 100-μL aliquot of the filtered extract was transferred to a tared aluminum weighing dish, evaporated to dryness under a heat lamp, and reweighed using a microbalance to determine the weight of low-volatility organic compounds. This method is suitable for the extraction of most phenols, anilines, and neutral semivolatile organic compounds. SEOM values determined for the contaminated soils are presented in Table 1.

Gas Chromatography Analysis of Solvent Extract

The MTBE extract of soil was analyzed by gas chromatography (GC) using a fused-silica capillary column and a [63]Ni electron capture detector (ECD). The GC conditions used were the following.

Column: 30 m $\times$ 0.25 mm i.d., 0.25 μm film thickness, J&W DB-5
Column temperature: 80 to 300°C at 4°/min, 300°C for 10 min
Injector temperature: 250°C
Detector temperature: 350°C
Carrier gas: helium at 30 cm/sec
Makeup gas: nitrogen at 30 mL/min
Injection: 2μL splitless.

The combined GC-ECD peak area of all chromatographic peaks eluting after 5 min, referred to as peak area in Table 1, was obtained using a computerized integration and data processing system. The combined peak area response was used as an indicator of the total amount of chlorinated semivolatile organic compounds present. Although the electron capture detector detects a variety of compounds other than chlorinated compounds, e.g., brominated compounds, nitro compounds, and phthalates, chlorinated compounds were expected to account for the majority of the electron capture response from the extracts generated in this study.

RESULTS AND DISCUSSION

Soil Properties

Soil 2,3,7,8-TCDD levels ranged from 8 to 26,000 ng/g (Table 1). Missouri samples as a group were lower in 2,3,7,8-TCDD content than those from New Jersey. This difference reflects two sources of soil contamination. The New Jersey soils were contaminated by leakage from 2,4,5-TCP reactor vessels or storage tanks, while samples from Missouri were contaminated by spraying with still bottoms diluted with used oil.

Soil organic carbon content ranged from 1.5 to 8.0% (Table 1). This wide range in organic carbon content produced a greater basis for the regression of this parameter versus K_p values. Solvent-extractable organic matter (SEOM) content ranged from 0.3 to 22 mg/g in the contaminated soils. This parameter ranged from 0.5 to 1.5 mg/g in background soils collected from sites in Missouri. Seven of the ten soil samples included in this study had SEOM values greater than background levels. This parameter represents the concentration of organic compounds which are soluble in a moderately polar organic solvent and relatively insoluble in water. The peak area of chlorinated semivolatile organic compounds for the soils ranged from 0.4 to 270. Peak area and SEOM were considered indicative of the relative concentrations of cocontaminants which were presumably applied to the soil as a mixture with 2,3,7,8-TCDD.

Selected GC-ECD chromatograms are presented in Figure 1. Although specific compounds were not identified, considerable differences were apparent in the chromatographic patterns obtained from these soils. The chemical plant soil contained a variety of discrete compounds distributed throughout the midrange of the chromatogram. The soil from Shenandoah Stables, Minker, and Bubbling Springs Stables gave chromatographic patterns indicative of a polyhalogenated mixture such as a polychlorinated biphenyl (PCB). The chromatographic patterns from the Minker and Bubbling Springs Stables soils are nearly identical, and represent a somewhat higher degree of chlorination than that from the Shenandoah Stables soil. The soil from Piazza Road-2 was nearly devoid of chlorinated components, even though this sample was relatively high in 2,3,7,8-TCDD. Electrolyte concentration as determined by EC in soil leachate (Table 2) was reduced 51% in the second leaching of the SWLP-R relative to the first fraction. Soils exhibiting this general trend contained the greatest levels of 2,3,7,8-TCDD and halogenated organic content as determined by peak area (Table 1). These results imply that the lower electrolyte level of the second extract allows a higher concentration of 2,3,7,8-TCDD to come into solution. These observations are supported by Karickhoff,[30] who discussed a relationship between solution electrolyte concentration and solubility of hydrophobic compounds.

Table 2. Electrical Conductivity (EC) of Soil Leachates.

Site Number	EC (μmhos/cm)	
	Extract 1	Extract 2
1	30.8	15.6
2	76.5	31.0
3	47.8	22.4
4	32.1	17.1
5	93.4	43.2
6	78.1	40.3
7	59.7	31.8
8	119	74.3
9	171	74.1
10	76.6	44.5
Mean	78.5	40.3

Soil Leachates

The 2,3,7,8-TCDD concentration in aqueous soil extracts ranged from
0.1 ng/L to 55 ng/L; 73% of the concentrations were 2.0 ng/L or less. The
greatest concentration of 2,3,7,8-TCDD in soil extracts, 55 ng/L (Figure 2),
was found in the second leaching fraction of the extraction procedure for
the chemical plant soil and represented a tenfold greater concentration than
the first leaching fraction. Relatively greater concentrations of 2,3,7,8-
TCDD were found in the second leaching fraction of the batch procedure in
six of the ten soils analyzed. Soils exhibiting this general trend contained the
greatest levels of 2,3,7,8-TCDD and chlorinated organic content as determined by peak area (Table 1).

Additional information on leachability of 2,3,7,8-TCDD from the Missouri and New Jersey soils using intact cores and another batch procedure
has been reported recently.[31]

Soil Partition Coefficients

A soil partition coefficient (K_p) was computed using Equation 1 for soil
extract produced by the batch extraction method. The coefficients, presented in Figure 3, ranged from 0.4×10^5 to 45.8×10^5, with an overall
mean of 9.6×10^5.

Log K_{oc} values were computed for each soil using Equation 3. Values of
log K_{oc} were computed in order to make comparisons between these values
and those predicted from Equation 3 using the log K_{ow} of 7.16 provided by
Dragun.[12] The log K_{oc} for 2,3,7,8-TCDD predicted by this method is 6.95.
The mean log K_{oc} value determined from experimental K_p values using
Equation 3 for the batch extraction procedure was 7.39, which was within
one standard deviation of the predicted value.

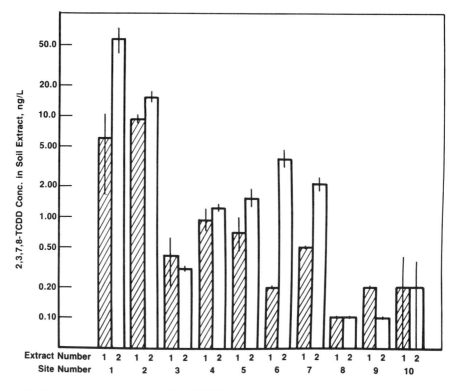

Figure 2.   Concentrations of 2,3,7,8-TCDD in soil leachates from ten contaminated soils.

Statistical Analyses

Correlation and regression analyses were used to investigate relationships between the soil property variables (independent variables) presented in Table 1 and the concentrations of 2,3,7,8-TCDD in soil extracts and soil partition coefficients (dependent variables) presented in Figures 2 and 3. The soil property variables were not experimentally manipulated to assess their effect on the dependent variables. However, the soils used in the study exhibited a sufficiently wide range of chemical and physical properties (Table 1) that correlation and regression analyses yielded information about the effect of soil properties on partition coefficients and 2,3,7,8-TCDD concentrations. Results of the correlation and regression analyses were used to develop hypotheses about mechanisms involved in 2,3,7,8-TCDD solubility in contaminated soils. Only soils from Missouri were included in the statistical analyses because most of the data were from Missouri, and because of the differences in soil type and pattern of contamination between the Missouri and New Jersey sites.

The correlation matrix presented in Table 3 was constructed to examine

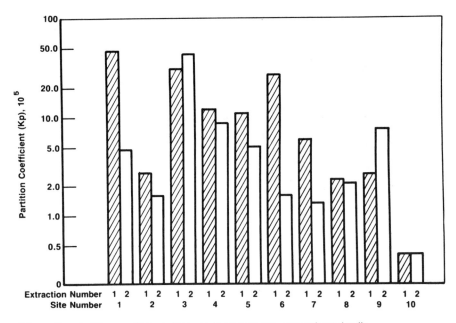

Figure 3. Partition coefficients (K_p) obtained from ten contaminated soils.

the significance of relationships between selected soil property variables and dependent variables. Observed significance levels (OSLs), which indicate the significance of the reported correlation coefficients, are provided in parentheses. For example, an OSL value of 0.05 indicates that the corresponding correlation coefficient is significantly different from zero at a confidence level of 95%.

The log soil concentration of 2,3,7,8-TCDD was positively correlated with the leachate concentrations. However, the leachate concentrations were most highly correlated with the log peak area parameter. The relationships between log peak area and the soil and extract 2,3,7,8-TCDD concentrations are illustrated graphically in Figure 4. These figures are log-log plots of 2,3,7,8-TCDD concentration versus peak area. The least squares regression line has been superimposed on each plot. The intercept, slope, percentage of explained variation, and observed significance level for each regression line are provided in Table 4. These relationships imply that chlo-

Table 3. Correlation Matrix for Soil Property Variables Versus TCDD Concentrations in Bulk Soil and Soil Extracts and K_p Values.[a]

Dependent Variable	Log 2,3,7,8-TCDD$_{Soil}$	Log$_{PeakArea}$	Log (SEOM)	% Organic Carbon
Log 2,3,7,8-TCDD$_{Soil}$		0.84 (0.01)	0.76 (0.03)	0.14 (0.74)
Log 2,3,7,8-TCDD$_{Extract}$	0.66 (0.08)	0.95 (0.00)	0.61 (0.11)	−0.16 (0.71)

[a]Observed significance levels in parentheses.

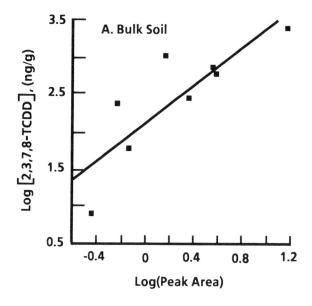

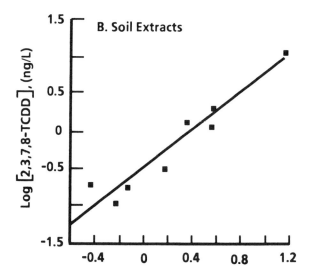

Figure 4. Log-log plots of 2,3,7,8-TCDD concentrations in soil and soil leachates versus peak area for eight contaminated soils from Missouri.

rinated semivolatile organic compounds were cocontaminants in the waste applied to these soils and may be a most important determinant for the relative aqueous extractability 2,3,7,8-TCDD in the Missouri soils.

This effect of the cocontaminants is most strikingly illustrated by com-

Table 4. Least Squares Regression Analysis of $Log_{2,3,7,8\text{-}TCDD}$ Versus $Log_{Peak\ Area}$.

Dependent Variable	Intercept	Slope	R^2	Observed Significance Level
Log 2,3,7,8-TCDD$_{Soil}$	2.14	1.26	0.70	0.01
Log 2,3,7,8-TCDD$_{Leachates}$	−0.49	1.25	0.91	0.00

paring the results from Bubbling Springs and Piazza Road-2. These soils contain comparable amounts of 2,3,7,8-TCDD (234 and 275 ng/g). Bubbling Springs has much greater amounts of cocontaminants (compare Figure 1–C and Figure 1–E) and the 2,3,7,8-TCDD concentrations in aqueous soil extracts from Bubbling Springs were about twice as great as those from Piazza Road-2, even though both soils contained comparable amounts of 2,3,7,8-TCDD.

The relationship between soil organic carbon content and the K_p values for the eight soils was investigated to determine the significance of computing K_{oc} values for the soils. The partition coefficients for these contaminated soils were not significantly related to soil organic matter content ($r = 0.36$). Thus, normalization of the K_p values for soil organic carbon content did not reduce the variability associated with the nonnormalized K_p values.

SUMMARY AND CONCLUSIONS

Log K_{oc} values based on the 2,3,7,8-TCDD concentration in bulk soil and soil leachates and soil organic carbon content were determined for ten contaminated soils from Missouri and New Jersey. The calculated mean log K_{oc} value of 7.39 was within one standard deviation of a predicted value of 6.95. The log K_{oc} parameter is based on a linear relationship between K_p and soil organic carbon content.[19] This relationship was not apparent for the soils analyzed in this investigation. However, a strong association was found between 2,3,7,8-TCDD concentration in soil extracts and log GC-ECD peak area of organic solvent extract. This relationship suggests that the relative aqueous extractability of 2,3,7,8-TCDD in these soils is regulated primarily on the basis of soil levels of organic solvent-extractable chlorinated compounds. This conclusion is compatible with the theory that bonding of hydrophobic compounds in soil is analogous to partitioning of the compounds between water and immiscible solvent.[29]

Studies by Karickhoff et al.[19] and Hassett et al.[10] are based on partitioning of hydrophobic compounds between initially clean soils and aqueous solutions containing only the hydrophobic compound. In these cases indigenous organic matter is the most compatible substrate in the soil matrix for sorption of the hydrophobic compounds.

Conditions in the soils used in this study were considerably different than those in the studies cited above. The 2,3,7,8-TCDD applied to the soils from the Missouri region is alleged to have originated from chemical still bottoms which would most likely contain chlorinated benzenes and trichlorophenols. These still bottoms were generally mixed with used engine crankcase oil and possibly transformer oil prior to soil application. Soil contamination at the New Jersey sites can be attributed primarily to storage tank leakage of still bottoms from pesticides manufacturing. In both cases chlorinated and other organic compounds which are insoluble in water and resistant to soil microbial degradation were added to the soil as cocontaminants. We conclude that these organic compounds are sorbed to the soil matrix, and that these chlorinated compounds as a group greatly influence the partitioning of 2,3,7,8-TCDD between the soil solid-phase and soil water extracts in the contaminated soils.

ACKNOWLEDGMENTS

The research described in this article has been funded wholly by the United States Environmental Protection Agency under Contract 68-03-3100 to Acurex Corporation and under Subcontract EC 59606A to Battelle Memorial Institute. It has been subject to the Agency's peer review and has been approved for publication. Mention of trade names or commercial products does not constitute endorsement or recommendation for use.

REFERENCES

1. Dryden, F. E., H. E. Ensley, R. J. Rossi, and E. J. Westbrook. "Dioxins: Volume III. Assessment of Dioxin-Forming Chemical Processes." U. S. EPA Report–600/2–80–158 (1980).
2. des Rosiers, P. E. "Remedial Measures for Wastes Containing Polychlorinated Dibenzo-p-dioxins (PCDDs) and Dibenzofurans (PCDFs): Destruction, Containment or Process Modification," *Ann. Occup. Hyg.* 27(1):57–72 (1983).
3. Kearney, P. C., E. A. Woolson, A. R. Isensee, and C. S. Helling. "Tetrachlorodibenzodioxin in the Environment: Sources, Fate, and Decontamination," *Environ. Health Pers.* 5:273–277 (1973).
4. Young, A. L. "Long-Term Studies on the Persistence and Movement of TCDD in a Natural Ecosystem," *Environ. Sci. Res.* 26:173–190 (1983).
5. Bartleson, F. D., Jr., D. D. Harrison, and J. D. Morgan. "Field Studies of Wildlife Exposed to TCDD Contaminated Soils," U. S. Air Force Arament Laboratory, Eglin Air Force Base, FL, AFATL-TR-75–49 (1975).
6. Bolton, L. "Seveso Dioxin: No Solution in Sight," *Chem. Eng.* 85:78 (1978).
7. Crummett, W. B., and R. H. Stehl. "Determination of Chlorinated

Dibenzo-p-dioxins and Dibenzofurans in Various Materials," *Environ. Health Pers.* 5:15–25 (1973).

8. Di Domenico, A., G. Viviano, and G. Zapponi. "Environmental Persistence of 2,3,7,8-TCDD at Seveso." in *Chlorinated Dioxins and Related Compounds*, O. Hutzinger, Ed. (New York: Pergamon Press, 1982).

9. Weber, W. J., Jr., T. C. Voice, M. Pirbazari, G. E. Hunt, and D. M. Ulanoff. "Sorption of Hydrophobic Compounds by Sediments, Soils and Suspended Solids. II. Sorbent Evaluation Studies," *Water Res.* 17:1433–1452 (1983).

10. Hassett, J. J., J. C. Means, W. L. Banwart, S. G. Wood, S. Ali, and A. Khan. "Sorption of Dibenzothiophene by Soils and Sediments," *J. Environ. Qual.* 9:184–186 (1980).

11. Moreale, A., and R. Van Bladel. "Adsorption and Migration of Lindane (1,2,3,4,5,6-Hexachlorocyclohexane) in Soil," *Parasitica* 34:233–255 (1978).

12. Dragun, J. Personal communication. E. C. Jordan Co., Southfield, MI (1984).

13. Leo, A. J. "Calculation of Partition Coefficients Useful in Evaluation of the Relative Hazards of Various Chemicals in the Environment," G. D. Keith and D. F. Konasewish, eds. in *Structure-Activity Correlations in Studies of Toxicity and Bioconcentration with Aquatic Organisms.* (International Joint Commission, Windsor, Ontario. NTISPB275670 1975).

14. Hansch, C., and A. Leo. *Substituent Constants for Correlation Analysis in Chemistry and Biology.* (New York: John Wiley & Sons, Inc., 1979).

15. "Ambient Water Quality Criteria for 2,3,7,8-Tetrachlorodibenzo-p-dioxin," Office of Water Regulations and Standards, Washington, D. C., U. S. EPA Report–440/5–84–007.

16. Means, J. C., S. G. Wood, J. J. Hassett, and W. L. Banwart. "Sorption of Aromatic Hydrocarbons by Sediment and Soils," *Environ. Sci. Technol.* 14:1524–1528 (1980).

17. Brown, D. S., and E. W. Flagg. "Empirical Prediction of Organic Pollutant Sorption in Natural Sediments," *J. Environ. Qual.* 10:382–386 (1981).

18. Schwarzenbach, R. P., and J. Westall. "Transport of Nonpolar Organic Compounds from Surface Water to Groundwater, Laboratory Sorption Studies," *Environ. Sci. Technol.* 15:1360–1367 (1981).

19. Karickhoff, S. W., D. S. Brown, T. A. Scott, and A. Trudy. "Sorption of Hydrophobic Pollutants on Natural Sediments," *Water Res.* 13:241–248 (1979).

20. Senesi, N., C. Testimi, and D. Mette. "Binding of Chlorophenoxy Alkonoic Herbicides From Aqueous Solution by Soil Humic Acid," in *Environmental Contamination* (Imperial College, London, 1984), pp. 96–101.

21. *Dioxin: Quandary for the '80s* (St. Louis, Missouri: St. Louis Post-Dispatch, 1983).

22. McLean, E. O. "Recommended pH and Lime Requirement Tests," in *Recommended Chemical Soil Test Procedures for the North Central Region*. North Dakota Agricultural Experiment Station, NCR Publ. No. 221 (rev.). (Fargo, ND: 1980).
23. Watson, M. E. "Soil Testing Procedures," Ohio Agricultural Research and Development Center, Research Extension Analytical Laboratory, Wooster, OH. (1978).
24. Allen, S. E., H. M. Grimshaw, J. A. Parkinsen, and C. Quarmby. *Chemical Analysis of Ecological Materials*. (New York: John Wiley and Sons, Inc., 1974).
25. *Standard Method for Particle-Size Analysis of Soils*, ASTM D422-63, Washington, DC. ASTM D422-63. (Washington D.C.: American Society of Testing and Materials, 1972).
26. Garrett, B. C., D. R. Jackson, W. E. Schwartz, and J. S. Warner. "Solid Waste Leaching Procedure," Office of Solid Waste and Emergency Response, U. S. EPA Technical Resource Document for Public Comment SW-924 (1984).
27. Ham, R., M. A. Anderson, R. Stegmann, and R. Stanforth. "Background Study on the Development of a Standard Leaching Test," U. S. EPA Report-600/2-79-109 (1979).
28. "Determination of 2,3,7,8-TCDD in Soil and Sediment," Region VII Laboratory, U. S. Environmental Protection Agency (September revision, 1983).
29. Warner, J. S. "Chemical Characterization of Marine Samples," Publication 4307, American Petroleum Institute (1978).
30. Karickhoff, S. W. "Organic Pollutant Sorption in Aquatic Systems," *J. Hydraul. Eng.* 110:707-735 (1984).
31. Jackson, D. R., M. H. Roulier, H. M. Grotta, S. W. Rust, J. S. Warner, M. F. Arthur, and F. L. DeRoos. "Leaching Potential of 2,3,7,8-TCDD in Contaminated Soils," in *Land Disposal of Hazardous Waste—Proceedings of the Eleventh Annual Research Symposium*, Cincinnati, Ohio. U. S. EPA Report-600/9-85-013 (1985) pp. 153-168.

CHAPTER 14

The Presence of Hexachloroxanthene at Missouri Dioxin Sites

Tenkasi S. Viswanathan and Robert D. Kleopfer

INTRODUCTION

Polychlorinated dibenzodioxins and dibenzofurans are a group of highly stable compounds present or produced as a contaminant in chlorinated phenols, phenoxyacids, chlorobenzenes, and the emissions and fly ash from power plants and municipal refuse incinerators. The high toxicity of these compounds, coupled with their apparent widespread distribution in the environment, has led to public concern and intense scientific research on these compounds. The state of Missouri has more than 40 hazardous waste sites containing the toxic compound 2,3,7,8-tetrachlorodibenzo-p-dioxin (TCDD). Most of these contaminated areas are related to the waste disposal activities of a 2,4,5-trichlorophenol and hexachlorophene producer (NEPACCO) in southwestern Missouri.

Since hexachlorophene production requires the intermediate production of 2,4,5-trichlorophenol, the waste materials from the hexachlorophene manufacturing process often contain two chlorinated by-products, 1,2,3,5,7,8-hexachloro-9H-xanthene (HCX) and TCDD. We have previously observed the presence of HCX in environmental samples containing TCDD, and there are published reports[1] about its presence in the contaminated soil from the Shenandoah horse arena in eastern Missouri.

Chlorinated 9H-xanthenes are nearly planar molecules like dibenzodiox-

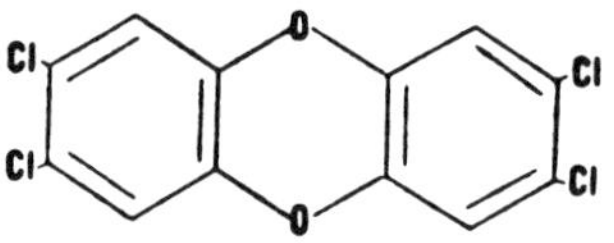

2,3,7,8-tetrachloro
dibenzo-p-dioxin
(TCDD)

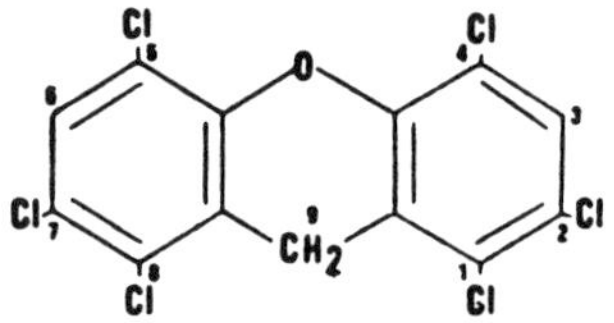

1,2,4,5,7,8, Hexachloro(9-H)xanthene
(HCX)

Figure 1. Structural formulas of 2,3,7,8-tetrachlorodibenzo-p-dioxin and 1,2,4,5,7,8-hexachloro-9H-xanthene.

ins (Figure 1) and dibenzofurans,[2] and they are expected to be quite toxic if the number and position of chlorines are similar to the toxic PCDD and PCDF isomers. Consequently, an effort was made to determine the extent of the known hexachloroxanthene contamination of sites known to be contaminated with significant quantities of TCDD. Since HCX is a unique by-product in the manufacture of hexachlorophene it was also hoped that the present study would indicate, at least in part, the source of the environmental pollution associated with dioxin sites.

EXPERIMENTAL

Materials

The analytical standard of 1,2,4,5,7,8-hexachloroxanthene (> 99% pure) used in quantitation was kindly provided by Dr. Fred Hileman of Monsanto Company. The standard of HCX used for toxicity determination and the unlabeled and labeled TCDD were provided by the U. S. EPA Quality Assurance Materials Bank in Research Triangle Park, North Carolina. Other equipment, reagents, and chemicals were obtained from commercial sources.

Analysis of Solid Samples for TCDD and HCX

Soil samples (10 g) were spiked with $^{13}C_{12}$-2,3,7,8-TCDD and $^{37}Cl_4$-2,3,7,8-TCDD, extracted with methanol/hexane in a jar, cleaned up using chromatography on silica gel, alumina and carbpack columns and analyzed on high resolution gas chromatography-low resolution mass spectrometry as described in detail elsewhere.[3,4]

Analysis of Liquid Samples for TCDD and HCX

The liquid samples generally contained ppm (mg/kg) levels of TCDD and HCX. About 1 g of the liquid was weighed accurately and diluted to 10 mL in a volumetric flask. An additional elevenfold dilution of the sample was achieved by diluting 10 μL of this stock solution with 100 μL of a solution normally containing $^{13}C_{12}$-2,3,7,8-TCDD (0.5 μg/mL) and $^{37}Cl_4$-2,3,7,8-TCDD (0.1 μg/mL). This extract, containing about 1 mg of the sample, often contained too many matrix interferences for the direct analysis of HCX and TCDD by GC/MS. Consequently, a partial cleanup was achieved using chromatography on silica gel and alumina columns[5] prior to GC/MS analysis.

GC/MS Analysis of TCDD and HCX

These analyses were performed using a Finnigan 4000 GC/MS/DS system. Hexachloroxanthene and TCDD were chromatographed on a 30M DB-5 (i.d. = 0.32 mm; film thickness 0.25 μm) capillary column under the following conditions: initial phase, isothermal at 70°C for 1 min; first temperature program, to 190° at 25°C/min; second temperature program, to 210° at 2°C/min; third temperature program, to 295° at 10°C/min. TCDD elutes during the second temperature program with a retention time of 980 ± 7 sec, while HCX elutes during the third temperature program with a retention time of 1411 ± 5 sec under these conditions.

The mass spectrometer was operated in the selected ion monitoring mode scanning ions of m/z 257, 320, 322, 328, 332 and 334 during the first 1100 sec and the ions of m/z 351, 353, 355, and 385 to 390 during the next 500 sec using a scan rate of 1 sec/scan. Positive ions produced by electron impact (70 eV) were monitored.

Quantitation of TCDD

The qualitative identification and quantitation of 2,3,7,8-TCDD in environmental samples was performed following the U. S. EPA region VII protocol.[3,4] The concentration of 2,3,7,8-TCDD in μg/kg (ppb) in samples was calculated using the following equations.

$$[\text{TCDD}] = \frac{(R_{320} + R_{322})}{(R_{332} + R_{334})} \circ \frac{W^{13}C - \text{TCDD}}{\text{RRF}_D} \circ \frac{1}{Ws} \tag{1}$$

and

$$[\text{RRF}_D] = \frac{(R_{320} + R_{322})\ std}{(R_{332} + R_{334})\ std} \circ \frac{W^{13}C - \text{TCDD, std}}{W - \text{TCDD,\quad std}} \tag{2}$$

where R_{320}, R_{322}, R_{332} and R_{334} are GC area or height responses for the ions of m/z 320 and 322 for unlabeled TCDD and m/z 332 and 334 for the $^{13}C_{12}$-labeled TCDD
$W^{13}C$-TCDD is the amount of $^{13}C_{12}$-2,3,7,8-TCDD in ng spiked to Ws g of sample prior to extraction.

A multipoint calibration curve[5] was obtained by determining the response factor (RRF_D) of 2,3,7,8-TCDD using five calibration solutions containing varying concentrations of 2,3,7,8-TCDD (0.2 ng/μL to 40 ng/μL) relative to a fixed concentration of $^{13}C_{12}$-2,3,7,8-TCDD (1.0 ng/μL).

Quantitation of HCX

The hexachloroxanthene was identified by the coelution at the correct GC retention time of two chlorine clusters from this compound: the M and M-H cluster in the mass range 385–390 and the M-Cl cluster (m/z 351, 353, and 355) with the basepeak as m/z 353 (Figure 2). The M-Cl cluster, which was relatively free of interferences from other matrix components, was used for quantitation, while the M and M-H clusters were used for confirming the qualitative presence of HCX. The amount of hexachloroxanthene in samples was calculated as follows:

$$[HCX] \ = \ R \circ [TCDD] \tag{3}$$

$$R \ = \ \frac{[HCX]}{[TCDD]} \ = \ \frac{(A_{351} + A_{353} + A_{355})}{(A_{320} + A_{322})} \circ \frac{1}{RRF_x} \tag{4}$$

and

$$[RRF_X] \ = \ \frac{(A_{351} + A_{353} + A_{355})\, std}{(A_{320} + A_{322})std} \circ \frac{W-TCDD, std}{W-HCX, \ std} \tag{5}$$

where A_{351}, A_{353}, and A_{355} are the GC area responses for the ions of m/z 351, 353 and 355 for hexachloroxanthene
A_{320} and A_{322} are similiar parameters for TCDD
RRF_x is the relative response factor for hexachloroxanthene defined with respect to 2,3,7,8-TCDD

The RRF_x was determined using multiple injections of a calibration solution that contained 2,3,7,8-TCDD (0.67 ng/μL), $^{13}C_{12}$-2,3,7,8-TCDD (0.67ng/μL), $^{37}Cl_4$-2,3,7,8-TCDD (0.080 ng/μL) and HCX (1.67) ng/μL). The mean RRF_x from four determinations was 0.458 with a standard deviation of 0.043. The RRF_D (relative response factor of 2,3,7,8-TCDD) was very close to unity. Consequently, the ions of m/z 332 and 334 from the $^{13}C_{12}$-2,3,7,8-TCDD were used, instead of ions of m/z 320 and 322, to determine the amount or detection limits of HCX in samples that did not contain any 2,3,7,8-TCDD. The GC retention times of TCDD and HCX were determined on a daily basis using the above calibration standard solution and control limits of $\pm$ 10% and $\pm$ 25% were placed on RRF_D and RRF_x,

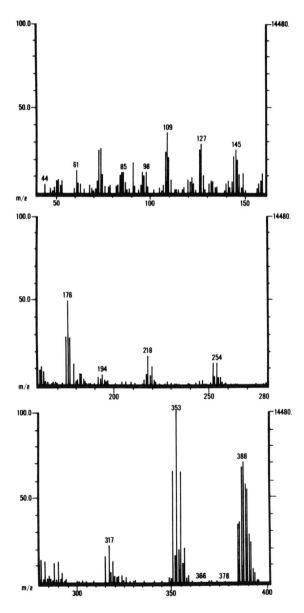

Figure 2. The mass spectrum of 1,2,4,5,7,8-hexachloro-9-H-xanthene.

Table 1.   Hexachloroxanthene and TCDD Contents of Samples from the Denney Farm.

No. of Samples	No. of GC/MS Analyses	TCDD Content		Hexachloroxanthene Content		[HCX]/[TCDD]	
		ppm	R.S.D.%[a]	ppm	R.S.D.%	Ratio	R.S.D.%
6	1	312	19.0	582	30	1.86	24
6	1	236	11.0	509	24	2.21	31
5	1	198	23.0	378	22	1.99	35
1	1	132		357		2.70	
2	1	46	6.1	127	8.1	2.77	14
1	3	70	17.0	173	26	2.45	10
1	2	229	9.9	452	25	1.96	16
1	2	1177	4.0	1193	44	1.02	48
1	3	595	13.0	1760	23	2.96	14
1	3	297	4.7	613	16	2.06	12
1	4	9648		28,600		2.54	27
1	4	619	9.2	1,663	28	2.66	22

[a]R.S.D.%, relative standard deviation percent.

respectively for correct qualitative and quantitative identification of these compounds.

RESULTS

The TCDD and HCX contents of waste oil, "still bottom," "filter clay," and soil samples taken from the James Denney farm in rural Barry County about 7 miles from Verona, Missouri are shown in Table 1. The samples listed in the first three lines were viscous, heterogeneous liquid samples that were quite similar in the nature of their matrix. A tenfold dilution of these liquids in toluene often led to light brown solutions with a black precipitate at the bottom. Both the precipitate and the supernatant contained TCDD and HCX. Representative samples of the two phases were used for further dilution to minimize error in sampling and subsampling.

Samples from this site contained various amounts of TCDD ranging from 46 ppm to 9648 ppm (the highest ever reported), and the concentration of HCX ranged from 127 ppm to 28,600 ppm. The ratio (Equation 4) of TCDD and HCX concentrations in 27 samples analyzed from this site averaged to 2.16, with a standard deviation of 0.58.

The TCDD and HCX contents of soil samples from some of the other dioxin sites are shown in Table 2. Samples with various concentrations of TCDD were chosen for this investigation, and it reveals the presence of HCX in all of them except one. This site, Economy Products, represents a known 2,4,5-T herbicide formulation site, while most of the other sites were recipients of waste originating from the NEPACCO facility. The last column in Table 2 shows that the [HCX]/[TCDD] ratio varies considerably,

Table 2.   Distribution of Hexachloroxanthene in Missouri Dioxin Sites.

Site Name	[TCDD] μg/kg	[HCX] μg/kg	[HCX] / [TCDD]
Denney Farm	46×10^3 to 9.6×10^6	1.3×10^3 to 29×10^6	2.16
Economy Products	1500	ND[a]	
Sullins	780	72	0.092
Minker	558	239	0.43
Quail Run	547	179	0.33
Cashel	153	13	0.083
Sontag Road	126	8,396	66.6
Stout	100	49	.49
Bull Moose	21	14	.67
Overnite Transport	9.2	13	1.4
Bliss Tank Site	5.5	5.2	0.94
Hamill Transfer Co.	2.6	6.4	2.5
Piazza Road	2.2	3.8	1.7

[a]ND, none detected.

and that values above and below the mean of 2.16 observed in Denney Farm are found for other sites.

DISCUSSION

The presence of hexachloroxanthene in most of the Missouri dioxin sites, although at levels that vary considerably from site to site and also within the site, implicates a hexachlorophene producer as the source of dioxin. A brief description of the commercial process[6] for the production of hexachlorophene (Figure 3) is useful for this discussion. The first step in the manufacture of hexachlorophene involves the hydrolysis to 2,4,5-trichlorophenol of 1,2,4,5-tetrachlorobenzene, which is the principal isomer produced by rechlorination of o-dichlorobenzene. The hydrolysis with caustic soda is a batch reaction conducted at elevated temperatures (160°C to 300°C) and pressures (20 to 1500 psi) depending on the solvent (ethylene glycol, propylene glycol, methanol, or water) used for the reaction. The high temperature, high pressure and the strongly alkaline conditions promote a continuation of the reaction, in which 2,4,5-trichlorophenol combines with itself to form 2,3,7,8-TCDD.

Hexachlorophene is formed by reacting two molecules of 2,4,5-trichlorophenol with one molecule of formaldehyde at elevated temperatures in the presence of an acid catalyst. The patented processes differ in conditions, and one process uses oleum as the catalyst and concentrated sulfuric acid is recovered as a by-product. Under these dehydrating conditions, hexachlorophene loses a molecule of water to produce hexachloroxanthene, with yields that may be highly variable depending on the reaction conditions.

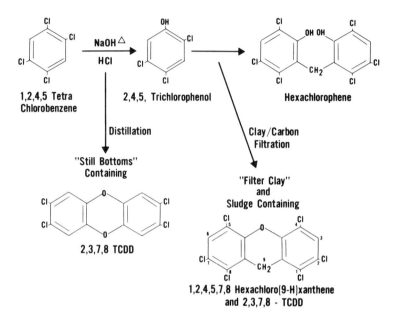

Figure 3. Hazardous waste produced in the manufacture of hexachlorophene.

Hazardous Wastes in Hexachlorophene Production

Hexachlorophene is used as an active constituent in several cosmetics and some over the counter and prescription drugs. Pharmaceutical-grade hexachlorophene requires purification of 2,4,5-trichlorophenol prior to its reaction with formaldehyde. This is often achieved using distillation, leading to a waste referred to as "still bottoms" that contains large amounts of dioxin, but no hexachloroxanthene. A second waste product containing dioxin is produced when the mixture, after completion of the reaction with formaldehyde, is treated with clay or activated carbon or a combination filter to remove colored impurities. This "filter clay" waste is expected to contain large amounts of hexachloroxanthene and hexachlorophene and moderate amounts of 2,3,7,8-TCDD. The "still bottom" and "filter clay" are dioxin-containing wastes that were distributed in the environment.

The [HXC]/[TCDD] ratio of the waste (Table 2) is thus expected to be highly variable, depending not only on the differences in reaction conditions from batch to batch, but also on the still bottom filter clay ratio in the waste. A possible additional variation to this waste product was introduced by the practices of the Bliss Salvage Oil Company, which mixed waste oil from other sources with the waste oil from a hexachlorophene producer in storage tanks near St. Louis, Missouri. Recent studies have also shown considerable differences in the volatitility and the relative mobilities in the soil of hexachloroxanthene and 2,3,7,8-TCDD.[7] It is conceivable that the spontaneous degradation rates of the two compounds in the soil are also quite different.

In conclusion, it appears that the Missouri dioxin sites are also contaminated with hexachloroxanthene at levels comparable to that of dioxin. As a corollary, it appears that the hexachlorophene producer, at least in part, was the source of the dioxin. Preliminary toxicological studies on HCX[8] indicate, however, that the HCX contamination does not pose any additional danger to the public in excess of the assessed risk from the known dioxin levels in these sites.

ACKNOWLEDGMENT

We would like to acknowledge the help of Dr. Fred Hileman of Monsanto, who kindly provided an analytical standard of HCX; Dr. Kimbrough of the Center for Disease Control and Dr. Gierthy of New York State Department of Health for the preliminary results on the toxicological information on this compound; and Dr. Cliff Kirchmer, Dr. Kenneth Yue, Kathy Bogges, Allen Smith, William Bunn and George Yeargens from our facility, for their contribution to this project.

REFERENCES

1. Buser, H. R., and C. Rappe, "High-Resolution Gas Chromatography of the 22 Tetrachlorodibenzo-p-dioxin Isomers," *Anal. Chem.* 52 (14):2257–2262 (1980).
2. Soderholm, M., U. Sonnerstram, R. Norrestam, and T.-B. Palm. "Structural Studies of Polychlorinated Hydrocarbon. II Hexachloroxanthene and Hexachloroxanthone," *Acta Cryst.* B32:3013–3018 (1976).
3. Kleopfer, R. D., K. T. Yue, and W. W. Bunn. "Determination of 2,3,7,8-Tetrachlorodibenzo-p-dioxin in Soil," in *Chlorinated Dioxin and Dibenzofurans in the Total Environment II*, L. H. Keith, C. Rappe, and G. Choudhary, Eds. (Stoneham, MA: Butterworth Publishers, 1985), pp. 367–375.
4. "Determination of 2,3,7,8-TCDD in Soil and Sediment," U. S. EPA Region VII Laboratory protocol (Revised September 1983).
5. "Dioxin Analysis — Soil/Sediment Matrix — Multi Concentration," U. S. EPA document IFB-WA-A002, statement of work, pp. 1–66.
6. Esposito, M. D., T. O. Tiernan, and F. E. Dryden. "Dioxins," U. S. EPA Report # EPA-600/2-80-197 (1980) pp. 89–93 and pp. 106–108.
7. Freeman, R. A., J. M. Schroy, and F. E. Hileman. "Environmental

Mobility of 2,3,7,8-TCDD and Companion Chemicals in a Roadway Soil Matrix," Chapter 12, this volume.
8. Gierthy, J. F. and D. Crane. "In Vitro Bioassay for Dioxin Like Compounds," Chapter 19, this volume.

CHAPTER 15

Disposition of Tetrachlorodibenzo-p-Dioxin in Soil

J. Palausky, J. J. Harwood, T. E. Clevenger, S. Kapila, and A. F. Yanders

INTRODUCTION

The contamination of many sites in Missouri with 2,3,7,8-tetrachlorodibenzo-p-dioxin (TCDD) is without doubt the single most publicized and pressing environmental issue in Missouri. This contamination is centered around the past improper disposal of waste from a hexachlorophene production facility at Verona, Missouri.[1] Due to its extremely high toxicity to a number of mammals, TCDD has drawn considerable attention and has been the focus of extensive research in many parts of the world.[2-4] Large data bases for the compound in terms of toxicity and incidences of environmental contamination are available[5,6]; however, certain important basic data are still required for realistic assessment of hazards posed by this compound in soil to human health and environment. Furthermore, there exists a need for fundamental environmental physiochemical studies directed toward gaining an understanding of the chemical processes and fates of TCDD and similar molecules in the environment.

The detection of TCDD residues in fish and sediment samples downstream from contaminated sites raises a special concern for the potential contamination of ground and surface water. This is particularly significant due to the extremely low solubility of TCDD in water. Clearly, an understanding of bioavailability of particle-bound TCDD and related molecules is required. Very little information is currently available in the literature on TCDD's interaction with sediments. Knowledge of a contaminant's equilibrium partition coefficient (K_{oc}) is an important part of environmental hazard assessment. However, the recent findings that some compounds may require as long as six months to reach equilibrium with sediment-water phases[7] suggests that the K_{oc} values may not adequately predict the bioavailability of TCDD through time. Several other variables may affect TCDD bioavailability, and these could include sediment mass, particle size distribution, concentration, and the nature of dissolved as well as colloidal organic species. Furthermore, an important parameter for the TCDD contamination in Missouri is the role of the dispersing agent and its effect on binding, mobility and volatilization from contaminated soil. A review of published data shows that TCDD is more tightly bound in soils having high organic content and that the TCDD becomes more tightly bound to soil components over time.[8,9] Loss of TCDD from soil through biodegradation and transport other than erosion has generally been considered low.[10] However, a recent model presented by Freeman and Schroy predicts that 90% of the TCDD at Times Beach has been lost through volatilization and subsequent photolysis during the ten years since initial contamination; furthermore, 57% of the loss was estimated to have occurred during the first summer.[11,12] The loss of TCDD through volatilization has also been reported by Nash and Beall.[13] It is apparent that the validity of the model can have a significant effect on the assessment of TCDD contamination in Missouri and similar contamination elsewhere. Similar models to predict the volatilization losses for pesticides with low water solubility and low vapor pressure have been developed by Jury and co-workers.[14,15] These models have been used for describing the volatilization losses from soil surface. However, due to the lack of direct experimental data on vapor phase movement of TCDD in soil beds, a number of assumptions have been made in formulating the models. The two important parameters governing the loss from the soil bed are the initial depth profile, which is dependent on the carrier solvents (copollutants), and vapor phase diffusion within the bed. These two aspects were experimentally explored in the present study, with the objectives of monitoring the effects of carrier medium on depth of migration of TCDD in soil, and determining the extent of vapor phase diffusion in TCDD in soil.

EXPERIMENTAL

Soil Column Studies

All the experiments for the study were conducted with the representative uncontaminated soil from Times Beach, Missouri. The soil was air dried

Table 1.   Soil Characteristics.

Soil Sample	% Sand	% Silt	% Clay	OM	Texture
Times Beach	11.4	55.2	33.4	2.0	Silty clay loam
ETSRC Columbia	15.2	57.4	27.4	1.4	Silty clay loam

and passed through a 60-mesh sieve. The sieved soil was characterized for typical soil parameters listed in Table 1. Then 600 g of the sieved soil was packed in 15-cm × 7.85-cm i.d. aluminum columns. All columns were packed to the same density, which was 80% of the natural soil density at Times Beach.[16] For vapor phase diffusion studies, known amounts of TCDD were introduced into the soil column with a microsyringe. In order to achieve better precision, an aluminum syringe-positioning block was employed in later studies. For these experiments 100 ng of TCDD was introduced into the soil column in 5 µL of toluene. Spike solution was introduced slowly over a 30–sec period, to minimize the spread of TCDD. The schematic of the soil spiking system is shown in Figure 1. The columns prepared in the described manner were incubated in temperature-controlled chambers set at 0°,10°,20°,30°,40°, and 100°C. The temperatures of the chambers were monitored daily. The columns were sampled after 30, 60, and 90 days. The columns at 100° were sampled after 18 and 38 days.

The soil columns for studies dealing with TCDD mobility in various dispersing mediums were conducted using six solvents ranging from acetone/water to o-xylene. For these experiments, a low dead-volume shower-

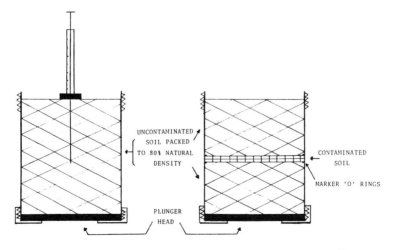

Figure 1.   Schematic of soil column used for 2,3,7,8-TCDD migration studies.

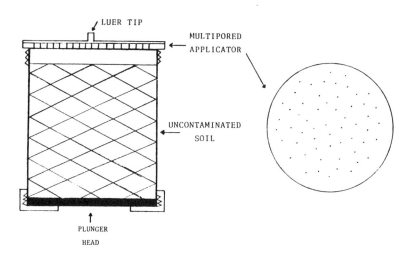

Figure 2. Schematic of soil columns used for TCDD-solvent migration studies.

head spreader was designed to fit over the aluminum soil columns, and 900 ng of TCDD was introduced onto the soil column in 9 mL of the appropriate solvents. The schematic of the shower-head application is shown in Figure 2. The solvent amounts used corresponded to 0.4 gal/yd^2, which is within the waste-oil loadings of 0.1 – 1.0 gal/yd^2 used for dust control.[11] The six solvents were selected to cover a wide range of polarity, volatility, aromaticity, and TCDD solubility. Since o-xylene has been used in hexachlorophene synthesis and thus would have been a likely constituent of the wastes used for dust control, it also was added to the list.

SAMPLING

In order to make an accurate assessment of TCDD mobility with solvents and vapor phase diffusion, a system capable of sampling small sections of the soil column was constructed at the Science Instrument Shop, University of Missouri, Columbia. The schematic of the sampler is shown in Figure 3. The system allowed sampling soil sections down to 1 mm thickness. The system allowed precise sampling of the soil sections; reproducibility of the soil sections obtained is given in Table 2.

ANALYSIS

The analytical scheme consisted of the extraction of soil samples with a 1:1 mixture of cyclohexane and methylene chloride, cleanup of extract with acidic, basic and neutral silica-gel, followed by analysis by gas chromatography with an electron capture detector (GC-EC) or, in selected cases, by

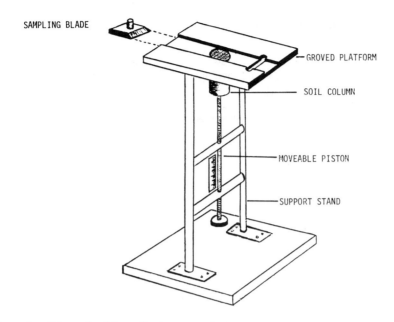

Figure 3. Schematic of the soil column sampler.

gas chromatography mass spectrometry (GC-MS). C–13 and Cl–37 labeled TCDD or p, p-DDD were used as the internal standards. Extensive validation studies were performed prior to the soil sample analysis of experimental soil samples.

The initial determinations were carried out with a Tracor Model 220 gas chromatograph equipped with an electron capture detector. The chromatographic column consisted of a 30-m DB–5 (J&W Scientific, Inc.) fused-silica capillary or a 180-cm × 0.2-cm i.d. borosilicate glass tube packed with 1.5% OV–17 + 2% QF–1 on chromsorb W(H. P.) 80–100 mesh. The confirmatory analyses were carried out with a Finnigan Model 30B GC-MS; a 30–m DB–5 fused-silica capillary column was used for these determinations. The mass spectrometer was operated in the multiple ion monitoring modes; ions monitored were m/e 257, 320, 322, 328, 332, and 334.

Table 2. Soil Sampling Precision.

Soil Section Width (mm)	Soil Wt/Section (g)
1	5.02
1	5.45
1	5.25
2	10.10
2	10.17
2	10.20
	% RSD = 1.5

RESULTS AND DISCUSSION

The models dealing with the fate of trace organic compounds in soil are based on a chemical mass balance approach and are expressed by the following general equation.[14]

$$C_T = \alpha C_S + \beta C_L + \gamma C_G$$

where C_T = Total Concentration
C_S = Concentration in the solid phase
C_L = Concentration in the liquid phase
C_G = Concentration in the vapor phase
α, β and γ are proportionality constants

For compounds with low water solubility in areas where particle movement is slow, migration and loss in the vapor phase can be significant.

Relatively rapid losses through volatilization from soil and other surfaces have been demonstrated for a number of compounds with low water solubility and vapor pressures.[16-17] The rate of these losses has been shown to be moderated by the boundary layer conditions.[15] However, direct evidence of volatilization and subsequent migration through vapor phase diffusion or convection has been more limited. The macropore spaces in soil constitute a significant portion (16–32%) of the total soil column volume,[18] providing significant volume for vapor phase migration to occur within the soil column. Parts of this study were designed to monitor the extent of this vapor phase migration. The concentration of TCDD was determined in the horizontal soil sections and plotted against the distance from the point of application. The TCDD concentration profiles obtained after 30–day incubation at various temperatures are shown in Figure 4. The TCDD concentration profile obtained after an initial equilibration period of 18 hr is also represented. This profile shows that the TCDD application procedure resulted in a small but measurable degree of band-broadening with the spiking solvent, toluene.

The chemical transport through soil has been expressed by Jury et al.[15] as a solute flux equation.

$$J_S = -D_G(\gamma)(\partial C_G/\partial Z) - D_L(\beta)(\partial C_L/\partial Z) + J_w C_L$$

where J_S is the solute flux in amt/cm^2/sec
D_G is the diffusion coefficient in gas phase
D_L is the diffusion coefficient in liquid phase
β and γ are proportionately constant
Z is the migration distance

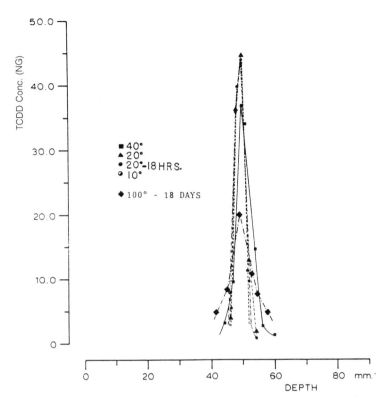

Figure 4. Concentration profiles of 2,3,7,8-TCDD in soil column incubated at various temperatures.

Under dry conditions in a soil column, the migration would be predominantly in the vapor phase. An expression for the change in vapor phase fraction in the soil column has been derived by Freeman and Schroy.[11]

$$\frac{Ca}{\partial t} = h\,Dab\,\frac{\partial^2 Ca}{\partial Z^2} + \frac{R}{\xi} + R_1$$

where CA is concentration of solute in air
  h is the hindrance factor
  Dab is the diffusivity of solute in air
  R is the rate of volatilization into the gas phase
  R_1 is the rate of disappearance due to chemical or biochemical processes
  ξ is the void fraction in soil column

In the present study, since soil columns were maintained under closed conditions and under constant temperature, it can be assumed that any

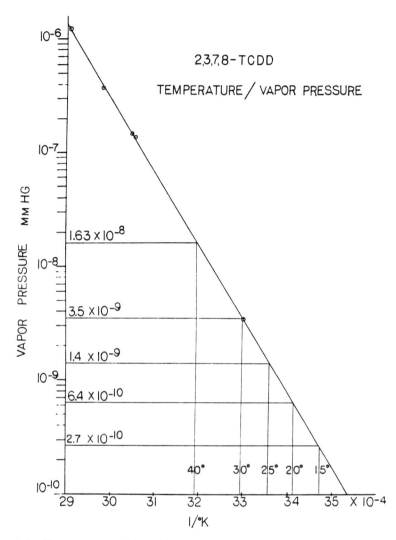

Figure 5. Vapor pressure of 2,3,7,8-TCDD.

additional migration of the TCDD zone would be determined primarily by movement in the vapor phase, either through diffusion or through convection with soil moisture. The concentration profile curves illustrate that no noticeable change in the TCDD concentration profile occurred after 30 days incubation at temperatures ranging from 0° to 20°C; however, a measurable change was observed at 40°C. These results fit well with the vapor pressure data for TCDD (Figure 5). The vapor pressure of TCDD at 40°C is approximately one order of magnitude higher than at 25°C. These results are consistent with the volatilization model of Freeman and Schroy,[11] which

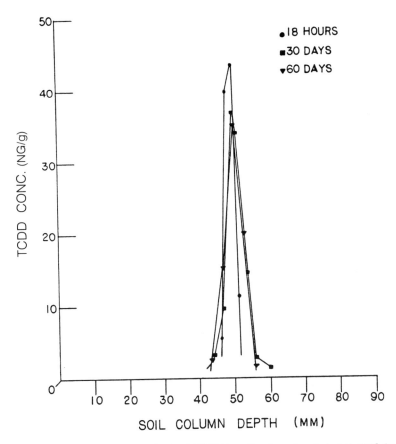

Figure 6.   Concentration profile of 2,3,7,8-TCDD in soil column incubated at 40°C for 30 and 60 days.

predicted that the volatilization losses after the initial application at Times Beach occurred primarily during the summer months, when the temperature rises as high as 30°C down to a depth of 2 cm. Similar concentration profiles were obtained after 60–day incubation periods. The 60–day 40°C profile was similar to the one obtained after 30 days (Figure 6). The lack of migration after the initial 30 days can be attributed in part to the loss of moisture from the soil, which would lower the migration rate. The gravimetric determination of moisture content in the soil samples showed an average moisture content of 3.6% prior to incubation. Moisture content of soil after incubation at 40°C for 60 days was found to be 2.2%. A rough estimate of diffusion flux for TCDD in air made from the 40°C profile was found to be approximately 30% higher than the value given by Freeman and Schroy.[11]

An important factor in the fate of TCDD at the contaminated site is the

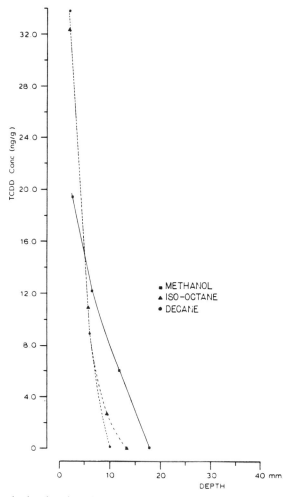

Figure 7. Graph showing the migration of 2,3,7,8-TCDD with methanol, isooctane and decane.

initial depth of migration of the TCDD in the soil. This parameter will not only affect the volatilization losses, but also the losses through erosion. It can be assumed that greater depths of initial contamination would lead to slower losses of the contaminant. The dispersing medium (solvent) would be the controlling parameter in the initial disposition and binding of TCDD in the soil, thus affecting its environmental fate. One of the objectives of our study was to monitor the effect of carrier copollutant on the extent of penetration into the soil. The results obtained are shown as TCDD concentration profiles in Figures 7 and 8. Little or no migration was observed with the acetone/water mixture, and more than 90% of TCDD applied with this

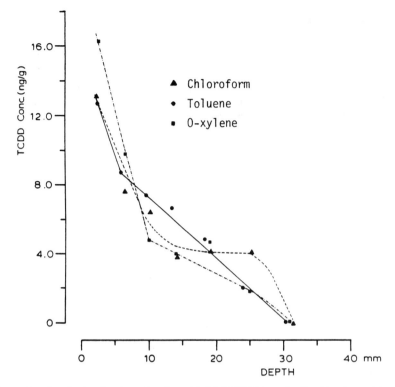

Figure 8.   Graph showing the migration of 2,3,7,8-TCDD in soil with chloroform, toluene, and o-xylene.

solvent mixture was confined to the first 1-mm section. The migration of TCDD, with organic solvents, was found to be slowest with the saturated hydrocarbons; 90% of the TCDD was confined to the first 5 mm of the column. The migration was somewhat greater with methanol; in this case approximately 20% of the applied TCDD moved to a depth greater than 10 mm. The extent of migration was found to be highest with aromatic solvents and chloroform. With these solvents, approximately 50% of TCDD moved to depths of more than 20 mm. In general, the extent of migration was related to the solubility of TCDD in the solvent; however, a direct correlation between solubility and extent of migration was not found. For example, the solubility of TCDD is approximately an order of magnitude higher in xylene than chloroform,[12] but the movement was found to be somewhat greater with chloroform. These results indicate that solvent volatility, viscosity and interaction of soil organic matter as well as other absorption sites also play a significant role.

ACKNOWLEDGMENTS

The study was supported in part by a grant from the U. S. Department of the Interior, administered through the Missouri Water Resources Center, University of Missouri. The assistance of Science Instrument Shop, University of Missouri, Columbia, is acknowledged. The graphical assistance of Mr. Carl Orazio is also acknowledged.

REFERENCES

1. "Final Report of the Missouri Dioxin Task Force," submitted to Governor Christopher S. Bond (October 31, 1983).
2. Slater, R. W. "Opening Address: 6th International Symposium on Chlorinated Dioxins and Related Compounds," *Chemosphere* 14:(6/7):575–579 (1985).
3. Young, A. L. "Dibenzo-p-dioxin: Public Concerns and National Science Policy," paper presented at the 189th Meeting of the American Chemical Society, Miami Beach, FL, April 29, 1985.
4. Somers, E., and V. M. Douglas. "Dioxins and Related Compounds as Issues of International Concern," in *Human and Environmental Risks of Chlorinated Dioxins and Related Compounds* (New York: Plenum Press, 1985).
5. Hutzinger, O., M. J. Blumich, M. van den Berg, and K. Olie. "Source and Fate of PCDDs and PCDFs: An Overview," *Chemosphere* 14(6/7):581–600 (1985).
6. Hutzinger, O. "Concluding Summary—Source Determination and Environmental Fate," *Chemosphere* 14(6/7):645–647 (1985).
7. Karickoff, S. W. "Sorption Kinetics of Hydrophobic Pollutants in Natural Sediments," paper presented at the 186th meeting of the American Chemical Society, 1983.
8. Kearney, P. C., E. A. Woolson, and C. P. Ellington. "Persistence and Metabolism of Chlorodioxins in Soil," *Environ. Sci. Technol.* 6(12):1017–1019 (1972).
9. Matsumura, F., and J. H. Benzet. "Studies on Bioaccumulation and Microbial Degradation of 2,3,7,8 Tetrachlorodibenzo-p-dioxin," *Environ. Health Pers.* 5:253–258 (1973).
10. Esposito, M. P., I. O. Tiernan, and F. E. Dryden. Dioxins, U. S. E. P. A., IERL, Office of Research and Development, Cincinnati, Ohio, EPA–600/2-80-197, pp. 230–246 (November 1980).
11. Freeman, R. A., and J. M. Schroy. "Environmental Mobility of Dioxins," paper presented at the 8th ASTM Aquatic Toxicology Symposium, Fort Mitchell, Kentucky, April 15–17, 1984.
12. Schroy, J. M., F. D. Hileman, and S. C. Cheng. "The Uniqueness of Dioxins? Physical/Chemical Characteristics," paper presented at the 8th ASTM Aquatic Toxicology Symposium, Fort Mitchell, Kentucky, April 15–17, 1984.

13. Nash, R. G., and M. L. Beall. "Distribution of Silvex, 2,4-D and TCDD Applied to Turf in Chambers and Field Plots," *J. Agric. Food Chem.*, 28:614–623 (1980).
14. Jury, W. A., W. F. Spencer, and W. J. Farmer. "Use of Model for Assessing Relative Volatility, Mobility and Persistence of Pesticides and Other Trace Organics in Soil Systems," in *Hazard Assessment of Chemicals—Current Developments*, J. Saxena, Ed. (New York: Academic Press, 1983).
15. Spencer, W. F., W. A. Jury, and W. J. Farmer. "Movement of Pesticides and Other Trace Organics Across to the Soil-Air Interface," paper presented at the 189th meeting of American Chemical Society, Miami Beach, Florida, April 19–May 31, 1985.
16. Phillips, F. T. "Persistence of Organochlorine Insecticides on Different Substrates under Different Environmental Conditions," *Pesticide Sci.* 2:255–266 (1971).
17. Nash, R. G., "Comparative Volatilization and Dissipation Rates of Several Pesticides from Soil," *J. Agric. Food Chem.* 31:210–217 (1983).
18. Thibodeaux, L. J. *Chemodynamics* (New York: John Wiley & Sons, Inc., 1979).

CHAPTER **16**

Studies on the Absorption of TCDD by Plant Species

S. Facchetti, A. Balasso, C. Fichtner, G. Frare, A. Leoni,
C. Mauri, and M. Vasconi

INTRODUCTION

This chapter presents the results of cultivations of maize and bean in pots containing soil doped with 2,3,7,8-TCDD in the concentration range between 1 and 752 ppt. The aim of the research is to establish the capacity of the higher plants to absorb, translocate, accumulate and possibly eliminate TCDD. Maize and bean plants were chosen on the basis of four considerations.

1. the possibility of having available results on two plants with morphophysiologically different characteristics: maize is a monocotyledon, and bean is a dicotyledon; this leads to considerable differences in roots, leaves, and cation exchange capacity
2. erect position, which minimizes contamination with the sublayer;
3. relative ease of cultivation in soil and rapid growth
4. these crops are widely grown in the Seveso, Italy* area

---

*See Chapter 30 for a description of the contaminated area in Seveso, Italy.

Table 1. Maize and Bean Cultivations in Pots: 2,3,7,8-TCDD Concentrations in Soils.

Sample	2,3,7,8-TCDD Concentrations (ppt)	Coefficient of Variation (%)	Total Amount of 2,3,7,8-TCDD per Pot (ng)
3[a]	22±2	10	220
6[a]	47±9	19	470
12[a]	80±3	4	800
24[a]	160±35	22	1600
48[a]	243±55	23	2430
24A[b]	478±105	22	4780
48A[b]	752±32	4	7520

[a]Doped with native 2,3,7,8-TCDD.
[b]Doped with seveso soil (Zone A).

EXPERIMENTAL

The doped soil was prepared by mixing soil from uncontaminated Seveso land with contaminated soil from Seveso zone A (the most highly contaminated zone) or with other soil which was previously spiked with acetonic TCDD solutions. The two types of soil were mixed by stirring in a rotating steel cylinder (diameter = 35 cm, height = 60 cm) with two transverse blades, for times between 10 and 30 min. The substrate was previously dried out at ambient temperature to make it friable without damaging the structure, and sieved to remove stones and lumps larger than 2 mm.

To encourage the growth of the plants, the following mineral salts were added to each 10 kg of soil: 5 g of ammonium nitrate, 5 g of calcium monophosphate, and 6 g of potassium sulfate. The contamination levels and the degree of homogenization of the soils prepared in this way are presented in Table 1.

Cultivation

Pots were 30 cm in diameter and contained 10 kg of soil. In each pot 12 pairs of seeds were buried at a depth of 5 cm. In the definition of the harvesting times for each crop, reference was made to the germination times. The crops were grown both outdoors and in an indoor environment with controlled conditions.

Greenhouses

The pots of the outdoor crops were placed in two greenhouses (A and B) 1.4 × 4 × 1.75 m, with polyethylene walls and a mobile roof to protect them from excessive rain and cold. The pots were placed in the greenhouses according to the following criteria.

A : Crops in soils with 22 ppt (1982) and 478 ppt (1983) of 2,3,7,8-TCDD
B : Crops in soils contaminated with 47 ppt (1982) and 752 ppt (1983) of 2,3,7,8-TCDD

Possible contamination between crops in the same outdoor greenhouse was controlled by cultivating plants growing in soil with low contamination. The temperature and humidity were recorded throughout the experiment.

The internal greenhouse was set up in a laboratory with air-conditioning equipment, which had an additional humidification system, and white light and UV illumination (3000 lux at the pot surface). In this way the temperature and humidity were kept sufficiently constant for the duration of the experiment while the artificial illumination, maintained for 17 hr per day, approximately reproduced the summer conditions.

It was observed that the plants grown in the inside greenhouse grew three times higher than those in the outside greenhouse. This phenomenon may be due to the positioning of the light source above the crops and to the strong aspiration, which encouraged evapotranspiration. The aspiration was therefore stopped.

Sample Collection

The plant samples were collected at 15–day intervals to study the possible transfer of 2,3,7,8-TCDD from the soil to the plant with the plant's age.

The last samples (at 75 days) corresponded in general to the maximum development of a plant grown in a pot and also to the appearance and/or to the ripening of the fruits. The characteristics of the samples subjected to analysis depended on the age of the plants: 50 g of sample (quantity needed for an analysis) at 15 days from germination were made up of shoots, while at 75 days the individual plant weighed more than 50 g, and the sample thus contained a high percentage of lignin.

The plants were sampled by cutting at 5 cm from the soil surface to avoid contamination, weighed, washed in running water and kept at -20°C until the moment of analysis

Cleanup

The cleanup procedure was based on an extraction performed first with a polar solvent (methanol) and then with a nonpolar one (methylene chloride), followed by a purification, and separations on columns.[1] High-resolution GC-MS was used for analysis, and the detection limit was lower than 1 ppt.

PERSISTENCE OF 2,3,7,8-TCDD IN THE SOIL

Some authors[2] have shown, with tests in the field and by studies in special chambers, that TCDD is volatile, detecting it in fact in the air.

The capacity of TCDD to sublimate at 200°C could be an indirect confirmation of this. Schroy et al.[3] measured the vapor tension of TCDD, which increases from 3.46×10^{-9} to 1.19×10^{-6} mm Hg in the temperature range from 30 to 71°C. Thus, they believe that TCDD is volatile at ambient temperature and that in the range mentioned above it behaves like an ideal gas.

Freeman and Schroy[4] have developed a mathematical model to describe the vaporization and diffusion of organic products with low volatility through a soil column, finding good agreement between the values predicted by the model and the experimental values obtained for TCDD at Times Beach, Missouri. In confirming the environmental volatility of 2,3,7,8-TCDD, they affirm that as the process is regulated by temperature, most of the vapor transport through the soil column occurs in the summer months, while at the soil surface both vaporization and photodegradation processes occur. As photodegradation of 2,3,7,8-TCDD is a very rapid process (10-min half-life) it predominates during the day. The low solubility of TCDD in water and its adsorption by soil colloids, however, reduce its migration in the soil because of rain, irrigation, or floods.

It seems opportune to add that experimental tests with Silvex with an addition of 7.5 and 15 ppm of tritiated dioxin[2] have shown that the volatilization rate decreases nonlinearly with time, which is an indication of the occurrence of complex physical phenomena.

The similarity of TCDD and of DDT, particularly as regards their vapor tension values and their solubility in water (although this is lower for TCDD), should allow us to attribute environmental behavior to them which is similar in terms of transport.

The numerous experimental studies which have been carried out on DDT[5-8] confirm what has been found for TCDD. Thus, on the basis of the literature one may conclude that a chemical product with low solubility in water and low vapor tension may volatilize from the soil at a rate which may be important for the fate of these products.

Various checks have been carried out on the soil of our crops. During plant growth the samples were taken by coring or by digging with a spoon, while at the end of the cultivation the vertical distribution of the TCDD concentration was studied by cutting the pot contents, which had been frozen beforehand, at the heights shown in Figure 1 and analyzing the layers marked with the letters A, C, and E. The results are given in Tables 2 and 3.

A loss of TCDD over time was evident, probably mainly by volatilization. The differences found even for pots with the same initial concentration may perhaps be attributed to different volatility rates due to different humidity contents in the soil, to the homogenization of the soil and to analytical precision. It would be advisable, in order to gain a better understanding of the process, to plan additional experiments.

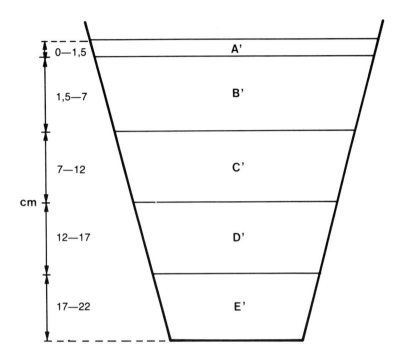

Figure 1. Pot for crop cultivation.

TRANSFER OF 2,3,7,8-TCDD TO THE PLANTS

Studies carried out in the 1970s[9-11] indicated absorption of 2,3,7,8-TCDD by the upper portions of plants, but reduced accumulation. Absorption must be interpreted as the metabolic entry, adsorption as the retention on surfaces.

The analysis of samples collected in the contaminated zone of Seveso showed a quantity of TCDD in the parts of the plant below ground that was double that measured in the parts above ground, from which the dioxin disappeared when the plants were transplanted into uncontaminated soil.[12,13] This disappearance could have occurred by photodegradation, transpiration and/or metabolization.

At the end of our cultivation of maize and bean plants we gathered some root samples to check whether there was any adsorption of dioxin; the results are given in Table 4. The samples were analyzed only after thorough washing in running water to remove the soil. Some samples were also washed with n-hexane. In most cases in the roots and the sediments created by suspension in water from the already washed roots, TCDD levels were found which were higher than those of the soil in which the crop was grown.

Table 2. Maize Cultivations: 2,3,7,8-TCDD Concentration in Soils at the End of Cultivation (About 90 Days).

Sample	Initial	2,3,7,8-TCDD Concentrations (ppt) — Final				Covering	Location	Temperature (°C)			Relative Humidity (%)		
		Whole Pot	Layer A (0–1.5 cm)	Layer C (7–12 cm)	Layer E (17–22 cm)			$\bar{X}$	Min	Max	$\bar{X}$	Min	Max
3	25		20	20	15		Outdoor	21	8	38	70	15	95
3	25		20	18	20		Outdoor	21	8	38	70	15	95
6	56		33	41	38		Outdoor	15	2	27	90	43	100
24	124	72					Indoor	24	18	29	80	30	90
24	127		15	95	104	Plastic Chips	Indoor	24	18	29	80	30	90
48	201		118	136	262	Plastic Chips	Indoor	24	18	29	80	30	90
24A	432		375	438	350		Outdoor	22	15	30	76	30	85
24A	471		362	471	503	Plastic sheet with holes	Outdoor	22	15	30	76	30	85
48A	720		778	727	581		Outdoor	22	15	30	76	30	85
3	22		26 (25 days) 22 (40 days) 51 (55 days)				Outdoor	21	8	38	70	15	95

Table 3. Bean Cultivations: 2,3,7,8-TCDD Concentration in Soils at the End of Cultivation (About 90 Days).

Sample	Initial	2,3,7,8-TCDD Concentrations (ppt) Final				Covering	Location	Temperature (°C)			Relative Humidity (%)		
		Whole Pot	Layer A (0–1.5 cm)	Layer C (7–12 cm)	Layer E (17–22 cm)			X̄	Min	Max	X̄	Min	Max
3	25		19	26	21		Outdoor	15	2	27	90	43	100
6	55		33	36	37		Outdoor	15	2	27	90	43	100
24	105	77					Indoor	24	18	29	80	30	90
24	184		147	70	112	Plastic chips	Indoor	24	18	29	80	30	90
48	188		234	115	121	Plastic chips	Indoor	24	18	29	80	30	90
48	264	190					Indoor	24	18	29	80	30	90
24A	452		305	251	341	Plastic sheet with holes	Outdoor	22	15	30	76	30	85
48A	734		355	456	420	Plastic sheet with holes	Outdoor	22	15	30	76	30	85
48A	1066	845					Outdoor	22	15	30	76	30	85

Table 4. **Maize and Bean Cultivations: 2,3,7,8-TCDD Concentrations in Roots and Sediment at the End of Cultivation (About 80 Days).**

| Sample | Cultivation | | 2,3,7,8-TCDD Concentrations (ppt) | | | | |
	Type	Location	Soil[a]	Roots[b]	Sediment	Roots[c]	Sediment
Blanc	Maize	Outdoor	1.5	16	37	11	58
24	Maize	Indoor	160	156	673	86	245
48	Maize	Indoor	243	254	181		
24A	Maize	Outdoor	478	849	856	578	906
48A	Maize	Outdoor	752	1278	1649	1019	1432
Blanc	Bean	Outdoor	1.5	37	52		
24A	Bean	Outdoor	478	705	1271		
48A	Bean	Outdoor	752	1807	3702		

[a]Initial values.
[b]Previously washed with water.
[c]Previously washed with water, then with n-hexane.

In this context it is interesting to remember the studies carried out at the University of Milan[14] on the same varieties of maize and bean put in a nutrient solution containing tritiated TCDD. That study showed that the roots adsorbed 90% of the dioxin previously added. In our crops the adsorption was obviously lower due to the different transport conditions between water and soil, because of the importance of the evaporation through the soil, and because of the lower availability of dioxin resulting from the competitive processes between the soil and the roots themselves.

Our soil-grown crops did not show, however, a significant increase of TCDD in the parts above ground, either as a function of time or with the increase in concentration of the pollutant in the soil. The results obtained are given in Table 5 for maize and Table 6 for beans.

The variations in dioxin concentration in the plants grown in the soil with pollutant concentration of 1 ppt are very interesting. In fact, the TCDD concentration varies in these plants with their location, being greater if they are near the pots containing contaminated soil or if they were grown in the indoor greenhouse with no air aspiration. Similar results[14] were obtained with bean and maize plants grown in uncontaminated soil placed near the pots with tritiated TCDD: in the blanks the radioactivity increased with time. We believe these results are a proof of the contamination of the aboveground parts of the plants by evaporation of the TCDD from the polluted soil. In the case of crops grown in soils with contamination levels up to several hundred ppt evaporation is likely the predominant process and could mask other contamination pathways such as adsorption through the roots.

The absence of proportionality between TCDD concentration in soil (1–752 ppt) and the leaf contamination is confirmed by an independent experiment on beans grown in soil contaminated with tritium-labeled TCDD.[14] The measurements done on the leaves did not show an increase

Table 5. Maize Cultivations: 2,3,7,8-TCDD Concentration in Plants (ppt fresh material).

		Soil Concentration (ppt)					
		Average Temperature = 22°C; Average Relative Humidity = 40%			Average Temperature = 24°C; Average Relative Humidity = 80%		
Location	Days	1	47	80	1	160	243
Indoor no air aspiration	15	0.05[a]	0.05[b]	0.1[b]	1.5[a]	1.5	1.6
	30	0.5[a]	0.6	0.5	1.6[a]	3.4	1.0
	45	0.3[a]	0.5	0.3	2.5[a]	2.2	1.6
	60	1.0[a]	0.9	0.7			
	75				1.9[a]	1.7	1.2

		Soil Concentration (ppt)				
		Average Temperature = 21°C; Average Relative Humidity = 70%		Average Temperature = 22°C; Average Relative Humidity = 76%		
Location	Days	1	22	1	478[c]	752[c]
Outdoor	15	3	8	0.1/0.5[a]	0.6	1.2
	30	4	5	0.4[a]	0.2	0.9
	45	4	5 (leaves = 5)	1.4/1.8[a]	1.2	0.6
	60	3				
	75			0.9/2[a]	0.6	1.2

[a]Grown together with doped cultivation.
[b]With air aspiration.
[c]Doped with Seveso soil (Zone A).

Table 6.   Bean Cultivations: 2,3,7,8-TCDD Concentration in Plants (ppt fresh material).

		Soil Concentration (ppt)					
		Average Temperature = 22°C; Average Relative Humidity = 40%			Average Temperature = 24°C; Average Relative Humidity = 80%		
Location	Days	1	47	80	1	160	243
Indoor no air aspiration	15	1.3[a]/0.1[ab]	0.1[b]	0.1[b]		0.8	1
	30		2.2	2.7	1.2[a]	1.3	0.9
	45	1[a]	2.1	3.4	2.2[a]	1.2	
	60	1.4[a]	0.6	0.6			
	75		1.8	0.8	2.2[a]	1.7	2.1

		Soil Concentration (ppt)					
		Average Temperature = 21°C; Average Relative Humidity = 70%			Average Temperature = 22°C; Average Relative Humidity = 76%		
Location	Days	1	22		1	478[c]	752[c]
Outdoor	15	5	5		0.7[a]		1.3
	30	4	6 (5 leaves; 4 stem)		0.7[a]	0.8	0.5
	45	7	6 (6 leaves; 7 stem)		0.8/1.3[a]	0.6	0.8
	60						
	75					1.4	

[a] Grown together with doped cultivation.
[b] With air aspiration.
[c] Doped with Seveso soil (Zone A).

with radioactivity levels in crops grown in soils containing from 14 ppt up to 1000 ppt dioxin.

Such a lack of proportionality between the pollution levels of the soil and the radioactivity measurements in the leaves confirms, in our opinion, the process whereby the parts of the plant above ground are mainly contaminated by evaporation of TCDD from the soil.

To conclude, the high adsorption of TCDD by the roots leads us to recommend precautions in the consumption of some root vegetables such as carrots, potatoes, etc. The contamination of the aboveground parts of the higher plants suggests that it occurs mainly through the evaporation of TCDD from contaminated soils. At the contamination levels of our experiment the plant uptake is on the order of a few ppt.

REFERENCES

1. Leoni, A., C. Fichtner, G. Frare, A. Balasso, C. Mauri, S. Facchetti. *Chemosphere* 12(4/5):493–497 (1983).
2. Nash, R. G., and M. L. Beall, Jr. *J. Agric. Food Chem.* 28(3):614 (1980).
3. Schroy, J. M., F. E. Hileman, and S. C. Cheng. "The uniqueness of dioxins ? Physical-Chemical Characteristics," 8th ASTM Aquatic Toxicology Symposium, Fort Mitchell, Kentucky, April 15–17, 1984.
4. Freeman, R. A., and J. M. Schroy. "Environmental Mobility of Dioxins," 8th ASTM Aquatic Toxicology Symposium, Fort Mitchell, Kentucky, April 15–17, 1984.
5. Nash, R. G., and E. A. Woolson. *Soil Sci. Soc. Am. Proc.* 32:525 (1968).
6. Guenzi, W. D., and W. E. Beard. *Soil Sci. Soc. Am. Proc.* 34:443 (1968).
7. Farmer, W. J., K. Igue, W. F. Spencer, and J. P. Martin. *Soil Sci. Soc. Am. Proc.* 36:443 (1972).
8. Hee, S. S. Q., K. S. McKinlay, and J. G. Soha. *Bulletin of Environmental Contamination and Toxicology* (1975), p. 284.
9. Isensee, A. R., and G. E. Jones. *J. Agric. Food Chem.* 19:1210 (1971).
10. Kearney, P. C., E. A. Woolson, A. R. Isensee, and C. S. Helling. *Environ. Health Pers.* (September 1973), p. 273.
11. Young, A. L., C. E. Thalken, E. L. Arnold, J. M. Cupello, and L. G. Cockerham. "Fate of 2.3.7.8-Tetrachlorodibenzo-p-dioxin (TCDD) in the Environment: Summary and Decontamination Recommendation," United States Air Force Academy, Colorado (October 1976).
12. Cocucci, S., F. Di Gerolamo, A. Verderio, A. Cavallaro, G. Colli, A. Gorni, G. Invernizzi, L. Luciani. *Experientia* 35:482 (1979).
13. Cocucci, S. "Adsorption, Translocation and Elimination of TCDD by Plants in Contaminated Soil." International Steering Committee, 3rd Meeting, Segrate, Italy, March 30-April 1, 1980.
14. Cocucci, S. Personal communication.

CHAPTER 17

Dioxin: Field Research Opportunities at Times Beach, Missouri

Armon F. Yanders, Shubhender Kapila, and Robert J. Schreiber, Jr.

INTRODUCTION

There are more than forty sites in Missouri which are known to be contaminated with 2,3,7,8-tetrachlorodibenzo-p-dioxin, the most toxic of the 75 possible dioxin isomers. The sites are located principally in two general regions: the first is in the east central part of the state, near St. Louis, and the second is in the southwest part of the state, near Verona, where the chemical plant which produced the dioxin is located. The dioxin occurred as one of the unwanted by-products formed in the production of hexachlorophene, and was at its greatest concentration in the residues remaining after distillation. During the period of February to October, 1971, the contractor who was paid to dispose of these wastes hauled six truckloads to a storage site near St. Louis, where most of the material was mixed with used oil in a large storage tank. In 1971 and 1972, this mixture, and in a few cases the undiluted wastes themselves, were used as sprays for the control of dust on roads, parking lots, and horse arenas. The majority of the affected soils in Missouri received their contamination as a result of such sprayings. Estimates of the amount of contaminated soil range between 150,000 and 400,000 m³. It is a problem of immense proportions, and has become a matter of much concern to the people of Missouri.

In response to this concern, in 1983 the Governor of Missouri established a Task Force on Dioxin with the charge to "recommend a practical and effective plan of action for implementing comprehensive and permanent solutions to the public health and environmental problems caused by dioxin contamination in Missouri." In its final report to the Governor,[1] one of the Task Force's recommendations was that further research be conducted on methods for destruction of dioxin. To implement this, the Missouri Department of Natural Resources (MDNR) and the University of Missouri cooperated in establishing a dioxin research group which includes representatives of governmental agencies (MDNR, U. S. Environmental Protection Agency, Missouri Division of Health, Missouri Department of Conservation), industry, and the University. This group concluded that in situ studies at Times Beach, Missouri, would be of great help in determining methods for treating dioxin-contaminated soils.

The city of Times Beach, which is located in eastern Missouri near St. Louis, is the most extensively contaminated area in the state. It is estimated that some 60% of all known dioxin-contaminated soil in Missouri is located in Times Beach, and as much as 90% is within 60 mi of that city. Times Beach includes approximately 640 acres of residential, commercial and park areas, with some 20 mi of streets. Most of the streets are now paved, but in 1972 all but a few were dirt and gravel, and virtually all of them were sprayed two or more times with waste oil as a dust control measure.[2] Some of the waste oil was heavily contaminated with dioxin, and levels as high as 1500 ppb have been found in the roads.

The U. S. Environmental Protection Agency and the State of Missouri are spending over $33 million to purchase all Times Beach property and relocate the 800 families living there. When the buyout is complete, the State of Missouri will hold title to the land. Two blocks of the most heavily contaminated street have been set aside to be used for research. In this section of Laurel Road, the dioxin concentration in the soil under the asphalt averages about 300 ppb. To prepare the research plots, the asphalt has been removed, and the underlying soil and gravel have been excavated to a depth of about 8 in. (dioxin is rarely found at greater depths in this street). The soil and gravel were screened to remove the larger gravel and rocks, and homogenized by mixing them thoroughly in a cement mixer. The homogeneous, screened material was placed in stainless steel bins 6 ft × 8 ft × 2 ft deep and compacted to 70% of the original density. A bottom liner was installed to drain liquids seeping through the soil. Water and electrical power outlets have been provided at each plot, and an onsite soils laboratory equipped for routine work is available. Security arrangements, including lockers, and a personnel decontamination facility are provided. A full-time MDNR onsite coordinator has been assigned to oversee operations and ensure that security is maintained. Emergency services are also available.

A comprehensive sampling and analysis program was conducted to determine initial reference levels of dioxin, and the plots are now available for in situ investigations. The units are normally assigned in sets of three, giving the investigators the capability of establishing one unit as a standard reference control and varying the parameters in the other two. Standardized soil can also be made available for offsite research if the applicant demonstrates the capability for proper in-house management of dioxin.

A major objective of this project is to identify methods which have the greatest potential to detoxify dioxin-contaminated material, and to evaluate those methods which appear to be successful. Several studies of in situ degradation of dioxin in soil began in the summer of 1984, and more are expected to begin this year. Other types of studies which can be accommodated at Times Beach include the following:

> Biological studies: biological uptake and bioconcentration; movement of dioxin through short food chains; biodegradation by microbial and/or mycorrhizal agents; animal toxicology; etc.
>
> Chemical studies: distribution, concentration and mobility of dioxin in soils and sediments; physical properties of dioxin in relation to matrix material and particle size; solubility; volatilization rate; etc.

The first-year cost for leasing a research plot (set of three units) is $16,500, which covers the cost of preparing the plots and providing the supporting facilities and includes the cost of sampling and analyzing one composite sample per unit before and after the experimental treatment (six GC/MS analyses). Costs for subsequent years will be based on the expense of maintaining the facility and providing the analyses, and should be much lower.

REFERENCES

1. "Final Report of the Missouri Dioxin Task Force," submitted to Governor Christopher S. Bond (October 31, 1983).
2. Leistner, M. "Public and Media Perspective on Times Beach, Missouri," in *Proceedings of the Third Annual Hazardous Materials Management Conference* (Wheaton, IL: Tower Conference Management Co., 1985), pp. 361–369.

SECTION IV

Bioassays

CHAPTER 18

Studies on the Molecular Basis of TCDD-Caused Changes in Proteins Associated with the Liver Plasma Membrane

Fumio Matsumura, D. W. Brewster, D. W. Bombick, and B. V. Madhukar

INTRODUCTION

Polychlorinated dibenzo-p-dioxins (PCDD) and dibenzofurans (PCDF) are serious environmental pollutants. From a toxicological point of view, this group of chemicals presents quite an enigma to scientists, since their toxicological action patterns are very unusual and do not fit into any of the known classes of poisons.

The characteristics of their toxic actions may be summarized as follows: lack of violent acute poisoning symptoms, slowness of development of their toxic effects (e.g., even at lethal doses death occurs after 10 to 40 days), loss of appetite, loss of body weight in most animal species, and skin lesions in several mammalian species.[1-3] Among PCDD and PCDF, the most toxic congener has been acknowledged to be 2,3,7,8-tetrachlorodibenzo-p-dioxin (TCDD).

TCDD has been used as a model and a probe for toxicological studies for this family of halogenated aromatics. The oral LD_{50}'s of TCDD in the male guinea pig and the rat are cited as 0.6 µg/kg and 50 µg/kg, respectively.[4] There are marked species differences in sensitivity to TCDD, with the Golden Syrian hamster being tolerant, having an $LD_{50} > 5000$ µg/kg.[5]

Toxic manifestations of this dioxin also vary qualitatively from species to species.

One of the most conspicuous TCDD-caused biochemical changes is "induction" of hepatic enzymes in many animal species. For instance, Kitchin and Woods[6] found that microsomal enzymes in the rat liver are induced at doses as low as 0.002 µg/kg, the most sensitive parameter in any animal species recorded. This induction effect is mediated by a cytosolic receptor which binds with TCDD, and is transported into the nucleus where changes in DNA activity trigger various pleiotropic responses.[7,8] Progress in this area has been remarkable, but there are indications that the phenomenon of induction per se is not directly related to TCDD's lethal action.[1] For example, some very effective AHH inducers, such as β-naphthoflavone, show little toxicity. Furthermore, hepatic induction by TCDD is not significant at lethal doses in the most sensitive species (guinea pigs), although it does induce hepatic enzymes in the hamster at sublethal doses. Available evidence indicates the affinity of the cytosolic receptor for TCDD and the rate of TCDD metabolism are not significantly different among the hamster, rat and guinea pig,[9] suggesting that such differences alone may not explain the enormous species differences in susceptibility. These observations point to the need for further studies on biochemical effects of TCDD, since many unexplainable toxic manifestations remain.

Another unique aspect of TCDD's action is that its toxic manifestations vary among different tissues.[1] Such a phenomenon immediately indicates that its action cannot be explained by a simple cause, such as inhibition of an enzyme that is present mainly in one type of tissue. An important observation which prompted our current investigation was that TCDD's effect seemed to be more pronounced in young growing animals than in mature animals. Therefore, the site(s) of action could be related to developmental processes. Another phenomenon that drew our attention was that some of TCDD's effects resemble those caused by hormones and bioactive peptides which have diverse effects in different tissues and species. Based upon these observations we have proposed a working hypothesis[10] that some of the TCDD-caused cellular changes may be explained if one assumes that TCDD causes drastic changes in the cell surface membrane (plasma membrane) which influences the functions of its constituents, including enzymes, receptors, ion channels, and surface glycoproteins.

Two preliminary observations provided stimuli to propose such a hypothesis. One is our observation in a collaborative study with Dr. R. E. Peterson that ATPases in the hepatic plasma membrane from TCDD-treated rats were much lower than those found in untreated rats.[11] The second observation was that protein profiles in electrophoretograms of hepatic plasma membrane from control and TCDD-treated rats were qualitatively and quantitatively different.[12] Also, the level of ^{3}H-concanavalin A binding was lower in the hepatic plasma membrane from the treated animals than that from the control. Concanavalin A (Con A) is a lectin which is known to bind with glycoproteins. This finding indicated to us that levels of some of the surface glycoproteins of the hepatocytes were reduced as a result of the action of TCDD. In addition, isolated rat hepatocytes cultured in vitro from TCDD-treated animals do not readily adhere to each other or to glass surfaces like those cultured from control animals.[13]

In subsequent studies we have found that TCDD treatment in vivo caused changes not only in several plasma membrane enzymes, but in several key receptors as well.[14,15] Two of the most conspicuous changes observed were those of epidermal growth factor (EGF) and low-density lipoprotein (LDL) receptor activities as measured with corresponding ligand binding assays. In this chapter we present experimental evidence for the above conclusion and results of recent work addressing the question of the possible mechanism by which treatment with TCDD evokes those simultaneous pleiotropic changes in the hepatocyte plasma membrane.

All methodological details have already been given in several recent studies done in this laboratory.[12,13,15-17]

RESULTS

Effects of TCDD on Hepatic Plasma Membrane in Rats

To study the biochemical characteristics of the plasma membrane, rats were treated with 25 µg/kg of TCDD (single dose) through intraperitoneal (i.p.) injection, and their hepatic plasma membrane isolated after 10 days as before.[11]

In the first series of tests the activities of several membrane-bound enzymes and receptor proteins were examined (Tables 1 and 2). The differences in Na–K and Ca ATPase were highly significant, but the difference in γ-glutamyl transpeptidase (γ-GT) activity between TCDD-treated and untreated rats was only marginally significant. The level of protein kinase in the plasma membrane from treated rats was significantly higher than that of the control, indicating that TCDD treatment does not always cause a reduction in enzyme activities. In this case, both c–AMP-stimulated and nonstimulated protein kinases from the TCDD-treated animals showed higher enzyme activities than those from controls.

As for receptor activities (Table 2), EGF-binding was the most severely affected at 25 µg/kg (single i.p. dose) TCDD. In agreement with our previous observation, Con A binding was also reduced; however, the effects of TCDD on glucagon receptors were significant only at 115 µg/kg. The effect on insulin binding was unusual, in that at low doses it was stimulated and at high doses it was inhibited (data not presented). Under our experimental conditions, TCDD treatment caused no appreciable change in prostaglandin E_1 binding.

Table 1.   Difference in Various Enzyme Activities Between the Hepatic Plasma Membrane Preparation from Untreated Control and In Vivo TCDD-Treated Rats.[a]

| Enzyme | Enzyme Activities (nmol product/mg protein/hr) | |
	Control	TCDD-Treated[b]
Na-K ATPase[c]	1496 + 142 (3)	890 + 178 (3)***
Mg ATPase[c]	1820 + 10 (2)	1110 + 149 (2)**
Ca ATPase[c]	608 + 65 (4)	340 + 140 (8)***
γ-Gluta myl transpeptidase[d]	887 + 231 (4)	658 + 35 (3)*
Protein kinase[e]		
in the absence of c-AMP	55 + 31 (6)	147 + 67 (7)**
in the presence of c-AMP	58 + 25 (4)	217 + 110 (7)**

[a]The results are expressed as mean + standard error, and the number in parentheses indicates the number of different membrane perparations tested. Each preparation was tested twice. Data were analyzed using Student's t test: * $p < 0.1$; ** $p < 0.05$; *** $p < 0.01$.
[b]TCDD at 25 μg/kg; single dose intraperitoneal injection. Plasma membrane collected at 10 days posttreatment.
[c]nmol P_i liberated/mg plasma membrane protein/hr at pH 7.1, 37°C.
[d]nmol p-nitroaniline liberated/mg plasma membrane protein/hr at pH 8.0,37°C.
[e]nmol P_i incorporated/mg plasma membrane protein/hr at pH 8.0,37°C.

The time course of TCDD effect was studied following a single i.p. dose (25 μg/kg, Figure 1). During the 40–day observation period, TCDD-treated rats gained consistently less body weight than did the control rats. Insulin binding was unaffected at the beginning, but 20 days after treatment it was significantly reduced ($P < 0.05$). Con A binding was continuously suppressed. The time course of TCDD effects on the membrane enzyme activities was studied at the same dose regimen. The results shown in Figure 2 clearly indicate that, with the exception of γ-GT activities, these enzyme activities follow a reduction pattern similar to that of the membrane-bound

Table 2.   Effect of In Vivo TCDD Treatment on Ligand Binding to Cell-Surface Membrane Receptors in the Rat Liver.[a]

| Ligand | Dose (μg/kg; single i.p.) | Specific Binding | |
		Control	TCDD-Treated
³H-Con A[b]	25	15.2 ± 2.5 (4)	11.7 ± 0.4 (3)*
¹²⁵I-EGF[c]	25	32.0 ± 8.6 (5)	3.8 ± 0.3 (3)**
¹²⁵I-Glucagon[c]	25	43.5 ± 10.1 (6)	43.7 ± 6.8 (4)**
	115		4.4 ± 4.1 (8)
¹²⁵I-Insulin[c]	25	8.1 ± 1.2 (7)	11.7 ± 0.9 (3)**
³H-Prostaglandin E₁[d]	25	3.6 ± 0.3 (3)	3.3 ± 2.3 (6)

[a]TCDD was intraperitoneally administered to rats at indicated doses. Plasma membrane was fractionated 10 days after treatment. Ligand binding was studied as described earlier.[13] The results are expressed as mean ± S.E. and the number of animals tested is given in parentheses. Data were analyzed using Student's t test; * $p < 0.05$; ** $p < 0.01$.
[b]ng of ³H-Con A bound/25 μg protein/10 min.
[c]pg of ¹²⁵I-EGF, glucagon- or insulin-bound/50 μg protein/20 min.
[d]pg of ³H-PGE₁ bound/150 μg protein/20 min.

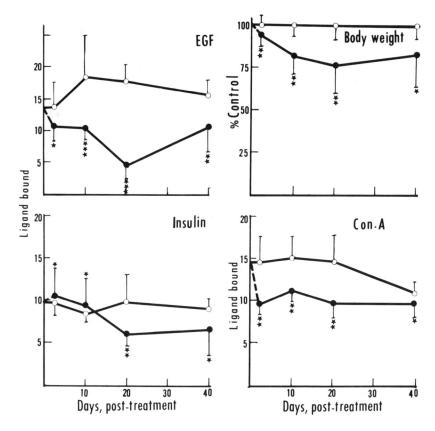

Figure 1. Time course of changes in specific binding of EGF insulin or Con A to liver plasma membranes and body weight changes of TCDD-treated rats (●) relative to untreated control rats (○). Rats were given a single i.p. dose of TCDD at 25 μg/kg and plasma membranes prepared at 2, 10, 20 and 40 days after treatment. Control rats received an appropriate volume of the vehicle (corn oil/acetone, 9:1). Binding was expressed as pg of ^{125}I-insulin or ^{125}I-EGF-bound/50 μg protein or ng of ^{3}H-Con-A bound/25 μg protein. Changes in the body weight of the TCDD-treated rats were expressed as percentage of body weight of controls at each time point. Values for controls are: 108 + 2, 140 + 8, 170 + 15 and 206 + 18 g at 2, 10, 20 and 40 days, respectively. Each point is a mean + S. E. of 4 to 8 animals. The data are analyzed using Student's t test. Asterisk designations are the same as in Table 1.

receptors: The decline becomes significant at day 10 and reaches the minimum level at day 20, followed by an apparent recovery. There is the possibility that this apparent recovery does not represent a true biochemical recovery, since by day 40 approximately 20 to 30% of the treated rats had died, while at other times little mortality was observed. In terms of the percentage of reduction, it is concluded that among the criteria examined, EGF binding was the most sensitive parameter to in vivo TCDD treatment

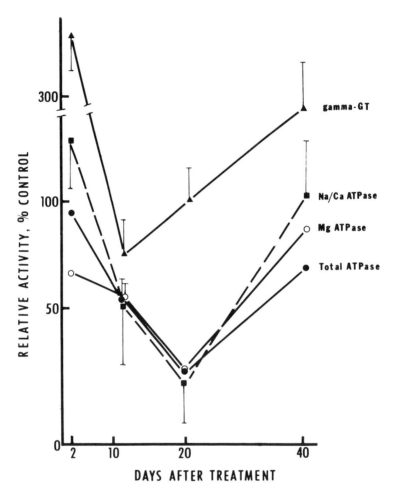

Figure 2.   Time course of changes in plasma membrane-associated enzyme activities in the rat liver as a result of in vivo TCDD exposure. TCDD was administered as a single i.p. dose at 25 μg/kg. Hepatic plasma membrane was isolated at indicated days of posttreatment and the various enzyme activities were assayed in the isolated membrane as described by Matsumura et al.[13] Values are expressed as percentage control activities and are represented (where indicated) by mean + S. E. of at least three different membrane preparations. Control values for the enzymes assayed at 2, 10, 20 and 40 days respectively are: Total ATPase, 972, 1479 + 142, 1768 and 1137 nmol of P_i-liberated/mg protein/hr; Mg-ATPase: 1211, 1820 + 10, 1076, and 1230 nmol of P_i/mg protein/hr; Ca-ATPase, 564 + 119, 609 + 65, 603 + 105, and 638 + 152 nmol P_i mg protein/hr; and γ-glutamyltranspeptidase (γ-GT) 884 + 116, 887 + 231, 935 + 223 and 666 + 68 nmol of p-nitroaniline produced mg protein/hr at 37°C and pH 8.0. It should be noted that an apparent recovery is seen on day 40 which only reflects the levels of these biochemical parameters in the surviving population. By day 40, 20–30% of the population had died at this dose (25 μg/kg).

(Table 2) and was observed on the second day after treatment (Figure 1), as is body weight loss.

At this stage, a question was raised as to whether this phenomenon of induced changes in plasma membrane functions is associated with any toxicants that cause general stress, or whether it is confined to TCDD. To answer this question, the rats were treated (all i.p. daily for 10 days) with 9:1 corn oil-acetone or with 3-methylcholanthrene (20 mg/kg), Aroclor® 1242 (50 mg/kg), phenobarbital (120 mg/kg), DDT (0.3 mg/kg) and Firemaster® B-6 (i.e., PBB, 50 mg/kg). At the end of the in vivo treatment, animals were sacrificed, hepatic plasma membrane was isolated, and the extent of Con A binding was studied. The levels of specific binding as expressed in terms of ng/50 µg plasma membrane protein were: control 17.0 ± 2.0, 3-methylcholanthrene 21.0 ± 4.9, Aroclor® 1242 18.4 ± 2.1, phenobarbital 10.6 ± 1.8, DDT 13.9 ± 1.2 and Firemaster® B-6 19.5 ± 6.6 (mean ± standard deviation; 3 animals, except for DDT and Firemaster® experiments, where 2 animals were used). Only the results of the phenobarbital experiment were statistically different from the control value (at $P \geq 0.05$); therefore such changes in the plasma membrane are caused only by rather specific chemicals.

Studies on the Cause for Hypercholesterolemia in Guinea Pigs

The above studies have clearly established that many functions on the hepatic plasma membrane are altered as a result of TCDD administration in vivo. However, in the case of the rat liver, other changes, such as induction of microsomal monooxygenases, are known to occur. Thus, it is not easy in this species to disseminate all of the information and ascribe one biochemical change to a certain toxic lesion in vivo. With guinea pigs, effects of TCDD in the liver are not extensive as judged by morphological[18] and biochemical[19] criteria. Rather, in this species the most noticeable sign of TCDD's toxic effect is the unusual accumulation of serum cholesterol, cholesteryl ester (hypercholesterolemia), and triglyceride (hypertriglyceridemia) carrying lipoproteins.[20,21]

Male guinea pigs (about 200 g) were treated with 1 µg/kg (single i.p.) of TCDD dissolved in corn oil-acetone. Low-density lipoprotein (LDL) was obtained by a centrifugation technique[22] and radiolabeled with $Na^{125}I$ by a Bio-Rad enzymobead method[23] and internalization of ^{125}I–LDL was studied using isolated guinea pig hepatocytes. The details of the methodologies used have been published elsewhere.[16]

When slices of liver tissues from treated and control animals were incubated with the medium containing ^{125}I-labeled bovine serum albumin or ^{125}I-labeled LDL for up to 20 min, slices from treated guinea pigs showed reduced total uptake of these labeled materials from the medium (Figure 3). The rate of uptake was also observed to be different. Accumulation of

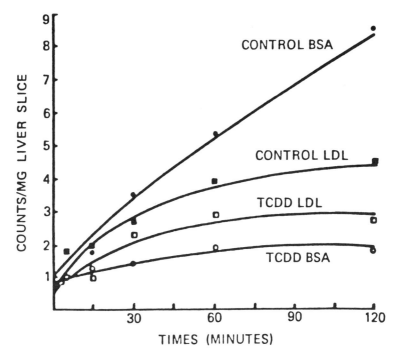

Figure 3. Uptake of ^{125}I-LDL and ^{125}I-BSA in liver slices from treated (1.0 μg/kg single i.p. injection) guinea pigs and controls ten days after dosing. Specific activities of ^{125}I-BSA and ^{125}I-LDL were 708 cpm/μg protein and 80 cpm/ng protein, respectively.

radiolabel by the control slices showed a steady increase from 30 to 120 min, whereas with slices from the treated guinea pigs no significant increase could be observed over the same time period.

In the following series of experiments with LDL, binding characteristics were directly studied using isolated hepatic plasma membrane fractions. The results summarized in Table 3 clearly show that the isolated plasma membranes from the TCDD-treated animals have a lower capacity to bind exogenously added ^{125}I–LDL. Since LDL receptor activities are known to be influenced by the nutritional states of the animals, the same experiment was repeated with carefully matched pair-fed controls, where control animals were fed with the same amount of food as the treated animals. Results were similar in that plasma membranes from the treated animals always showed a lower LDL-binding capacity than that of the paired control animals. Since it has been reported by several researchers[24-26] that TCDD causes dose-related increases in serum cholesterol levels in rats, a similar experiment was also conducted with the male rats treated with TCDD (25 µg/kg; single i.p.). The plasma membrane from the liver of the treated animals (sacrificed at

Table 3. Differences in the Levels of Specific Binding of ^{125}I-LDL to the Hepatic Plasma Membrane and Isolated Hepatocytes from Control and TCDD-Treated Guinea Pigs and Rats.[a]

Source	Treatment (μg/kg; single i.p.)	^{125}I-LDL binding (pg/μg protein)	
		Control	TCDD-Tested
Guinea pigs			
Plasma membrane	1.0	210.4 ± 21.1 (3)	78.7 ± 7.4 (3)*
Plasma membrane (control, pair-fed)	1.0	320.0 ± 9.7 (3)	84.9 ± 18.8 (4)*
Hepatocytes	1.0	1.630[b] ± 2.21 (1)	0.269[b] ± 0.14 (1)
Rats			
Plasma membrane	25	1633 ± 386 (3)	873 ± 116 (3)

[a]Data are expressed in mean ± standard error (number of independent tests with different animals for eath test); * p < 0.05.
[b]μg of LDL bound per 10^6 cells. Each value represents three replicates of a single animal.

day 10) showed a significantly lower level of ^{125}I-LDL binding than the ones from control rats.

Bound LDL is subsequently internalized through an endocytotic process and eventually metabolized within the hepatocyte.[27] Therefore, the reduced binding of LDL to its receptor in the TCDD-treated animals could mean a reduction in LDL metabolism. To test this possibility, a primary hepatocyte culture was established. Using a fixed number of cells (i.e., 200,000 cells per test) ^{125}I-LDL uptake was studied. The results (Table 3) are in good agreement with the previous observations with tissue slices and isolated plasma membranes that the hepatic cells from the treated animals are not taking up the same amount of LDL as controls. Reduced receptor activities at the cellular surface can explain the decreased internalization in treated hepatocytes, since LDL binding is integral to LDL internalization and metabolism.[28]

The time course of internalization was also studied using the same hepatocyte preparations and by the method developed by Soltys and Portman.[29] The results (Figure 4) show that hepatocytes from guinea pigs definitely have a lower ability to internalize LDL than those of untreated controls.

One of the most prominent signs of TCDD toxicity in the guinea pig is hyperlipidemia.[20,21] This phenomenon is unusual because serum LDL and VLDL levels are elevated without changes in the levels of HDL and free fatty acids.[25] According to the general body of knowledge on the biochemical processes of serum lipoprotein homeostasis,[27] VLDL is produced by the liver and converted to LDL. LDL, the final form of all the cholesteryl ester-carrying lipoproteins, is bound by LDL receptors, concentrated in the hepatocytes in many animals, then metabolized. Therefore, the reduced receptor

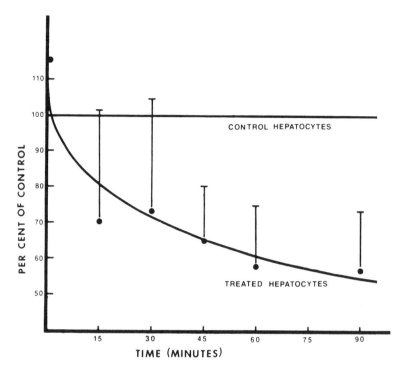

Figure 4.   Comparison (as a percentage) of ^{125}I-LDL internalization in primary hepatocytes from treated guinea pigs (1.0 μg/kg single i.p. injection) and control primary hepatocytes. Control values (μg^{125}I-LDL/200,000 cells for 0-, 15-, 30-, 45-, 60- and 90-min incubation times are 0.85, 2.77, 2.78, 2.95, 3.22 and 3.53, respectively. An lsd test ($\alpha = 0.05$) after a completely randomized factorial analysis of variance showed significant differences in the means at incubation times of 15, 45, 60 and 90 min.

activity in the treated animal can potentially explain the simultaneous increase in LDL levels in the serum.

Moreover, recent results on this subject indicate that the LDL receptor may have a dual role to degrade both LDL and VLDL remnants.[30] Rabbits lacking the LDL receptor (i.e., "the Watanabe rabbit") have LDL degradation hindered but also have VLDL production rates stimulated. This may be due to an inability of this strain's receptor to degrade VLDL remnants which serve as a precursor for LDL.[30] Thus, the decrease in LDL receptor could explain the increase in VLDL levels as well. A good example of such a cause-effect relationship is the genetic disorder in humans known as familial hypercholesterolemia (FH), in which the LDL level in the serum is elevated as a result of a lowered LDL receptor activity.

A potentially serious aspect of the current finding is that there is a possibility that TCDD causes a similar effect in humans. For instance, Oliver[31] has reported cases of hypercholesterolemia in humans who have been exposed to TCDD three years prior to examination time. Pazderova-Vejlupkova and coworkers[32] have also reported that about one-half of the 55 workers involved in 2,4,5-T production showed increased levels of serum cholesterol and phospholipids. Walker and Martin[33] have examined eight workers who showed chloracne due to occupational exposure to TCDD-containing products, and found serum triglyceride and cholesterol levels higher than in age-matched controls (n = 100). Though these epidemiological data do not provide an absolute proof for a causal relationship, it points to the need of further studies in this area.

Reduced LDL receptor activity could explain the hypercholesterolemia seen in TCDD-treated guinea pigs. Cholesterol-metabolizing systems including the LDL receptor are scattered among various tissues in the guinea pig. This is in contrast to other animals, such as rats, which have a more centralized system in the liver. It is possible the guinea pig is more susceptible to TCDD because of such an unusual distribution of systems handling cholesterol metabolism coupled with a susceptibility to elevated levels of cholesterol (e.g., via diet). This results not only in increases of LDL but also reduced LDL-receptor activities, as in the case of the rabbit.

Nature of TCDD-Caused Changes in EGF Receptor

The above studies have shown that, at least in one case, TCDD-caused changes in plasma membrane receptor function could lead to deleterious physiological consequences. Certainly the guinea pig and the rabbit are good models in this regard. At the same time, we are aware of many other biochemical changes and lesions occurring simultaneously in animals exposed to TCDD, and the problems of not being able to explain prominent and consistent toxic manifestations, such as body weight loss, thymic involution, and hyperplastic responses of epithelial tissues. During the survey phase of our investigation we noted that the hepatic EGF receptor is very sensitive to TCDD's effect. To gain further insights to the cause of TCDD's action, we have decided to study the nature of EGF receptor changes in several rodent species in more detail.

We have examined the effects of in vivo-administered TCDD on three receptors on the rat hepatic plasma membrane at various doses to see which one of them is most sensitive (Figure 5). It is clear from the results that the effect of TCDD on EGF binding was most pronounced followed by that on Con A and insulin under these experimental conditions. It appears most significant that the effect of EGF binding is observable even at such a low does as 1.0 µg/kg (i.p.). Since the LD_{50} value (40 days after single application) of TCDD in the male rat via intraperitoneal application is on the order of 50 µg/kg (as compared with 25 µg/kg oral LD_{50}), such an effect must be considered a very sensitive parameter.

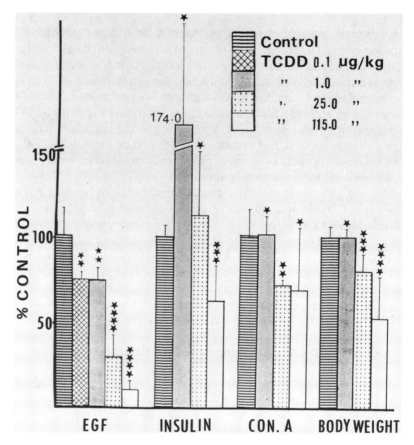

Figure 5. Dose-response relationship of specific binding of ^{125}I-EGF, ^{125}I-insulin or ^{3}H-Con A to liver plasma membrane from control and TCDD-treated rats. TCDD was administered as a single i.p. dose at 0.1, 1.0, 25 or 115 μg/kg. Controls received the same volume of the vehicle (corn oil/acetone, 9:1). Ligand binding was studied in the plasma membrane prepared 10 days after treatment. Data are expressed as percentages of control values. Specific bindings in control for EGF, insulin and Con-A are 24.5 + 3.6 pg, 8.4 + 0.5 pg/50 μg protein and 15.2 + 2.5 ng/25 μg protein, respectively. Body weight gains or losses at these doses are also represented in the graph and are expressed as percentages of the final body weight of the control rats. Mean + S. E. of 4 to 8 rats. Asterisks indicate p values of Student's t test: **** p <0.01, *** p <0.05, ** p <0.1 and * p <0.1.

One of the characteristic toxic manifestations of TCDD is the enormous species difference in susceptibility and symptoms. Comparison of the toxicological parameters between the hamster and the guinea pig has been made by other researchers.[9] For this purpose, the dose response of EGF binding was examined. Ten days after single i.p. treatments the plasma membranes were isolated from each animal, and the levels of EGF binding were quantified. The results indicate that the guinea pig system was most sensitive

($ED_{50} \simeq 0.5 \, \mu g/kg$), followed by that of the rat ($ED_{50} = 10 \, \mu g/kg$) and the hamster ($ED_{50} = 30 \, \mu g/kg$). If one adopts the I_{50} (i.e., the dose of TCDD to suppress EGF binding to 50% of the control level) as a critical dose, the rat and the hamster may be considered to be about 20 and 60 times less sensitive than the guinea pig. At I_{75} the species difference was much greater.

Next, several strains of mice known to show different degrees of tolerance to TCDD were examined.[34,35] The results (Table 4) clearly indicate that all the sensitive strains have shown reduction in EGF receptors, while the tolerant strain shows less of a response. One of the strains, CBA/J, is known to possess a high-affinity TCDD receptor in their cytosol, though it is tolerant as judged by their response to TCDD in terms of cleft palate formation (i.e., teratogenic parameter).[35] It is noteworthy that this CBA/J strain is sensitive to TCDD as judged by the EGF receptor assay as well as body weight loss and thymic involution.

To study whether some of the TCDD-caused toxic effects are similar to those produced by excess EGF, mouse neonates were treated postnatally with TCDD through mothers' milk, and various developmental parameters were examined. The most recognized specific in vivo effects of EGF are early eye opening and tooth eruption in mouse neonates.[36,37] TCDD's action was clear (Table 5) in that both events occur at earlier dates in the treated mice than in controls. Other parameters examined also show that the lesions caused by TCDD are remarkably similar to those occurring in EGF-treated animals.

To study the cause for the changes in EGF receptors, hepatic plasma membrane preparations from TCDD-treated and control rats were incubated with gamma-^{32}P-ATP as described by Rubin et al.[38] Resulting phosphoproteins clearly indicate that the general pattern of changes in protein kinase activity is similar to the rest of the membrane biochemical parameter, starting from day 2, reaching the maximum level by day 20, and showing some recovery by day 40, except that in the case of kinases, TCDD's effect was stimulation rather than reduction (Figure 6).

To study the nature of native protein substrates for increased protein kinases, hepatic plasma membrane preparations from TCDD-treated and control rats and guinea pigs were incubated with gamma-^{32}P-ATP without histone, and resulting phosphoproteins were analyzed with SDS polyacrylamide gel-electrophoresis. It was noted immediately that intensities of a number of bands in preparations from TCDD-treated animals were high in the resulting electrophoretogram/autoradiogram (Figure 7). The bands showing increased intensity were: for guinea pigs 270 K, 170 K, 118 K, 90–95 K, 80–85 K, 68 K, 58–64 K, 53 K, 41 K, and 35 K, and for rats 270 K, 170 K, 150 K, 130 K, 115 K, 98 K, 87 K, 77 K, 67 K, 55–58 K, 53 K, 48–50 K, 37–40 K, 33 K, 27 K daltons. On the other hand, the 54 K band intensity was decreased for guinea pigs, and the 84 K region was decreased for rats.

As stated previously, the band region at 170 K corresponds to the EGF

Table 4.   Changes in ^{125}I-EGF Binding to Hepatic Plasma Membrane, Body Weight and Thymus Weight in Mouse Strains 10 Days after TCDD Treatment.[a]

Strain	EGF Bound (pg/50 μg Protein)		Body Weight (% Initial)		Thymus (mg)	
	Control	Treated[b]	Control	Treated[b]	Control	Treated[b]
C57BL/6J	102.7 + 5.8(7)	7.1 + 3.3(3)***	95.7 + 2.9(3)	89.8 + 4.6(3)*	46.0 + 1.7(3)	10.7 + 1.2(3)***
CBA/J	122.0 + 14.0(7)	2.7 + 0.1(3)***	103.1 + 2.8(3)	88.8 + 3.9(3)**	42.7 + 2.5(3)	11.0 + 2.7(3)***
AKR/J	102.8 + 11.9(4)	55.5 + 15.2(3)***	98.3 + 1.1(3)	96.8 + 5.7(3)	82.0 + 7.6(3)	23.0 + 0.9(3)***

[a]Mice were treated with a single i.p. dose of TCDD in corn oil/acetone (9:1) at 115 μg/kg. Controls received an appropriate volume of the vehicle alone (0.5 ml/100 g body weight). Plasma membrane was prepared from individual mice and ^{125}I-EGF binding was studied as described earlier.[15]

[b]Values are significantly different from respective controls; * $p < 0.05$; ** $p < 0.01$; *** $p < 0.0005$.

Table 5. Effect of TCDD on Eyelid Opening, Incisor Eruption and Hair Growth on Neonatal BALB/C Mice.[a]

| | Treatment | | |
	Control	TCDD	EGF
Development			
Eyelid opening (days)	13.7 ± 0.5(14)	11.4 ± 0.5(13)**	10.7 ± 0.5(10)**
Tooth eruption (days)			
lower	9.9 ± 0.5(14)	9.0 ± 0.4(13)**	7.5 ± 0.5(10)**
upper	11.0 ± 0.0(14)	10.2 ± 0.4(13)**	8.2 ± 0.1(10)**
Hair[b]			
length (mm)	7.3 ± 0.9(24)	4.9 ± 0.7(36)**	5.1 ± 0.7(24)**
diameter (μm)	17.5 ± 0.3(30)	12.3 ± 0.2(30)**	13.2 ± 0.3(25)**
Weights[c]			
Body Weight (g)	10.0 ± 0.7(10)	7.9 ± 0.8(10)**	8.2 ± 0.5(5)*
Thymus (mg)	74.0 ± 12.6(10)	41.7 ± 6.8(10)**	54.5 ± 2.5(5)*

[a]Neonatal mice were exposed to TCDD through mothers' milk. The dams received a single i.p. dose of TCDD (10 μg/kg) within 3 hr after delivery. Control mothers received vehicle only (acetone/corn oil, 1:9). EGF at 2 μg/g body weight was injected subcutaneously to the neonatal mice daily for 14 days. * $p < 0.1$; ** $p < 0.0005$.
[b]Hair measurements were taken on 14-day-old neonatal mice.
[c]Measured at 22 days of age. EGF treatment was continued for 21 days. All data are expressed as mean ± standard error (number of animals).

receptor, which is often accompanied with 150 K band representing its degradation product.[41] We have previously shown that TCDD, when administered in vivo, causes a reduction in the number of EGF receptors in the hepatic plasma membrane in guinea pigs and rats.[13,15] One of the agents known to cause a reduction in the number of EGF receptors is TPA (12-0-tetradecanoyl phorbol-13-acetate). This cancer-promoter acts in this manner by activating the protein kinase C-TPA receptor,[42] which in turn increases phosphorylation of EGF receptor proteins. Since there are a number of similarities between the biological effects caused by TCDD and TPA, an attempt was made to study whether some of the stimulatory effects of TCDD on phosphorylation activities are due to its activation on protein kinase C.

Protein kinase C activity was defined as the portion of kinase activity stimulated by calcium and phosphatidyserine.[42-44] Since the properties of protein kinase C in the liver tissues have not been well documented, we have first examined the effects of agents known to stimulate protein kinase C in other tissues on the EGTA-EDTA-solubilized enzyme preparation from the rat hepatic plasma membrane. It is clear from the results (Figure 8) that protein kinase C in this hepatic preparation responds to these agents in the manner consistent with the description of this enzyme from other tissues.[42,44] As expected, TPA's stimulatory effect was particularly significant.

To study whether the pattern of increases in phosphorylation among

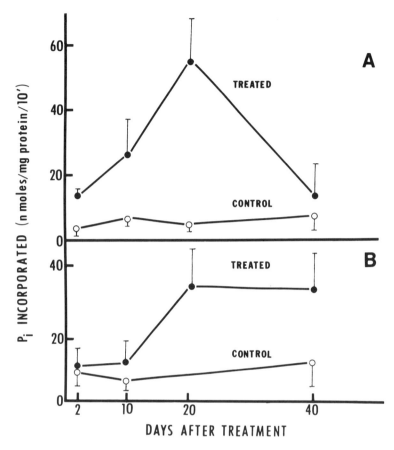

Figure 6. Time course of changes in protein kinase activities in the hepatic plasma membranes from male Sprague-Dawley rats treated with 25 μg/kg, (single, intraperitoneal injection) of TCDD. (A) c-AMP-independent protein kinases, and (B) c-AMP-dependent protein kinases. Control rats received the same volume of vehicle (corn oil/acetone, 9:1) only. At given time periods, the animals were sacrificed and their livers were processed for plasma membrane. Protein kinase activity was measured according to Corbin and Reimann.[39]

native plasma membrane proteins by TCDD administration is similar to that evoked by that of TPA, hepatic plasma membrane preparations from TCDD treated (1 μg/kg; single i.p.; tested on day 10) and control guinea pigs were incubated with TAP (1 μg added to 100 μg membrane protein) for 10 min at 0°C in vitro and then with gamma-^{32}P-ATP. The resulting phosphoproteins were analyzed through SDS-PAGE and radioautography as before. TPA was found to increase phosphorylation of proteins at 35 K, 41 K, 82 K, 150 K, and 170 K, in addition to low-molecular-weight proteins (< 20 K). EGF alone (2.5 μg for 10 min at 0°C) increased the band intensity at

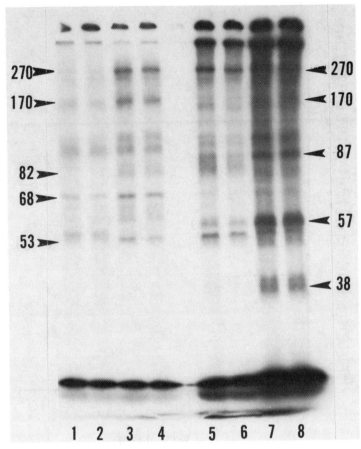

Figure 7. SDS-electrophoretogram/radioautogram of hepatic plasma membrane proteins phosphorylated with gamma ^{32}P-ATP from control (Lanes 1,2) and TCDD-treated (1 μg/kg single i.p.) guinea pigs (Lanes 3,4) and control (Lanes 5,6) and treated (25 μg/kg single i.p.) rats (Lanes 7,8). To obtain the radioautogram, the hepatic plasma membrane preparations were isolated from the control and treated animals after 10 days from the time of TCDD injection, and incubated directly with gamma-^{32}P-ATP without histone as described by Rubin et al.[38] The reaction was stopped with addition of SDS, the level of total protein for each electrophoresis test was adjusted and electrophoresis was developed.[40] Thereafter, the gel was dried and exposed to Fuji X-ray film for radioautography for 7 to 14 days. The position of phosphorylated EGF receptor 170 and 150 kD and insulin receptor (subunit, approximately 90 kD) were determined by incubating fresh rat hepatic plasma membranes with unlabeled EGF and insulin, respectively, in a separate experiment.

64 K, 150 K, and 170 K. TCDD increased band intensities of all those proteins.

We have also examined the levels of protein kinase C in various preparations from TCDD-treated and control animals. The results (Table 6) clearly indicate that the levels of protein kinase C are higher in the membrane

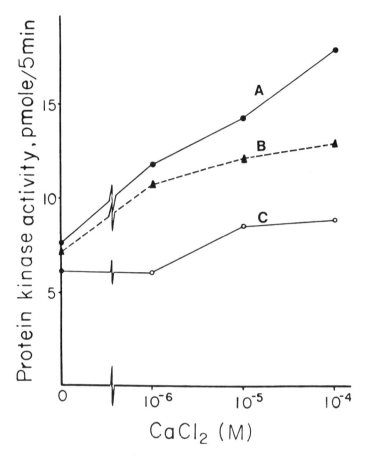

Figure 8. Stimulatory effects of Ca^{2+} on the protein kinase C activity in the rat hepatic plasma membrane in the presence of (A) TPA + phosphatidylserine (PS), (B) diolein + PS, and (C) PS alone. Protein kinase C in the plasma membrane was solubilized with EGTA-EDTA and assayed as described by Niedel et al.[42] Kinase activity was expressed as pmol Pi incorporated into histone per 5 min per tube, each tube containing 40 μg membrane protein. The buffer contained 20 mM tris-HCl (pH 7.5), 50 MM β-mercaptoethanol and 2 mM of phenylmethylsulfonylfluoride. The amounts of agents used were: 20 μg/mL PS, 1 μg/mL diolein and 10 μg/mL of TPA. The assay was performed in the presence of 0.7 mM EGTA or added CaCl$_2$ at 10 μM to 0.1 mM incubated at 37°C.

fractions from TCDD-treated rats and guinea pigs than from control animals.

DISCUSSION

In this chapter, we have summarized our recent work concerning TCDD's effects on hepatic plasma membrane associated proteins. The most notice-

Table 6. Levels of Protein Kinase C in the Hepatic Plasma Membrane Preparations from Control and TCDD-Treated Rats (25 μg/kg single i.p. assayed on day 10) and Guinea Pigs (1 μg/kg single i.p. assayed on day 10).[a]

Assay Conditions	Protein Kinase Activity (pmol Pi/5 min/40 μg protein)	
	Control	TCDD-treated
Rats (6 animals)		
Basal kinase activity (a)	10.49 ± 2.29	15.65 ± 9.09
Kinase activity in the presence of Ca^{2+} and phosphatidyl-serine (b)	13.64 ± 2.17	23.30 ± 8.90*
Protein kinase C activity (i.e., b – a)	3.15 ± 0.99	7.62 ± 3.60*
Guniea pigs (6 animals)		
Basal kinase activity (c)	3.02 ± 0.72	3.87 ± 0.56*
Kinase activity in the presence of Ca^{2+} and phosphatidyl-serine (d)	4.25 ± 0.47	6.19 ± 0.76*
Protein kinase C activity (i.e., d – c)	1.23 ± 0.69	2.32 ± 1.18

[a]Significant difference against corresponding control values; * $p < 0.05$. All data are presented as mean ± standard deviation determined by using six animals for each test.

able changes occurring as a result of a single administration of TCDD are reduction in receptor activities for EGF, LDL, and insulin, and decreased enzymatic activities such as various ATPases as well as increases in protein kinases. All these changes become significant on about day 2 TCDD administration in vivo in rats and the difference continues to magnify until day 20.

The important question is how a single administration of TCDD causes such simultaneous changes in these surface membrane proteins. Such a phenomenon resembles closely the pleiotropic changes observed among cells receiving a message through a hormone, an inducer, a genetically predetermined DNA activity change such as a developmental sequence, or even a physical stimulus such as partial hepatectomy.

To investigate the cause for such TCDD-evoked pleiotropic changes, we have used EGF receptor changes as a molecular probe. This parameter was the most sensitive to TCDD, and there is a wealth of information regarding the mode of modulation of this particular receptor with respect to its association with cellular changes leading to ultimate carcinogenic expression.

In the past, several agents have been identified as being able to reduce EGF receptor binding.[45] Examples include: phorbol acid esters such as TPA,[46,47] Rous sarcoma virus and other viruses causing neoplastic transformation,[48,49] hormones such as vasopressin,[50,51] and the solvent dimethylsulfoxide.[41]

At least the first three groups of agents are known to cause or promote neoplastic growth in various tissues. The most intriguing aspect of this phenomenon is that the tissues and the cells exhibiting loss of EGF receptors by these treatments start behaving as though they have received a large dose of EGF.[45-47] For instance, cells which have been treated with TPA show a decreased number of EGF receptors and exhibit a rise in ornithine decarboxylase, stimulation of plasminogen activator, enhancement of sugar transport and prostaglandin synthesis, and stimulation of mitogenic activities. All these cellular changes are observed in cells and tissues receiving excess EGF.[46,47] Rose et al.[52] have shown that EGF itself promotes skin tumors in mice treated with methylcholanthrene, as in the case of TPA.

The cause of this "EGF-like" effect of agents which reduce EGF receptor activity has been proposed by DeLarco and Todaro[53] and Weinstein and coworkers[46,47,54,55] to be due to production of endogenous growth factor(s) by the affected cells and tissues. In the case of transformation viruses such growth factors (termed TGF for "transformation growth factors") have been found, isolated, and characterized.[56,57] An alternative explanation for these events may be that these agents cause phosphorylation of the EGF receptor, as shown in the case of TPA, dimethysulfoxide, Rous sarcoma virus, etc., and that this phosphorylation action alone could activate a chain of events normally triggered by the external ligand, EGF. In turn, these cells with phosphorylated receptors may no longer require external ligands to process the messages of transaction. After a prolonged phosphorylation of the receptor-associated protein, the receptor may lose the ability to bind with the external ligand (i.e., "down regulation") and in due course disappear altogether.

Whatever the cause for the EGF-like effects of these agents, there is indeed good evidence to indicate that the decrease in number of the EGF receptors, whether caused by chemicals or physical treatments, is causally related to physiological changes of tissues. For example, after partial hepatectomy, EGF binding starts declining in male rats, and this decline precedes the onset of DNA synthesis and eventual cell division. Phosphorylation of the EGF receptor protein is intimately associated with the initial event of decline of EGF binding.

In this study, we have found that TCDD does indeed increase phosphorylation of the EGF receptor in the rat liver plasma membrane, and that it has the same in vivo effects as exogenously administered EGF with regard to the early eye opening and the tooth eruption in mouse neonates.

The pleiotropic nature of protein kinase-induced cellular responses has been pointed out by many scientists. Particularly affected are cell surface receptors. Changes in the cell surface characteristics induced by chemical cancer promoters and transforming viruses are well known. Thus it is reasonable to assume that stimulation of protein kinase C by TCDD produces end results similar to the ones caused by TPA and EGF.

However, it must be pointed out that there are some differences between the in vivo effects of TCDD and TPA. TCDD has been shown to be a promoter of both the skin and liver tumors, while TPA is known as a promoter of skin tumors only. Dermal application of TCDD does not cause necrotic skin damage such as that caused by TPA. Therefore, it is wrong to assume that all of the cellular changes brought by TCDD or TPA administration are related to the increase in protein kinase C. However, reduction in receptor activities of EGF and insulin, and changes in lipid metabolism, etc., are events which are potentially attributable to changes in protein kinase C. Particularly with the EGF receptor, recent reports clearly indicate that the increased protein kinase C activity, due to administration of TPA, results in changes in phosphorylation of the EGF receptor protein.

The cause for the stimulation of protein kinases by TCDD is not known at this time, though it is unlikely to be identical to TPA, since it takes a much longer time for TCDD to cause the same end result. Preliminary data from this laboratory show that reduction in EGF receptor activity by TCDD occurs only in mice strains which have the specific cytosolic receptor for TCDD (i.e., B57BL/6J and CBA/J strains with Ah^b locus). Less of an effect was seen in the AKR/J strain, which lacks such a receptor. In that case, the stimulation of protein kinases is likely to be mediated by changes in DNA activity. Another possibility is that TCDD causes excess production of hormones in other cellular regulatory factors such as growth factors by cells, which in turn stimulate intracellular protein kinase activities.

Whatever the cause for TCDD-evoked stimulation of protein kinases in hepatocytes, it seems worthwhile to pursue this line of investigation. There is increased awareness of the importance of various protein kinases, including that associated with the EGF receptor, in regulating cellular functional and differentiational expression.

In conclusion, we have been able to demonstrate that many biochemical functions on the hepatic plasma membrane are altered as a result of a single treatment of TCDD. Particularly significant are the changes in insulin, LDL, and EGF receptor properties. It may be pertinent to point out that these receptors are very similar to each other in terms of their amino acid composition,[58] molecular arrangements, and receptor activation processes (e.g., protein kinase activation, internalization and receptor metabolism). Particularly important is the fact that they are under tight homeostatic controls, involving down regulation and protein kinase activation. Thus our observation that there is a change in a whole series of protein kinases as a result of TCDD administration raises the possibility that such an increase is causally related to the changes observed in various membrane receptors and enzymes in hepatocytes.

One must be careful in determining which of these changes are the result of the initial and the primary action of TCDD and which are secondary and tertiary effects of the primary ones, in view of the whole scale of changes observed. Also, one must be extremely cautious in relating any of the above biochemical changes to in vivo lesions and symptoms caused by TCDD.

Within such constraints, we were able to point out two likely TCDD-caused biochemical changes, EGF and LDL receptors, which are likely to contribute to in vivo symptoms and lesions. The accompanying increase in protein kinases raises the possibility that these two phenomena are causally related.

ACKNOWLEDGMENTS

This research was supported by the Michigan Agricultural Experiment Station, Journal Article No. 11677, Research Grant ES01963 and ES03575 from the National Institute of Environmental Health Sciences, Research Triangle Park, North Carolina, and by the Center for Environmental Toxicology, Michigan State University, East Lansing, Michigan.

REFERENCES

1. Poland, A., and J. Knutson. "2,3,7,8-Tetrachlorodibenzo-p-dioxin and related halogenated aromatic hydrocarbons: Examination of the mechanisms of toxicity," *Annu. Rev. Pharmacol. Toxicol.* 22:516-554 (1982).
2. Tucker, R. E., A. L. Young, and A. P. Gray. *Human and Environmental Risks of Chlorinated Dioxins and Related Compounds*, (New York: Plenum Publishing Corporation, 1983).
3. Lawrance, W. W. *Public Health Risks of the Dioxins* (New York: Rockefeller University, (1984).
4. Schwetz, B. A., J. M. Norris, G. L. Sparschu, V. K. Rowe, P. J. Gehring, J. L. Emerson, and C. G. Gerbig. "Toxicology of chlorinated dibenzo-p-dioxins," *Environ. Health Perspec.* 5:87-99 (1973).
5. Olson, J. R., T. A. Gasiewicz, and R. A. Neal. "Tissue distribution, excretion, and metabolism of 2,3,7,8-tetrachlorodibenzo-p-dioxin (TCDD) in the Golden Syrian Hamster," *Toxicol. Appl. Pharmacol.* 56:78-85 (1980).
6. Kitchin, K. T., and J. Woods. "2,3,7,8-tetrachlorodibenzo-p-dioxin (TCDD) effects on hepatic microsomal cytochrome P-448-mediated enzyme activities," *Toxicol. Appl. Pharmacol.* 47:537-546 (1979).
7. Poland, A., and A. Kende. "The genetic expression of aryl hydrocarbon hydroxylase activity: Evidence for a receptor mutation in non-responsive mice," In *Origins of Human Cancer*, H. H. Hiatt, S. D. Watson, and J. A. Winston, Eds. (Cold Spring, New York: Cold Spring Harbor Laboratory, 1977), p. 847.
8. Nebert, D. W. "Genetic differences in the induction on monooxygenase activities by polycyclic aromatic compounds," *Pharmacol. Ther.* 6:395-417 (1979).
9. Gasiewicz, T. A., and G. Rucci. "Comparison of the binding properties of cytosolic receptors for 2,3,7,8-tetrachlorodibenzo-p-dioxin (TCDD) in various mammalian species," *Toxicologist* 2:145 (1982).
10. Matsumura, F. "Biochemical aspects of action mechanisms of

2,3,7,8–tetrachlorodibenzo-p-dioxin (TCDD) and related chemicals in animals," *Pharmacol. Ther.* 19:195–209 (1983).

11. Peterson, R. E., B. V. Madhukar, K. H. Yang, and F. Matsumura. "Depression of adenosine triphosphatase activities in isolated liver surface membranes of 2,3,7,8–tetrachlorodibenzo-p-dioxin-treated rats," *J. Pharmacol. Exp. Ther.* 210:275–282 (1979).

12. Brewster, D. W., B. V. Madhukar, and F. Matsumura. "Influence of 2,3,7,8–TCDD on the protein composition of the plasma membrane of the hepatic cells from the rat," *Biochem. Biophys. Res. Commun.* 107:68–74 (1983).

13. Matsumura, F., D. W. Brewster, B. V. Madhukar, and D. W. Bombick. "Alteration of rat hepatic plasma membrane functions by 2,3,7,8–tetrachlorodibenzo-p-dioxin (TCDD)," *Arch. Environ. Contam. Toxicol.* 13:509–515 (1984).

14. Matsumura, F., B. V. Madhukar, D. W. Bombick, and D. W. Brewster. "Toxicological significance of pleiotropic changes of plasma membrane functions, particularly that of EGF receptor, caused by TCDD," In *Banbury Report* 18: *Biological Mechanisms of Dioxin Action.* A. Poland and R. D. Kimbrough, Eds. (Cold Spring, New York: Cold Spring Harbor Laboratory, 1984) p. 267.

15. Madhukar, B. V., D. W. Brewster, and F. Matsumura. "Effects of in vivo-administered 2,3,7,8–tetrachlorodibenzo-p-dioxin on receptor binding of epidermal growth factor in the hepatic plasma membrane of rat, guinea pig, mouse, and hamster," *Proc. Natl. Acad. Sci. U.S.A.* 81:7407–7411 (1984).

16. Bombick, D. W., F. Matsumura, and B. V. Madhukar. "TCDD (2,3,7,8–tetrachlorodibenzo-p-dioxin) causes reduction in the low density lipoprotein (LDL) receptor activities in the hepatic plasma membrane of the guinea pig and rat," *Biochem. Biophys. Res. Commun.* 118:548–554 (1984).

17. Bombick, D. W., B. V. Madhukar, D. W. Brewster, and F. Matsumura. "TCDD (2,3,7,8–tetrachlorodibenzo-p-dioxin) causes increases in protein kinases especially protein kinase C in the hepatic plasma membrane of the guinea pig and rat," *Biochem. Biophys. Res. Commun.* 127:296–302 (1985).

18. Gupta, B. N., J. G. Vos, J. A. Moore, J. G. Zinkl, and B. C. Bullock, "Pathologic effects of 2,3,7,8–tetrachlorodibenzo-p-dioxin in laboratory animals. *Environ. Health Perspect.* 73:125–140 (1973).

19. Hook, G. E. R., J. K. Haseman, and G. W. Lucier. "Induction and suppression of hepatic and extra-hepatic microsomal foreign-compound-metabolizing enzyme systems by 2,3,7,8–TCDD," *Chem. Biol. Interact.* 10:199–214 (1975).

20. Swift, L. L., T. A. Gasiewicz, G. D. Dunn, P. D. Soule, and R. A. Neal. "Characterization of the hyperlipidemia in guinea pigs induced by 2,3,7,8–tetrachlorodibenzo-p-dioxin," *Toxicol. Appl. Pharmacol.* 59:489–499 (1981).

21. Gasiewicz, T. A., and R. A. Neal. "2,3,7,8–tetrachlorodibenzo-p-

dioxin tissue distribution, excretion, and effects on clinical parameters in guinea pigs," *Toxicol. Appl. Pharmacol.* 51:329–339 (1979).

22. Havel, R. J., H. A. Eder, and J. F. Bragdon. "The distribution and chemical composition of ultracentrifugally separated lipoproteins in human serum," *J. Clin. Invest.* 34:1345–1353 (1955).

23. Bio-Rad Laboratories, Richmond, California. *Technical Bulletin 1071 on Enzymobeads.*

24. Zinkl, J. G., J. G. Vos, J. A. Moore, and B. N. Gupta, "Hematologic and clinical chemistry effects of 2,3,7,8-tetrachlorodibenzo-p-dioxin in laboratory animals," *Environ. Health Perspec.* 5:111–118 (1973).

25. Albro, P. W., J. T. Corbett, M. Harris, and L. D. Lawson. "Effects of 2,3,7,8-tetrachlorodibenzo-p-dioxin on lipid profiles in tissue of the Fischer rat", *Chem. Biol. Interact.* 23:315–330 (1978).

26. Poli, A., G. Grancescini, L. Puglisi, and C. R. Sirtori. "Increased total and high density lipoprotein cholesterol with apoprotein changes resembling Streptozotocin diabetes in tetrachlorodibenzodioxin (TCDD) treated rats," *Biochem. Pharmacol.* 28:835–838 (1980).

27. Brown, M. S., P. T. Kovanen, and J. Goldstein. "Regulation of plasma cholesterol by lipoprotein receptors," *Science* 212:628–635 (1981).

28. Brown, M. S., P. T. Kovanen, and J. Goldstein. "Receptor mediated uptake of lipoprotein cholesterol and its utilization for steroid synthesis in the adrenal cortex," *Recent Prog. Horm. Res.* 35:215–249 (1979).

29. Soltys, P., and O. W. Portman. "Low density lipoprotein receptors and catabolism in primary cultures of Rabbit hepatocytes," *Biochim. Biophys. Acta* 617:335–346 (1979).

30. Goldstein, J., T. Kita, and M. Brown. "Defective lipoprotein receptors and atherosclerosis. Lessons from an animal counterpart of fumilial hypercholesterolemia," *N. Engl. J. Med.* 309:288–296 (1983).

31. Oliver, R. M. "Toxic effects of 2,3,7,8 tetrachlorodibenzo 1,4 dioxin in laboratory workers," *Brit. J. Ind. Med.* 32:49–53 (1975).

32. Pazderova-Vejlupkova, J., E. Lukas, M. Nemcova, M. Spacilova, L. Jorasek, J. Kalensky, J. John, J. Jirasek, and J. Pickova. "Chronic intoxication by chlorinated hydrocarbons produced during the manufacture of sodium 2,4,5-trichlorophenoxyacetate. *Pracok.Lek* 26:332–339 (1974).

33. Walker, A. E., and J. V. Martin. "Lipid profiles in dioxin-exposed workers," *Lancet*, February 24:446–447 (1979).

34. Poland, A., and E. Glover. "Genetic expression of aryl hydrocarbon hydroxylase activity by 2,3,7,8-tetrachlorodibenzo-p-dioxin: Evidence for receptor mutation in genetically non-responsive mice," *Mol. Pharmacol.* 11:389–398 (1975).

35. Poland, A., and E. Glover. "2,3,7,8-Tetrachlorodibenzo-p-dioxin: Segregation of toxicity with the Ah locus," *Mol. Pharmacol.* 17:86–94 (1980).

36. Heinberg, M., I. Weinstein, V. S. LeQuire, and S. Cohen. "The induction of fatty liver in neonatal animals," *Life Sci.* 4:1625–1633 (1965).

37. Cohen, S., and G. A. Elliott. "The stimulation of epidermal keratiniza-

tion by a protein isolated from the submaxillary gland of the mouse," *J. Invest. Dermatol.* 40:1–5 (1962).

38. Rubin, R. A., E. J. O'Keefe, and H. S. Earp. "Alteration of epidermal growth factor-dependent phosphorylation during rat liver regeneration," *Proc. Natl. Acad. Sci. U.S.A.* 79:776–780 (1982).
39. Corbin, J. D., E. M. Reimann. "Assay of c-AMP dependent protein kinases," In *Methods in Enzymology*, J. G. Hardman and B. W. O'Malley, Eds. (New York: Academic Press, vol. 38, 1974) p. 287.
40. Laemmli, U. K. "Cleavage of structural proteins during the assembly of the head of bacteriophage T_4," *Nature* 277:680–685 (1970).
41. Rubin, R. A., and H. S. Earp. "Dimethylsulfoxide stimulates tyrosine residue phosphorylation of rat liver epidermal growth factor receptor," *Science* 219:60–63 (1983).
42. Niedel, J. E., L. J. Kuhn, and G. R. Vandenbark. "Phorbol diester receptor co-purifies with protein kinase C," *Proc. Natl. Acad. Sci.* 80:36–40 (1983).
43. Kishimoto, A., Y. Takai, T. Mori, U. Kikkawa, and Y. Nishizuka. "Activation of calcium and phospholipid-dependent protein kinase by diacylglycerol, its possible relation to phosphatidylinositol turnover," *J. Biol. Chem.* 255:2273–2276 (1980).
44. Takai, Y., A. Kishimoto, Y. Iwasa, Y. Kawahara, T. Mori, and Y. Nishizuka. "Calcium dependent activation of a multifunctional protein kinase by membrane phospholipids," *J. Biol. Chem.* 254:3692–3695 (1979).
45. Schlessinger, J., A. B. Schreiber, A. Levi, I. Lax, T. Libermann, and Y. Yarden. "Regulation of cell proliferation by epidermal growth factor," *CRC Crit. Rev. Biochem.* 14:93–111 (1982).
46. Lee, L. S., and I. B. Weinstein. "Tumor-promoting phorbol esters inhibit binding of epidermal growth factor to cellular receptors," *Science* 202:313–315 (1978).
47. Lee, L. S., and I. B. Weinstein. "Mechanism of tumor promotor inhibition of cellular binding of epidermal growth factor," *Proc. Natl. Acad. Sci. U.S.A.* 76:5168–5172 (1979).
48. Todaro, G. J., J. B. DeLarco, and S. Cohen, "Transformation by murine and feline sarcoma virsues specifically block binding of epidermal growth factor to cells," *Nature* 264:26–31 (1976).
49. Erikson, E., P. Shealy, and R. L. Erikson. "Evidence that viral transforming gene products and epidermal growth factor stimulate phosphorylation of the same cellular protein with same specificity. *J. Biol. Chem.* 256:11381–11384 (1981).
50. Rosengurt, E., K. D. Brown, and D. Pettican. "Vasopressin inhibition of epidermal growth factor binding to cultured mouse cells," *J. Biol. Chem.* 256:716–722 (1981).
51. Hayden, L. J., and D. L. Severson. "Correlation of membrane phosphorylation and epidermal growth factor binding to hepatic membranes isolated from triidothyronine-treated rats," *Biochim. Biophys. Acta* 730:226–230 (1983).
52. Rose, S., P. R. Stahn, D. S. Passovoy, and H. Herschman. "Epidermal

growth factor enchancement of skin tumor induction in mice," *Experientia* 32:913–914 (1976).

53. DeLarco, J. E., and G. J. Todaro. "Growth factors from murine sarcoma virus-transformed cells," *Proc. Natl. Acad. Sci. U.S.A.* 75:4001–4005 (1978).

54. Weinstein, I. B., M. Wigler, and C. Pietropaolo. In *Carcinogenesis*, T. J. Slaga, A. Sivak and R. K. Boutwell, Eds. (New York: Raven Press, Vol 2, 1977a), p. 313.

55. Weinstein, I. B., M. Wigler, and C. Pietropaolo. "The action of tumor promoting agents in culture," In *Origins of Human Cancer*, H. H. Hiatt, J. D. Watson and J. A. Winston, Eds. (Cold Spring, New York: Cold Spring Harbor Laboratory, 1977b), p. 751.

56. Todaro, G. J., J. E. DeLarco, "Growth factors produced by sarcoma virus-transformed cells," *Cancer Res.* 38:4147–4154 (1978).

57. Todaro, G. J., J. E. DeLarco, H. Marquardt, M. L. Bryant, S. A. Sherwin, and A. H. Sliski, "Polypeptide growth factors produced by tumor cells and virsus-transformed cells," In *Hormones and cell culture* G. Sato and R. Ross Eds. (Cold Spring Harbor, New York: Cold Spring Harbor Laboratory, 1979), p. 113.

58. Sudhof, T. C., D. W. Russell, J. L. Goldstein, M. S. Brown, R. Sanchez-Pescador, and G. I. Bell. "Casette of eight exons shared by genes for LDL receptor and EGF precursor," *Science* 228:893–895 (1985).

CHAPTER 19

Development and Application of an In Vitro Bioassay for Dioxinlike Compounds

John F. Gierthy and Deborah Crane

INTRODUCTION

Environmental contamination by 2,3,7,8–tetrachlorodibenzo-p-dioxin (TCDD) has become a major public health issue. TCDD is the most toxic synthetic substance known.[1,2] The toxicity of TCDD varies greatly from species to species, with the LD_{50} for guinea pigs, the most sensitive animal, being about 1 µg/kg body weight, while that for the hamster is 5 mg/kg body weight. TCDD toxicity is pleomorphic and, in animals, includes prolonged wasting syndrome prior to death; lymphoid involution; embryotoxicity and/or teratogenicity; hyperkeratosis; edema; hyperplasia of the epithelium of the stomach, intestines, and urinary bladder; hepatocellular damage; and thymic involution.[3-5] The ultimate cause of death in animals exposed to TCDD is unknown.[2] TCDD has also been shown, in animal studies, to be a potent carcinogen and tumor promoter.[6,7]

Few details of the toxicity of TCDD in humans, either acute or chronic, are currently known, and the topic is highly controversial. Acneform lesions (chloracne) are one of the most common known toxic manifestations of TCDD exposure in humans,[3,4] and evidence exists which shows possible links between human exposure to TCDD and the occurrence of cancer (soft tissue sarcoma)[8,9] although this has been disputed.[10]

Another class of halogenated aromatic hydrocarbon pollutants in the environment is the polychlorinated dibenzofurans (PCDFs), which have been found as the result of polychlorinated biphenyl (PCB) pyrolysis in heating fluids[11,12] and in soot resulting from a PCB-containing transformer fire.[13,14] In some cases PCDF exposure has induced acute toxicity symptoms in humans similar to those of TCDD (e.g., chloracne).[4]

Hyperkeratinization of the epithelium is considered to be responsible for the occurrence of chloracne in humans exposed to polychlorinated dioxins (PCDDs) and PCDFs. This effect is thought to be caused by an induced differentiation of the squamous epithelium.[2] In this regard, TCDD is considered to produce a sustained stimulation of a normal physiological response which leads to chloracne. Knutson and Poland have developed and used an in vitro cell keratinization system to study this effect of PCDD isomers and congeners.[15] We have demonstrated the use of this keratinization system for the detection of these compounds in soot extracts from a fire which involved a PCB-containing electrical transformer.[16] This investigation was performed using the XB epithelial cell line cloned by Rheinwald and Green[17] from a mouse teratoma cocultured with lethally irradiated mouse fibroblasts (3T3) as described by Knutson and Poland for their in vitro keratinization studies.[15] Subsequent experiments in our laboratory led to the identification of a nonkeratinizing derivative of the XB line which we designated XBF. This variant of the XB line, when cocultured with lethally irradiated 3T3 cells, grew to high saturation density as compared to the original XB line grown under similar conditions. TCDD, at a minimal concentration of $10^{-11}M$, was shown to induce a morphological change and reversible inhibition of postconfluent cell proliferation in the XBF/3T3 system.[18] The morphological change induced by TCDD in these cultures is characterized by the appearance of flat cobblestonelike cells as compared to the fusiform, high-density control cultures. We will refer to this TCDD-induced morphological change as the flat-cell effect.

There are 75 possible PCDD congeners, and 135 possible PCDF congeners. A correlation has been demonstrated between the structures of these compounds and their biological potency with regard to toxicity, enzyme induction, aryl hydrocarbon (A*H*) receptor binding, and hyperkeratinization.[2] The potency of these compounds is modulated by the positions chlorinated. Since TCDD has been shown to be the most potent member of this class of compounds, we will refer to this general toxicity as dioxinlike activity.

Because of the possible human health hazards associated with TCDD and PCDF exposure, great importance has been placed on their detection in the environment. PCDDs and PCDFs are detected and quantitated in environmental samples by high-resolution gas chromatography and mass spectrometry (GC/MS). While these methods are highly sensitive, accurate and isomer specific, they are also costly and time consuming. Current capabilities may not be adequate to handle the large numbers of samples generated by testing programs.[19] A rapid, inexpensive screen assay for semiquantitative determination of dioxinlike activity would allow priority ranking of large numbers of environmental samples for subsequent isomer-specific quantitation by high-resolution chemical analysis. Perhaps more important, such an assay would also identify those samples with no dioxinlike activity at a predetermined level, thereby eliminating the need to subject these samples to the more rigorous and costly chemical analysis.

This chapter describes our studies performed to test the specificity and sensitivity of the TCDD-induced flat-cell effect in the XBF/3T3 culture system and the application of this effect as a bioassay for the detection of PCDDs and/or related compounds in soot extracts, surface swipes, fish, and other environmental samples.

METHODS

Chemicals

TCDD was obtained from Dow Chemical (Midland, MI). Purity was determined by mass spectrometry to be >99%. Other PCDDs and PCDFs (purity > 99%) were obtained from the National Institute of Environmental Health Sciences, the Illinois Institute of Technology, and Analabs (North Haven, CN). Single PCB isomers were obtained from Analabs at 99% purity. Commercial Aroclor 1254 was from Monsanto (St. Louis, MO). Polynuclear aromatic hydrocarbons (PAHs) were obtained from Aldrich Chemical (Milwaukee, WI), Eastman Organic (Rochester, NY), and K and K Laboratories (Plainview, NY) and recrystallized by Dr. B. Bush in these laboratories.[20] Pesticides were obtained as reference standards from the Environmental Protection Agency at > 99% purity. Analytical grade dimethysulfoxide (DMSO) was obtained from the Aldrich Chemical Co. (Milwaukee, WI). Samples of soot were collected from the upper surfaces of ceiling panels from various floors of the Binghamton State Office Building (BSOB) in Binghamton, New York after it was involved in a PCB-containing transformer fire. These soot samples were subjected to Soxhlet extraction in benzene, as described by Gierthy et al.[16] The surface swipes were collected dry from vinyl walls on the 7th and 14th floors of the BSOB using Whatman No. 40 filter paper discs. Benzene extracts of the surface swipes were supplied for testing. Fish from various sources were subjected to a hexane-benzene extraction followed by sulfuric acid treatment to remove oil and subsequent neutralization.

Cell Culture

The XBF cells were derived from the cloned XB mouse epithelial cell line which, along with the 3T3 feeder cells, were a gift from H. Green, at Harvard University.[17] The derivation of the XBF line from the XB line has been described in detail elsewhere.[18] XBF cells were routinely propagated every one to two weeks, when confluency was reached, by trypsinization (0.25%) and replating at a concentration of 3×10^4 cells per cm^2. These cells were grown in Dulbecco's Modified Eagle Medium (DMEM) obtained from Gibco, supplemented with 20% fetal bovine serum (Flow, Rockville, MD), 100 U of penicillin/mL, and 100 μg of streptomycin/mL in a humidified atmosphere of 5% CO_2. The 3T3 feeder cell stocks were grown in the same incubation conditions in DMEM supplemented with 10% calf serum (Flow).

Assay Procedure

XBF cells were suspended by trypsinization and seeded into 24–well plates (16–mm-diameter wells, 5×10^4 cells per mL per well) in medium conditioned by 24–hr exposure to confluent cultures of 3T3 cells (25 mL of medium per 75–cm^2 flask) with lethally irradiated (6000 rads) 3T3 cells (5×10^5 per mL per well). After overnight incubation (37°C, 5% CO_2, humidified), the cultures were refed with a series comprising tenfold dilutions of a stock solution of test chemical or soot extract in DMSO, or with DMSO alone, in nonconditioned DMEM supplemented with 20% fetal bovine serum. Extracts of various samples were solvent exchanged to DMSO as described elsewhere[16] and were first diluted 1:1000 in the culture medium before subsequent tenfold dilutions. Medium replacement was repeated every 3 or 4 days using freshly prepared dilutions. The highest cumulative DMSO concentration was 0.1%. After 14 days, the culture was washed with phosphate-buffered saline (PBS), fixed with formalin in PBS and stained with either Giemsa or 1% Rhodamine B in water.

The cultures were evaluated for the flat-cell effect by phase-microscopic assessment and confirmation that the tested cultures had grown to form a confluent monolayer of morphologically flat cells as compared to the high-density control cultures comprised of multilayered fusiform cells described in detail previously.[18] Macroscopic evaluation of staining intensity, i.e., an indication of cell culture density, was then made on the fixed and stained cultures as shown previously for TCDD[18] and in Figure 1 for TCDF, as compared to the TCDD calibration standard.

RESULTS

Validation of the Flat-Cell Assay

A comparison of Figures 2A and 2B illustrates the dioxin-induced morphological alteration from controls, referred to as the flat-cell effect, in XBF/3T3 cultures seeded and exposed to 10^{-9} M TCDD for 14 days. The

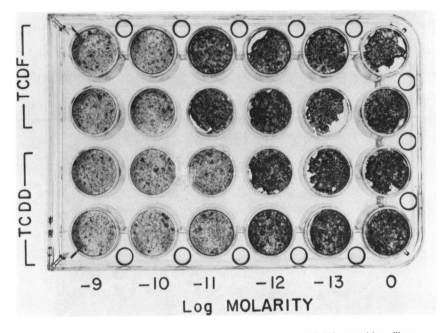

Figure 1. Dose-response relation of flat-cell induction by 2,3,7,8-tetrachlorodibenzo-p-dioxin (TCDD) and 2,3,7,8-tetrachlorodibenzofuran (TCDF). XBF/3T3 cultures were exposed to various concentrations of 2,3,7,8-TCDD and 2,3,7,8-TCDF as described in text. After 14 days the cultures were fixed and stained with Giemsa stain. The less intensely stained low-density cultures (10^{-9} to 10^{-10} M for 2,3,7,8-TCDD and 10^{-9} to 10^{-10} M for 2,3,7,8-TCDF) are similar to that shown in Figure 2B, while the remaining, intensely staining high-density cultures (10^{-12}, 10^{-13}, and 0 M) are similar to that seen in Figure 2A. The flat-cell induction is first evident with 10^{-11} M 2,3,7,8-TCDD and 10^{-10} M 2,3,7,8-TCDF.

TCDD-treated culture comprises a monolayer of flat, nondividing cells, as compared to the denser control cultures. Control cultures were treated with 0.01% DMSO and were populated by fusiform cells in a parallel fashion. The concentration of the treated cells after 14 days of incubation was routinely found to be 15×10^4 cells/cm^2, while that of the control culture was 30×10^4 cells/cm^2.

Figure 1 shows the effect of a series of tenfold dilutions of TCDD and 2,3,7,8–tetrachlorodibenzofuran (TCDF) on XBF/3T3 cultures exposed and stained with Giemsa stain. The light staining wells correspond to the flat-cell appearance seen in Figure 2B. The intensity of the staining can be used as an indicator of the flat-cell effect. Here, the TCDF exhibits an endpoint at a concentration of $10^{-10}M$, while of TCDD is $10^{-11}M$, indicating an approximate tenfold greater potency for the TCDD.

For a series of 24 chemicals, including PCDDs, PCDFs, PCBs, polycyclic aromatic hydrocarbons (PAHs), and pesticides, there was at least a 63 millionfold range in the potential of these compounds for inducing the flat-

Figure 2A. Induction of the flat-cell response by 2,3,7,8-TCDD. Figure shows the dense packing and multilayering of control XBF/3T3 cultures. Phase microscopy, 100X magnification.

cell effect (Table 1). These data show that the flat-cell effect, like the keratinization response,[15] is most sensitive to the more toxic PCDDs and PCDFs, with TCDD being the most potent. Specifically, 1,2,4,7,8–penta-CDD, which lacks chlorination in a lateral position of one of the benzene rings, shows 100 times less activity than TCDD. TCDF was shown to be the most potent PCDF congener tested and, as in the keratinization system,[15] had a flat-cell-inducing activity an order of magnitude lower than that of TCDD. Other PCDFs tested gave a range of activity, with the hexa-CDF being more potent than the octa- or di-CDFs. The range of activity observed for the PCDDs and PCDFs tested, relative to TCDD was about one thousandfold. This is contrasted by the activity of the various PCBs tested. Here

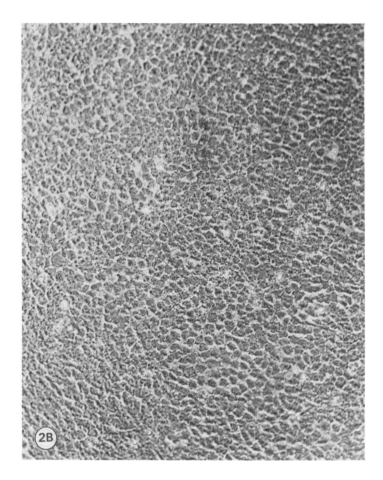

Figure 2B. The apparent induction of density-dependent inhibition of cell proliferation and flat-cell morphology characteristic of XBF/3T3 cultures after 14 days of exposure to 10^{-9} M 2,3,7,8-TCDD. Phase microscopy, 100X magnification.

the flat-cell-inducing activity was 10^4–10^6 times less potent than TCDD. Similar relative activity was reported in the keratinization system for another halogenated biphenyl, 2,3,4,2′,3′,4′–hexabromobiphenyl.[15] The PAHs varied in activity in relation to TCDD. Dibenzo(a,h)anthracene and benz(a)anthracene were about 10^3 and 10^4 times less potent, respectively, than TCDD. This correlates well with the reported activities of these compounds in the cell keratinization system.[15] 3–Methylcholanthrene and benzo(a)pyrene were both toxic at concentrations insufficient to induce a flat-cell effect. These compounds were found to be inactive as inducers of keratinization in the routine XB/3T3 system.[15] The pesticides were all inactive in inducing the flat-cell response at the concentrations tested with the

Table 1.   Induction of the Flat-Cell Effect by Various Chemicals.

Compound	Minimum Detectable Concentration (ppb)
2,3,7,8-Tetrachlorodibenzo-p-dioxin	0.0032
1,2,4,7,8-Pentachlorodibenzo-p-dioxin	0.359
2,3,7,8-Tetrachlorodibenzofuran	0.032
2,3,4,6,7,8-Hexachlorodibenzofuran	0.378
Octachlorodibenzofuran	4.48
2,6-Dichlorodibenzofuran	> 2.38
3,4,3',4'-Tetrachlorobiphenyl	100
2,4,5,2',4',5'-Hexachlorobiphenyl	1,000
2,5,2',5'-Tetrachlorobiphenyl	> 10,000
2,3,4,2',4',5'-Hexachlorobiphenyl	> 10,000
2,3,4,2',3',4'-Hexachlorobiphenyl	> 10,000
Aroclor 1254	10,000
Dibenzo(a,h)anthracene	10
Benz(a)anthracene	100
3-Methylcholanthrene	> 100[a]
Benzo(a)pyrene	> 100[a]
β-Naphthoflavone	1,000
Pyrene	> 10,000
Mirex	10,000
Dieldrin	> 10,000
Aldrin	> 10,000
o,p'DDT	> 10,000
Lindane	> 10,000
α-BHC	> 200,000

[a]Toxic concentration was 1000 ppb.

exception of Mirex, which had the highest potency of this group (10^6 times less active than TCDD).

Application to Binghamton State Office Building Soot Samples

Benzene extracts of 10 soot samples taken from above the ceiling tiles of different floors of the BSOB were tested for their ability to induce the flat-cell effect. The results of triplicate assays (Table 2) show that flat-cell-inducing activity was detected in all ten samples. However, the minimum amount of soot equivalent of the extract needed to induce this effect at greater than background levels varied from 0.3 to 114.2 μg. The activity of each sample relative to that of the 1st-floor sample was calculated (Table 2).

In order to determine if these results reflected actual variation between samples, these values were compared to the total PCDF concentrations determined by GC/MS analysis for these samples.[14] Total PCDF values were used in this comparison, since, as predicted by pyrolysis studies of PCBs,[11,12] PCDF concentrations in the soot were found to be much greater than those of the PCDDs.[13,14] As with the flat-cell-inducing activity, the total PCDF concentrations varied greatly from floor to floor.[14] Comparison of the relative flat-cell-inducing activity (Table 2) with relative total PCDF

Table 2.   Flat-Cell-Inducing Activity of Binghamton State Office Building (BSOB) Soot Extracts.

BSOB Floor Number	Endpoint (µg soot/mL)[a]	Relative Activity[b]
1	114.2	1
4	1.4	82
6	15.9	7
7	0.3	381
8	10.0	11
9	1.5	76
10	0.5	228
14	12.9	9
15	0.4	284
17	0.4	284

[a]Lowest concentration of a series of tenfold dilutions of soot extract capable of inducing a flat-cell effect greater than background. Each endpoint is an average of two evaluations of three replicates.
[b]Relative to activity of 1st-floor sample.

concentrations of the various soot extracts demonstrated a good correlation between the results of the two methods (correlation coefficient 0.82) (Fig. 3).

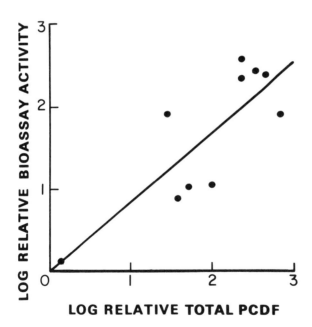

Figure 3.   Comparison of the data for relative flat-cell induction (from Table 2) and relative mass spectrometric analysis for PCDFs[14] for the various soot extracts from the Binghamton State Office Building. The relative flat-cell-inducing activity and relative total PCDF in each sample is plotted as a ratio to the values on Floor 1. The correlation coefficient for these two sets of data is 0.82.

Application to Surface Swipes

The surface swipe extracts were tested to determine whether: 1) they were toxic to cultured cells, 2) they interfered with the induction of dioxinlike activity in spiked samples, 3) dioxinlike activity could be detected in the unspiked samples, and 4) any activity would correlate with the chemical analyses for PCDDs and PCDFs performed on the same samples.

The samples were divided in half; 30 μL of DMSO were added to the one half and 30 μL of 10^{-6} M 2,3,7,8–TCDD in DMSO to the other. These spiked and unspiked samples were then solvent exchanged to the DMSO by evaporation of the benzene under a flow of dry nitrogen.

Two of the samples were cytotoxic at the highest test concentration but not at 1/10th of that concentration. This toxicity raised the level of detection for these samples tenfold. None of the spiked samples showed interference with the induction of the flat-cell effect. Those unspiked samples which were not cytotoxic induced a fairly consistent level of flat-cell induction, with the exception of the field blank and the reagent blank, which had no activity.

The chemical analysis of these samples showed that the tetra- through hexachlorinated dibenzodioxins, which would be the most biologically active components, were below the limits of detection. The 2,3,7,8–TCDF concentrations were relatively consistent throughout the samples, with a range of 0.5 to 1.1 ng/m^2, as were the concentrations of total PCDF congeners and isomers. The biological activity of most PCDFs, other than 2,3,7,8–TCDF, is unknown. However, it is assumed that the 2,3,7,8–tetra isomer is, as is the case with the PCDDs, the most potent isomer of this group.

The results showed that dioxinlike activity (the flat-cell effect) was detected in six of the ten samples. Two of the negative samples were also toxic at the highest concentration tested. 2,3,7,8–TCDF has about 1/20 the activity of 2,3,7,8–TCDD. If this factor is taken into account, the concentrations of total PCDFs were shown by chemical analysis to be within the limits of the activity range predicted by the bioassay. The additive effects of all the PCDF isomers in the flat-cell assay (FCA) are unknown, therefore an exact quantitative correlation between the FCA results and chemical analysis cannot be made at this time. Most significantly the field blank and the reagent blank had no dioxinlike activity as measured by the flat-cell assay.

Application to Fish

Preliminary experiments were conducted to determine if the flat-cell assay was applicable to detection of dioxinlike activity in fish. Fish which were shown to have measurable levels of 2,3,7,8–TCDD, along with negative controls (undetectable TCDD), were tested after the extensive cleanup procedure needed for gas chromatographic/mass spectrometric (GC/MS) analysis. Results indicated that the two methodologies gave consistent results. An effort was then made to determine the minimum level of cleanup necessary to produce accurate results using the flat-cell assay. It was noted that a simple organic extract resulted in cell death when applied to the flat-cell assay cultures. However, when this crude extract was treated with concentrated H_2SO_4 followed by neutralization, the samples were no longer toxic. A total of 12 fish samples were tested in the flat-cell assay, and the results were consistent with the GC/MS analysis for 2,3,7,8–TCDD in all cases. Significantly, three of the samples had no detectable contamination, as shown by both the chemical analysis and bioassay methods.

Discussion

The potential health hazards associated with human exposure to TCDD are not well understood, but are of major concern. Furthermore, as indicated in the introduction of this chapter, other dioxin isomers and congeners as well as dioxinlike compounds, which also occur in the environment, also exhibit the toxic effects associated with TCDD (dioxinlike activity). The detection of these less potent compounds is as important as the detection of TCDD, since at sufficiently high concentrations, their overall biological potency may equal or exceed that of TCDD in a particular sample. The currently used GC/MS analysis, while providing the concentrations of specific components of the environmental sample, does not indicate the overall dioxinlike activity. Furthermore, these methods are expensive, and are of limited capability in regard to mass screening programs.

Less costly assays, used to detect the biological activity caused by dioxin and dioxinlike compounds, rather than a specific isomer, would supplement GC/MS analysis. This would allow priority ranking of samples for subsequent GC/MS analysis, thereby removing samples with no dioxinlike activity from the pool of samples to be tested for specific isomers.

The development of proposed broad screen biological assays has generally followed the elucidation of the mechanism of dioxin toxicity. A proposed mechanism for the toxicity of TCDD and other dioxinlike compounds is that the compound first binds to a high-affinity receptor, followed by translocation to the nucleus[21] and nuclear interaction, presumably with chromatin, which results in the expression of a number of gene products, including aryl hydrocarbon hydroxylase (AHH).[2] This high affinity TCDD receptor has been identified in cytosolic preparations of liver and other tissues.[22,23] Studies of the affinity of other compounds, including dioxinlike chemicals and PAHs, for their ability to bind to these receptors indicate a good correlation between the binding affinity of compounds for the cytosol receptor and their potencies to induce AHH activity.[2] PAHs, which also induce AHH activity, had a binding affinity of between 1/3 and 1/30 that of TCDD.[24] This system has been and will continue to be very useful for the study of the molecular basis of the proposed initial events, i.e., receptor binding, of dioxinlike toxicity. Some work has been done in application of this system to fly ash.[19] However, the apparent lack of specificity of this system for dioxinlike compounds, compared to nonhalogenated PAHs, may pose problems for application to mass screening of complex environmental samples that may contain other compounds which demonstrate high affinity but exhibit relatively little dioxinlike toxicity.

Other in vitro cell culture systems have been investigated and proven to be useful in the detection of planar polychlorinated organic compounds. The in vitro induction of AHH in rat hepatoma cell cultures[25] has been one of the most extensively studied. AHH is rapidly induced by a wide variety of PAHs. While TCDD is the most potent known inducer of AHH activity, other unrelated compounds, such as 3–methylcholanthrene and benzo(a)pyrene, also induce this enzyme.[26] Thus the in vitro AHH assay, while very useful for the study of the mechanism and structure function relationships between dioxinlike compounds, may, like the cytosolic receptor system, have difficulty distinguishing between dioxinlike compounds and other potent AHH inducers. We have shown that these compounds are much less potent inducers of the flat-cell effect. This allows the flat-cell assay to have a greater specificity for the detection of polychlorinated dioxin isomers and congeners.

We have previously demonstrated the feasibility of using the keratinization system developed by Knutson and Poland[15] as an assay for dioxinlike compounds, e.g., PCDFs, in extracts from soot produced by a fire which involved a PCB-containing electrical transformer.[16] However, we also noted, as had others, a change in phenotype which was apparently associated with extended serial culture. In our experience, this change was characterized by a decline in the magnitude of the keratinization response to TCDD exposure, making use of low-passage frozen cell stocks necessary for extensive use of this system as a screen assay.[16]

This work led to the evaluation of a derivative of the XB cell line used by Knutson and Poland. This subline, designated XBF, was shown, when cocultured with irradiated 3T3 cells, to exhibit a dioxin-induced reversible postconfluent inhibition of cell proliferation.[18] This XBF/3T3 phenotype has been stable during more than a year of routine culturing, unlike the XB/3T3 culture system, which lost its dioxin-induced keratinization characteristic under similar conditions. The keratinization of XB/3T3 cultures is suggested to be an example of epithelial differentiation.[2] The dioxin-induced flat-cell effect seen in the XBF/3T3 culture may also represent a differentiation response analogous to the cessation of cell division and cell flattening seen in vivo as the proliferating cells of the stratum basale become the quiescent adjacent cells of the stratum spinosum.

The sensitivity and stability of the XBF/3T3 flat-cell effect for TCDD suggested its application as an assay for dioxin. Previous studies using inhibition of macromolecular synthesis and mitosis indicated that the morphological change seen in postconfluent cultures was not simply a consequence of inhibition of cell division.[18] The experiments described in this chapter indicate that, for the classes of chemicals tested, this effect is most sensitive to TCDD and TCDF. Other compounds, such as PCBs and certain PAHs, demonstrate relatively little potential for inducing this activity. The ranking of flat-cell-inducing potential of these chemicals also suggests a structure-function relationship which is consistent with that of the keratinization system, cytosolic receptor binding, arylhydrocarbon hydroxylase induction and animal toxicity.

This report also demonstrates the ability of the in vitro flat-cell assay to detect this activity in relevant environmental samples and to discriminate between samples having relatively high and low levels of PCDDs and PCDFs as determined by GC/MS analysis. It has been demonstrated that for the induction of AHH and cytosolic receptor binding, the potency of a compound is dependent on its structure.[27] We have shown that TCDF, while exhibiting one-tenth the activity of TCDD, is about 10–times more potent in the flat-cell assay than 1,2,4,7,8–pentachlorodibenzofuran and 2,3,4,6,7,8–hexachlorodibenzofuran. Other PCDF congeners tested are even less potent. Thus, an important advantage of the XBF/3T3 in vitro flat-cell system for detection of dioxinlike activity is its apparent sensitivity to the most toxic members of the PCDDs and PCDFs.

Dioxinlike activity is defined here by the induction of the flat-cell effect by 2,3,7,8–TCDD, the most potent and toxic PCDD isomer known. As such, the aggregate toxicity of a sample, whether comprised of high concentrations of a less toxic isomer or low concentrations of highly toxic isomer, could be determined and expressed as TCDD equivalents rather than as specific concentrations of isomers and congeners whose biological activity and relevance are unknown. This approach would also take into account possible synergisms and antagonisms which may be associated with complex environmental mixtures as may be found in toxic waste dump sites or in used PCB mixtures found in electrical transformers and capacitors.

The flat-cell assay for dioxinlike activity has potential usefulness in a number of applications. It can be used as a supplement to the more expensive high-resolution chemical analysis by identifying those samples with no dioxinlike activity. These samples could then be removed from the positive samples scheduled for subsequent chemical analysis for definite identification and quantitation of specific congeners and isomers. This application would be particularly useful in delineating the specific locations of dioxin hot spots in a land area which is suspected of being contaminated or in the mass screening of large numbers of fish or other biota to track the source of contamination, e.g., along a bifurcated stream system. The data presented in this chapter demonstrate the ability of the flat-cell assay to discriminate between various samples such as soot, surface swipes, and fish, with regard to lack of activity and low or high levels of activity. The semiquantitative determination of dioxinlike activity would allow priority ranking of positive samples for subsequent instrument analysis, thus making more efficient use of this limited resource.

Another use of this bioassay is the detection of compounds which exhibit dioxinlike activity but are not the subject of chemical analysis for a particular sample. An example of this is the recent detection, in this laboratory using the flat-cell assay, of 2,3,7,8–tetrachlorothianthrene, a sulfur analog of TCDD, whose activity was detected in a sediment sample from a municipal sanitary sewer. This sample contained very low levels of TCDD, as shown by GC/MS, which could not account for the high activity seen in the bioassay. This resulted in further chemical analysis which allowed the detection, quantitation and presumable determination of the source of this compound.

Another use for this bioassay is the examination of suspect compounds, such as PCDD analogs, for dioxinlike activity. An example of this is the detection, by chemical analysis, of 1,2,4,5,7,8–hexachloroxanthene (HCX) at dioxin sites in Missouri. Like TCDD, HCX is a by-product of hexachlorophene production. Risk assessment was impossible, since no toxicity data were available on this compound. Synthesized HCX was subjected to the flat-cell assay. The results indicated that dioxinlike activity associated with this compound, if any, was about 10^6 less than TCDD. Based on these data, the preliminary determination was made that HCX posed no additional health hazard, since it occurred mainly at dioxin sites. Animal toxicity studies of HCX are currently being performed in this laboratory.

Since the flat-cell assay may represent a TCDD-induced differentiation of epithelium, similar to in vitro induction of keratinization, which is thought to result in chloracne in humans, the flat-cell assay may be useful as a screen for acnegenic potential in commercial products, especially those of the cosmetics industry. It could also be used to detect possible cutaneous hazards in the workplace.

ACKNOWLEDGMENTS

The authors thank Ms. Victoria Lamberton for preparation of the typescript. Portions of this chapter were published in *Fundamental and Applied Toxicology*, copyright Academic Press (1985), and are reproduced here by permission. This work was supported in part by a NIOSH grant OHO1533, NIEHS grant ES03561, and Electric Power Research Institute contract RP2028–4.

REFERENCES

1. Poland, A., and A. Kende. "2,3,7,8-Tetrachlorodibenzo-p-dioxin: Environmental Contaminant and Molecular Probe," *Fed. Proc.* 35:2404–2411 (1976).
2. Poland, A., and J. Knutson. "2,3,7,8-Tetrachlorodibenzo-p-dioxin and Related Halogenated Aromatic Hydrocarbons: Examination of the Mechanism of Toxicity," *Ann. Rev. Pharmacol. Toxicol.* 22:517–554 (1982).
3. Kimbrough, R. "The Toxicity of Polychlorinated Polycyclic compounds and Related Chemicals," *CRC Critical Reviews in Toxicology* (January 1974) pp. 445–498.
4. Huff, J. E., J. A. Moore, R. Saracci, and L. Tomatis. "Long Term Hazards of Polychlorinated Dibenzodioxins and Polychlorinated Dibenzofurans," *Environ. Health Pers.* 36:221–240 (1980).
5. Kimbrough, R. D., H. Falk, and P. Stehr. "Health Implications of 2,3,7,8-Tetrachlorodibenzodioxin (TCDD) Contamination of Residential Soil," *J. Tox. Environ. Health* 14:47–93 (1984).
6. Kociba, R. J., D. G. Keyes, J. E. Beyer, R. M. Carreon, E. C. Wade, D. Dittenber, R. Kalnins, L. Frauson, C. M. Park, S. Bernard, R. Hummel, and C. G. Humiston. "Results of a Two Year Chronic Toxicity and Oncogenicity Study of 2,3,7,8-Tetrachlorodibenzo-p-dioxin (TCDD) in Rats," *Toxic. Appl. Pharmacol.* 46:279–303 (1978).
7. Pitot, H. C., T. Goldsworthy, H. A. Campbell, and A. Poland. "Quantitative Evaluation of the Promotion by 2,3,7,8-Tetrachlorodibenzo-p-dioxin of Hepatocarcinogenesis from Diethylnitrosamine," *Cancer Res.* 40:3616–20 (1980).
8. Cook, R. R. "Dioxin, Chloracne and Soft Tissue Sarcoma," *Lancet* 1:618–619 (1981).
9. Hardell, L., and A. Sandstorm. "Case-Control Study: Soft Tissue Sarcomas and Exposure to Phenoxyacetic Acids or Chlorophenols," *Brit. J. Cancer* 39:711–717 (1979).
10. Fingerhut, M. A., W. E. Halperin, P. A. Honchar, A. B. Smith, D. H. Groth, and W. O. Russell. "Review of Exposure and Pathology Data for Seven Cases Reported as Soft Tissue Sarcoma Among Persons Occupationally Exposed to Dioxin-Contaminated Herbicides," in *Public Health Risks of the Dioxins*, W. W. Lawrance, Ed. (New York: Rockefeller University Press, 1984) p. 187.
11. Buser, H. R., H. P. Bosshardt, and C. Rappe. "Formation of Polychlorinated Dibenzofurans (PCDFs) from the Pyrolysis of PCBs," *Chemosphere* 7:109–119 (1978).
12. Buser, H. R., H. P. Bosshardt, C. Rappe, and R. Lindahl. "Identification of Polychlorinated Dibenzofuran Isomers in Flyash and PCB Pyrolyses," *Chemosphere* 7:419–429 (1978).
13. Smith, R. M., P. W. O'Keefe, D. L. Hilker, B. L. Jelus-Tyror, and K. Aldous. "Analysis for 2,3,7,8-Tetrachlorodibenzofuran and 2,3,7,8-Tetrachlorodibenzo-p-dioxin in a Soot Sample from a Trans-

former Explosion in Binghamton, New York," *Chemosphere* 11:715–720 (1982).

14. Smith, R. M., D. L. Hilker, P. W. O'Keefe, S. Kumar, K. Aldous, and B. L. Jelus-Tyror. "Determination of Polychlorinated Dibenzofurans and Polychlorinated Dibenzodioxins in Soot Samples from a Contaminated Office Building," *New York State Department of Health Report* (March, 1982).

15. Knutson, J. C., and A. Poland. "Keratinization of Mouse Teratoma Cell Line XB Produced by 2,3,7,8–Tetrachlorodibenzo-p-dioxin: An In Vitro Model of Toxicity," *Cell* 22:27–36 (1980).

16. Gierthy, J. F., D. Crane, and G. D. Frenkel. "Application of an In Vitro Keratinization Assay to Extracts of Soot from a Fire in a PCB-Containing Transformer," *Fund. Appl. Pharmacol.* 4:1036–1041 (1984).

17. Rheinwald, J. G., and H. Green. "Formation of a Keratinizing Epithelium in Culture by a Cloned Cell Line Derived from a Teratoma," *Cell* 6:317–330 (1975).

18. Gierthy, J. F., and D. Crane. "Reversible Inhibition of In Vitro Epithelial Cell Proliferation by 2,3,7,8–Tetrachlorodibenzo(p)dioxin," *Toxicol. Appl. Pharmacol.* 74:91–93 (1984).

19. Hutzinger, O., K. Olie, J. W. A. Lustenhouwer, A. B. Okey, S. Bandiera, and S. Safe. "Polychlorinated dibenzo-p-dioxins and dibenzofurans: A bioanalytical approach," *Chemosphere* 10:19–25 (1981).

20. Choudhury, D. R., and B. Bush. "Chromatographic-Spectrometric Identification of Airborne Polynuclear Aromatic Hydrocarbons," *Anal. Chem.* 53:1351–1356 (1981).

21. Greenlee, W. F., and A. Poland. "Nuclear Uptake of 2,3,7,8–Tetrachlorodibenzo-p-dioxin in C57BL/6J and DBA/2J Mice," *J. Biol. Chem.* 254:9814–9821 (1979).

22. Carstedt-Duke, J. M. B. "Tissue Distribution of the Receptor for 2,3,7,8–Tetrachlorodibenzo-p-dioxin in the Rat," *Cancer Res.* 39:3172–76 (1979).

23. Whitlock, J. P., and D. R. Galeazzi. "2,3,7,8–Tetrachlorodibenzo-p-dioxin Receptors in Wild Type and Variant Mouse Hepatoma Cells," *J. Biol. Chem.* 259:980–985 (1984).

24. Poland, A., E. Glover, and A. S. Kende. "Stereospecific, High Affinity Binding of 2,3,7,8–Tetrachlorodibenzo-p-dioxin by Hepatic Cytosol," *J. Biol. Chem.* 251:4936–46 (1976).

25. Bradlaw, J. A., and J. L. Casterline. "Induction of Enzyme Activity in Cell Culture: A Rapid Screen for Detection of Planar Polychlorinated Organic Compounds," *J. Assoc. Off. Anal. Chem.* 62:904–916 (1979).

26. Poland, A., and E. Glover. "Genetic Expression of Aryl Hydrocarbon Hydroxylase by 2,3,7,8–Tetrachlorodibenzo-p-dioxin: Evidence for a Receptor Mutation in Genetically Nonresponsive Mice," *Mol. Pharmacol.* 11:389–398 (1975).

27. Knutson, J., and A. Poland. "2,3,7,8–Tetrachlorodibenzo-p-dioxin: Toxicity In Vivo and In Vitro," in *Halogenated Hydrocarbons*, H. Khan, Ed. (New York: Pergammon Press, 1981), pp. 187–201.

CHAPTER **20**

Alterations in Lipid Parameters Associated with Changes in 2,3,7,8-Tetrachlorodibenzo-p-Dioxin (TCDD)-Induced Mortality in Rats

Carol M. Schiller, Margaret W. King, and Ramsey Walden

INTRODUCTION

2,3,7,8-Tetrachlorodibenzo-p-dioxin (TCDD) is a highly toxic chlorin-ated dioxin that is the unwanted by-product of several industrial syntheses, e.g. 2,4,5-trichlorophenoxyacetic acid and hexachlorophene.[1] TCDD has become an increasing environmental concern because it has been found as a contaminant in fly ash.[2,3] Several areas of research have been pursued to examine the biological responses to TCDD. These include: the wasting syndrome, hypophagia, liver damage, serum changes and, in humans, chloracne. Epidemiologic studies reveal that no increase in heart disease results from accidental TCDD exposure; however, animal studies indicate it is a potential teratogen, hepatotoxin and carcinogen. TCDD is thought to exert its affects through a common mechanism, and this commonality has been ascribed to altered epithelial cell functions.[4] It is generally held that TCDD alters normal lipid metabolism. Reported observations describe changes in lipid metabolism in humans accidentally exposed to TCDD and in a wide range of laboratory species, especially rats. This chapter reviews the current literature describing alterations in lipid metabolism after TCDD exposure, and focuses on lipid changes in laboratory rats.

Clinical studies indicate increases in blood triglyceride[5] as well as increases in blood cholesterol.[6-8] Some of the clinical reports have been concerned with the association of cardiovascular disease in populations exposed to TCDD.[9,10] A recent clinical epidemiologic study examining the long-term (30–yr) health effects of workplace exposure to TCDD indicated clinical evidence of chloracne persistance in 55.7% of the exposed workers, but no evidence of increased risk for cardiovascular disease.[11]

Although lipid metabolism is generally studied in laboratory rats, there are several reports indicating TCDD-induced changes in the serum lipids in other species. In hamsters, both intraperitoneal (i.p.) and peroral (p.o.) administration of TCDD (1000 µg/kg) increased the serum cholesterol levels, but not the serum triglyceride levels.[12] Guinea pigs, a TCDD-susceptible species, exhibited both markedly increased serum cholesterol and triglyceride levels after i.p. administration of TCDD (1.0 µg/kg).[13] Mice strains have demonstrated a 14–fold difference in oral LD_{50}–30 values[14]; however, at equivalent lethal doses of TCDD, p.o. exposure increased serum cholesteral but not serum triglyceride levels in each strain.[14]

Several earlier studies examined the lipid contents of rat liver after TCDD exposure.[15,16] A frequently quoted report was designed to determine rates of hepatic synthesis of lipid.[16] In that study, TCDD did not increase the rate of incorporation of ^3H-acetate into liver lipids expressed as dpm ^3H/mg liver. Rather, the dpm ^3H/mg liver decreased with increased TCDD dose and time after TCDD exposure. Decreases in the rate of ^3H-acetate incorporation into epididymal tissue were also observed.[16] Presumably, since large quantities of lipid accumulate in the liver, this result is perplexing without dpm ^3H-acetate/mg lipid. In addition, the male Wistar rats used in these liver lipid studies were administered TCDD orally (10 µg/kg) and the body weight gain fluctuated one week after exposure, but no significant change in body weight was observed.

A more recent study examined serum and liver lipids in female Fischer rats[17] by differential solvent extraction followed by gravimetric measurements, or by transmission electron microscopy. Generally, the lipids were analyzed 13 days after oral administration of TCDD (50 µg/kg). Although no increase in liver triglyceride was observed, total liver lipids did increase. Both cholesterol ester and free fatty acid increased in the liver. Only serum cholesterol ester was increased in this study.[17] However, these results should be used cautiously, since no chemical or biochemical determination of the lipid classes was performed.

In addition to increased serum cholesterol ester in female Fischer rats,[17] a more biochemical study reported twofold increases in both serum triglyceride and serum total cholesterol.[18] In this later study, pair-fed rats were administered TCDD (100 µg/kg) or corn oil p.o. 14 days before experimentation. Serum glucose markedly decreased in this study. The serum lipoproteins from female Fischer rats were separated by cellulose acetate electrophoresis and revealed decreases in the pre-albumin, α_1-globulin bands and an increase in the α_1-globulin bands.[17] Another study, which monitored increases in serum cholesterol and high-density lipoprotein (HDL) cholesterol in male Sprague-Dawley rats, administered TCDD (20 μg/kg) i.p. for up to 60 days, also examined the serum HDL apolipoprotein composition by isoelectric focusing on polyacrylamide gel electrophoresis.[19] The altered distribution of C-III apolipoproteins (↑C-III–3 and ↓C-III–0) in the TCDD-exposed rat serum HDL[19] is similar to that previously reported for rats with streptozotocin-induced diabetes.[20]

The laboratory rat has been well studied in both basic and toxicological investigations. The normal rat diet is low in fat (1.5 to 4.5% by weight) and, therefore, there is minimal postprandial hyperlipidemia. However, the rat has been utilized routinely as an animal model for studying lipid absorption with extrapolation to humans. Average (normal) values of serum and liver lipids vary with rat strain. For example, fasting serum triglyceride values range form 39 to 87 mg/dL[21] and serum cholesterol values range from 44 to 78 mg/dL.[22] When 10 + % of the rate diet is fat (weight/weight), increases in serum cholesterol values are observed,[23,24] but serum triglyceride values rarely reach 200 mg/dL.[25] In the nonfed state, circulating triglycerides decrease,[26] while serum cholesterol responses in fasted rats are known to fluctuate.[27-30] It has been suggested that this variation in serum cholesterol may be due to differences in sex, age or strain of rat examined.[31]

It has been suggested that TCDD acts through a common mechanism. However, this commonality is difficult to discern from the literature, in part because of the variations in experimental protocols. More recently it has been observed that rats exhibit (sub)strain variation in susceptibility to TCDD-induced mortality[32] as previously suggested for strains of mice.[33,14] Thus, it is important to denote the precise dose, strain of rat, and time after dosing in reporting the effects of TCDD on a particular metabolic response. These notations should also be observed when comparing the results obtained from various laboratories.

To observe changes in lipid parameters resulting from TCDD exposure, several initial determinations have been made. The preparation of the reagents is described briefly below. TCDD was obtained from the Laboratory of Chemistry, NIEHS, and was 99% pure, as determined by gas chromatographic and mass spectrophotometric analysis. The concentrations of the stock and diluted solutions of TCDD in corn oil were evaluated by neutron activation analysis of chlorine. Protocols approved by the NIEHS Safety Office were employed, and required use of high-hazard laboratory facilities. Adenine hemisulfate (Ad) and 4–aminopyrazolo[3,4–d]pyrimidine (4APP) were obtained from Sigma Chemical Company (St. Louis, MO). Fresh solutions were prepared in 0.155 M NaCl immediately prior to use.

Adult male Fischer (F/344N) rats were obtained from Charles River

Table 1. Time Course of Effects of 2,3,7,8-Tetrachlorodibenzo-p-Dioxin on Serum Parameters in Rats.[a]

Days After Dosing	Glucose		Triglyceride		Cholesterol (Free)		Cholesterol (Esterified)	
	Control	Treated	Control	Treated	Control	Treated	Control	Treated
0	100		100		100		100	
2	102	88[b,c]	89[b,c]	99	113	140[b,c]	88	132[b,c]
4	103	84[b,c]	106	283[b,c]	106	181[b,c]	102	208[b,c]
7	95	57[b,c]	106	314[b,c]	133	269[b,c]	122	234[b,c]
12	97	58[b,c]	104	293[b,c]	120	317[b,c]	95	241[b,c]
17	98	43[b,c]	93	272[b,c]	131	211[b,c]	101	249[b,c]
21	92	53[b,c]	99	254[b,c]	95	204[b,c]	107	243[b,c]

[a]Fischer rats were treated with a single oral dose of 60 µg TCDD/kg body wt, and were sacrificed at various times after exposure. Control rats were given a comparable volume of corn oil. Day 0 rats received no treatment. All animals were fasted overnight before sacrifice. Values are expressed as percentage of control (Day 0) and represent the mean of 5 animals analyzed separately.
[b]Significantly different from control values (Day 0), $p < 0.05$.
[c]Significantly different from control values at similar time after dosing, $p < 0.05$.

Breeding Laboratories (Wilmington, MA) at 11 to 12 weeks of age (240–290 g). The animals were housed one per cage and maintained under controlled temperature, lighting (12 hr light and 12 hr dark) and humidity. Animals were allowed free access to water and NIH–31 mouse and rat ration.

In some experiments, it was necessary to pair-feed controls. In such cases, the TCDD-exposed animals were allowed to feed ad libitum from a weighed portion of ration, and the next day the ration was reweighed and the amount removed was given to the control animals. The TCDD and TCDD + Ad animals were also monitored for spillage to correct for the ration given to the control animals. The serum and liver parameters were measured by methods described previously in detail.[32,34]

Since the oral LD_{50} value for Charles River/Fischer (CR/F) rats is 160 µg TCDD/kg, a single oral dose of 45 or 60 µg/kg was administered by gavage. This nonlethal dose readily induces changes in serum parameters (Table 1). Two days after treatment serum glucose levels declined, decreased to approximately 50% of control values at day 7, and then remained at this lower level for 3 weeks. Serum lipid parameters, triglyceride, cholesterol (free) and cholesterol (esterified), all increased more than twofold, with a peak at 1 week after exposure (Table 1). The dose-response effects of TCDD on these serum parameters were then determined 1 week after exposure at doses of 30 to 360 µg TCDD/kg (Table 2). At the lowest dose, 30 µg/kg, decreased serum glucose and increased serum cholesterol (free) and cholesterol (esterified) levels were observed. The serum triglyceride levels demonstrated an 88% increase at 90 µg TCDD/kg. For comparative purposes, groups of control rats were pair-fed, with the treated rats given 60, 180 and 360 µg TCDD/kg (Table 2). The serum parameters in the pair-fed

Table 2.   Dose-Response Effects of 2,3,7,8-Tetrachlorodibenzo-p-Dioxin on Serum Parameters in Rats.[a]

| Dose (μg/kg) | Serum Parameters | | | |
	Glucose	Triglyceride	Cholesterol Free	Cholesterol (Esterified)
0	100	100	100	100
30	74[b]	127	143[b]	176[b]
60	54[b]	141[b]	244[b]	268[b]
	(84)[c]	(82)	(108)	(100)
90	30[b]	188[b]	255[b]	316[b]
180	43[b]	267[b]	274[b]	323[b]
	(90)	(117)	(86)	(91)
270	38[b]	274[b]	307[b]	381[b]
360	39[b]	217[b]	330[b]	351[b]
	(91)	(107)	(88)	(97)

[a]Data represent groups of 5 animals given various doses of TCDD in corn oil (1 mL/250g) and sacrificed 7 days later. All animals were fasted overnight before sacrifice. Values are expressed as percentage of control (Dose 0).
[b]Significantly different from control values (Dose 0), $p < 0.05$.
[c]Values in parentheses were obtained from control animals pair-fed to animals treated with the various doses of TCDD.

control groups fluctuated by $\pm$ 15 of the control values, but not by the –60 to + 200% observed in the TCDD-treated rats.

At low doses of TCDD, both the control (corn oil) and treated (TCDD, in corn oil) rats continue to eat ad libitum as pair-fed animals (Figure 1). Only at times longer than 1 week after treatment are differences in feed consumption (net) observed (Figure 1C). Significant body weight loss occurs 5 days after treatment (Figure 1F). Three weeks after exposure to 60 μg/kg, the body weight and epididymal fat pads have decreased by 20%, while the liver weight has decreased by 10% (Table 3). At higher doses of TCDD, rats lose as much as 39% of their body weight in one week, which is more than that lost by the pair-fed control rats (Table 4).

Preliminary fasting experiments established that CR/F rats lost approximately 10% body weight after a 72–hr fast. Thus, TCDD-exposed rats and 72–hr fasted control rats, which exhibit a 10% body weight loss, were examined and compared for changes in lipid parameters (Table 5). The exposure to TCDD alone (0–hr fast) increased the liver total lipid by 95% (60.5 vs 118 mg/g) and the liver triglyceride concentration by 208% (12.5 vs 38.5 mg/g) as compared to the control values. These TCDD-induced increases in lipid concentration were evident after the 72–hr fast. The 72–hr fast alone (control values) increased the liver total lipid and the liver triglyceride concentration also, but these increases were not of the same magnitude as the TCDD-induced increases.

There were marked increases in serum triglyceride and serum cholesterol levels in the TCDD-exposed groups (Table 5): at 0 hr of fasting, increases of 63 and 112% triglyceride and cholesterol, respectively, and at 72 hr of

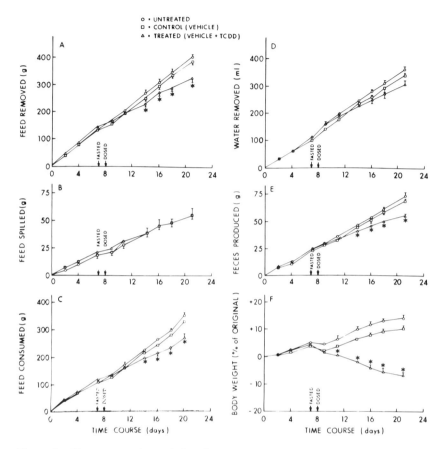

Figure 1. Time course of cumulative changes induced by TCDD in adult, male Fischer rats. Groups of rats (n = 7) were administered nothing (untreated), a single oral dose of corn oil (control), or corn oil with TCDD (60 μg/kg body wt, treated). Parameters were monitored for 1 week prior and 2 weeks after treatment. Treated values statistically different from other two values (p < 0.05). Standard error bars were omitted where smaller than symbol used.[34]

fasting, increases of 154 and 84% triglyceride and cholesterol, respectively. Fasting alone did not increase serum lipids in the control group; rather, the serum triglycerides decreased. There was no change in serum fatty acid levels in any of the groups. Comparison of the two groups with a 10% body weight loss, 0-hr fasted TCDD-exposed group and the 72-hr fasted control group, indicated that only the 0-hr fasted TCDD-exposed group demonstrated serum hyperlipidemia, with differences of 127 and 67%, triglyceride and cholesterol, respectively. The 72-hr fasted TCDD-exposed group was also hyperlipidemia.

The time course of lipid accumulation in the liver indicates a twofold increase in triglyceride concentration within two days after exposure to only

Table 3. Time Course of Changes in Body and Organ Weights in Rats After Exposure to 2,3,7,8-Tetrachlorodibenzo-p-Dioxin.[a]

Days After Dosing	Body Weight		Liver Weight		Epididymal Fat Pad Weight	
	Control	Treated	Control	Treated	Control	Treated
0	100[b]		100[b]		100[b]	
2	101	98	95	114[c,d]	113[c]	100
4	101	96[d]	97	131[c,d]	110	108
7	106[c]	94[c,d]	101	129[c,d]	110	115[c]
12	106[c]	90[c,d]	109	120[c,d]	127[c]	98[c,d]
17	106[c]	84[c,d]	89	107[c,d]	120[c]	88[d]
21	112[c]	80[c,d]	117[c]	91[d]	146[c]	81[d]

[a]Fischer rats were treated with a single oral dose of 60 µg TCDD/kg body wt, and were sacrificed at various times after exposure. Control rats were given a comparable volume of corn oil. Day 0 rats received no treatment. All animals were fasted overnight before sacrifice. Body wt measurements were made prior to beginning the fast. Values are expressed as percentage of control (Day 0) and represent the mean of 5 animals.

[b]Absolute body and organ weights are expressed in g.

[c]Significantly different from control values (Day 0), $p < 0.05$.

[d]Significantly different from control values at same time after exposure, $p < 0.05$.

Table 4.   Dose-Response Effects of 2,3,7,8-Tetrachlorodibenzo-p-Dioxin on Body and Organ Weights in Rats.[a]

Dose (μg/kg)	Body Weight			Liver Weight	Epididymal Fat Pad Weight
	Initial	Final	Change		
0	100	100	+ 1	100	100
30	100	92[b]	− 5	128[b]	89[b]
60	100	86[b]	− 14	119[b]	68[b]
	(100)[c]	(90)	(− 7)	(81)	(82)
90	100	86[b]	− 13	109	80[b]
180	98	59[b]	− 39	104	70[b]
	(100)	(82)	(− 17)	(76)	(70)
270	99	80[b]	− 18	105	79[b]
360	99	65	− 34	105	67[b]
	(99)	(82)	(− 17)	(77)	(82)

[a]Data represent groups of 5 animals given various doses of TCDD in corn oil (1 mL/250g) and sacrificed 7 days later. All animals were fasted overnight before sacrifice. Values are expressed as percentage of control (Day 0).

[b]Significantly different from control values (Dose 0), $p < 0.05$.

[c]Values in parentheses were obtained from control animals pair-fed to animals treated with the various doses of TCDD.

Table 5.   Effect of TCDD and Fasting on Body Weight and Lipid Parameters in Rats.[a]

Group	0-hr Fast		72-hr Fast	
	Control	Treated	Control	Treated
A. Body Weight				
prefast (g)	294±3	248±4[c]	307±3	241±5[c]
postfast (g)			281±4	217±5[c]
change (%)	(+7.9±1.4)[b]	(−11.4±1.3)[b,c]	−8.4+0.3	−10.3±0.1[c]
B. Liver Weight				
total (g)	8.7±0.1	8.1±0.4	6.2±0.2	6.2±0.2
relative to body wt (mg/g)	29.5±0.3	32.2±1.3	22.0±0.5[d]	27.8±0.7[c,d]
C. Liver Lipid				
total lipid (mg/g)	60.5±2.7	118±10[c]	82.4±2.3[d]	143±9[c]
triglyceride (mg/g)	12.5±1.5	38.5±4.3[c]	15.5±2.3[d]	45.8±6.5[c]
D. Serum				
triglyceride (mg/dL)	66.2±8.8	108±15[c]	47.6±5.7[d]	121±9[c]
cholesterol (mg/dL)	44.6±5.9	94.4±6.9[c]	56.4±4.5	104±9[c]
fatty acid (mg/dL)	16.0±0.8	18.8±2.9	19.5±1.0	20.5±1.7

[a]Groups of 14 animals were assigned randomly to control and treated groups before the initial exposure to corn oil (control) or TCDD in corn oil (treated). Each group was subdivided and 7 animals from each group were fasted for 0 or 72 hr. The initial body wt were, respectively, 290 ± 3 and 290 ± 2g for control and treated groups.

[b]Values in parentheses are the percentage body wt changes resulting during the 1 week between the initial exposure and the beginning of the fast for the 0-hr fast group.

[c]Significantly different ($p < 0.05$) from other group within the same duration of fasting. (Used for all parameters.)

[d]Significantly different ($p < 0.05$) from other group within the same exposure at 0 hr of fasting. (Used for comparing relative parameters, mg/g and mg/dL).[34]

Table 6. Time Course of Effects of 2,3,7,8-Tetrachlorodibenzo-p-Dioxin on Liver
Lipid Parameters in Rats.[a]

| Days After Dosing | Liver Lipid Parameters | | | |
| | Total Lipids | | Triglyceride | |
	Control	Treated	Control	Treated
0	100		100	
2	105	195[b,c]	105	229[b,c]
4	98	585[b,c]	92	437[b,c]
7	103	714[b,c]	105	479[b,c]
12	118	725[b,c]	133	447[b,c]
17	117	749[b,c]	124	493[b,c]
21	108	631[b,c]	87	401[b,c]

[a]Fischer rats were treated with a single oral dose of 60 μg TCDD/kg body wt and were sacrificed at various times after exposure. Control rats were given a comparable volume of corn oil. Day 0 rats received no treatment. All animals were fasted overnight before sacrifice. Values are expressed as precentage of control (Day 0) and represent the mean of 5 animals analyzed separately.
[b]Significantly different from control values (Day 0), $p < 0.05$.
[c]Significantly different from control values at similar time after dosing, $p < 0.05$.

60 μg TCDD/kg (Table 6). This increase reaches a peak 7 days after exposure to TCDD, and plateaus for at least 3 weeks. At only 30 μg/kg, the liver lipids are increased twofold 1 week after exposure, but reach even greater levels at higher levels of TCDD (Table 7). These increases in liver lipids are reflected in increased liver triglyceride.

The relationship between the hyperlipidemic or fatty liver formation and TCDD-induced mortality is not known. The approach used to examine such a relationship was to decrease serum or liver lipids with the use of 4-aminopyrazola(3,4-d) pyrimidine (4APP) or adenine (Ad), respectively (Figure 2).

Table 7. Dose-Response Effects of 2,3,7,8-Tetrachlorodibenzo-p-Dioxin on Liver
Lipid Parameters in Rats.[a]

| Dose (μg/kg) | Liver Lipid Parameters | |
	Total Lipids	Triglyceride
0	100	100
30	258[b]	160[b]
60	692[b]	440[b]
	(82)[c]	(114)
90	778[b]	511[b]
180	760[b]	548[b]
	(95)	(107)
270	817[b]	604[b]
360	807[b]	901[b]
	(77)	(103)

[a]Data represent groups of 5 animals given various doses of TCDD in corn oil (1 mL/250g) and sacrificed 7 days later. All animals were fasted overnight before sacrifice. Values are expressed as percentage of control (Dose 0).
[b]Significantly different from control values (Dose 0), $p < 0.05$.
[c]Values in parentheses were obtained from control animals pair-fed to animals treated with the various doses of TCDD.

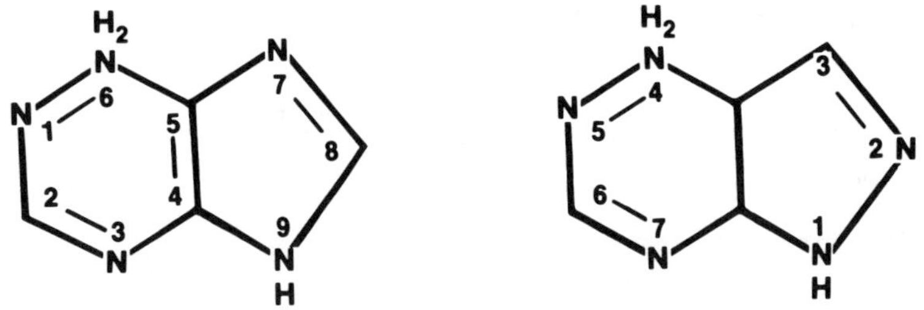

Adenine

4-Aminopyrazolo (3,4-d) pyrimidine

Figure 2. Structures of metabolite (Ad) and antimetabolite (4APP) used in mortality studies.

4APP is known to decrease the release of triglyceride from the liver[35] by blocking the synthesis and/or release of triglyceride-rich, very low-density lipoproteins from the liver,[36] thereby decreasing serum triglyceride and very low-density lipoproteins while increasing liver lipid content.[37] Ad prevents the formation of fatty liver by a number of agents, including ethionine,[38] orotic acid[39-41] and 4APP,[38] without affecting normal liver lipid concentrations.[38] By treating rats with 4APP or Ad simultaneously with TCDD or corn oil alone, the role of hyperlipidemia and fatty liver formation in TCDD-induced mortality can be evaluated. The effects of 4APP and Ad on TCDD-induced mortality were examined by monitoring body weight loss and mean time to death in rats. Body weight changes are a good indicator for predicting nonsurvival after lethal doses of TCDD.[42]

Changes in serum trigylceride levels were measured with time and with increasing doses of 4APP to determine the most effective dose. A dose-related decrease in serum triglyceride concentrations was observed 8 days after initiation of treatment. After 12 days, the serum triglyceride decreased from 69 ± 14 to 17 ± 4 mg/dL (0 and 12 mg 4APP/kg, every other day). At 8 days after the initiation of treatment, the serum very low-density lipoproteins and low-density lipoproteins also decreased in the 4APP animals, in a dose-dependent manner, as demonstrated on acrylamide electrophoresis gels stained for lipid with Sudan Black B (data not shown). These range-finding experiments are important because it is known that the susceptibility to 4APP varies with rat (sub)strains,[35] and that the metabolically important forms of triglyceride are carried as serum lipoproteins. The highest dose, 12 mg 4APP/kg, was then used in future experiments.

Lipid changes associated with 4APP and Ad were compared 10 days after the rats were exposed to a single oral dose of corn oil or TCDD in corn oil (325 µg/kg) (Table 8). The 12 mg 4APP/kg dose was taken from the range-finding experiments and the Ad dose was based on studies with semipureed diets and 4APP.[37] The TCDD exposure increased the serum triglyceride and

Table 8. Serum and Liver Lipid Changes 10 Days After Exposure to TCDD.[a]

Group:	4APP	Adenine	Serum Triglyceride	Serum Cholesterol	Liver Triglyceride
Control	–	–	100	100	100
Control	+	–	75	52	268
Control	–	+	125	91	76
TCDD	–	–	210	186	653
TCDD	+	–	86	62	708
TCDD	–	+	183	202	479

[a]Groups of 7 animals were given corn oil (control) or TCDD in corn oil (325 μg TCDD/kg body wt) (TCDD) by gavage. Some animals were also given either 4APP (12 mg 4APP/kg body wt) or Ad (38 mg Ad/kg body wt) by i.p. injection every other day. Values are expressed as the percentage of control (no additional treatment) and represent the mean of 7 animals analyzed separately.
[b]Values are significantly different from control values (p < 0.05).

cholesterol levels from 111 and 90 mg/dL to 233 and 167 mg/dL, respectively, and the liver triglyceride concentrations from 38 to 248 mg/g wet weight. 4APP decreased serum triglyceride and cholesterol levels in both control and TCDD-exposed animals while increasing the liver triglyceride. In contrast, the major Ad effect was decreased liver triglyceride, which prevented formation of fatty liver in the TCDD-exposed rats. Based on these results, several mortality studies were performed.

In the first mortality study, the effect of decreasing serum lipids was monitored by giving groups of rats a single oral dose of TCDD (325 μg/kg) equal to two times the LD_{50}[32] and 4APP (12 mg/kg, every other day, i.p.). Since mortality, mean time to death and body weight loss were monitored, it was necessary to pair-feed the control and TCDD groups, and the 4APP and TCDD + 4APP groups throughout the observation. Although the body weight loss was similar (p > 0.05), within 12 days, between the TCDD and TCDD + 4APP groups (29 and 27%, respectively), only the TCDD + 4APP animals died (0/7 vs 7/7). The mean time to death of the TCDD + 4APP group was 9.7 ± 0.5 days. None of the other animals died within 12 days, and the study was discontinued. In separate experiments, 4APP animals were demonstrated to survive more than 30 days on this treatment.

Similar mortality studies were performed with control and TCDD groups of pair-fed animals given Ad (38 mg/kg, every other day, i.p.) or Ad (76 mg/kg, daily, i.p.). The results of the pair-feeding indicated that the TCDD + Ad animals removed, spilled and consumed more feed than the TCDD animals. The effect of Ad is evident, in that both the Ad and TCDD + Ad animals lost less body weight than the appropriate controls, i.e., Ad vs control, and TCDD + Ad vs Ad. Based on the percentage mortality (100% for TCDD and TCDD + Ad) and the mean time to death, 22 ± 1 day (19–26 day range) for TCDD rats and 21 ± 1 day (16–24 day range) for TCDD + Ad rats, however, Ad treatment, even at 76 mg/kg, was not effective in prolonging life.

A single nonlethal dose of TCDD produced marked serum hyperlipidemia and fatty liver in rats within 1 week after treatment. At this low dose of TCDD (60 μg/kg), the TCDD animals continued eating ad libitum at the same level as control animals, i.e., they were naturally pair-fed. Despite the similarity in feed consumption, the TCDD animals lost approximately 10% body weight during the 1 week after exposure to TCDD.

The serum and liver lipid profiles of TCDD-exposed rats and fasted rats are not similar. Although the TCDD-exposed rats lost 10% body weight within 1 week after treatment and 72–hr fasted, control rats lost 10% body weight, only the TCDD-exposed rats demonstrate the serum hyperlipidemia and marked increase in liver lipid concentrations. Thus, the marked changes in lipid parameters after exposure are indicative of altered metabolism unrelated to feed consumption or body weight loss.

Ad stimulated feed consumption and decreased body weight loss in the pair-fed control and TCDD animals without prolonging the life span of the TCDD-treated animals given two times the LD_{50} dose of TCDD. Thus, stimulation of lipid metabolism within the liver did not prolong the longevity of the TCDD-exposed rats. In contrast, 4APP markedly shortened the mean time to death of the TCDD-exposed rats without increasing body weight loss. It is possible that only a small portion of the serum triglyceride (present as serum lipoproteins) was being utilized by the TCDD-exposed rat, and that the 4APP-induced decrease in serum triglyceride affected the available energy sources and prompted death.

In situ experiments, in which rats were pretreated for 2 days with 4APP (20 mg/kg/day), indicated that duodenally infused emulsified [^{14}C]-triglyceride had decreased the appearance of radioactivity in the plasma.[26] In our experiments the rats were not pretreated with 4APP for 2 days prior to the administration of the TCDD (see above). However, if there was a decrease in absorbed dose of TCDD, then the 4APP further enhanced the toxicity of TCDD. It has been reported previously that dietary hexadecane potentiates the toxicity of TCDD, while enhancing the elimination of the orally administered dose of TCDD.[42] The mechanism of the potentiated toxicity of TCDD by hexadecane is unknown, but it has been suggested that hexadecane alters the disposition of the TCDD as it does that of hexachlorobenzene (HCB).[42] An alternative explanation is that hexadecane acts similarly to squalane, which is known to stimulate the excretion of HCB and, more important, decreases the plasma triglyceride concentration by a factor of 2.[43] Thus, both agents which potentiate TCDD-induced toxicity may act by lowering the serum concentration of triglyceride, which is essential for energy metabolism. The TCDD-induced increased sensitivity to 4APP implies that blocking the release and/or synthesis of triglyceride-rich lipoproteins by the liver may play an important role in the TCDD-induced mortality separate from body weight loss. The reduction of serum triglyceride by 4APP in the TCDD rats indicates that the elevated level of circu-

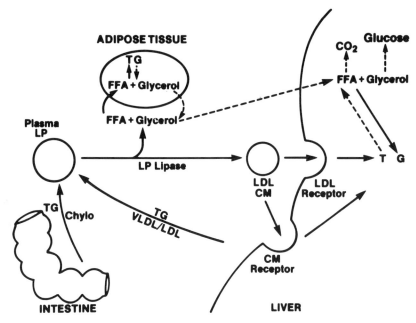

Figure 3.   Outline of lipid metabolism during fed (→) and fasting (——→) states. TG = triglyceride, Chylo = chylomicra, VLDL = very low-density lipoproteins, LDL = low-density lipoproteins, LP = lipoproteins, CM = chylomicron remnant, FFA = free fatty acid.

lating triglyceride is essential to the energy metabolism of the TCDD-stressed rats.

The importance of lipid as an energy source arises between feedings, during periods of fasting, and during times of metabolic and hormonal imbalance. The flow of dietary (exogenous) lipid absorbed from the intestinal lumen and released via the mesentive lymph into the blood as chylomicra and very low-density lipoproteins and endogenous (hepatic) lipid secreted into the blood as very low-density lipoproteins and low-density lipoproteins is illustrated in Figure 3. During the fed state, serum triglyceride levels are maintained by intestinal absorption, and between meals and during fasting by the endogenous production of triglyceride by the liver. The enzyme lipoprotein lipose hydrolyzes some of the circulating triglyceride, forming glycerol and free fatty acids. The glycerol is water soluble and the free fatty acid complexes with albumin for transport. The peripheral tissues utilize the fatty acid for energy by oxidation, or, during the fed state, the fatty acid is resynthesized into triglyceride for storage, especially in the adipose tissue. During periods of fasting and metabolic or hormonal imbalance, the stored triglyceride is hydrolyzed by a hormone-sensitive lipase in the adipose tissue, and the fatty acid and glycerol are again released into the serum as an energy source for the peripheral tissues.

After exposure to TCDD, the serum and liver triglyceride levels increase without changes in serum free fatty acid levels. The epididymal fat pads (adipose tissue) decrease in wet weight, while the liver weight decreases and then declines after exposure to TCDD. These results indicate that after TCDD exposure there is a net accumulation of triglyceride in the serum and liver despite a decline in body and organ weight. In addition, the serum glucose levels decrease markedly. There is no glycogen apparent in the liver tissue in the TCDD-exposed rats (personal observation). A factor contributing to the decreased serum glucose is the utilization of the glycerol (a glucose precursor) for resynthesis of triglyceride via the α-glycerolphosphate pathway in the liver. As demonstrated in Figure 3, a futile metabolic cycle of triglyceride synthesis, hydrolysis and resynthesis seems to occur after exposure to TCDD. After the exhaustion of serum and hepatic glucose stores, the TCDD-exposed rat relies on lipid for metabolic energy, but the triglyceride recycles through the liver, limiting the fatty acid and glycerol available for oxidation and gluconcogenesis, respectively. Thus, lowering serum triglyceride levels, by an agent such as 4APP, in TCDD-exposed animals is critical to survival, and maintaining the increased serum triglyceride levels is a beneficial form of metabolic compensation.

REFERENCES

1. Milnes, H. H. "Formation of 2,3,7,8–Tetrachlorodibenzo-p-dioxin by Thermal Decomposition of Sodium Trichlorophenate," *Nature* (London) 232:395–396 (1971).
2. Bumb, R. R., W. B. Crummett, and S. S. Cutie. "Trace Chemistries of Fire: A Source of Chlorinated Dioxins," *Science* 210:385–390 (1980).
3. Buser, H. R., H. P. Losshardt, and C. Rappe. "Identification of Polychlorinated Dibenzo-*p*-dioxin Isomers Found in Flyash," *Chemosphere* 7:165–172 (1978).
4. Poland, A., and J. C. Knutson. "2,3,7,8–Tetrachlorodibenzo-p-dioxin and Related Halogenated Aromatric Hydrocarbons: Examination of the Mechanism of Toxicity," *Ann. Rev. Pharmacol. Toxicol.* 22:517–554 (1982).
5. May, G. "Chloracne from the Accidental Production of Tetrachlorodibenzo-p-dioxin," *Brit. J. Ind. Med.* 30:276–283 (1973).
6. Poland, A. P., D. Smith, G. Nietter, et al. "A Health Survey of Workers in a 2,4–D and 2,4,5–T Plant: With Special Attention to Chloracne, Porphycia Cutanea Tarda, and Psychologic Parameters," *Arch. Environ. Health* 22:316–327 (1971).
7. Oliver, R. M. "Toxic Effects of 2,3,7,8–Tetrachlorodibenzo-p-dioxin in Laboratory Workers," *Brit. J. Ind. Med.* 32:49–53 (1975).

8. Walker, A. E., and J. V. Martin. "Lipid Profiles in Dioxin-Exposed Workers," *Lancet* 1:446–447 (1979).

9. Axelson, O., and L. Sundell. "Herbicide Exposure, Mortality and Tumor Incidence: An Epidemiological Investigation on Swedish Railroad Workers," *Work Environ. Health* 11:21–28 (1974).

10. Hardell, L. "Malignant Lymphoma in Histiocytic Type and Exposure to Phenoxyacetic Acids or Chlorophenols," *Lancet* 1:55–56 (1979).

11. Suskind, R. R., and V. S. Hertzberg. "Human Health Effects of 2,4,5-T and its Toxic Contaminants," *J. Am. Med. Assn.* 251 (18):2372–2380 (1984).

12. Olson, J. R., M. A. Holscher, and R. A. Neal. "Toxicity of 2,3,7,8-Tetrachlorodibenzo-p-dioxin in the Golden Syrian Hamster," *Toxicol. Appl. Pharmal.* 55:67–78 (1980).

13. Gasiewicz, T. A., and R. A. Neal. "2,3,7,8-Tetrachlorodibenzo-p-dioxin Tissue Distribution, Excretion, and Effects on Clinical Chemical Parameters in Guinea Pigs," *Toxicol. Appl. Pharmacol.* 51:329–339 (1979).

14. Chapman, D. E., and C. M. Schiller. "Dose-Related Effects of 2,3,7,8-Tetrachlorodibenzo-p-dioxin (TCDD) in C57BL/6J and DBA/2J Mice," *Toxicol. Appl. Pharmacol.* 78:147–157 (1985).

15. Campbell, T. C., and L. Friedman. "Chick Edema Factor: Some Tissue Distribution Data and Toxicologic Effects in the Rat and Chick," *Proc. Soc. Expt. Biol. Med.* 121:1283–1287 (1966).

16. Cunningham, H. M., and D. T. Williams. "Effect of Tetrachlorodibenzo-p-dioxin on Growth Rate and the Synthesis of Lipids and Proteins in Rats," *Bull. Environ. Contam. Toxicol.* 7(1):45–51 (1972).

17. Albro, P. W., J. T. Corbett, M. Harris, and L. D. Lawson. "Effects of 2,3,7,8-Tetrachlorodibenzo-*p*-dioxin on Lipid Profiles in Tissue of the Fischer Rat," *Toxicol. Appl. Pharmacol.* 23:315–330 (1978).

18. Gasiewicz, T. A., M. A. Holscher, and R. A. Neal. "The Effect of Total Parentical Nutrition on the Toxicity of 2,3,7,8-Tetrachlorodibenzo-p-dioxin in the Rat," *Toxicol. Appl. Pharmacol.* 54:469–488 (1980).

19. Poli, A., G. Franceschini, L. Puglisi, and C. R. Sirtori. "Increased Total and High Density Lipoprotein Cholesterol With Apoprotein Changes Resembling Streptozotocin Diabetes in Tetrachlorodibenzodioxin (TCDD) Treated Rats," *Biochem. Pharmacol.* 29:835–838 (1980).

20. Bar-On, H., P. S. Roheim, and H. A. Eder. "Serum Lipoproteins and Apolipoproteins in Rats With Streptozotocin-Induced Diabetes," *J. Clin. Invest.* 57:714–721 (1976).

21. Mills, G. L., and C. E. Taylaur. "The Distribution and Composition of Serum Lipoproteins in Eighteen Animals," *Comp. Biochem. Physical.* 40B:489–501 (1971).

22. McNamara, D. J., A. Proca, and K. D. G. Edwards. "Cholesterol Homeostasis in Rats Fed a Purified Diet," *Biochem. Biophys. Acta* 711:252–260 (1982).

23. Reiser, R., M. C. Williams, M. F. Soriels, and N. L. Murtz. "Dietary

Myristate and Plasma Cholesterol Concentration," *J. Am. Oil Chem. Soc.* 42:1155–1162 (1965).

24. Feldman, E. B., B. S. Russell, F. H. Schnare, et al. "Effects of Diets on Homogeneous Saturated Triglycerides on Cholesterol Balance in Rats," *J. Nutr.* 109:2237–2246 (1979).

25. Carroll, R. M., and E. B. Feldman. "Lipids and Lipoproteins," in *Clinical Chemistry of Laboratory Animals*, W. F. Loeb, Ed. (Baltimore: University Park Press, 1985).

26. Krause, B. R., S. H. Sloop, and P. S. Roheim. "Lipid Absorption in Unanesthetized Rats: Effects of 4–Aminopyrazolopyrimidine and Ethinyl Estradiol," *Biochem. Biophys. Acta* 665:165–169 (1981).

27. Hutchens, T. T., J. T. Van Bruggan, P. M. Cockburn, and E. S. West. "Effect of Fasting Upon Tissue Lipogenesis in Intact Rat," *J. Biol. Chem.* 208:115–122 (1954).

28. Bragdon, J. H., R. J. Havel, and R. S. Gordon. "Effects of Carbohydrate Feeding on Serum Lipid and Lipoproteins in the Rat," *Am. J. Physiol.* 189:63–67 (1957).

29. Andersen, J. M., and J. M. Dietschy. "Regulation of Sterol Synthesis in 15 Tissues in Rat. II. Role of Rat and Human High and Low Density Plasma Lipoproteins and of Rat Chylomicron Remnants," *J. Biol. Chem.* 252:3652–3659 (1977).

30. Nordby, G., L. Helgeland, O. D. Mjos, and K. R. Norum. "Lecithin: Cholesterol Acyltransferase Activity in Rats Treated With 4–Aminopyrazolopyrimidine," *Scand. J. Clin. Lab. Invest.* 38:643–647 (1978).

31. Krause, B. R., M. Balzer, and A. D. Hartman, "Adipocyte Cholesterol Storage: Effect of Starvation," *Proc. Soc. Exptl. Biol. Med.* 167:407–411 (1981).

32. Walden, R., and C. M. Schiller. "Comparative Toxicity of 2,3,7,8-Tetrachlorodibenzo-p-dioxin (TCDD) in Four (Sub) strains of Adult Male Rats," *Toxicol. Appl. Pharmacol. 77:490–495 (1985).*

33. Nebert, D. W. "Genetic Differences in Susceptibility to Chemically Induced Myelotoxicity and Leukemia," *Environ. Health Perspect.* 39:11–22 (1981).

34. Schiller, C. M., C. M. Adcock, R. A. Moore, and R. Walden. "Effect of 2,3,7,8–Tetrachlorodibenzo-p-dioxin (TCDD) and Fasting on Body Weight and Lipid Parameters in Rats," *Toxicol. Appl. Pharmacol.*, in press.

35. Scholler, J., F. S. Philips, and S. S. Sternberg, "Production of Fatty Livers by 4–Aminopyrazolo [3,4–*d*] pyrimidine. Toxicological and Pathologic Studies," *Proc. Soc. Exp. Biol. Med.* 93:398–402 (1956).

36. Henderson, J. F. "Studies on Fatty Liver Induction by 4–Aminopyrazolopyrimidine," *J. Lipid Res.* 4:68–74 (1963).

37. Shiff, T. S., P. S. Roheim, and H. A. Eder. "Effects of High Sucrose Diets and 4–Aminopyrazolopyrimidine on Serum Lipids and Lipoproteins in the Rat," *J. Lipid Res.* 12:596–603 (1971).

38. Farber, E., B. Lombardi, and A. E. Castillo. "The Prevention by Adenosine Triphosphate of the Fatty Liver Induced by Ethionine," *Lab. Invest.* 12:873–883 (1963).

39. Handschumacher, R. E., W. A. Creasey, J. J. Jaffee, et al. "Biochemical and Nutritional Studies on the Induction of Fatty Livers by Dietary Orotic Acid," *Proc. Nat. Acad. Sci. U.S.* 46:178–186 (1960).

40. Windmueller, H. G. "An Orotic Acid-Induced Adenine-Reversed Inhibition of Hepatic Lipoprotein Secretion in the Rat," *J. Biol. Chem.* 239:530–537 (1964).

41. Novikoff, A. B., P. S. Roheim, and N. Quintana. "Changes in Rat Liver Cells Induced by Orotic Acid Feeding," *Lab. Invest.* 15:27–49 (1966).

42. Rozman, K. "Hexadecane Increases the Toxicity of 2,3,7,8–Tetrachlorodibenzo-p-dioxin (TCDD): Is Brown Adipose Tissue the Primary Target in TCDD-Induced Wasting Syndrome?" *Biochem. Biophys. Res. Commun.* 125(3):996–1004 (1984).

43. Richter, E., S. G. Schafer, and B. Fichtl. "Stimulation of the Faecal Excretion 2,4,5,2',4',5'-Hexachlorobiphenyl in Rats by Squalane," *Xenobiotica* 13(6):337–343 (1983).

SECTION V

Analytical

CHAPTER 21

Development of High-Performance Liquid Chromatographic Methods for Separation and Automated Collection of Tetrachlorodibenzo-p-Dioxin Isomers

Daniel Wagel, Thomas O. Tiernan, Michael L. Taylor,
B. Ramalingam, John H. Garrett, and Joseph G. Solch

INTRODUCTION

The formation, environmental distribution, fate and toxicological properties of the 22 tetrachlorodibenzo-p-dioxins (TCDDs) have been the subject of intense interest in recent years. Investigations of the properties of the TCDDs have been hampered by the availability of only limited quantities of the individual isomers. As part of a continuing project to prepare larger quantities of these compounds, our laboratory has developed new high-performance liquid chromatographic (HPLC) procedures to isolate the individual TCDD isomers from synthetic mixtures. The primary objectives of this research were the development of higher-capacity HPLC separation techniques and the full automation of the mixture injection and fraction collection operations. By carefully adjusting the chromatographic parameters, more stable and efficient reversed-phase HPLC (RP-HPLC) and normal-phase HPLC (NP-HPLC) methods have been developed and employed to separate components of seven synthetic mixtures of the TCDD isomers.

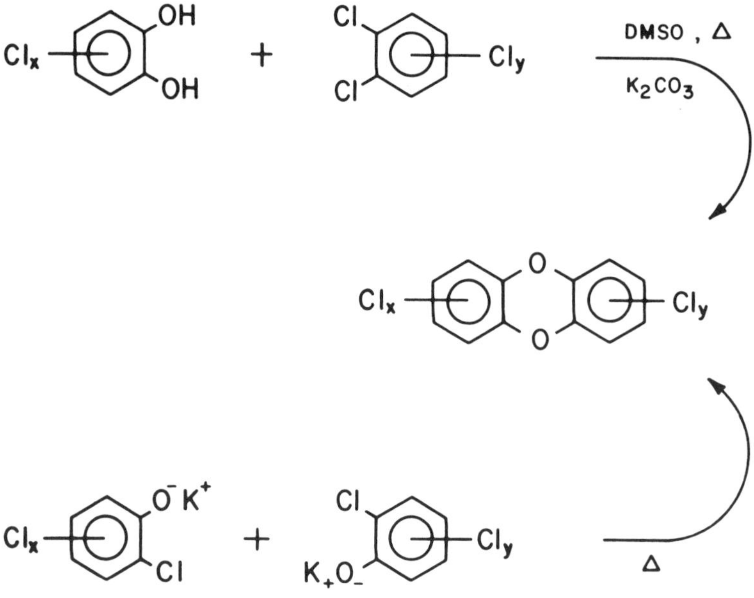

Figure 1.   General TCDD synthesis procedures.

SYNTHESIS OF TCDD ISOMERS

The synthesis procedures initially employed in our laboratory to prepare the isomer mixtures have been reported,[1] and are summarized by the reaction scheme shown in Figure 1. These syntheses resulted in the isomer mixtures listed in Table 1. Two of the isomers, 1,2,3,4- and 2,3,7,8-TCDD, were prepared as single isomer products. Gas chromatographic-mass spectrometric (GC-MS) analysis was used to confirm the identity and purity of the TCDD isomers isolated from each of the seven reaction mixtures. The operating conditions and parameters for the GC-MS system have been described previously.[1]

Table 1.   Mixtures of TCDD Isomers Resulting from Syntheses by the Brehm Laboratory.

Mixture Identification	Isolated TCDD Isomers			
A	1,2,3,7–	1,2,3,8–	1,2,4,7–	1,2,4,8–
B	1,2,3,6–	1,2,3,9–	1,2,4,6–	1,2,4,9–
C	1,2,7,9–	1,4,7,8–	1,3,7,8–	
D	1,2,6,7–	1,2,8,9–		
E	1,2,6,8–	1,3,6,9–	1,2,7,8–	
F	1,3,6,8–	1,3,7,9–		
G	1,4,6,9–	1,2,6,9–		
–	1,2,3,4–			
–	2,3,7,8–			

DESCRIPTION OF THE HPLC SYSTEM AND PARAMETERS

The automated HPLC system used for this research was configured to accomplish repetitive injections of aliquots of a TCDD mixture and collect the separated isomers. This HPLC system included a DuPont Instruments (Wilmington, DE) Series 8800 Gradient Controller. The two component gradients developed with this system can be varied, and both linear and exponential changes in the mobile-phase composition are attainable. A modified DuPont Model 834 Automatic Sampler was connected to the Gradient Controller. The Automatic Sampler was configured to provide automated, repetitive introduction of aliquots of a single TCDD isomer mixture into the HPLC. A DuPont Model 8800 Pump Module and Manually Operated Column Compartment were also used in the HPLC systems. The detector employed here was a Varian Instruments (Palo Alto, CA) Varichrom Model VUV–10, UV-Visible Detector.

An LKB Model 2211 Super Rac (LKB Produkter, Bromma, Sweden) was used to collect fractions of the eluate containing the separated TCDDs. The Super Rac incorporates a microprocessor-controlled three-way valve which shunts all waste to a single container, and the fractions to be collected are directed to a movable drop head. Up to nine time windows may be established, with collection occurring only during each window, while all eluate produced in the time periods outside the windows is dumped to a waste container. The LKB Fraction Collector uses a sophisticated program to control the collection of components within a given chromatographic window. The signal response from the UV detector is monitored and integrated by the Fraction Collector system. The slope of this signal is then compared to a set of parameters describing the peak shape, and the portion of the eluate corresponding to a peak is collected. The Gradient Controller is used to reset the Fraction Collector at the completion of a chromatographic separation, and to restart its operation following the next injection. These components are therefore connected in such a manner that the entire HPLC system may be operated automatically to inject, separate and collect the TCDD isomers.

The chromatographic columns used in the course of this work are listed below:

Zorbax ODS	6.2 mm × 25.0 cm
	9.4 mm × 25.0 cm
Zorbax SIL	6.2 mm × 25.0 cm
Zorbax PSM60	6.2 mm × 25.0 cm
Zorbax Phenyl	4.6 mm × 25.0 cm
Zorbax CN	4.6 mm × 25.0 cm

The experimental parameters used for the RP-HPLC separations are as follows:

Columns   2-Zorbax ODS columns in
                                       series
Injection solvent   Benzene
Injection volume   10μL
Column temperature   40°C
Detector wavelength   235 nm
Mobile phase flow rate   1.50 to 5.00 mL/min
Gradient profiles   Exponential increase of organic
                                       phase relative to water phase
Mobile phases   MA = methyl alcohol
                                       IPA = isopropyl alcohol
                                       ACN = acetonitrile
                                       THF = tetrahydrofuran

EVALUATION OF PREVIOUSLY REPORTED HPLC PROCEDURES FOR SEPARATING TCDD ISOMERS

Following the synthesis of Mixture A (Table 1), the HPLC separation procedures reported by Nestrick et al.[2] and O'Keefe et al.[3] for isolating TCDD isomers were evaluated. Preliminary GC-MS analysis of Mixture A indicated that the 1,2,4,7- and 1,2,4,8-TCDD isomer pair was present at three times the concentration of the 1,2,3,7- and 1,2,3,8-TCDD isomer pair. After establishing HPLC conditions essentially identical to those reported by Nestrick et al.,[2] this mixture was analyzed. The predicted HPLC retention time for the 1,2,4,7- and 1,2,4,8-TCDD isomers, based on the relative retention time reported by Nestrick et al.,[2] and the retention time for 2,3,7,8-TCDD in this study, was 17 min. As shown in Figure 2, a single TCDD peak was observed at 16.8 min and no resolution of the 1,2,4,7- and 1,2,4,8-TCDD isomers was obtained in the RP-HPLC separation.

Following isolation of the single RP-HPLC fraction containing the four TCDD isomers in Mixture A, the NP-HPLC methods described by Nestrick et al.[2] were employed in an attempt to achieve separation of the Mixture A isomer components. The previously reported results of these authors for the separation of the four isomers in this mixture list relative retention times between 1.100 and 1.199. When the silica columns utilized in the previous study were activated, 2,3,7,8-TCDD exhibited a retention time of approximately 15 min. The TCDDs in Mixture A thus eluted between 16.5 and 18.0 min, producing four isomer peaks having retention times within an interval of 1.5 min. The NP-HPLC analysis of Mixture A in the present study, using HPLC conditions identical to those reported by Nestrick et al.,[2] showed that the peaks tailed markedly, and the individual isomer components could not be effectively collected without severe cross-contamination. During the course of several successive NP-HPLC analyses accomplished in the present

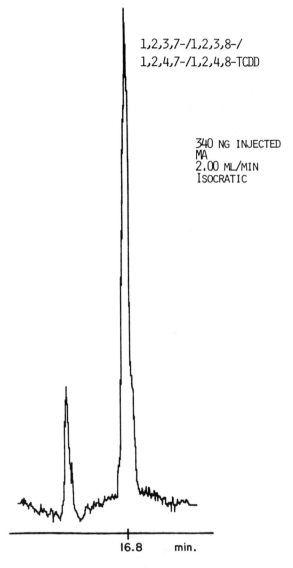

1,2,3,7-/1,2,3,8-/
1,2,4,7-/1,2,4,8-TCDD

340 NG INJECTED
MA
2.00 ML/MIN
ISOCRATIC

16.8   min.

Figure 2.   Chromatogram of TCDDs contained in Mixture A using published conditions.

study, the retention times of the isomers were also observed to shift because of changes in the activity of the silica packing material. These results indicated that while the NP-HPLC procedures may be useful for collecting small quantities of the separated TCDDs, more efficient and reproducible separation procedures were necessary to permit development of an automated separation and collection system. Therefore, further studies were undertaken to improve the resolution of the TCDD isomers which could be achieved using the more stable RP-HPLC separations.

FACTORS INFLUENCING
LIQUID CHROMATOGRAPHIC RESOLUTION

The resolution of two symmetrical liquid chromatographic peaks is given by the equation

$$R_s = 2(t_{R2} - t_{R1})/(W_1 + W_2) \tag{1}$$

where t_{R1} and t_{R2} represent the retention times of the two peaks and W_1 and W_2 are the corresponding peak widths. The resolution between the two peaks, as defined above, is a function of the capacity factor, k', the number of theoretical plates, N, and the selectivity, α, as shown in the following equation.[4]

$$R_s = 1/4[(\alpha - 1)/\alpha] [k_2'/(1 + k_2')] (N_2)^{1/2} \tag{2}$$

Obviously, the resolution can be increased by appropriate changes in the three parameters shown in Equation 2. The number of theoretical plates, N, is defined by the relation,

$$N = 16(t_R/W)^2 \tag{3}$$

and is a measure of the chromatographic column efficiency. Two methods which may be used to increase the number of theoretical plates for a particular column are: first, to decrease the mobile-phase flow rate, to allow for more efficient mass transfer; and second, to increase the length of the chromatographic column. As can be seen from Equation 2, however, doubling the column length increases the resolution only by a factor of 1.4. The capacity factor is defined as

$$k' = (t_R - t_o)/t_o \tag{4}$$

which shows that k' is a measure of the net retention of a component on the column, relative to that of a "nonretained" peak. This term is indicative of how strongly a component is retained on the column. The optimum retention usually involves a k' value between one and ten.[5] Changes in the composition of the stationary and mobile phases are an effective means of adjusting the k' value. For example, the use of a more polar mobile phase may increase the retention time of a weakly retained component in a RP-HPLC separation.

The relative retention, α, is defined as:

$$\alpha = (t_{R2} - t_o)/(t_{R1} - t_o) \tag{5}$$

During a chromatographic separation, two components are separated because they interact differently with the mobile and stationary phases, and thus migrate at different rates through the column. The α term is used to quantify this difference. When $\alpha = 1$, there is no separation of the components, because the chromatographic conditions are not adjusted to utilize the differences which exist between the two compounds. From Equation 2, it is seen that R_s is proportional to $(\alpha-1)/\alpha$, which means that a very small change in α (such as from 1.10 to 1.20) will nearly double the resolution. Therefore, α is a very important parameter in determining the resolution. Changes in α may be produced by changing either the stationary phase or the temperature. Significant changes in the selectivity may also result from changes in the composition of the mobile phase. The almost infinite number of mobile phase possibilities extends the range of HPLC selectivity over a very broad range.[6]

As already noted, optimization of an HPLC separation involves appropriate selection of the three parameters in Equation 2. A reasonable approach entails optimizing the number of theoretical plates and the capacity factor for a given liquid chromatographic separation, followed by an actual experimental test of the separation attainable for a given pair of isomer components. If acceptable resolution of these sample components is not obtained, then changes in the selectivity are required. In cases where the selectivity is low, adjustments in the composition of the mobile phase are an important means of increasing the selectivity, while maintaining an acceptable value for the capacity factor.[6]

DEVELOPMENT AND AUTOMATION OF A RP-HPLC PROCEDURE FOR SEPARATING THE COMPONENTS OF ISOMER MIXTURE A

Utilizing the approach described in the preceding section, various experiments were accomplished in an effort to increase the resolution of the TCDD isomers achieved by the RP-HPLC separations. The procedure of Nestrick et al.[2] involved the use of two Zorbax ODS columns in series, and isocratic elution with methanol at a flow rate of 2.00 mL/min. The capacity factor for the isomers in Mixture A under these conditions was approximately 1.6, which is a low value, in view of the fact that the optimum range of k' values extends from one to ten. Using an isocratic mobile phase containing 90% methanol and 10% water, we were able to increase the capacity factor to 7.9, and the separation illustrated by the chromatogram shown in Figure 3 was obtained. To increase the number of theoretical plates, the flow rate was slowed from 2.00 mL/min to 1.50 mL/min. At the same time, an exponential gradient ranging from 90% to 100% methanol in water was employed, and with these conditions, the chromatogram shown

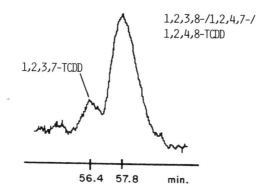

Figure 3.   Chromatogram of TCDDs contained in Mixture A (k' increased).

in Figure 4 was obtained. These experiments indicated that adjusting the capacity factor and the number of theoretical plates did indeed increase the resolution achieved for the isomers, but the separation obtained was still not great enough to permit effective isolation of the individual isomers.

As discussed earlier, changes in the mobile phase can produce significant increases in the resolution because of the selectivity factor. The three chromatograms shown in Figure 5 illustrate the effects of mobile-phase changes on the separation of the isomer components of Mixture A. All of these separations involved use of an exponential gradient elution with a mobile-phase flow rate of 1.5 mL/min. The mobile phase composed of 75% isopropanol and 25% methanol in an exponential gradient with water (abbreviated as: methanol, isopropanol/water gradient) yielded the most effective separation. The individual peaks from a separation using this mobile phase were collected manually and analyzed by GC-MS. Analysis of the first peak showed it contained only 1,2,3,7-TCDD. The second peak contained 1,2,3,8-TCDD and approximately 2% of 1,2,4,7- and 1,2,4,8-TCDD, while the third peak contained only 1,2,4,7- and 1,2,4,8-TCDD. Therefore, this new RP-HPLC procedure is useful for isolating 1,2,3,7- and 1,2,3,8-TCDD in nearly pure form. In these initial analyses of Mixture A, no separation of the 1,2,4,7- and 1,2,4,8-TCDD isomer pair was observed. Since RP-HPLC did not appear to be applicable for separation of the latter

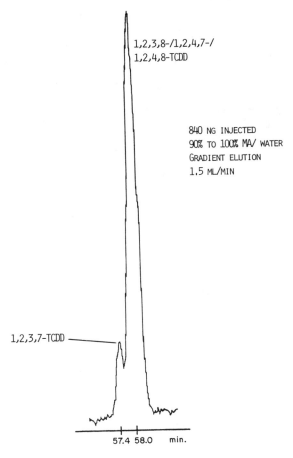

1,2,3,8-/1,2,4,7-/
1,2,4,8-TCDD

840 NG INJECTED
90% TO 100% MA/ WATER
GRADIENT ELUTION
1.5 ML/MIN

1,2,3,7-TCDD

57.4 58.0   min.

Figure 4. Chromatogram of TCDDs contained in Mixture A (gradient elution and slower flow rate).

isomer pair, it was decided to utilize simplified NP-HPLC procedures for this purpose, in a second HPLC run.

The RP-HPLC separations of the TCDD isomer components of Mixture A demonstrate the potential for "fine-tuning" the mobile-phase composition to obtain improved resolution of other TCDD isomers. Because the retention times are reproducible and the peak shapes are highly symmetrical for the RP-HPLC separations described here, these procedures are quite amenable to automation. Consequently, the fraction collector was programmed for a chromatographic retention time window during which the three components of Mixture A would elute. The peak detector was programmed to collect peaks having a width of 15 sec and a threshold signal response greater than 4 mV. Under these conditions, 8.4-µg quantities of Mixture A were injected, and the separated components were collected. GC-

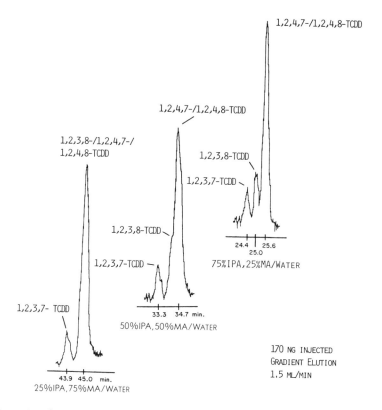

Figure 5. Chromatograms of TCDDs contained in Mixture A demonstrating effects of charges in mobile phase selectivity.

MS analysis of these fractions gave results similar to those reported for the manually collected HPLC fraction discussed above.

METHANOL, ISOPROPANOL/WATER GRADIENT SEPARATION OF THE 22 TCDD ISOMERS

As described above, the methanol, isopropanol/water gradient HPLC procedure permitted the separation and collection of 1,2,3,7- and 1,2,3,8-TCDD (as well as the unseparated 1,2,4,7- and 1,2,4,8-TCDD isomer pair). The results of initial attempts to separate the components of Mixture B using this newly developed gradient procedure showed that this solvent system was superior to the method of isocratic elution with methanol. The successful separation of the components of Mixtures A and B using the new RP-HPLC gradient prompted further attempts to separate and collect each of the 22 TCDD isomers using this solvent system. The data from these separations are shown in Table 2 and are compared to the results

Table 2. Comparison of RP-HPLC Retention Data Obtained for TCDD Isomer Mixtures Using a New Gradient Elution Technique with Those Obtained with an Isocratic Elution Technique.[a]

Mixture	TCDD Isomer Contents	Retention Time MA (Isocratic)	Retention Time IPA, MA/Water (Gradient)
A	1237	16.8	33.0
	1238	17.2	33.9
	1247	17.0	34.8
	1248	17.0	34.8
B	1236	16.8	32.3
	1239	16.2	29.6
	1246	16.2	30.2
	1249	16.2	30.2
C	1279	15.7	30.8
	1478	15.8	33.2
	1378	17.5	38.4
D	1267	14.4	25.6
	1289	14.4	24.0
E	1268	15.5	30.7
	1369	15.8	33.5
	1278	16.7	33.5
F	1368	18.6	42.8
	1379	17.7	38.9
G	1469	14.2	25.2
	1269	14.0	25.4
	1234	18.8	34.2
	2378	16.5	35.5

[a]MA, methyl alchohol; IPA, isopropyl alcohol; ACN, acetonitrile.

obtained using isocratic elution with methanol as the mobile phase. Comparison of the two sets of data shows that the methanol, isopropanol/water gradient system permits the separation of a greater number of TCDD isomers. However, these procedures do not permit efficient isolation of all the isomers in pure form. For example, 1,4,6,9- and 1,2,6,9–TCDD are separated by only 0.2 min even with this gradient elution procedure. Therefore, even more effective HPLC separation procedures were sought.

DEVELOPMENT OF MIXTURE SEPARATION PROCEDURES

The syntheses employed for production of the TCDD isomer mixtures shown in Table 1 obviously yielded specific groups of isomers. In order to take maximum advantage of the selectivity in separation which is possible using specially developed mobile phases, each of these mixtures of isomers was investigated separately, and a set of optimum HPLC separation conditions was developed for each.

For most of the mixtures indicated in Table 1, it was possible to develop a single mobile-phase system which would result in separation of the isomers in that mixture. However, if the resolution achieved was marginal, then the

Table 3.   Results of Various HPLC Separation Experiments Accomplished in an Effort to Separate the TCDD Isomers Contained in Mixture A.

| | | TCDD Isomers Producing Observed HPLC Peaks | |
Mobile Phase	Rs	Peak 1	Peak 2
75% IPA, 25% MA/Water	1.2	1237	1238
(Mobile phase previously	2.3	1237	1247/1248
discussed)	1.0	1238	1247/1248
ACN/Water	2.0	1237/1238	1247/1248
25% IPA, 75% MA/Water	1.6	1237	1238

injection of highly concentrated solutions made the collection of pure isomer fractions more difficult, and often reinjection of the collected fractions was necessary in order to remove cross-contamination. In these cases, multiple mobile-phase separation procedures were developed to increase the efficiency of the isolation process. These procedures entailed the collection of HPLC fractions resulting from application of one mobile phase, followed by reinjection of each collected fraction containing more than one isomer using a different mobile phase, which yielded separation of the previously unresolved isomers. The resolution obtained with the more specialized mobile phases permitted the isolation of greater quantities of the TCDD isomers from each injection, because the broadened peaks resulting from overloading of the columns could still be collected without cross-contamination. Several multiple mobile-phase separation schemes were developed for the TCDD mixtures during this research. In the following sections, the development and application of separation schemes for isolating isomerically pure TCDDs from four of the seven synthesis mixtures are briefly discussed.

Separation Procedure for Mixture A

As discussed in a preceding section, the methanol, isopropanol/water gradient was used to isolate the 1,2,3,7- and 1,2,3,8-TCDD isomers as well as the unseparated 1,2,4,7- and 1,2,4,8-TCDD isomer pair. As shown in Table 3, the resolution of the isomers using this mobile phase was poor, especially between the 1,2,3,8-TCDD peak and the 1,2,4,7- and 1,2,4,8-TCDD composite peak. Following an evaluation of 16 different mobile phases, two were selected as being useful in the development of a two-step isolation procedure for this mixture. Referring to Table 3, it is seen that use of an acetonitrile and water mobile phase yielded a resolution of 2.0 for the separation of the 1,2,3,7- and 1,2,3,8-TCDD composite isomer pair from the 1,2,4,7- and 1,2,4,8-TCDD composite isomer pair. The chromatogram corresponding to this separation is shown in Figure 6. The 1,2,3,7- and 1,2,3,8-TCDD isomer pair isolated in this first step was then effectively separated using a mobile phase composed of 25% isopropanol

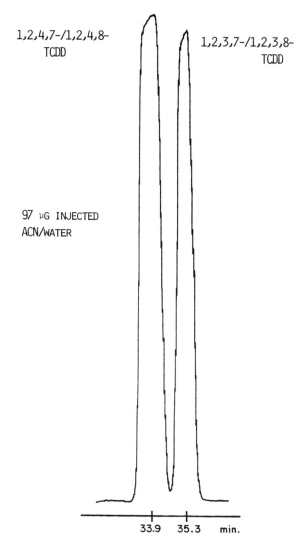

1,2,4,7-/1,2,4,8-
TCDD

1,2,3,7-/1,2,3,8-
TCDD

97 µG INJECTED
ACN/WATER

33.9   35.3   min.

Figure 6. Chromatogram of the first step in the separation of the TCDDs contained in Mixture A.

and 75% methanol in a gradient with water. This separation is shown by the chromatograph presented in Figure 7. This two-step isomer isolation process is more efficient than the single-step procedure discussed earlier because the greater resolution of the isomers achieved in each step permits the injection of significantly higher concentrations of Mixture A without peak overlapping. The separation of the 1,2,4,7- and 1,2,4,8-TCDD isomer pair will be discussed in a later section.

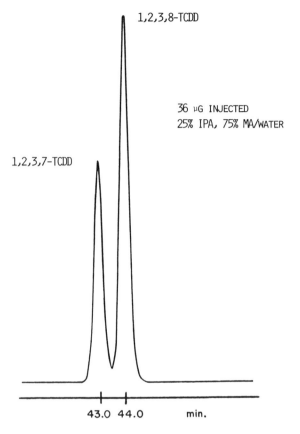

1,2,3,8-TCDD

36 μG INJECTED
25% IPA, 75% MA/WATER

1,2,3,7-TCDD

43.0 44.0   min.

Figure 7. Chromatogram of the second step in the separation of the TCDDs contained in Mixture A.

Separation Procedure for Mixture E

Use of the methanol, isopropanol/water mobile phase originally developed for Mixture A yielded the separation shown in Figure 8 when used with Mixture E. The two well-resolved peaks from this separation were collected and GC-MS analysis showed that the first peak was attributable to pure 1,2,6,8–TCDD, while the second peak contained a mixture of the 1,3,6,9– and 1,2,7,8–TCDD isomers. Further, experiments were then conducted to develop methods to separate this isomer pair using several other mobile phases. The most complete resolution of the two isomers was obtained using an acetonitrile and water mixture. The chromatogram illustrating this separation is shown in Figure 9. Using this two-step process, the three isomers in Mixture E were separated and collected in pure form.

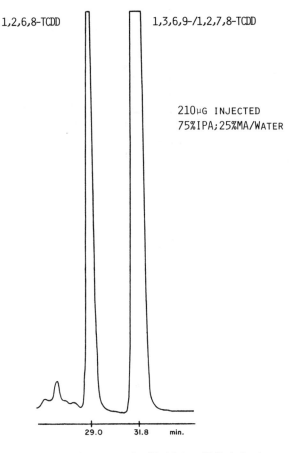

Figure 8.   Chromatogram of TCDDs contained in Mixture E; first step in separation procedure.

Separation Procedure for Mixture C

The three isomers present in Mixture C were 1,2,7,9– 1,4,7,8– and
1,3,7,8–TCDD. The methanol, isopropanol/water mobile phase used in the
earlier experiments was very effective in separating the components of this
mixture. The chromatogram illustrating this separation is shown in Figure
10. In this mixture, the most critical separations involved the resolution of
organic contaminants from the TCDD isomer peaks. Using the above conditions, a total of 240 µg of TCDD isomers was isolated from each injection.

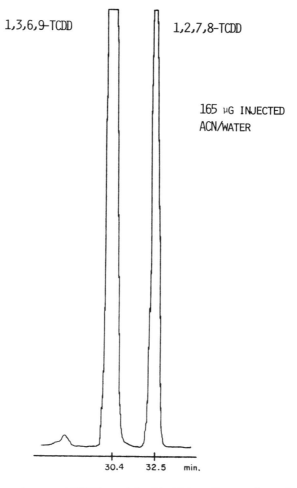

Figure 9. Chromatogram of TCDDs contained in Mixture E; second step in separation procedure.

Separation Procedure for Mixture B

An attempt to separate the TCDD isomers in Mixture B using a methanol mobile phase yielded almost no apparent separation of the four isomer components. The methanol, isopropanol/water gradient mobile phase was more effective in separating these components and was used to isolate nanogram quantities of these isomers during a preliminary separation. The major problem encountered with this mobile phase was the relatively low resolution of the TCDD isomers from other residual components of the reaction product. Evaluations of several mobile phases showed that a mixture of 50% acetonitrile and 50% methanol in a gradient with water was the

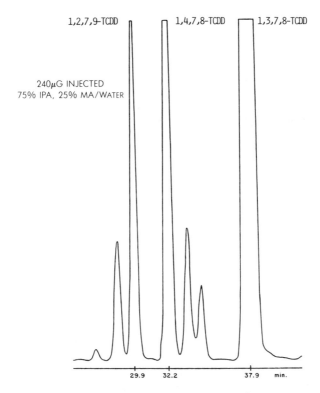

Figure 10. Chromatogram of TCDDs contained in Mixture C.

most effective solvent mixture for separating the various synthesis products. The chromatogram shown in Figure 11 illustrates the separation obtained using this mobile phase. The 1,2,4,6– and 1,2,4,9–TCDD isomer pair was not separated under any of the RP-HPLC conditions tested in this study.

NORMAL-PHASE SEPARATION PROCEDURES

The initial evaluations of the NP-HPLC separations of the isomers showed that this was not an efficient or easily automated method for isolating the TCDD isomers. The RP-HPLC procedures were therefore developed to accomplish the majority of the separations described herein, and 18 of the isomers were finally isolated using these techniques. During this portion of the research, none of the RP-HPLC conditions investigated here were found to separate the 1,2,4,6– and 1,2,4,9–TCDD pair (as noted earlier), or the 1,2,4,7– and 1,2,4,8–TCDD pair. In addition to the columns described above, Zorbax-Phenyl and Zorbax-CN stationary phases were also investigated in this study, and were found to be ineffective in separating

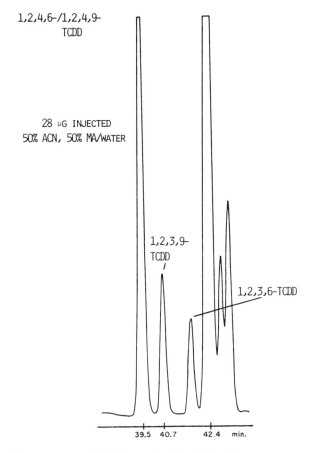

Figure 11. Chromatogram of TCDDs contained in Mixture B.

the indicated isomer pairs. Consequently, since RP-HPLC did not appear to be applicable for these separations, it was decided to reevaluate the NP-HPLC procedures reported by Nestrick et al.[2] and O'Keefe et al.[3] From the previous NP-HPLC analyses, the poor peak shapes and marginal resolution were judged to be the major chromatographic problems limiting effective isomer separation, since manual fraction collection could be used to compensate for the problem of drifting retention times observed with the NP-HPLC. Using the unresolved pairs of isomers isolated from Mixtures A and B, experiments were undertaken in an attempt to optimize the NP-HPLC separations. The optimum NP-HPLC separation conditions ultimately identified are summarized below.

Columns	2–Zorbax SIL and 2–Zorbax PSM60 columns in series
Mobile phase	0.5% toluene in hexane

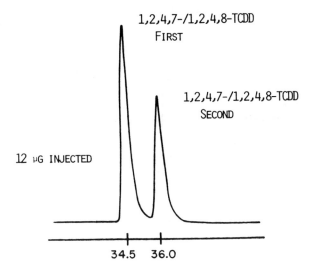

Figure 12.   Chromatogram of 1,2,4,7- and 1,2,4,8-TCDD using NP-HPLC.

Flow rate	1.50 mL/min
Injection volume	10 µL of hexane
Temperature	24°C
Detector Wavelength	313nm

As previous researchers have observed,[3] the use of 0.5% toluene as a modifier in the hexane mobile phase is effective in reducing the peak tailing frequently observed in NP-HPLC. An evaluation of the Zorbax PSM60 columns showed that the separations attainable on these columns were comparable to those obtained using the Zorbax SIL columns. The need for greater resolution prompted the coupling of four NP-HPLC columns in series. This configuration was ultimately effective in separating the previously unresolved TCDD isomer pairs, as shown by the chromatograms presented in Figures 12 and 13. The identification of the order of elution of the four isomers in these figures remains unknown.

PURITY OF ISOLATED ISOMERS

Using the automated HPLC procedures and the specialized mobile phases developed during this research program, microgram to milligram quantities of each TCDD isomer have been isolated in pure form. The isomeric purity of each of these TCDD isomers was determined by GC-MS analysis using two different columns. In the case of the isomers separated by NP-HPLC, the isolated components were reinjected to quantify any cross-contamination. Figure 14 shows two of the GC-MS chromatograms which were obtained for the isomers in Mixture D. Table 4 lists the purity determined

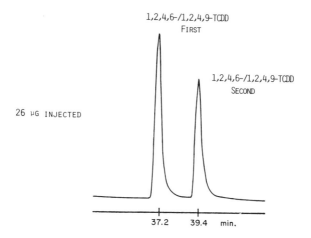

Figure 13. Chromatogram of 1,2,4,6- and 1,2,4,9-TCDD using NP-HPLC.

for each isolated isomer obtained in this study. The data in Table 4 indicate that the isomeric purity of each TCDD isomer obtained here was greater than 98%, except in the case of the isomers which had to be separated using the NP-HPLC procedures, in which case the purity was at least 96%. Very pure TCDD isomer standards have therefore been prepared by applying the higher capacity RP- and NP-HPLC procedures developed during this research.

SUMMARY

Procedures for the automated isolation of the 22 TCDD isomers have been developed in our laboratory. Previously published methods were found to be ineffective when isolating larger quantities of the isomers because of the poor resolution achieved by the RP-HPLC procedures, and the poor stability and peak shapes obtained in the NP-HPLC separations. The procedures described herein were developed empirically by evaluating various RP-HPLC separation conditions and developing optimum mobile phases to separate synthetic mixtures of TCDD isomers. The automated collection of HPLC fractions was easily implemented because the retention time stability and peak shape obtained with the specialized HPLC mobile phases were superior to those observed under normal-phase conditions. The improved resolution of the RP-HPLC procedures developed here also permitted overloading of the columns and the collection of the broadened peaks without cross-contamination, facilitating more rapid separation and collection of larger quantities of the pure TCDD isomers. In the two cases where NP-HPLC separations were required, the separation of the isomers

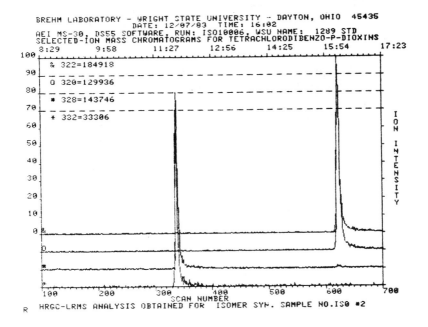

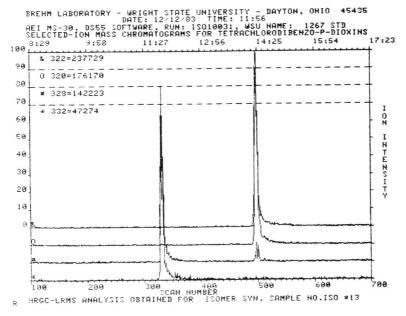

Figure 14.   GC-MS chromatograms of 1,2,8,9- and 1,2,6,7-TCDD.

Table 4.   Isomer Purity of Each TCDD Isomer Prepared and Isolated in the Present Study as Determined by GC-MS and NP-HPLC.

TCDD Isomer	Isomeric Purity (%)
1378	> 99
1289	99.5
1234	> 98
1478	> 99.5
1279	99
1369	> 99.5
1268	> 99.5
1239	99
1278	> 99
1236	> 99.5
1269	> 99.5
1469	> 99.5
1267	> 99.5
1368	> 99.5
1379	> 99.5
1237	> 98
1238	> 98
1246 /1249 First[a]	> 97
1246 /1249 Second[a]	> 97
1247 /1248 First[a]	> 97
1247 /1248 Second[a]	96
2378	> 99

[a]Isomers are designated with isomer pair and elution on NP-HPLC (Silica).

was improved by increasing the column length and reducing the peak tailing.

The development of these procedures has also yielded retention time and isomer resolution data for the 22 isomers under many different HPLC mobile and stationary phase conditions. Correlations of the TCDD isomer structure with the indicated optimum separation conditions are presently being studied. Further, experiments are also in progress to combine the newly developed mobile phases with higher-capacity preparative-scale HPLC techniques.

ACKNOWLEDGMENTS

This work was supported in part by United States Environmental Protection Agency Cooperative Agreement No. CR–809972–01–1, which was monitored by EPA/EMSL (Las Vegas). The earlier, generous contributions of TCDD isomer standards to our laboratory by the Dow Chemical Co., and by Dr. H. R. Buser of the Swiss Federal Research Agency, which greatly facilitated identification of isomers produced in the present program, are gratefully acknowledged.

REFERENCES

1. Taylor, M. L., T. O. Tiernan, B. Ramalingam, D. J. Wagel, J. H. Garrett, J. G. Solch, and Gerald L. Ferguson. "Synthesis, Isolation, and Characterization of the Tetrachlorinated Dibenzo-p-dioxins and Other Related Compounds," in *Chlorinated Dioxins and Dibenzofurans in the Total Environment*, G. Choudhary, L. H. Keith, and C. Rappe, Eds., (Stoneham, MA: Butterworth Publishers, 1985), p. 17.
2. Nestrick, T. J., L. L. Lamparski, and R. H. Stehl. "Synthesis and Identification of the 22 Tetrachlorodibenzo-p-dioxin Isomers by High Performance Liquid Chromatography and Gas Chromatography," *Anal. Chem.* 51:2273–2281 (1979).
3. O'Keefe, P. W., R. Smith, C. Meyer, et al. "Modification of a High-Performance Liquid Chromatographic-Gas Chromatographic Procedure for Separation of the 22 Tetrachlorodibenzo-p-dioxin Isomers," *J. Chromatog.* 242:305–312 (1982).
4. Karger, B. L. "The Relationship of Theory to Practice in High-Speed Liquid Chromatography," in *Modern Practice of Liquid Chromatography*, J. J. Kirkland, Ed. (New York: John Wiley and Sons, Inc., 1971), p. 3.
5. Snyder, L. R. "The Role of the Mobile Phase in Liquid Chromatography," in *Modern Practice of Liquid Chromatography*, J. J. Kirkland, Ed. (New York: John Wiley and Sons, Inc., 1971), p. 125.
6. McLander, W. R., and G. Horvath. "Reversed-Phase Chromatography," in *High Performance Liquid Chromatography: Advances and Perspectives, Vol. 2*, C. Horvath, Ed. (New York: Academic Press, Inc., 1980), p. 113.

CHAPTER **22**

Evaluation of XAD-2 Resin Cartridge for Concentration/Isolation of Chlorinated Dibenzo-p-Dioxins and Furans from Drinking Water at the Parts-per-Quadrillion Level

Guy L. LeBel, David T. Williams, John J. Ryan, and Ben P.-Y Lau

INTRODUCTION

Because of the toxicity and persistence of some isomers of polychlorinated dibenzo-p-dioxins (PCDD) and polychlorinated dibenzofurans (PCDF), methodology is required for the analysis of PCDD and PCDF at the parts-per-quadrillion (pg/L) level in water. Analysis at these extremely low levels requires large sample size, sensitive/selective detectors and/or very clean extracts. The methods usually used for the analysis of chlorinated dioxins in water employ liquid/liquid solvent extraction with extensive cleanup of the extracts prior to instrumental analysis, usually by GC/MS.

Adsorbents, particularly the macroreticular resins, i.e., Amberlite XAD-2 resin, have been extensively used for the concentration and isolation of trace levels of organics from water.[1-7] We have developed a sampling cartridge using the macroreticular XAD-2 resin for the concentration/isolation of trace organics from large volumes (200 L) of drinking water, and have evaluated the technique for organic phosphates (pesticides and triesters),[3,4] halogenated organic compounds (pesticides, PCB's and other environmental contaminants)[5,6] and polycyclic aromatic hydrocarbons.[7] The XAD extracts have also been used to determine the mutagenic activity (Ames test) in drinking water samples.[8,9]

The XAD resin methodology has so far been used only for compounds present at ng or μg/L levels in water. However, we have evaluated its potential for the extraction and analysis of pg/L PCDD and PCDF in drinking water and now present the results of this evaluation.

MATERIALS AND METHOD

All glassware, chemicals and laboratory apparatus were thoroughly rinsed with distilled-in-glass grade solvents prior to use. Disposable glassware was used whenever possible to avoid the possibility of cross-contamination.

Sampling Cartridge and Preparation of Extracts

The construction and preparation of the sampling cartridge have been described previously.[3] Although the reuse of regenerated cartridges is a positive feature of the technique, the cartridges used for the present study were all packed with freshly prepared resin. In summary, the sampling cartridges were connected to a drinking water tap (allowed to run about 15 min before sampling), and the water flow through the cartridge adjusted to 3 bed-volumes/min, i.e., 140 mL/min for 24 hr with appropriate flow rate monitoring. After sampling, the cartridge was disconnected, the residual water in the cartridge was allowed to drain, the cartridge was eluted with 300 mL of 15% acetone/hexane (v/v), and the eluate was collected in a separatory funnel. The aqueous phase was discarded and the organic phase dried by passing through anhydrous sodium sulfate. The funnel and filter were rinsed with the eluting solvent and the filtrate was concentrated to ~ 2-3 mL on a rotary evaporator. The concentrate was transferred to a 15mL centrifuge tube with 2-3 x 1.5 mL of hexane.

Fortification Techniques

Because it is impractical to adequately fortify the large volume of water sampled at the low levels required, two alternative fortification techniques were used,[4] i.e., direct on-column fortification and continuous-feed fortification. A standard mixture of several PCDD/PCDF congeners was prepared in acetonitrile at the 25pg/μL level for each compound evaluated.

For the continuous fortification technique, 4 L of three concentrations of aqueous PCDD/PCDF solutions were prepared so that 1 mL/min of each solution pumped into the tap water line for each sampling cartridge (i.e., 2 mL/min for duplicate cartridges) would give a final concentration of 1, 10

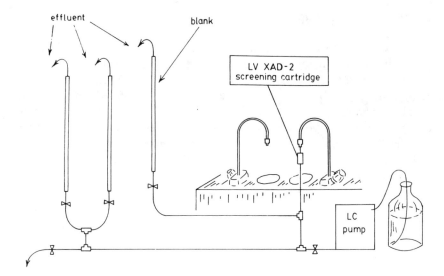

Figure 1. Schematic of continuous-feed fortification using LC pump.

and 50 pg/L respectively for each compound. The fortification experiment
was set up as shown in Figure 1. The tap water was passed through a large-
volume XAD cartridge prior to the fortification to remove any PCDD/
PCDF in the original water. The concentrated spiking solution was fed into
the XAD-screened water using an HPLC pump (Eldex Model B-100-S) with
at least 6 ft of 1/4 -in.-o.d. copper tubing between the tee and the cartridges
to provide adequate mixing. At the end of the fortification, i.e., after 200 L
of water passed through the XAD cartridges, 900 mL of spiking solution,
was pumped into a separatory funnel and back-extracted with dichloromethane (3 × 50 mL) to measure the actual concentration of the dioxins/
furans in the feed solution.

For the direct fortification procedure, an aliquot of the PCDD/PCDF
solution in acetonitrile was applied with a syringe directly to the head of the
cartridge and flushed down with ~ 250 mL of purified water. The aliquot
contained an amount of analyte equivalent to fortification levels of 1, 10
and 25 pg/L after the sampling of 200 L of XAD-2-screened tap water. The
fortified cartridges were then attached to a tap and 200 L of XAD-screened
water was passed through the cartridge. For all fortification experiments, a
blank cartridge was always run in parallel under the same sampling conditions (Figure 1).

Sulfuric Acid Cleanup

The acid cleanup step was adapted from a procedure used for the analysis of dioxins/furans in adipose tissue.[10] About 2 mL of concentrated sulfuric acid was added to the ~ 6–7 mL of the concentrated XAD extract in the 15mL centrifuge tube, and the mixture vortexed for ~ 30 sec and allowed to stand for phase separation. The lower acid layer was removed with a Pasteur pipette and the procedure repeated 4 to 6 times until the acid layer was colorless. The solution was then washed sequentially with 2 mL each of water, 1N KOH and water. The hexane layer was dried by passing through anhydrous sodium sulfate (~ 1 in. sulfate in a Pasteur pipet) and the sulfate was rinsed with hexane (3 ×). The rinses and filtrate were collected in a 15-mL centrifuge tube and concentrated to ~ 0.5 mL with a gentle stream of nitrogen.

Florisil Column Chromatography

The Florisil cleanup step is basically the same as that used in an earlier method for the determination of PCDD/PCDF in adipose tissue.[10] The column, a 6-mm-i.d. chromaflex column with a 50-mL reservoir, Kontes # K-420100-0021, was packed with 1 g of Florisil (activated overnight at 130° C) topped with ~ 5 mm of anhydrous sodium sulfate. The column was washed with 5 mL dichloromethane followed by 2 mL of hexane/dichloromethane (98:2). The extract from the sulfuric acid cleanup step was added to the column and eluted (a) with 10 mL of hexane/dichloromethane (98:2) using the first 3 × 0.5 mL to rinse/transfer the extract, and (b) with 15 mL of dichloromethane. The dichloromethane was collected in a 15-mL centrifuge tube and concentrated to ~ 0.3 mL with a jet of prepurified nitrogen with occasional washing of the tube walls. After transfer to a 1.0-mL conical reactivial with 2–3 washings and evaporation to near dryness, the extract was made up to 0.1 mL with isooctane for GC analysis or further cleanup.

Carbopack Column Cleanup

The carbopack column cleanup is an adaptation of a U.S. EPA protocol.[11] The column was prepared by packing 0.25 g of 18% Carbopack C (80/100 mesh) on Celite 545® in a 5-mL disposable borosilicate serological pipette. The column was washed sequentially with 2 mL of toluene, 1 mL of dichloromethane/methanol/toluene (75:20:5), 1 mL cyclohexane/dichloromethane (50:50) and 2 mL hexane. The extract from the sulfuric acid and Florisil cleanup steps was transferred to the column with 2 × 0.5 mL hexane and eluted sequentially (the first 0.5 mL of hexane was used to further rinse the vial) with the reverse order of washing solvent, except that 12 mL of toluene was used for the final elution. The toluene eluate was collected, evaporated on a rotary evaporator to near dryness, and transferred to a 1-mL conical reactivial with isooctane and finally reevaporated to 0.1 mL with a gentle stream of nitrogen.

Gas Chromatographic Determination

The XAD-2 extracts were all analyzed by capillary column gas chromatography with electron capture detector. The GC used was a Varian Model 4600 equipped with a Ni[63] ECD, a Varian on-column injector and interfaced to a Vista 402 chromatography data system with dual disk drives. The column parameters and operating conditions were: 30m × 0.32mm i.d. DB-5 (film thickness 0.25 μm) fused-silica capillary column (J & W); oven temperature: initial 80°C, hold 1 min, program at 10°/min to 200°C, hold 1 min then program at 5°/min to 280°C, hold 8 min; helium carrier gas set at 2.0 mL/min (ca 1 mL/min to purge) with nitrogen makeup gas at 25 mL/min; detector at 325°C; on-column injector program: initial 80°C, then program at 100°/min to 280°C, hold time 33 min. Aliquots (2 μL) of extract (or standard) solutions were injected and the chromatographic data were stored on floppy disk for subsequent data processing. The amount of unknown or spiked material was determined by comparison of peak area with the corresponding peak of standards injected under identical conditions.

Gas Chromatography/Mass Spectrometry Analysis

The extracts were analyzed for PCDD/PCDF by two GC/MS techniques: (1) The extracts with minimal cleanup, i.e., acid and Florisil column, were anlayzed by GC-tandem MS (TAGA 6000 Triple Quadrupole MS/MS system). The equipment, conditions and ions monitored have been previously described.[10,12] (2) The extracts that were processed through the additional Carbopack column cleanup stage were analyzed by capillary column GC/MS (VG-ZAB-2F). The technique, equipment, conditions and ions monitored were as previously described.[10,13] For estimation of percentage recovery, the extracts were analyzed by MS selected ion monitoring at a resolution of 2K.

RESULTS AND DISCUSSION

The analysis by GC/MS of organics present in XAD extracts obtained from 200-L tap water samples required no additional cleanup procedure when the organics were at ng/L or higher levels. However, application of the technique to low-pg/L levels of PCDD/PCDF clearly required suitable cleanup procedures. Several cleanup procedures have been used in our labo-

Table 1. ¹⁴C-TCDD Recoveries (%)[a] in Fortified Drinking Water (200 L) from XAD-2 Resin Cartridge.

Procedure	10 pg/L	100 pg/L
Cartridge elution (15% acetone/hexane)	97.7 / 103.3	96.6 / 99.4
Sulfuric acid	88.4 / 84.6	93.4 / 103.3
Florisil column	74.2 / 83.9	84.3 / 89.6
Cartridge elution (7 days)	100.5	96.2

[a] Two determinations.

ratories[10] for the determination of PCDD/PCDF in fish and human adipose tissues, and these were adapted to use with the XAD water extracts.

XAD extracts of ~ 200-L drinking water samples usually contain between 1-10 mg of residue depending on the water source.[9] Acid cleanup and Florisil column fractionation removed the majority of this residue (< 0.1 mg remaining). Before investigating the necessity for further cleanup procedures, a preliminary recovery experiment using 2378 ¹⁴C-TCDD at the 10 and 100 pg/L fortification level was carried out to monitor the recovery through the various extraction and cleanup procedures. The ¹⁴C-TCDD was determined by liquid scintillation counting techniques. The results (Table 1) indicate basically quantitative recoveries from the cartridge and following the sulfuric treatment and Florisil column cleanup stages. To measure possible loss due to storage, a recovery run at both spiking levels was also attempted with the cartridge elution done after storage of the cartridges at room temperature for 7 days. Again, the recovery of ¹⁴C-TCDD was essentially quantitative.

To evaluate the technique for a wider range of PCDD/PCDF isomers, a solution of at least one isomer of tetra- to octa-PCDD and PCDF was prepared in acetonitrile for fortification purposes. Because it is not practical to fortify the large volumes of tap water at these low fortification levels, the continuous fortification technique, in which a "concentrated" aqueous solution of the analytes is continuously pumped with an HPLC pump into a water tap line, was used for the fortification experiments at the 1, 10 and 50 pg/L for a 200-L sample of water. The generated extracts and appropriate blank extracts were processed through the acid and Florisil cleanup steps prior to GC analysis.

The extracts could be screened by GC/ECD for the analytes at the higher fortification levels and for the higher chlorinated congeners. However, this was clearly not adequate at the lower levels, or for the less chlorinated PCDD/PCDF congeners (see A in Figure 2).

The extracts were then analyzed by GC-tandem MS, which has been used previously[11] to determine trace levels of PCDD/PCDF analytes in extracts that had not been processed through rigorous cleanup procedures usually required for GC/MS confirmation. The percent recoveries, corrected for

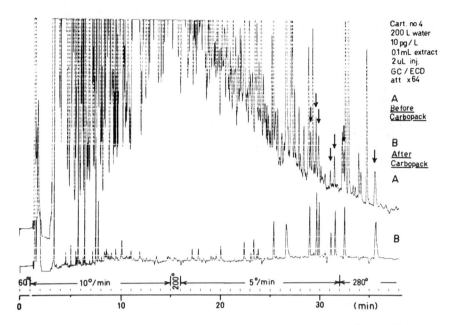

Figure 2. GC/ECD chromatogram of a 10-pg/L fortified water extract (200 L): A) before Carbopack column cleanup; and B) after Carbopack column cleanup. Arrows indicate PCDD/PCDF congeners.

loss in feed solution concentration, are shown in Table 2 and illustrate decreasing recoveries for the higher-chlorinated congeners.

Although most analytes were adequately determined in the fortified extracts by GC-tandem MS (Table 2), the presence of low-level interferences

Table 2. PCDD/PCDF recoveries (%)[a] by GC / Tandem MS in Fortified (Continuous Feed) XAD-2 Drinking Water Extracts (200 L).[b,c]

Analyte [d]	1 pg/L	10 pg/L	50 pg/L
1368-TCDD	67.5 / 35.4	61.4 / 68.0	56.8 / 77.3
2378-TCDD	230.8 / 163.4	67.1 / 73.5	45.1 / 62.5
2378-TCDF	106.7 / 77.8	56.6 / 60.9	51.6 / 69.5
12378-PnCDD	166.1 / 101.4	38.1 / 40.5	37.7 / 47.6
12478-PnCDF	75.3 / 64.6	40.9 / 45.2	50.7 / 65.6
123678-HxCDD	45.8 / 44.0	30.5 / 23.8	28.8 / 33.9
123789-HxCDD	43.2 / 38.7	31.5 / 21.0	28.6 / 33.6
123678-HxCDF	60.9 / 59.1	32.3 / 29.0	32.4 / 42.3
1234678-HpCDD	39.7 / 38.1	24.2 / 12.0	25.8 / 30.4
1234678-HpCDF	41.4 / 44.6	26.8 / 16.0	28.0 / 32.9
OCDD	79.9 / 80.5	34.4 / 10.3	31.5 / 26.9
OCDF	36.4 / 29.5	26.9 / 6.5	28.7 / 30.8

[a]Two determinations.
[b]Corrected for feed solution loss.
[c]After sulfuric acid and Florisil column cleanup.
[d]See Table 3 for ions monitored.

Table 3.   PCDD/PCDF Recoveries (%) by GC-Tandem MS in the Continuous Fortification Feed Solutions.

Analyte	m/z[a]	Fortification Experiment		
		1 pg/L	10 pg/L	50 pg/L
1368-TCDD	320/257	50.9	80.6	70.2
2378-TCDD	320/257	50.6	70.5	67.6
2378-TCDF	304/241	52.3	76.7	71.5
12378-PnCDD	356/293	44.0	73.5	65.6
12478-PnCDF	340/272	41.3	69.2	56.1
123678-HxCDD	390/327	43.9	74.8	68.6
123789-HxCDD	390/327	38.2	64.9	62.8
123678-HxCDF	374/311	36.9	66.8	58.2
1234678-HpCDD	424/361	36.5	73.0	62.8
1234678-HpCDF	408/345	33.6	65.1	57.8
OCDD	460/397	31.3	47.8	42.6
OCDF	444/381	26.1	56.6	46.2

[a]Parent ion/daughter ion.

was detected in the method blanks. Furthermore, the extract of a real tap water sample was considered "too dirty" for adequate determination of PCDD/PCDF by GC-tandem MS, and further cleanup was deemed necessary.

The analysis of the fortification feed solution, collected at $4\times$ the actual fortification pumping rate, illustrated loss of the analytes in the system prior to the XAD cartridge (see Table 3). The loss is most likely due to adsorption on the glass jug or on the Teflon® line used between the feed solution reservoir and the pump.

To avoid the adsorption problems, a second set of fortification studies was carried out using the direct fortification technique, in which the analytes are placed directly on the head of the resin bed to obtain an equivalent of 1, 10 and 25 pg/L after passing 200 L of tap water through the cartridge. The extracts were all processed as above and again screened by GC/ECD. The extracts were then further cleaned by the Carbopack column procedure prior to GC/ECD analysis. The pre-"Carbopack cleaned" extracts could again be screened for the higher chlorinated analytes fortified at the higher levels (see A in Figure 2). However, the Carbopack step was effective in removing nearly all interferences to allow GC/ECD determination in the extracts at all fortification levels. Figure 2 illustrates a GC/ECD chromatogram of a 10-pg/L fortified water extract before and after the Carbopack cleanup stage. The extract volumes were all 0.1 mL in isooctane. Figure 3 illustrates GC/ECD chromatograms of a 10-pg/L fortified water extract (B) along with the appropriate method blank (C). The extracts fortified at 1 pg/L were more readily determined when evaporated to 25 μL. However, at that concentration, artifacts rendered the ECD determination more uncertain (Figure 4). The results of analysis by GC/ECD (Table 4) show high-percentage recoveries for most PCDD/PCDF congeners.

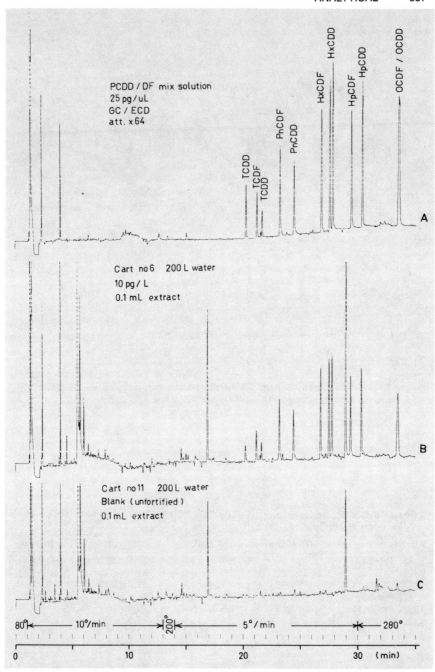

Figure 3. GC/ECD chromatograms of A) 25-pg/uL PCDD/PCDF mixture solution; B) 10-pg/L fortified water extract (200 L) after total cleanup procedures; and C) method blank (unfortified 200-L water extract). (See Table 4 for isomer identification and retention times.)

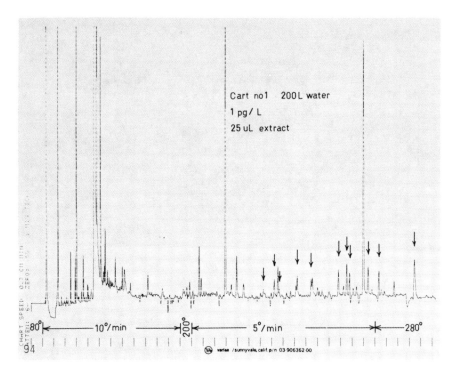

Figure 4. GC/ECD chromatogram of a 1-pg/L fortified extract (200 L) concentrated to 25 μL. Arrows indicate PCDD/PCDF congeners.

Table 4. PCDD/PCDF Recoveries (%)[a] by GC / ECD in Fortified (Direct) XAD-2 Drinking Water Extract (200 L).[b]

Analyte	RT (DB-5)	1 pg/L	10 pg/L	25 pg/L
1368-TCDD	20.26	39.4 ± 7.3	47.6 ± 16.0	40.0 ± 11.9
2378-TCDF	21.22	93.3 ± 6.6	83.6 ± 3.9	98.3 ± 11.3
2378-TCDD	21.67	41.8 ± 19.2	83.9 ± 20.0	94.0 ± 8.2
12378-PnCDF	23.22	86.8 ± 9.8	81.6 ± 10.5	100.0 ± 10.3
12347-PnCDD	24.45	65.9 ± 5.4	79.4 ± 10.7	96.0 ± 7.5
123678-HxCDF	26.81	64.2 ± 5.9	76.5 ± 6.5	102.2 ± 11.9
123678-HxCDD	27.52	75.8 ± 6.8	82.5 ± 9.4	106.8 ± 12.1
123789-HxCDD	27.77	57.4 ± 2.7	70.9 ± 10.8	102.2 ± 9.2
1234678-HpCDF	29.39	65.3 ± 9.9	72.9 ± 8.4	93.6 ± 13.6
1234678-HpCDD	30.34	67.9 ± 6.4	78.9 ± 8.9	92.7 ± 14.3
OCDF/OCDD	33.34	53.5 ± 8.8	47.6 ± 8.5	57.8 ± 14.4

[a]Mean ± SD (3 determinations); % recoveries include added IS.
[b]After acid, Florisil and Carbopack column cleanup.

Table 5.   PCDD/PCDF Recoveries (%)[a] by GC / MS in Fortified (Direct) XAD-2 Drinking Water Extract (200 L).[b]

Analyte	m/z	1 pg/L	10 pg/L	25 pg/L
1368-TCDD	320	34.2 / 51.1	67.3 / 25.8	64.1 / 28.2
2378-TCDF	304	97.2 / 88.3	91.7 / 89.5	105.1 / 101.0
2378-TCDD	320	77.4 / 73.2	67.3 / 95.9	116.4 / 105.7
12478-PnCDF	338	87.2 / 91.1	70.8 / 82.4	100.3 / 85.6
12347-PnCDD	354	106.3 / 128.6	85.1 / 76.9	128.3 / 103.9
123678-HxCDF	374	82.5 / 89.0	69.4 / 78.7	101.4 / 99.2
123678-HxCDD	390	95.6 / 86.7	72.6 / 95.7	113.2 / 110.9
123789-HxCDD	390	69.5 / 68.3	56.1 / 74.0	94.4 / 84.8
1234678-HpCDF	408	91.6 / 73.3	68.4 / 77.0	95.6 / 90.9
1234678-HpCDD	424	71.5 / 75.0	59.4 / 83.7	110.5 / 94.2
OCDF	444	46.7 / 50.8	42.2 / 70.5	91.9 / 77.6
OCDD	460	35.9 / 33.1	17.9 / 44.5	64.0 / 54.2

[a]Two determinations.
[b]After sulfuric acid, Florisil and Carbopack column cleanup.

Due to instrument breakdown at the time of analysis, the GC-tandem MS was unavailable, and the fortified extracts were analyzed by GC/MS using selected ion monitoring of one ion/analyte at relatively low (2K) MS resolution. The results of these analyses (Table 5) are in agreement with the GC/ECD results. Further, the GC/MS analysis also demonstrated that the ECD positive peaks found in the unfortified water (C in Figure 3) were not PCDD/PCDF congeners. The analytes in the 1-pg/L fortified extracts were all readily analyzed. Figure 5 illustrates reconstructed ion chromatograms for the studied PCDD/PCDF congeners at the 1-pg/L level. The instrumentation is easily capable of detecting lower levels in these extracts. The use of GC-tandem MS and high-resolution MS would give increased specificity.

Triplicate Ottawa tap water samples collected at a laboratory tap were analyzed, and no PCDD/PCDF congeners were detected at the screening level of 1 pg/L.

SUMMARY

The XAD-2 resin technique was evaluated and found to be effective in the extraction/concentration of PCDD and PCDF at levels down to 1 pg/L. The use of this technique permits sampling of large volumes (200 L) of water without tedious extraction with large volumes of solvent.

Although the rapid gas chromatography-tandem MS technique was effective in screening fortified XAD-2 water extracts following minimal cleanup of acid and Florisil stages, the real water extracts were not considered clean enough for screening purposes.

The additional Carbopack cleanup procedure produced extracts clean enough to be screened by capillary column GC/ECD even at low pg/L. The clean extracts facilitated GC/MS analysis and confirmation. The instru-

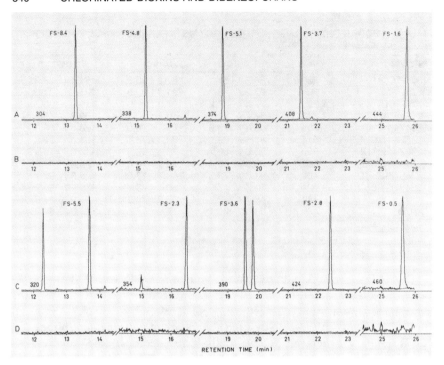

Figure 5. Selected ion chromatograms of furans (A) and dioxins (C) from a 200-L XAD-screened tap water, 1-pg/L fortified extract (cart #2) in 25 μL with the appropriate blank extract (cart #12) ion chromatograms (B—furans, D—dioxins) of 200-L unfortified XAD-screened tap water. All analyte chromatograms are normalized to the major peak for each selected ion. (See Table 5 for analyte screened.)

mentation is capable of screening at lower concentration, the limiting factor being the ability to produce suitable blanks and overcome carryover (cross contamination) problems.

No PCDD or PCDF was found in an Ottawa tap water sample at the screening level of 1 pg/L.

ACKNOWLEDGMENT

The CG-tandem MS analyses were done by Dr. Takeo Sakuma.

REFERENCES

1. Junk, J. A., J. J. Richard, M. D. Greiser, J. L. Witiak, M. D. Arguello, R. Vick, H. J. Svec, J. S. Fritz, and G. V. Calder. "Use of Macroreticular Resins in the Analysis of Water for Trace Organic Contaminants," *J. Chromatog.* 99:745–762 (1974).

2. Van Rossum, P., and R. G. Webb. "Isolation of Organic Water Pollutants by XAD Resins and Carbon," *J. Chromatog.* 150:381-392 (1978).

3. LeBel, G. L., D. T. Williams, G. Griffith, and F. M. Benoit. "Isolation and Concentration of OP Pesticides from Drinking Water at the ng/L Level, using Macroreticular Resin," *J. Assoc. Off. Anal. Chem.* 62:241-249 (1979).

4. LeBel, G. L., D. T. Williams, and F. M. Benoit. "GC Determination of TAAP in Drinking Water, Following Isolation using Macroreticular Resin," *J. Assoc. Off. Anal. Chem.* 64:991-998 (1981).

5. McNeil, E., R. Otson, W. Miles, and F. J. M. Rajabalee. "Determination of Chlorinated Pesticides in Potable Water," *J. Chromatog.* 132:277-286 (1977).

6. LeBel, G. L., and D. T. Williams. "Determination of Low ng/L Level of PCB in Drinking Water by Extraction with Macroreticular Resin and Analysis using a Capillary Column," *Bull. Environ. Contam. Toxicol.* 24:397-403 (1980).

7. Benoit, F. M., G. L. LeBel, and D. T. Williams. "The Determination of PAH at the ng/L Level in Ottawa Tap Water," *Intern. J. Environ. Anal. Chem.* 6:277-287 (1979).

8. Nestman, E. R., G. L. LeBel, D. T. Williams, and D. Kowbel. "Mutagenicity of Organic Extracts from Canadian Drinking Water in the Salmonella/Mammalian-Chromosome Assay," *Environ. Mutagenesis* 1:337-345 (1979).

9. Williams, D. T., E. R. Nestman, G. L. LeBel, F. M. Benoit, and R. Otson. "The Determination of Mutagenic Potential and Organic Contaminants of Great Lakes Drinking Water," *Chemosphere* 11:263-276 (1982).

10. Ryan, J. J., D. T. Williams, B. P.-Y. Lau, and T. Sakuma. "Analysis of Human Fat Tissue for 2378-TCDD and Chlorinated Dibenzofuran Residues," in *Chlorinated Dioxins and Dibenzofurans in the Total Environment II*, L. H. Keith, C. Rappe and G. Choudhary, Eds. (Stoneham, MA: Butterworth Publishers, 1983), pp. 205-214.

11. U. S. E. P. A. Region VII Dioxin Protocol. "Determination of 2378-TCDD in Soil and Sediment," Option D (September, 1983).

12. Sakuma, T., N. Gurprasad, S. D. Tanner, A. Ngo, W. R. Davidson, H. A. McLeod, B. P.-Y. Lau, and J. J. Ryan. "The Application of Rapid Gas Chromatography-Tandem Mass Spectrometry in the Analysis of Complex Samples for Chlorinated Dioxins and Furans." in *Chlorinated Dioxins and Dibenzofurans in the Total Environment II*, L. H. Keith, C. Rappe and G. Choudhary, Eds. (Stoneham, MA: Butterworth Publishers, 1983), pp. 139-152.

13. Lau, B. P.-Y., W. F. Sun, and J. J. Ryan. Paper presented at 32nd Annual Conference of the American Society for Mass Spectrometry, San Antonio, Texas, May 1984.

CHAPTER 23

Measurement of PCDD-PCDF from Incinerator and PCP Samples

R. C. Lao, R. S. Thomas, C. Chiu, R. Halman, and K. Li

INTRODUCTION

The primary concerns over polychlorinated dibenzo-p-dioxins (PCDDs) and polychlorinated dibenzofurans (PCDFs) in the Canadian environment are their adverse effects on human health, wildlife, and commercial and sport fishing.[1] The federal and provincial governments intend to focus their attention on commercial chemicals and formulations known to contain PCDD/PCDF as impurities, and the safe destruction/disposal of PCDD/PCDF in manufacturing process waters. Simultaneously, studies have been initiated into the formation of these compounds from sources such as refuse incineration, which is recognized to have high potential for PCDD/PCDF production.

As part of the assessment programs for combustion sources, PCDD/PCDF have been measured in the emission from typical municipal incinerators. The present work describes the application of a protocol for determination of PCDD/PCDF present both in gaseous phase and adsorbed on particulate matter.

In contrast with the continuous environmental burden provided by incinerators, we were frequently requested to examine hot-spot sources which represent short-term intrusion into the environment, often with long-lasting effects. A fire at a wood treatment plant which used pentachlorophenols

(PCPs) as preservatives aroused public concern because of the potential for exposure to toxic chemicals, and the contamination of adjacent properties and of waters downstream from runoff sources.

SAMPLE SOURCES

1. The incinerator monitored in this work is located in a major Canadian city, and was constructed in 1970.[2] It has four incinerator units, each consisting of a furnace, an energy recovery steam generator and a gas purification system. The original design capacity was 340,000 tonnes of waste a year, and currently the load, almost exclusively domestic waste, is 275,000 tonnes. Details on operational processing and characteristics are given by Environment Canada.[2] Two sampling programs were conducted — the first in November 1982, and a second in March 1983 — using Environment Canada.[3] The method was modified in order to collect PCDD/PCDF in particulate and gaseous form.

The distribution of samples in the sampling train, recovery of samples, and handlings are described in detail in the report cited above.[2]

2. A second group of samples was generated as the result of a postmortem survey of a wood treatment facility after an accidental fire. Initial concerns over creosote combustion by-product emissions were quickly refocussed with the discovery of high PCP concentrations and the potential for PCDD/PCDF formation from uncontrolled burning. Samples were taken to provide background levels in adjacent water bodies and exposed areas proximate to or runoff from the fire zone.

ANALYSIS

The sample preparation and cleanup procedures used in this laboratory have been described.[4] Details can be found in a separate report.[5]

Train samples collected from the incinerator were divided into two major portions: (1) the "front-half," comprising solvent washings of the nozzle, probe, and the filter with its holder; (2) the "back-half," including the contents of the condensation and adsorption unit, amberlite XAD absorbent, two impinger solutions, and washings of all glassware and connecting pieces. In addition, samples of fly ash and other solid material destined for landfill were analyzed.

The Finnigan 4500 HRGC-LRMS was used for the identification and quantification of PCDD/PCDF. Due to low concentrations found in this survey, negative chemical ionization techniques had to be employed. The technique permits the use of methane or oxygen percentage concentrations as a reactant gas and to enhance detection limits.[6] Water samples were taken from both the wood preserving plant and nearby communities, including the river, wastewater, condensate, flocculation outlet, inlet and outlet of water treatment facility, and cooling water effluent. Solid samples were from sludges, products, fresh (regenerated) and spent carbon, fresh and used PCP solutions (5%), and treated wood.

All wood preservative samples collected were divided into two aliquots: one for the analysis of PCDD/PCDF by GC/MSD (HP 5970A—mass selective detector [MSD]) and the other for chlorophenols (CPs). CPs were extracted into an aqueous solution of potassium carbonate, derivatized using acetic anyhdride and back-extracted into hexane. Quantitation analysis was performed by GC/ECD and GC/MSD.

To evaluate the extraction efficiency and overall recovery of the selected analytical protocol, the incinerator samples, prior to extraction, were spiked with known amounts of OCDD-^{13}C$_{12}$ and 2,3,7,8-TCDD-^{13}C$_{12}$. For the wood preserving plant survey, sixteen water samples were spiked with known amounts of the same surrogates together with 2,4,5-trichlorophenol-^{13}C$_6$ and PCP-^{13}C$_6$. Six carbon samples were also spiked with the same labeled TCDD and OCDD. Recoveries for each of these surrogates were determined from GC/MS analysis on an individual sample basis. Two blind samples from the incinerator were selected and analyzed by another laboratory as part of the QA/QC programs designed for all PCDD/PCDF projects conducted by this Center.

RESULTS AND DISCUSSION

Figure 1 illustrates the reconstructed multiple-ion detection (MID) chromatogram, using electron impact (EI) mode, for tetra to octa congeners of PCDD/PCDF in the Amberlite trap-rinse portion of a train sample drawn from the incinerator. Figure 2 gives the corresponding MID in chemical ionization (CI) mode for the same sample, as the EI response for the higher chlorine-containing species was below a level sufficient to allow positive identification. In Figure 1, the two large peaks at spectrum numbers 688 and 2220 are spiked isotope-labeled isomers, but the negative methane CI MID in Figure 2 fails to give the T$_{CDD}$ response. Details of the negative ion chemical ionization (NCI) spectra of PCDD/PCDF have been reported previously.[4,6]

Tables 1 and 2 give the laboratory results for PCDD/PCDF measured. Tables 3 and 4 summarize the emission concentrations for the tests in November 1982 and March 1983. The average concentration for all four tests conducted in 1982 was 12.8 pg/m^3 for PCDD and 19.4 pg/m^3 for PCDF. The results of the 1983 survey indicate that the average for the eight stack tests was 750 pg/m^3 for PCDD and 540 pg/m^3 for PCDF. For all the tests in two surveys, most of the compounds were found in gaseous form

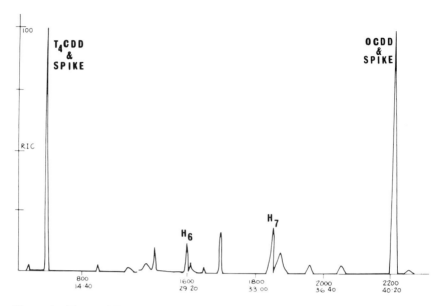

Figure 1.   The total EI MID chromatogram of an incinerator sample.

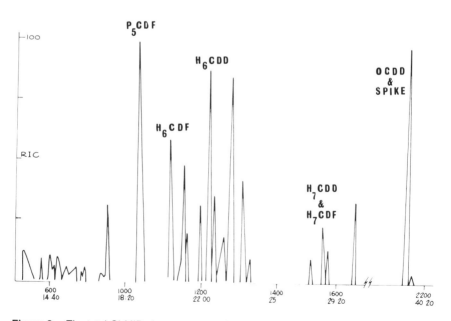

Figure 2.   The total CI MID chromatogram of the incinerator sample in Figure 1.

Table 1. PCDD and PCDF in Incinerator Stack Emissions (pg/sample), November 1982.[a]

Sample Type	Cl4 DD	Cl4 DF	Cl5 DD	Cl5 DF	Cl6 DD	Cl6 DF	Cl7 DD	Cl7 DF	Cl8 DD	Cl8 DF
Filters	0–3	0–2	0–13	0–20	0–3	0–12	0–3	0–8	0–5	0–5
Liquid Sample (Solvent)	0	0	0–2	0–3	0–4	0–5	0–5	0–4	0–2	0–2
Amberlite	0	0–4	0	0	0	0	0	0	0	0
Fly Ash[b]	3–22	15–68	4–16	7–32	13–26	9–33	17–37	8–18	14–106	1–10

[a]Detection limit: 1 pg per sample.
[b]In ng/g (ppb).

Table 2. PCDD and PCDF in Incinerator Stack Emissions (pg/sample), March 1983.

Sample Type	Cl4 DD	Cl4 DF	Cl5 DD	Cl5 DF	Cl6 DD	Cl6 DF	Cl7 DD	Cl7 DF	Cl8 DD	Cl8 DF
Filter, Probe, and Washings	0–120	3–451	0–140	0–344	3–504	2–190	0–370	0–62	4–430	0–93
Amberlite and Trap	0–48	0–220	0–72	0–212	0–123	0–435	0–306	0–197	0–333	0–71
Glassware	0	0	0	0	0	0	0	0	0	0
Bottom Ash	0–3	0–2	0–4	0–2	0–14	0–7	0–17	0–5	0–16	0–2

Table 3. PCDD-PCDF Emissions, November 1982.[2]

Test No.	Cl4 DD	Cl4 DF	Cl5 DD	Cl5 DF	Cl6 DD	Cl6 DF	Cl7 DD	Cl7 DF	Cl8 DD	Cl8 DF
1. A[a]	0	5.0	0	5.0	5.0	0	0	0	0	0
B[b]	0	3.3	0	3.2	3.3	0	0	0	0	0
2. A	0	10.0	6.7	13.4	0	0	0	0	0	0
B	0	5.6	3.7	7.5	0	0	0	0	0	0
3. A	6.3	6.2	47.3	75.7	28.4	60.0	34.8	50.6	22.1	12.2
B	3.5	3.1	26.6	42.5	15.9	33.7	19.5	28.4	12.4	7.1
4. A	9.1	18.3	12.2	15.2	18.3	15.2	12.2	12.2	6.1	6.0
B	5.3	10.6	7.1	8.8	10.6	8.8	7.1	7.1	3.5	3.4

[a]A. Average emission rate (ng/tonne) of waste burned over the duration of the test.
[b]B. Daily emission rate (μg/day) from both units in operation during each of the tests (based on stack gas flow).

Table 4.   PCDD/PCDF Emissions, March 1983.[2]

Test No.	Cl4 DD	Cl4 DF	Cl5 DD	Cl5 DF	Cl6 DD	Cl6 DF	Cl7 DD	Cl7 DF	Cl8 DD	Cl8 DF
1. A[a]	1065	2605	702	1899	1039	503	841	371	1078	266
B[b]	376	920	248	671	367	178	297	131	381	94
2. A	0	0	0	0	14	8	30	35	114	6
B	0	0	0	0	5	3	10	12	41	2
3. A	184	124	299	243	335	145	507	202	2540	243
B	60	40	98	47	109	47	166	66	831	80
4. A	251	250	431	378	295	173	514	254	1216	197
B	93	90	160	64	109	64	190	94	451	73
5. A	159	145	185	145	130	98	268	167	667	159
B	48	43	55	29	39	29	80	50	200	48
6. A	525	2097	663	1738	1950	1954	2114	834	2406	528
B	161	643	203	599	600	256	648	599	737	162
7. A	21	43	17	18	6	9	18	15	188	12
B	7	14	5	3	2	5	6	3	60	4
8. A	446	184	549	284	533	258	343	120	713	181
B	136	56	167	79	162	36	104	79	217	55

[a]A. Average emission rate (ng/tonne) of waste burned.
[b]B. Daily emission rate (μg/day).

and were trapped by the Amberlite cartridges. Some variations were observed and are considered to be a function of sampling conditions and incinerator operational parameters.

For the overall average concentrations of the twelve tests, the calculated PCDD emissions were about 2 μg/tonne.[2] These values are much lower than the reported data from other incinerators around the world.[2] Although the sampling methodology used in this survey has since been improved somewhat, it was considered satisfactory. The analytical protocols had been thoroughly tested and the results reported would seem to be reliable. Each sample had been spiked for recovery assay and split blind samples were prepared and analyzed by a well recognized laboratory. Results are compared in Table 5. The average differences are less than 15%, well within the experimental error.

The results obtained during March 1983 fail to indicate that the injection of pulverized lime at a rate of 81.6 kg/hr for the last four stack tests altered PCDD/PCDF emissions, although, as expected, the concentrations of HCl and SO$_2$ were reduced.

The average PCDD/PCDF concentrations in the fly ash were 130 and 115 ng/g, respectively. Their levels varied appreciably in the stack emissions, but they appeared to remain fairly constant for the fly ash samples collected in the electrostatic precipitator.

In one of the solid residue samples designated for landfill disposal, the

Table 5.   Quality Control: PCDD/PCDF Content (ng/g) of Ash Samples.

Sample No:		17		48	
Laboratory:		In-House	Outside	In-House	Outside
Cl4	DD	5	8	8	6
	DF	51	60	68	57
Cl5	DD	16	20	14	12
	DF	30	38	32	35
Cl6	DD	16	19	31	27
	DF	17	20	33	25
Cl7	DD	25	27	37	40
	DF	8	10	18	16
Cl8	DD	14	20	182	183
	DF	1	2	10	10
Total	DD	76	94	182	183
	DF	107	129	161	143

measured PCDD/PCDF levels were 1.7 and 0.4 pg/g. The solid materials from the surface of the incinerator sedimentation tank, which contains water mixed with incineration residues and fly ash, were also analyzed for PCDD/PCDF. The concentrations were 53 and 18 pg/g, respectively.

Given the limited number of tests carried out and the paucity of statistical information on the effect of varying operating conditions, it is practically impossible to draw any conclusion which might explain the variation among the different emission results. However, the fact remains that this particular incinerator produced lower PCDD/PCDF emissions than many other corresponding units previously surveyed. Further work to evaluate operating conditions and waste feed stock could clarify this apparent anomaly.

Figure 3 illustrates the total MID chromatogram of a water sample drawn from the inlet to the effluent treatment facility of the wood preserving plant. Figure 4 gives the MID monitored for compounds detected. Table 6 is the compilation of the results for PCDD/PCDF data in several samples from this survey. Only hepta and octa dioxins and traces of furans were present in measureable amounts in the PCP in oil samples. The concentration of these compounds was considerably higher than the values reported earlier.[6] The previous samples indicated a cross section of PCDD/PCDF congeners from penta to octa, and the data are probably indicative of the variations in composition of industrial PCP blends available to various users.

Samples were taken and analyzed at a number of locations next to the primary exposure zone, which included surface river water, main ditches, effluent from a water treatment facility and the regenerated carbon absorbent. Examination of these samples proved negative for PCDD/PCDF, indicating adequate containment at the site. Table 7 includes the corresponding CP concentrations detected in various water and solid samples.

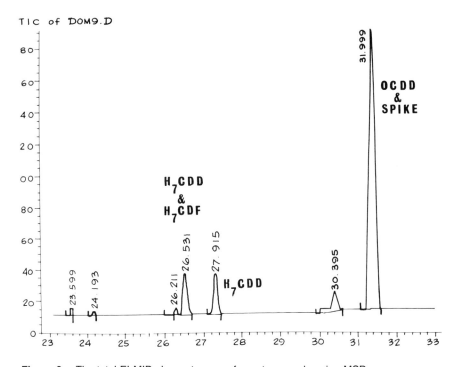

Figure 3. The total EI MID chromatogram of a water sample using MSD.

The analytical procedures for OCDD in "purified penta" in PCP in oil are well documented.[7] However, assays indicated that the spiked isotope-labeled OCDD levels measured were quite different for each method. In a comparative evaluation of published methods and procedures developed at this laboratory for solid PCP and PCP in oil samples, the inclusion of an acid-base column, used by the present work, prior to separation on basic alumina, gives the most consistent spiked OCDD recovery efficiency (95% ± 25%). Using other reported methods, PCP in oil samples could exhibit a different amount of OCDD by orders of magnitude, depending on the method selected. Small changes in the sample constituents could result in a different elution profile from the basic alumina column.[8] It is possible that the presence of the oil matrix can affect retention characteristics sufficiently to cause OCDD to elute in the other chromatographic fractions not normally analyzed for PCDD/PCDF.

This series of analyses of wood preserving plant samples demonstrated that the MSD system is a useful instrument for rapid screening of samples when ultimate sensitivities are not required and fast sample turnaround is desirable.[9]

A significant amount of intralaboratory QA/QC programs is built in to the method protocols. These procedures include use of isotopically labeled

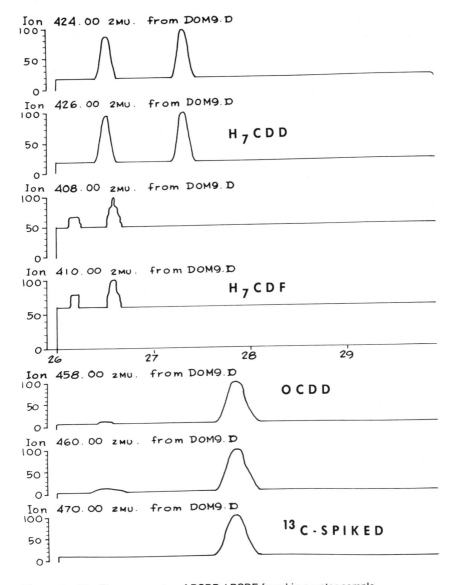

Figure 4. The EI mass spectra of PCDD / PCDF found in a water sample.

Table 6. Results of PCDD/PCDF in Wood Preservation Plant Survey.

Sample Type	Cl7 DD	Cl7 DF	Cl8 DD	Cl8 DF	Average Recovery (%) TCDD	Average Recovery (%) OCDD
Liquid (ng/L)						
1. Ditches	0–20	0–3	25–111	0	72	70
2. Flocculation	22–38	0	88–170	0	75	61
3. Flocculation Exit	0	0	5–3	0	59	62
4. Principal Ditch	0	0	2	0	86	73
5. Treatment Plant Inlet	37–40	0–2	170	0	87	85
6. Treatment Plant Outlet	0	0	14–28	0	80	68

Product (ng/g)					Cl6 DD	Cl6 DF
1. Fresh PCP Solution (5%)	9–14	0	29–54	0	0	0
2. Fresh Carbon (Regenerated)	0	0	0	0	0	0
3. Spent Carbon	1–2	0	6–11	0	0	0
4. Treated Wood	1.1–7.0	0	6.0–22.5	0	0	0
Sludges (ng/g)						
1. Cylinder for PCP Oil	0–17.3	0	50–161	0	0	0
2. Flocculation	0.8–0.9	0.1	0.6–0.7	0.1	0.1	0.1
3. Used PCP Oil	14–17	0.6–0.8	31.43	0	0	1–1.3
Solid (ng/g)						
1. Used Carbon	1–2	0	6–11	0	0	0
2. Flocculation	0.8–0.9	0.1	0.6–0.7	0	0.1	0
3. Treated Wood	1.1–7.0	0	6.0–22.5	0	0	0

internal standards, performance standards, method blanks, control samples, repeat analysis, etc.

Samples were spiked with labeled ^{13}C-TCDD and ^{13}C-OCDD to evaluate the recovery efficiencies from extraction to the analytical finish-ups. D-2-triphenylene was used as the performance standard to evaluate the injection variations and the global instrumental response.

Table 7. Chlorophenols in Wood Preservation Plant Survey.

Sample Type	T4CP	PCP	Average Recovery (%) T3CP	Average Recovery (%) PCP
Liquid (μg/L)				
1. Ditches	0	0–0.6	114	130
2. Flocculation	400–526	1000–1200		
3. Flocculation Exit	590	1037–4000		
4. Treatment Plant Inlet	600–980	4500–5700		
5. Treatment Plant Outlet	0	3–20	93	64
Solid (μg/g)				
1. Flocculation Sludge	40–390	250–2500		
2. Wood	160–380	3500–6400		
PCP in Oil (mg/g)	1–3	16–50		

With each batch (6-10) of samples processed, one blank and one standard mixture were included to monitor the performance of sample extraction and cleanup. Furthermore, a minimum of 20% of sample extracts were analyzed in duplicate to demonstrate the precision of analytical instrumentation.

REFERENCES

1. "Dioxins in Canada: the Federal Approach," Interdepartmental Committee on Toxic Chemicals (1983).
2. "Measurement of PCDD and PCDF Emissions from the Des Carrieres in Montreal," Air Pollution Division, Environmental Protection Service, Quebec Region, Environment Canada, EPS Report 5/UP/RQ-1 (December 1984).
3. "Standard Reference Methods for Source Testing: Measurement of Emission of Particulates from Stationary Sources," Environmental Protection Service, Environment Canada, EPS 1-AP-74-1 (1974).
4. Lao, R.C., et al. "Polychlorinated Hydrocarbons from Power Plants, Wood Burning and Municipal Incinerators," *Chemisphere* 12(3):607–616 (1983).
5. "Sample Preparation for the Analysis of PCDD-PCDF in Emission Samples," Analytical Services Division, Technical Services Branch, Environmental Protection Service, Environment Canada (1984).
6. Lao, R.C., et al. "Analysis of PCDD-PCDF in Environmental Samples," in *Chlorinated Dioxins and Dibenzofurans in the Total Environment II*, L. Keith, C. Rappe, and G. Choudhary, Eds. (Stoneham, MA: Butterworth Publishers, 1985), p. 65.
7. "PCDD: Analysis," National Research Council of Canada, Report NRCC no. 18574 (1981).
8. Rappe, C., and H. R. Buser, "Identification of PCDF in Commercial Chlorophenols," *Chemosphere* 7(4):981–991 (1978).
9. Lao, R., C. Chiu, and K. Li, "Evaluation of the MSD for PCDD-PCDF Analysis," Fourth International Symposium on Dioxin and Related Compounds, Ottawa, Canada (1984).

CHAPTER 24

Dioxin Residues in Fish and Other Foods

David Firestone, Richard A. Niemann, Louis F. Schneider, John R. Gridley, and
Darryl E. Brown

INTRODUCTION

Chlorinated dibenzo-p-dioxins (dioxins) can arise and enter the environment in a variety of ways: as by-products of chemical syntheses, as contaminants in commercial chlorophenol-based products, as products of combustion, and as constituents of waste disposal materials.[1] The presence of 2,3,7,8-tetrachlorodibenzo-p-dioxin (2,3,7,8-TCDD) in the environment is primarily associated with the production and use of 2,4,5-trichlorophenol (2,4,5-TCP) and its derivatives. The presence of more highly chlorinated dioxins in the environment is due largely to the extensive use of penta-chlorophenol (PCP) as a wood preservative and general bactericide and fungicide. For example, Neidert et al.[2] found low levels of PCP residues in all 1,072 liver and 723 fat samples collected from flocks of chickens slaughtered at Canadian packing plants (higher chlorinated dioxins were not found in any of the samples; PCP levels >0.1 µg/g were attributed to contamination of the litter by PCP). Residues of dioxins have also been reported in birds and fish from the Great Lakes region[3-5] and in food-grade gelatin[6] and beef fat.[7] Levels of 2,3,7,8-TCDD residues >100 pg/g were reported in fish collected from the Tittabawassee River and its estuary in Saginaw Bay, from Lake Huron,[8] and from the Arkansas River and a tributary, the Bayou Meto.[9]

To determine the presence of dioxin residues in food, several other food

355

commodities as well as fish have been analyzed for residues of 2,3,7,8-TCDD, and a number of foods of animal origin have been analyzed for residues of the higher-chlorinated dioxins present in PCP, hexa-, hepta-, and octachlorodibenzo-p-dioxin (HCDD, HpCDD, and OCDD). Particular attention has been given to the commercial fishing industry in Saginaw Bay, not far from the Dow Chemical Company facilities at Midland, Michigan, from which more than 1.5 million lb of commercial catch is harvested annually. Since 1979, fish have been collected from Saginaw Bay for determination of 2,3,7,8-TCDD residues. Also, fish have been collected from other Great Lakes sites as well as from various locations suspected to have been contaminated by industrial wastes from 2,4,5-TCP manufacturing plants or from the agricultural use of 2,4,5-T and related herbicides. A Food and Drug Administration (FDA) survey of 2,3,7,8-TCDD in fish from Great Lakes locations and a number of Michigan rivers has already been reported.[10] In addition, market basket samples collected under the FDA's Total Diet Program[11] with PCP residues >50 ng/g have been routinely analyzed for HCDD, HpCDD, and OCDD.

MATERIALS AND METHODS

Sample Preparation

Laboratory preparation followed standard procedures.[12] Fish samples were ground composites of the edible fillet portion of several fish. Frozen fish were allowed to thaw, and cut fillets were ground in a meat grinder at least three times with thorough mixing between each grinding. A medium screen (4- to 4.5-mm-diameter holes) was used for the first grinding and a fine screen (2- to 2.5-mm-diameter holes) was used for subsequent grindings.

Cleanup and Analysis

Test portions were cleaned up and analyzed for 2,3,7,8-TCDD residues according to the isomer-specific method described by Niemann et al.[13] Preliminary extraction and cleanup involved a basic digestion, hexane extraction, and sulfuric acid wash or a modification of the sulfuric acid column cleanup described by Lamparski and Nestrick.[14] Extracts were then subjected to three high-performance liquid chromatographic cleanups followed by quantitation using capillary column gas chromatography (GC) with electron capture (EC) detection. Multiple-ion detection gas chromatography/mass spectrometry (GC/MS/MID) was used to identify 2,3,7,8-TCDD residues in extracts. GC/MS/MID confirmation of identity was required to report findings of 2,3,7,8-TCDD. Method recovery was determined by fortification of a separate portion of each prepared sample with 2,3,7,8-TCDD, except for several samples examined more recently by a modification of the isomer-specific method,[15] which employs 1,3,7,8-TCDD as a recovery surrogate. The minimum amount of 2,3,7,8-TCDD that could be quantitated and the identity of which could be confirmed by MS was about 10 pg/g. Values below this level were considered as not present.

Samples were analyzed for residues of HCDD, HpCDD, and OCDD by a procedure[6,16] involving basic digestion, hexane extraction, sulfuric acid wash of the hexane extracts, chromatography on neutral alumina and Florisil minicolumns, and analysis of extracts by GC/EC. The presence of higher-chlorinated dioxins in several extracts was confirmed by negative ion chemical ionization GC/MS.

All results reported in this chapter have been adjusted for reagent blank and recovery. Most of the analyses for residues of 2,3,7,8-TCDD were carried out in the FDA's Detroit, Michigan laboratories. Determinations of HCDD, HpCDD, and OCDD residues were conducted in FDA's Dallas, Texas laboratories.

RESULTS AND DISCUSSION

Residues of 2,3,7,8-TCDD in Fish and Other Foods

Results of analyses of commercial fish from Saginaw Bay collected during 1979–1983 are summarized in Table 1. Residues of 2,3,7,8-TCDD were not observed in perch, sucker, walleye, bullhead, and whitefish. Carp and catfish collected in 1979–1981 contained up to about 70 pg 2,3,7,8-TCDD/g edible portion. However, reduced levels ($\leq$30 pg/g) were found in carp and catfish collected in 1983, and the catfish had 2,3,7,8-TCDD residues below the FDA's 25-pg/g "level of concern."[17]

Results of analysis of fish from various locations for 2,3,7,8-TCDD residues are shown in Table 2. No 2,3,7,8-TCDD above 10 ppt was found in carp, catfish, and buffalo fish collected in 1982 from the Mississippi River (St. Louis area). Kaczmar et al.[18] reported that carp and sucker from 19 major rivers in Michigan contained 2,3,7,8-TCDD residues ranging from 17 to 586 pg/g. However, FDA sampling and analyses found no 2,3,7,8-TCDD residues in carp from 13 Michigan rivers (Table 2), whereas a carp from a fourteenth river, the Tittabawassee, contained 93 pg 2,3,7,8-TCDD/g. The FDA findings were in agreement with data reported by the State of Michigan,[19] which conducted a comprehensive study of Michigan waterways.

The State of Michigan study found that residues of 2,3,7,8-TCDD in whole carp from ten rivers ranged from nondetectable to 8.6 pg/g. About 5 pg/g was found in a carp from Muskegon Lake, and levels ranging from nondetected ("detection limits" 0.2–2.5 pg/g) to 12 pg/g, with no detectable

Table 1.   Analytical Results for 2,3,7,8-TCDD Residues in Commercial Fish Collected from Saginaw Bay, Michigan.

Year Collected	Type of Fish	No. of Fish Samples	2,3,7,8-TCDD pg/g[a]	Recovery (%)[b]
1979	Sucker	9	ND	48–72
	Perch	8	ND	50–75
	Bullhead	2	ND	62
	Whitefish	1	ND	51
	Carp	6	ND	50–76
	Carp	1	21	58
	Carp	1	57	61
	Catfish	2	ND	57
	Catfish	1	60	58
	Catfish	1	19	59
	Catfish	1	52	60
	Catfish	1	43	58
	Catfish	1	34	68
1980	Carp	1	ND	48
	Carp	1	35	48
	Catfish	1	18	57
	Catfish	1	18	57
1981	Perch	1	ND	48
	Carp	1	ND	71
	Carp	1	28	52
	Carp	1	37	74
	Catfish	1	28	52
	Catfish	1	44	84
	Catfish	1	50	56
	Catfish	1	57	51
1983	Sucker	1	ND	48[c]
	Walleye	1	ND	32[c]
	Whitefish	1	14	79
	Whitefish	1	20	41
	Carp	3	ND	46–74
	Carp	1	15	80
	Carp	1	16	84
	Carp	1	18	55
	Carp	1	20	75
	Carp	1	30	76
	Catfish	4	ND	47–88
	Catfish	1	19	45
	Catfish	1	18	57
1984	Catfish	1	ND	84[c]
	Catfish	1	7	70[c]
	Catfish	1	13	76[c]

[a]ND, not quantitated or confirmed; if 2,3,7,8-TCDD is present, it is present at a level below about 10 pg/g. Values corrected for reagent blank (~ 3 pg/g) and recovery.
[b]Test portions fortified at level of 50–100 pg/g with 2,3,7,8-TCDD standard solution. Each test portion is a composite of from 1 to 10 fish.
[c]Test portions fortified with 1,3,7,8-TCDD recovery surrogate.[15]

Table 2.   Analytical Results for 2,3,7,8-TCDD Residues in Fish Collected from Various U.S. Locations in 1981–1983.

Type of Fish	No. of Fish Samples	Origin	2,3,7,8–TCDD pg/g[a]	Recovery (%)[b]
Buffalo Fish	5	Mississippi River, vicinity St. Louis	ND	58–64
Buffalo Fish	2	Mississippi River, Sauget, IL	ND	53,25[c]
Catfish	1	Mississippi River, vicinity St. Louis	10	61
Catfish	1	Mississippi River, vicinity St. Louis	ND	43
Carp	2	Mississippi River, vicinity St. Louis	ND	48, 84[c]
Carp	13	13 Michigan rivers[d]	ND	50–62
Walleye	1	Clinton River, MI	ND	56
Carp	1	Tittabawassee River, MI	93	74
Coho Salmon	1	Detroit River, MI	ND	86
Carp	1	Muskegon Lake, MI	10	84
Coho Salmon	1	Trout Run, PA	ND	95
Carp	1	Lake Pippin, WI	ND	48
Lake Trout	1	Lake Huron	6	81
Whitefish	1	Lake Huron, Thunder Bay	ND	69
Coho Salmon	1	Lake Huron, Tawas Bay	ND	60
Whitefish	4	Lake Michigan	ND	54, 78
Sucker	1	Lake Michigan	ND	—
Coho Salmon	3	Lake Michigan	ND	74–78
Lake Trout	1	Lake Michigan	ND	79
Carp	1	Lake Michigan	ND	54
Whitefish	1	Lake Superior, Whitefish Bay	ND	—
Lake Trout	1	Lake Superior	ND	96
Coho Salmon	3	Lake Erie	ND	55–82
Sucker	1	Lake Erie, West Seneca, NY	ND	56[c]
Sucker	1	Lake Erie, Angola, NY	ND	62[c]
Various Fish[e]	6	Lake Erie, Dunkirk, NY	ND	50–67[c]
Coho Salmon	1	Lake Ontario, Salmon River, NY	ND	88
Coho Salmon	1	Lake Ontario, Salmon River, NY	35	73
Sucker	2	Lake Ontario, Wilson, NY	ND	58, 94
Lake Trout	1	Lake Ontario, Wilson, NY	46	76
Brown Trout	1	Lake Ontario, Wilson, NY	8	73
Rainbow Trout	1	Lake Ontario, Wilson, NY	21	73
White Perch	1	Lake Ontario, Oswego, NY	25	96
Yellow Perch	1	Lake Ontario at Chaumont Bay	ND	—
Butterfish	1	Atlantic Ocean[f]	ND	85
Flounder	1	Atlantic Ocean[f]	ND	73
Hake	3	Atlantic Ocean[f]	ND	83–103
Herring	1	Atlantic Ocean[f]	ND	85

[a]ND, not quantitated or confirmed; if 2,3,7,8-TCDD is present, it is present at a level below about 10 pg/g. Values corrected for reagent blank (~ 3pg/g) and recovery.
[b]Test portions fortified at level of 50–100 pg/g with 2,3,7,8-TCDD standard solution. Each test portion is a composite of from 1 to 10 fish.
[c]Test portions fortified with 1,3,7,8-TCDD recovery surrogate.[15]
[d]Au Sable, Cass, Clinton, Flint, Grand, Huron, Kalamazoo, Muskegon, Pine, Raisin, Shiawassee, St. Clair, and St. Joseph.
[e]White bass, catfish, drum, yellow perch, chinook salmon, and brown trout.
[f]8–12 mi off Long Branch, NJ, collected in 1984.

2,3,7,8-TCDD or detectable levels <5 pg/g, were found in most skinless fillet samples of 25 carp from the Grand River at Grand Ledge. On the other hand, 2,3,7,8-TCDD levels in 26 carp fillets from the Tittabawassee River at Midland ranged from 12 to 530 pg/g, and 13 of the carp had >25 pg/g.

No residues of 2,3,7,8-TCDD were found by FDA in fish from Lakes Erie, Michigan, and Superior (Table 2). Residues in fish from Lake Ontario ranged from none to 46 pg/g. O'Keefe et al.[3] reported levels up to 162 pg/g in fish collected during 1978 from Lake Ontario, with lower values (up to about 50 pg/g) in fish collected in 1979–1980. No residues of 2,3,7,8-TCDD were found in several kinds of fish collected in the Atlantic Ocean off the coast of northern New Jersey (Table 2) in response to reports[20] that >50 pg 2,3,7,8-TCDD/g was found in fish from the Passaic River.

Mitchum et al.[9] reported the presence of 2,3,7,8-TCDD in fish collected from the Arkansas River and its associated tributary, the Bayou Meto. The Bayou Meto was closed to commercial fishing by the Arkansas Game and Fish Commission in April 1980 after the Arkansas State Department of Pollution Control and Ecology reported the presence in fish of 2,3,7,8-TCDD residues as high as 257 pg/g, apparently due to stream contamination from the Vertac 2,4,5-T plant in Jacksonville, Arkansas. The Rocky Branch flows through the Vertac property into the Bayou Meto about 132 mi upstream from the Bayou Meto's confluence with the Arkansas River. The State of Arkansas reported 2,3,7,8-TCDD residues from 82 to > 200 pg/g in fish caught 100–134 mi above the mouth of the Bayou Meto versus 46 pg/g at 50 mi and <13 pg/g at 1.5 mi above the mouth of the Bayou meto. Table 3 shows the analytical results from a number of fish collected from the lower Bayou Meto and Arkansas River in 1979–1980. The first three fish, collected from the Bayou Meto at or 1.5 mi above its confluence with the Arkansas River, contained up to 15 pg of 2,3,7,8-TCDD/g. The remaining fish, collected from the Arkansas River at various locations near the mouth of the Bayou Meto, did not contain residues of 2,3,7,8-TCDD.

Results of analysis of bovine fat, soybeans, rice, and crawfish from Missouri, Arkansas, and Louisiana are shown in Table 4. Kleopfer[21] reported residues of 2,3,7,8-TCDD up to about 40 pg/g in fish from the Spring River in the vicinity of a former 2,4,5-T/hexachlorophene plant at Verona, Missouri. No 2,3,7,8-TCDD was found in bovine fat collected from two nearby dairy farms. No 2,3,7,8-TCDD residues were found in soybeans, rice, and crawfish collected in 1979–1980 and 1983 (crawfish) from rice-growing areas in Arkansas and Louisiana that were sprayed with 2,4,5-T herbicide.

Table 3. Analytical Results for 2,3,7,8-TCDD Residues in Fish Collected from the Bayou Meto and Arkansas River in 1979–1980.

Type of Fish	No. of Fish Samples	Collection Site	2,3,7,8-TCDD pg/g[a]	Recovery (%)[b]
Carp, Catfish[c]	1	Bayou Meto[d]	ND	57
Short Nose Gar	1	Bayou Meto[d]	13	48
Short Nose Gar, Long Nose Gar	1	Bayou Meto[e]	15	61
Carp, Catfish[c]	1	Arkansas River[f]	ND	64
Buffalo Fish	3	Arkansas River[g]	ND	59
Bowfin	1	Arkansas River[g]	ND	50
Catfish	1	Arkansas River[g]	ND	67
Catfish	1	Arkansas River[g]	ND	57
Crappie	1	Arkansas River[g]	ND	–
Drum Fish	1	Arkansas River[g]	ND	–
Rock Bass	1	Arkansas River[g]	ND	–

[a]ND, not quantitated or confirmed; if 2,3,7,8-TCDD is present, it is present at a level below about 10 pg/g. Values corrected for reagent blank ($\sim$ 3 pg/g) and recovery.
[b]Test portions fortified at level of 50–100 pg/g with 2,3,7,8-TCDD standard solution. Each test portion is a composite of from 3 to 12 individual fish.
[c]Mixed types of fish.
[d]About 1.5 mi from confluence with Arkansas River, near Gillette, AR.
[e]At confluence with Arkansas River.
[f]3 mi downstream from confluence with Bayou Meto.
[g]Near Bayou Meto.

Table 4. Analytical Results for 2,3,7,8-TCDD Residues in Various Samples from Missouri, Arkansas, and Louisiana Collected in 1979–1980.

Product[a]	No. of Products	Collection Site	2,3,7,8-TCDD pg/g[b]	Recovery (%)[c]
Bovine Fat	2	farm A, Verona, MO	ND	50,63
Bovine Fat	1	farm B, Verona, MO	ND	50
Soybeans	3	AR[d]	ND	50–60
Soybeans	3	Monterey, LA[d]	ND	50–60
Rough Rice	3	Monterey, LA[d]	ND	60–70
Rough Rice	1	Bunkie, LA[d]	ND	60–70
Rough Rice	1	Crowley, LA[d]	ND	60–70
Crawfish	12	LA[d]	ND	70–90
Crawfish	8	LA[d,e]	ND	59–84

[a]Bovine fat was collected in 1983.
[b]ND, not quantitated or confirmed; if 2,3,7,8-TCDD is present, it is present at a level below 5–10 pg/g.
[c]Test portions fortified at level of 50–100 pg/g with 2,3,7,8-TCDD standard solution.
[d]Samples collected from rice growing areas that use 2,4,5-T herbicide.
[e]Additional crawfish collected in 1983.

Table 5. Higher-Chlorinated Dioxin Residues in Various Foods Collected in the U.S., 1979–1984.

Food	No. of Products	FDA Collecting District[a]	PCP (μg/g)	pg/g[b] 1,2,3,4,6,7,9- HpCDD	pg/g[b] 1,2,3,4,6,7,8- HpCDD	pg/g[b] 1,2,3,4,6,7,8- OCDD
Bacon	1	MIN	0.06	ND	46	160
Blue Crab	1	NSV	—	ND	ND	ND
Crab	1	NOL	0.0	ND	ND	ND
Catfish	1	NOL	0.18	ND	ND	ND
Trout	1	NOL	0.14	ND	ND	ND
Ground Beef	16	c	—	ND	ND	ND
Peanut Butter	1	SAN	0.10	ND	ND	ND
Milk	58	c	0.01–0.05	ND	ND	ND
Chicken	14	c	—	ND	ND	ND
Chicken	1	MIN	0.17	42	28	252
Chicken	1	NSV	—	ND	ND	76
Chicken	1	SEA	—	ND	ND	29
Eggs	17	c	—	ND	ND	ND
Eggs	1	HOU	0.29	39	21	304
Eggs	6	HOU	0.19–0.24	ND	ND	80–205
Eggs	5	NOL[d]	0.3– 1.2	40–60	88–588	295–1610
Eggs	6	NOL[e]	0.1– 1.4	ND–60	44–303	105–940
Pork Chops	16	c	—	ND	ND	ND
Pork Chops	2	SAN, SEA	—	ND	ND	53, 27
Liver, Calf	1	HFD	—	ND	ND	133
Liver, Beef	3	c	—	ND	ND	ND
Liver, Beef	1	SEA	0.1	ND	428	3830
Liver, Beef	1	ORL	0.07	ND	168	614
Liver, Beef	1	SEA	0.05	ND	136	818
Liver, Beef	22	c	—	ND-37	ND-64	ND-197

[a]HFD = Hartford; HOU = Houston; MIN = Minneapolis; NOL = New Orleans; NSV = Nashville; SAN = San Francisco; SEA = Seattle.
[b]ND, not measured; limit of measurement about 10–40 pg/g; presence of dioxin residues not routinely confirmed by GC/MS. Values corrected for recovery (70–90%).
[c]Samples collected at various locations in U.S.
[d]Samples collected from farms in Mena, AR area in 1983.
[e]Samples collected from farms in Mena, AR area in 1984.

Residues of Higher-Chlorinated Dioxins in Foods

Various foods are examined by FDA for residues of PCP as part of the agency's Total Diet Program. Individual foods or food ingredients found to contain ≥0.05 μg PCP/g are examined for higher-chlorinated dioxin residues. In addition, several portions of ground beef, pork chops, chicken, eggs, and beef liver from the FDA market basket are specifically analyzed for residues of higher-chlorinated dioxins regardless of PCP residues in the products. Results of analysis of the various foods collected in the 5-yr period beginning in 1979 are shown in Table 5. Low levels (<300 pg/g) of HpCDD and OCDD were found in bacon, chicken, pork chops, and beef liver. HCDD was not found in any of the foods. Several beef livers had higher levels of OCDD residues, and one beef liver contained about 400 and 3,800 pg 1,2,3,4,6,7,8-HpCDD and OCDD/g, respectively. No dioxins

Table 6. Higher-Chlorinated Dioxin and PCP Residues in Layer Feed Ingredients and Market Eggs (1982–1983 Market Egg Contamination Episode in Houston, Texas Area).

Product	No. of Products	PCP (μg/g)	ng/g[a]		
			1,2,3,4,6,7,9- HpCDD	1,2,3,4,6,7,8- HpCDD	OCDD
Sheepskin Fleshing (source of rendered feed ingredients)	1	1862	625	394	4950
Meat and Bone Meal (feed ingredient)	1	16.0	3.5	0.6	20.4
	1	41.8	8.9	0.7	27.5
Animal Fat (feed ingredient)	1	10.6	28.2	3.4	133
Eggs	1	0.29	0.04	0.02	0.30
	1	0.22	ND	ND	0.22
	1	0.24	ND	ND	0.16

[a]ND, not measured; limit of measurement for egg samples about 10–40 pg/g; presence of dioxin residues in sheepskin fleshings, meat and bone meal, and animal fat confirmed by GC/MS/MID. Values corrected for recovery (70–90%).

(limit of measurement in the range of 10–40 pg/g) were found in ground beef. A survey of milk for higher-chlorinated dioxin residues was carried out in 1981–1983. None (limits of measurement, 5–15 pg/g) were found in 58 milk samples collected in different parts of the United States. No dioxins were found in 17 egg products collected in various parts of the United States. PCP and dioxin residues in eggs from the Houston, Texas and Mena, Arkansas areas, collected in 1982 and 1983–1984, respectively, were due to local PCP contamination problems in these areas.

In 1982, PCP residues of 0.2–0.3 μg/g were found in market eggs in the Houston area. HpCDD and OCDD residues ranging from 30 to 300 pg/g were also found in the eggs. The source of the PCP contamination was traced to fleshings from sheep hides that had been treated with PCP. Rendered fleshings (protein meal and fat derived from the fleshings) used as feed ingredients were determined to be the source of the PCP contamination. PCP and dioxin levels in the fleshings, feed ingredients, and market eggs are shown in Table 6. Levels of dioxins are in proportion to the PCP levels in the products. The Texas State Departments of Health and Agriculture were alerted. They participated with the FDA in determining the sources of the PCP and dioxin contamination and in terminating the use of the PCP-contaminated ingredients in animal feeds.

During 1983, an investigation of chlordane-contaminated chicken farms in northwestern Arkansas led to finding PCP (0.3–1.2 μg/g) and dioxins (about 40–60 pg 1,2,3,4,6,7,9-HpCDD/g, about 90–600 pg 1,2,3,4,6,7,8-HpCDD/g, and about 300–1,600 pg OCDD/g) in eggs from several farms (see Table 5, eggs collected in Mena area). Soil from a farm in the area contained about 8 ng/g each of 1,2,4,6,7,9-, 1,2,3,6,8,9-, and 1,2,3,6,7,8-

HCDD; 394, 270, and 1,300 ng/g, respectively, of 1,2,3,4,6,7,9-HpCDD, 1,2,3,4,6,7,8-HpCDD, and OCDD; and about 250 µg PCP/g. The source of the chlordane and PCP has not been determined, and the investigation of chlordane and PCP contamination in flocks in the area is continuing.

CONCLUSIONS

Ultratrace residues of 2,3,7,8-TCDD in the edible portion of fish from certain Great Lakes areas and other locations appear to be associated with the production of chlorinated phenol products and the disposal of dioxin-containing wastes. Levels of 2,3,7,8-TCDD in Saginaw Bay fish appear to be declining steadily since 1979. This may be related to the cessation of 2,4,5-T production by the Dow Chemical Co. in 1979. No 2,3,7,8-TCDD residues were found in soybeans, rice, and crawfish from rice-growing areas in Arkansas and Louisiana sprayed with 2,4,5-T herbicide. The presence of 2,3,7,8-TCDD in fish does not appear to be widespread, but rather is localized in areas near 2,4,5-TCP and 2,4,5-T production sites. The presence of higher-chlorinated dioxin residues in eggs and other animal-derived foods collected as market basket samples appears to be associated with the use of PCP in agriculture and industry, including hide processing.

REFERENCES

1. Esposito, M. P, T. O. Tiernan, and F. E. Dryden. "Dioxins," EPA 600/2-80-197; Environmental Protection Agency, Washington, DC (November 1980), Section 3, pp. 37–132.
2. Neidert, E., P. Saschenbrecker, and J. R. Patterson. *J. Environ. Sci. Health* B19:579–592 (1984).
3. O'Keefe, P., C. Meyer, D. Hilker, K. Aldous, B. Jelus-Tyror, K. Dillin, R. Donnelly, E. Horn, and R. Sloan. *Chemosphere* 12:325–332 (1983).
4. Ryan, J. J., P.-Y. Lau, J. C. Pilon, and D. Lewis. In *Chlorinated Dioxins and Dibenzofurans in the Total Environment*, G. Choudhary, L. H. Keith, and C. Rappe, Eds. (Stoneham, MA: Butterworth Publishers, 1983), pp. 87–97.
5. Stalling, D. L., L. M. Smith, J. D. Petty, J. W. Hogan, J. L. Johnson, C. Rappe, and H. R. Buser. In *Human and Environmental Risks of Chlorinated Dioxins*, R. E. Tucker, A. L. Young, and A. P. Gray, Eds. (New York: Plenum Press, 1983), pp. 221–240.
6. Firestone, D. *J. Agric. Food Chem.* 25:1274–1280 (1977).
7. Tiernan, T. O., and M. L. Taylor. "Development and Application of Analytical Methodology for Determination of Hexa-, Hepta-, and Octachlorodibenzodioxins in Beef Samples." Final report. USDA Contract No. 12-64-4-378 (1978).
8. Harless, R. L., E. O. Oswald, R. G. Lewis, A. E. Dupuy, D. D. McDaniel, and H. Tai. *Chemosphere* 11:193–198 (1982).

9. Mitchum, R. K., G. F. Moler, and W. A. Korfmacher. *Anal. Chem.* 52:2278–2282 (1980).
10. Fehringer, N. V., S. M. Walters, R. J. Kozara, and L. F. Schneider. *J. Agric. Food Chem.* 33:626–630 (1985).
11. Pennington, J. A. T. *J. Am. Diet. Assoc.* 82:166–173 (1983).
12. Food and Drug Administration. "Pesticide Analytical Manual," Association of Official Analytical Chemists, Arlington, Virginia (1968 and revisions), Vol. I, 2nd ed.
13. Niemann, R. A., W. C. Brumley, D. Firestone, and J. A. Sphon. *Anal. Chem.* 55:1497–1504 (1983).
14. Lamparski, L. L., and T. J. Nestrick. *Anal. Chem.* 52:2045–2054 (1980).
15. Niemann, R. A. Unpublished data (1984).
16. Firestone, D., M. Clower, Jr., A. P. Borsetti, R. H. Teske, and P. E. Long. *J. Agric. Food Chem.* 27:1171–1177 (1979).
17. Food and Drug Administration. Talk Paper T-81-32, "Dioxin in Fish," FDA, Rockville, Maryland (August 28, 1981).
18. Kaczmar, S. W., M. J. Zabik, and F. M. D'Itri. "Part Per Trillion Residues of 2,3,7,8-Tetrachlorodibenzo-p-dioxin in Michigan Fish," paper presented at the 186th meeting of the American Chemical Society, Washington, DC, August 29, 1983.
19. Duling, L. State of Michigan Environmental Services Division, Lansing, Michigan, news release (October 4, 1984).
20. Belton, T. New Jersey Department of Environmental Protection, Trenton, New Jersey, Unpublished data (1983).
21. Kleopfer, R. D. Environmental Protection Agency, Kansas City, Missouri, Unpublished data (1982).

CHAPTER 25

Comparison of a New Rapid Extraction GC/MS/MS and the Contract Laboratory Program GC/MS Methodologies for the Analysis of 2,3,7,8-Tetrachlorodibenzo-p-Dioxin

James S. Smith, David Ben Hur, Michael J. Urban, Robert D. Kleopfer, Cliff J. Kirchmer, W. Allen Smith, and Tenkasi S. Viswanathan

INTRODUCTION

The objective of the project was to establish a new method for quick TCDD analysis that can be performed in the field, and yields results analogous to the conventional EPA method.[1] Secondarily, the project was aimed at developing QA/QC criteria on the basis of which it could be determined whether the result of any given analysis is acceptable or if sample cleanup is dictated.

The protocol adopted for this project called for analyses of eight samples, each in six replicates, by both the GC/MS/MS method and the conventional GC/MS technique. The samples for GC/MS/MS analysis were to be run with no cleanup, after one stage of cleanup, and after two stages of cleanup. The sample consisted of five Missouri soils, one New Jersey soil, and two synthetic Potter's clay mixtures impregnated with TCDD. Each sample was air dried and thoroughly homogenized by the EPA Region VII laboratory. The samples were submitted without identification revealing the nature of the sample.

Before extraction, each sample was adjusted to a water content of 20% as defined in the following equation. The samples with added water were allowed to equilibrate in tightly sealed containers for 24 hr.

$$\% \text{ water} = \frac{\text{g water added}}{\text{g sample as received} + \text{g water added}} \times 100$$

TECHNICAL DISCUSSION

Two distinct methods have been employed in this comparison. The GC/MS/MS technique is comparatively new, and has been designed to provide rapid answers in the field. The GC/MS method has been employed by the EPA laboratories and contract laboratories for a number of years. The GC/MS method has been assumed to be the standard method, and new techniques are therefore expected to be compared to that standard.

GC/MS/MS TECHNIQUE

Utilization of tandem mass spectrometers for the analysis of complex mixtures provides for a rapid technique by virtue of its high specificity. Since the first quadrupole can be set to transmit only prescribed parent ions, the selectivity provided by the technique eliminates the need for a well defined chromatographic separation of the components of interest. Thus, the duration of the chromatographic run could be decreased significantly by employing the MS/MS as a detector.

In order to achieve a rapid analysis of samples for dioxin, not only was it necessary to reduce the chromatographic run time, but even more important, the sample preparation time had to be reduced drastically. The protocol for the conventional GC/MS method calls for about two days of sample preparation, the first day being taken up with the extraction, and the second day with sample cleanup and concentration.

The rapid extraction method and chromatographic conditions were developed and applied to Missouri soils with apparent success.[2] Attempts to utilize the technique on New Jersey soils proved futile because the extent of organics contamination was such that interferences still existed in spite of the greater selectivity of the procedure.[3] To take advantage of the efficient extraction and extremely rapid analysis time, cleanup procedures were found to be needed. These procedures have been evaluated in this study.

Figure 1 illustrates the sample preparation method. Briefly, a 5-g sample is extracted with a mixed solvent consisting of acetonitrile and dichloromethane in a 2:1 ratio by volume. The choice of solvent mixture is dictated by several features. First, the two solvents are miscible in each other. Second, because of the presence of acetonitrile in the mixture, the solvent is a

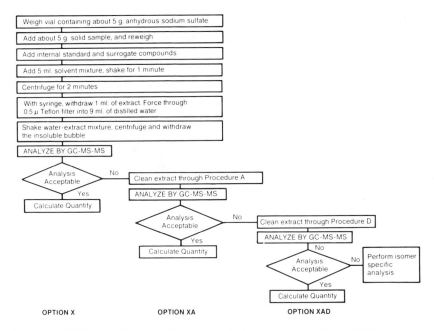

Figure 1.   TCDD sample preparation and analysis scheme using GC/MS/MS.

powerful wetting agent, making it possible to penetrate and loosen the organic components of the matrix. Third, the presence of dichloromethane permits the rapid and nearly complete dissolution of the dioxin in the solvent system. The major disadvantage of the method is the ready solubility of other organic components in the solvent mixture, leading to an extract that potentially contains many interferents. Additionally, should an extract cleanup be necessary, the solvent mixture would have to be exchanged to an essentially nonpolar solvent. Ordinary exchange is performed by adding a nonpolar solvent that has a higher boiling point than the extraction solvent, then boiling off the lower-boiling solvent.

In this case, however, acetonitrile has a fairly high boiling point, so that the conventional technique would be rather time consuming. To facilitate solvent exchange, the acetonitrile is partitioned between water and dichloromethane. The partitioning equilibrium greatly favors the transfer of acetonitrile to the aqueous phase. Thus, by mixing the extract with about a ninefold excess of water, acetonitrile is almost completely removed from the dichloromethane phase. The resulting TCDD extract, now present in the dichloromethane phase, is more concentrated than the original, but more important, it is in a form that is readily exchanged by conventional means. In addition, water-soluble soil components that were extracted with the solvent mixture are removed from the extract.

For most Missouri soils the original extract did not require any cleanup;

consequently, solvent exchange was not needed. For the New Jersey soil and the synthetic Potter's clay samples, cleanup was necessary. In this study, extracts were cleaned up regardless of whether they needed it. In this fashion, results from each step of the procedure could be compared to the other steps and to the GC/MS results.

The extracts, after undergoing acetonitrile removal, were subjected to solvent exchange from dichloromethane to cyclohexane. Cyclohexane is a convenient carrier solvent for the cleanup operation.

The first cleanup step is a modification of EPA Region VII Option A.[1] The extract is deposited on a silica gel column with cyclohexane. TCDD does not become absorbed on the silica gel, but many of the contaminants in the extract do. The column is eluted with cyclohexane, and the eluate is immediately deposited on an acid alumina column. TCDD is adsorbed on the alumina column, so that before recovering the TCDD, the column is washed with cyclohexane. A dichloromethane/cyclohexane solvent system permits the recovery of TCDD from the alumina column.

The second cleanup step is the EPA Region VII Option D[1] used without modification. This step involves the adsorption of TCDD on activated charcoal, followed by elution of less well adsorbed contaminants from the charcoal. TCDD is finally eluted using toluene.

Because the instrument is very sensitive, the extensive extract-concentrating step that is needed in the conventional GC/MS method is not needed. A 1-ppb level of TCDD is detected without the need of any concentration. This instrumental sensitivity allows work with small volumes of extract. Usually, 1 mL of extract is recovered from the extractions step. The ability to work with small volumes of extract makes it possible to miniaturize the cleanup procedure, thus permitting the performance of sample preparation in the field without slowing down the analytical process. The silica and alumina columns are each made from a Pasteur pipette, plugged with silanized glass wool, and containing 1-cm layer of adsorbent.

Beyond the sample preparation method, the technique of TCDD analysis relies on the specificity of the instrument to isolate the desired components, so that the chromatography preceding the instrument analysis is rapid and provides only very crude separation.

The instrumental specificity arises from the ability to set selectively the parent ions that would be transmitted by the first quadrupole and the daughter ions that would be transmitted to the detector by the third quadrupole. In tuning the instrument, the rod offsets and collision gas thickness are adjusted so that the preferential fragmentation resulting from the neutral loss of the group COCl is maximized. Under these conditions, the ions in Table 1 are monitored. Chromatography is performed on a 15-m DB-5 capillary column, 0.32 mm i.d., with a $1-\mu$ coating thickness. The temperature program is $20°C/min$ from $150°C$ to $240°C$. The TCDD retention time is approximately 5 min.

Table 1.   Instrumental Specificity.

Component	Parent Ion (m/z)	Daughter Ion (m/z)
Native TCDD	320	257
	322	259
Surrogate, $^{37}Cl_4$-TCDD	328	263
Internal Standard, $^{13}C_{12}$-TCDD	332	268

Quality Assurance Criteria for the GC/MS/MS Method

Quality assurance criteria are established so that each individual run could be examined for acceptance, without requiring reliance on sample spikes, replicate runs, or analysis of a split sample by a conventional technique.

The first two criteria are chromatographic functions. First, the four monitored ions must exhibit the same retention time; second, the peak shape of each trace must be symmetrical. Aside from indicating good chromatographic conditions, these two criteria are excellent means of observing interferences. Any interferent that does not precisely coelute with TCDD will cause the peak of the ion with which it interferes to be distorted or to exhibit an apparent shift in the retention time. These criteria have been employed as visual means of ascertaining that a particular run is acceptable; however, quantitative measures can be easily applied. The peak symmetry can be measured through the determination of the tailing factor. The retention times of the four monitored ion pairs should be within three scans.

The third criterion is that of surrogate recovery. The definition of a surrogate compound is that of a component added to the matrix in a known amount and allowed to equilibrate with the matrix so that its measured concentration in the extract, relative to an internal standard, is an indication of the extraction efficiency. The internal standard, on the other hand, is added to the extract at a known concentration, so that its concentration is precisely known. In this study, a cocktail consisting of both surrogate and internal standard is injected onto the sample matrix, and the sample is immediately extracted. In this situation, no equilibrium is established. Thus, the ratio of surrogate to internal standard should be a constant. To determine this ratio, and its range, the sample spiking mixture was run repeatedly, and the average and standard deviation were determined. The deviation from the average actually reflects counting errors. To determine this acceptable range of surrogate recovery, three times the standard deviation was taken. Based on the experimental runs of standards, the acceptable recovery of the surrogate, when the surrogate is added to the sample at the level of 1 ppb, is in the range 83.4–116.6%.

The fourth criterion is the ratio A_{257}/A_{259}. On a theoretical isotope distribution basis, this ratio should be about 1.04. The theoretical ratio should be obtainable if the instrument is run at 1 dalton(amu) resolution. In order to improve the signal intensity, however, the instrument is run at a lower resolution. Under such circumstances, the ratio of the two ions is expected to deviate from the theoretical. The ratio is determined empirically from the calibration runs, and in this project a window of three standard deviations was taken as the acceptable range. Thus, the following limits were set:

$$1.17 < \frac{A_{257}}{A_{259}} < 1.52$$

One exception must be noted: when the concentration of native TCDD in the sample is below the detection limit, the ion ratio does not apply.

Quality Control Measure for the GC/MS/MS Method

Calibration Curve Construction

At the beginning of the program, a calibration curve was constructed using the EPA-supplied standards covering the range of concentration of native TCDD from 1 ppb to 200 ppb and the range for surrogate from 0.3 ppb to 1 ppb. Each of the supplied standards was diluted 1:200 by volume with the extraction solvent. The diluted standard (1 mL) was extracted with water to remove the acetonitrile, and 2 μL of the remaining dichloromethane solution was injected.

The diluted standards, without going through the water treatment, were also injected to verify adequately the response factor for each of the ions.

Daily Calibration Verification

At least once each day a calibration standard was injected to verify that the instrumental response was still within an acceptable range. Frequently more than one standard was analyzed. These standards were treated as samples, and the concentration of native TCDD was calculated and compared with the known amount.

Surrogate Response

The spiking mixture was analyzed on several different days at various stages of sample treatment.

Spiked Samples

Several samples were spiked during the course of the program at irregular intervals.

GC/MS TECHNIQUE

The GC/MS method used in this project is the Contract Laboratory Program Method.[1]

RESULTS AND DISCUSSION

GC/MS/MS

Correction Factors

Native TCDD contains an innate amount of $^{37}Cl_4$-TCDD. Ignoring this contribution to the surrogate would lead to erroneous surrogate recoveries, especially when the native TCDD content of the sample is high. The theoretical contribution, based on natural isotope distribution, shows that the ratio of A_{263}/A_{257} in the native sample is 0.0108. For actual calculations, however, we chose to measure the contribution. The measurements were performed on the two calibration standards, supplied by the EPA, that contained no surrogate compound. Based on these results, the area count of the surrogate must be corrected as follows.

$$A'_{263} = A_{263} - 0.01113A_{257}$$

where A'_{263} = corrected area count of the surrogate
         A_{263} = raw area count of the surrogate
         A_{257} = raw area count of native TCDD measured at the fragment of mass 257

Experience has also indicated that the surrogate contributes to the native TCDD. In all likelihood, the surrogate, $^{37}Cl_4$-TCDD, is made by isotopic exchange from the native; hence, it is not unexpected that some native TCDD would still remain.

Furthermore, because of the exchange process, the normally expected ratio of A_{257}/A_{259} cannot hold; the level of A_{259} should be relatively greater. The contribution has been calculated on the basis of the surrogate peak because historically it has been the surrogate that has contributed to the native count, not the internal standard. Nonetheless, correction factors are not derived on the basis of the spiking mixture contribution to the native, because certainties concerning the source of contribution have not been established. It would be an easy matter to correct if the level of surrogate and internal standard in all samples and in the calibration runs were constant. The ratio of surrogate to internal standard, however, is variable in the calibration runs.

The contribution of the spiking mixture to the native TCDD level is low, and would not significantly affect samples with 1 ppb or more TCDD. Close assessment of this contribution should be determined when lower levels of TCDD are measured. In this project the contribution of the spiking mixture to the native TCDD has been ignored.

Table 2.   Calibration Curve Data.

C_n^a/C_{is}^b	A_{257}/A_{268}	A_{259}/A_{268}	RF_{257}^c	RF_{259}^d
0.2	0.2413	0.1644	1.206	0.822
1.0	1.1002	0.8080	1.100	0.808
5.0	5.2853	3.8789	1.057	0.776
20.0	18.7908	14.2171	0.940	0.711
40.0	32.6772	25.7554	0.817	0.644

$^a C_n$, concentration of native TCDD in the solution.
$^b C_{is}$, concentration of internal standard in the solution.
$^c RF_{257}$, the response factor for the ion of mass 257, determined from $(A_{257}/A_{268})/(C_n/C_{is})$.
$^d RF_{259}$, the response factor for the ion of mass 259, determined from $(A_{259}/A_{268})/(C_n/C_{is})$.

Surrogate Response

Although the surrogate response can be measured from the calibration curve samples, a better way is to determine the response by measuring the areas associated with the spiking mixture. This measurement would relate more closely to the determinations performed on the samples, because the same spiking mixture is used in dosing the samples. Thus, any error in preparing the spiking mixture would be canceled.

The surrogate has a measured response factor of 1.265. In calculating surrogate recovery in samples, the equation is

$$\% R = \frac{A_{263}/A_{268}}{0.253} \times 100$$

where A_{263} = the corrected area of the peak of mass 263
A_{268} = the area of the peak of mass 268, associated with the internal standard

Initial Calibration Curve

The initial calibration curve was obtained by diluting each of the EPA-supplied calibration standards 1:200 with acetonitrile/dichloromethane (2:1 by volume). The diluted standards (1 mL of each) were mixed with 9 mL of distilled water, shaken well, centrifuged, and the dichloromethane phase recovered. This procedure was followed in order to simulate the steps taken in the sample extraction procedure. Table 2 summarizes the results.

Examination of the RF values indicates a distinct deviation from linearity as the concentration of the active TCDD increases. Because of this, only the three lowest points have been taken to determine the calibration line. These points have been used to determine the following equations for the calculation of native TCDD.

$$\frac{C_n}{-C_{is}} = \frac{\dfrac{A_{257}}{A_{268}} - 0.0402}{1.0494}$$

$$\frac{C_n}{-C_{is}} = \frac{\dfrac{A_{259}}{A_{268}} - 0.0219}{1.7719}$$

Regression line parameters were used in calculating the concentration because the intercept is a necessary correction factor in interpreting the data.

Daily Calibration Runs

Calibration standards, diluted 1:200 with the extraction solvent, were run on a daily basis to verify that the instrument calibration was still valid. These calibration runs were treated as samples, and the concentrations of native and surrogate TCDD were calculated. The results show that the instrument has performed satisfactorily throughout the period of the project. The relative deviations are at their highest in the low-concentration calibration standard, but this is to be expected, since small errors in counting would be relatively more significant at the low level than at the high-concentration level.

Quality Control Limits

In the absence of established guidelines for the GC/MS/MS technique, numerical values for the acceptance of data were determined on the basis of standard deviation of the corresponding values in the calibration runs.

For the surrogate recovery, the standard deviation of the ratio A_{263}/A_{268} is 0.013, and the mean is 0.253. Using three standard deviations as the acceptance limits, experimental values falling within the range 0.253 ± 0.039 are acceptable. This corresponds to a surrogate recovery in the range 83.4–116.6%, when the sample is spiked with 1 ppb surrogate.

For the ratio A_{257}/A_{259}, the data from the calibration curve runs are used, again applying the three standard deviations rule to determine the acceptable range. For this ratio, the acceptable range is 1.3473 ± 0.1770. Peak appearances, their coelution and shapes, were judged visually.

Detection Limit

There are several means of determining a theoretical detection limit. The conventional technique employed by the IFB method[1] is dependent upon measurement of noise level in the scans immediately adjacent to the peak of interest. Employing this approach, the detection limit for this project is 0.07 ppb native TCDD (it is assumed that a signal 2.5 times the noise level is detectable).

Table 3.   Sample Soil Type.

Missouri Soil Number 1	New Jersey Soil
Missouri Soil Number 2	EMSL-LV Low-Clay
Missouri Soil Number 3	EMSL-LV High-Clay
Missouri Soil Number 4	
Missouri Soil Number 5	

The detection limit calculated on the basis of signal-to-noise ratio is purely a theoretical limit, not realizable in real situations. It ignores any experimental variability, other than counting errors of the detector.

To estimate a detection limit, we have chosen to base it on the standard deviation of repeated runs of the 1-ppb standard solution. Taking three times the standard deviation as a measure of the detection limit, the obtained value is 0.13 ppb.

Another possible approach is to calculate the concentration equivalent to the intercept in the calibration curve. Using this approach, the detection limit based on the daughter ion of mass 257 is 0.19 ppb, and based on the daughter ion of mass 259 it is 0.14 ppb. While the detection limit is stated in terms of concentration units, which is applicable to this project, the actual detection limit is dependent upon absolute amounts of TCDD. Thus, by concentrating the extract further, or by taking a larger sample, the detection limit can be lowered.

Results of Sample Analysis

The individual samples were identified by the EPA after data of the analyses were submitted. The soil identifications are in Table 3.

Based on quality assurance criteria, Missouri Soils Numbers 1, 2 and 4 required no cleanup. The data with no cleanup met the requirements, and no significant differences in values are found after cleanup. Missouri Soil Number 3 showed significant interference with the surrogate ion before any cleanup. However, it met all criteria after the first cleanup step. Missouri Soil Number 5 marginally required the first cleanup steps (if one used the EPA IFB criteria for surrogate recovery, Missouri Soil Number 5 would require no cleanup at all). The New Jersey sample exhibited interferences with both the surrogate and the internal standard ions. This sample required both steps of cleanup.

The Potter's clays contain several interferents that contribute to the native TCDD peaks. Most of these interferents are removed by the two-step cleanup. However, one of the interferents is 1,2,3,4-TCDD, which can only be separated completely from 2,3,7,8-TCDD by changing the chromatographic conditions. This interference is visible, upon examination of peak shapes. In the low-clay sample, the levels of the two dioxin isomers are about the same, and the quantitation of 2,3,7,8-TCDD singly is not possible. In the high-clay sample, the level of 2,3,7,8-TCDD is substantially

Table 4.   Analysis of Spiking Mixture 1.

Cleanup Step	Nominal Concentration (ppb)		Found Concentration (ppb)	
	Native	Surrogate	Native	Surrogate
X	1	1	0.928	0.967
XA	1	1	0.949	0.940
XAD	1	1	0.948	0.924

greater than that of 1,2,3,4-TCDD, making it possible to estimate the quantity of 2,3,7,8-TCDD. In both cases, the presence of a second component is detectable. Hence, in actual sample analysis, such occurrences can be caught and the extract subjected to a reanalysis under conditions more favorable to the separation of the two isomers.

It was not the aim of this study to demonstrate isomer separation capabilities; because of this, reanalysis of the EMSL-LV clays was not attempted.

Spike Recovery

In conventional field analysis it would be expected that a certain percentage of the samples would be spiked, and the spike recovery determined. Two samples were chosen for spike recovery in this study. Before attempting to calculate the recovery of the spike, the concentration in the spiking solution (supplied by the EPA) was determined. The spiking solution was also carried through the cleanup procedure to assure that its behavior throughout the process is the same as for the standard calibration solutions. The data are summarized in Table 4.

The results of recovery of the two spiked samples are shown in Table 5. The data indicate that the spike recovery in this method is the same as the spike recovery ordinarily found in the conventional mode.

Analysis of Single Extract by Both Methods

The extract of sample ACO8027-M3376 of Missouri Soil Number 5 prepared for the MS/MS method was analyzed also by GC/MS. The results are shown in Table 6.

Table 5.   Spike Recovery.

Sample Number	Cleanup Step	Native TCDD Added (ppb)	Native TCDD Found		Spike Recovery	
			Spiked	Unspiked	ppb	%
ACO808-M6381 (Missouri Soil Number 2)	X	1	2.18	1.42	0.76	76
	XA	1	2.18	1.41	0.77	77
ACO0806-M2252 (Missouri Soil Number 1)	X	1	0.972	ND	0.972	97

Table 6.   Sample Extract Analyses by GC/MS/MS and GC/MS.

Mean TCDD (ppb)		Surrogate Recovery (%)	
GC/MS/MS	GC/MS	GC/MS/MS	GC/MS
15.22	17.77	96.6	85.4

Isomer Separation

Although the method is not meant to be isomer specific, some idea of the capabilities of the technique to separate TCDD isomers is necessary. The standard isomer mixture supplied by the EPA was injected and analyzed under the same conditions as those used for sample analysis. The isomer mixture was mixed with the sample spiking mixture, so that the surrogate and internal standard could be employed to locate the 2,3,7,8-TCDD peak. With the exception of 1,2,3,4-TCDD, all the isomers appear well separated from the target compound 2,3,7,8-TCDD. The chromatogram is shown in Figure 2.

Although 1,2,3,4-TCDD is not adequately resolved from 2,3,7,8-TCDD, this should not pose severe limitations on the application of the technique. First, the separation of the two isomers is sufficient to warn the operator that a more isomer-specific approach is needed. Second, the 1,2,3,4-TCDD is not a common impurity in any process; usually it must be synthesized in order to introduce it into the sample. Thus, the occurrence of interference from 1,2,3,4-TCDD occurs only when the compound is intentionally spiked into a sample.

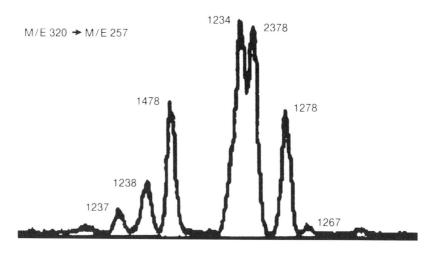

Figure 2.   Tentative assignments for TCDD performance check solution using a 30-m DB-5 fused-silica capillary column and MS/MS.

Table 7. Summary of Results of TCDD Analysis by GC/MS/MS and GC/MS (in ppb).

Sample	GC/MS/MS Results						GC/MS Results	
	Step X[a]		Step XA[b]		Step XAD[c]			
	Mean	Std. Dev.	Mean	Std. Dev.	Mean	Std. Dev.	Mean	Std. Dev.
Missouri Soil No. 1	< 0.13	—	< 0.13	—	< 0.13	—	< 0.07	—
Missouri Soil No. 2	1.72	0.39	1.74	0.35	1.76	0.39	2.07	0.39
Missouri Soil No. 3	—	—	2.86	0.38	2.95	0.42	3.11	0.24
Missouri Soil No. 4	6.39	0.30	6.37	0.38	6.53	0.59	6.97	0.60
Missouri Soil No. 5	16.0	0.63	15.8	0.69	15.7	0.63	29.8[d]	1.3
New Jersey Soil	—	—	—	—	9.4	2.0	8.08	0.53
EMSL-LV PE No. 1	—	—	—	—	—	—	0.605	0.066
EMSL-LV PE No. 2	—	—	—	—	5.25	0.84	3.70	0.28

[a]Extract with no cleanup.
[b]Extract after the first cleanup stage.
[c]Extract after the first and the second cleanup stages.
[d]Missouri Soil Number 5 contained 18.1 ppb TCDD, based on EPA Region VII Analysis. The values obtained by GC/MS shown above are anomalous.

SUMMARY OF RESULTS

The average results of the analysis of each sample type are summarized in Table 7. For the values obtained by GC/MS/MS, the summary table shows only those measurements that met the internal QA/QC criteria—i.e., if an "as is" sample extract clearly exhibited a need for sample cleanup, the results obtained without sample cleanup were excluded from the table.

CONCLUSIONS

Table 8 shows a comparison of the data obtained by the GC/MS/MS method to those obtained by the GC/MS method, the latter having been determined both by Roy F. Weston, Inc. and by the EPA Region 7 Laboratory. From the practical aspects of obtaining results in a timely fashion, the three sets of data exhibit a high degree of correspondence, with some exceptions. These exceptions are as follows:

1. Weston's GC/MS results for Missouri Soil Number 5 are anomalously high. Examination of the data obtained, and rerunning every sample in the set, did not help in explaining the discrepancy. In view of the EPA GC/MS results' close correspondence to Weston's GC/MS/MS results, it is suspected that the wrong sample was accidentally submitted to Weston for the GC/MS analysis. Since the samples have been completely consumed, this cannot be verified.
2. Both EMSL-LV Performance Evaluation samples contain 1,2,3,4-TCDD in addition to 2,3,7,8-TCDD. The GC/MS/MS method is designed for rapid analysis, recognizing that these two isomers are not separated under the experimental conditions. On Performance Evaluation Sample Number 1, the levels of the two isomers are approximately equal, making quantitation without separation virtually impossible. In Performance Evaluation Sample Number 2, the level of 2,3,7,8-TCDD is significantly higher than

Table 8. Comparison of Results of TCDD Analysis (in ppb).

GC/MS/MS[a] (Weston)		GC/MS (Weston)		GC/MS (EPA)		Sample Identity
Cleanup Step						
	Mean	Std. Dev.	Mean	Std. Dev.	Mean	Std. Dev.	
X	< 0.13	—	< 0.07	—	< 0.08	—	Missouri Soil No. 1
X	1.72	0.39	2.07	0.39	2.01	0.36	Missouri Soil No. 2
XA	2.86	0.38	3.11	0.24	3.07	0.15	Missouri Soil No. 3
X	6.39	0.30	6.97	0.60	6.99	0.51	Missouri Soil No. 4
X	16.0	0.63	29.8	1.3	18.2	0.51	Missouri Soil No. 5[b]
XAD	9.4	2.0	8.08	0.53	6.79	0.48	New Jersey Soil
—	—	—	0.605	0.066	0.939	0.130	EMSL-LV PE No. 1[c]
XAD	5.25	0.84	3.70	0.28	4.36	0.38	EMSL-LV PE No. 2

[a]Under the GC/MS/MS heading, the results reported are for the first stage of cleanup that produced runs that meet the internal QA requirements.
[b]Missouri Soil Number 5 shows an anomalously high value by the Weston GC/MS determination. No explanation could be gleaned from the data.
[c]EMSL-LV PE Number 1 showed the presence of 1,2,3,4-TCDD in amounts about equal to the 2,3,7,8-TCDD, but without producing sufficient separation of the peaks to quantitate. Therefore, no results are reported by the GC/MS/MS method.

that of 1,2,3,4-TCDD, making it possible to estimate the amount of 2,3,7,8-TCDD. Even though the GC/MS/MS method does not separate the two TCDD isomers from each other, the method is still applicable because 1,2,3,4-TCDD is a purely synthetic isomer not occurring as a natural by-product of any reaction. Thus, in natural samples, the 1,2,3,4-TCDD isomer is not expected to be found. Were it to be found in natural samples, the method is sufficiently sensitive to show the presence of another isomer, so that the sample could be reanalyzed using isomer-specific methodology.

Rigorous statistical analysis of all the data was applied by the EPA Region 7 Laboratory. This study has clearly demonstrated that the rapid GC/MS/MS method produces results with the accuracy and precision of the more cumbersome and time-consuming GC/MS method.

REFERENCES

1. Invitation for Bid, WA 84-A002 U.S. E.P.A. "The Analysis of 2,3,7,8-TCDD in Soils" (January 1984).
2. "Analysis of Dioxin in Soil in Rosati, Missouri," prepared for the Environmental Response Team, U.S. EPA, Edison, New Jersey, by Roy F. Weston, Inc., West Chester, Pennsylvania. (September 1983).
3. "Analysis of New Jersey Soils for TCDD to Determine Method Equivalency," prepared for the Environmental Response Team, U.S. EPA, Edison, New Jersey, by Roy F. Weston, Inc., West Chester, Pennsylvania. (September 1983).

CHAPTER 26

Some Analytical Considerations Relating to the Development of a High-Resolution Mass Spectrometric Dioxin Analytical Protocol

J. R. Donnelly, G. W. Sovocool, Y. Tondeur, S. Billets, and R. K. Mitchum

INTRODUCTION

As a part of the National Dioxin Strategy,[1] the U. S. Environmental Protection Agency is investigating sites where production, processing, or disposal of herbicides and pesticides could result in contamination of the environment by dioxins and/or related compounds. This National Dioxin Strategy provides a multitiered approach, ranking sites according to anticipated probability and severity of contamination. The seven Tiers identified under this Strategy may be summarized as follows:

Tier 1: 2,4,5-trichlorophenol production and disposal sites. There are believed to be about twenty production sites, but the total number of Tier 1 sites is not known.

Tier 2: Sites where 2,4,5-trichlorophenol was used as a chemical precursor for another product, such as hexachlorophene or 2,4,5-T herbicide. While the total number of disposal sites is unknown, approximately eighty production sites are known.

Tier 3: Sites where 2,4,5-trichlorophenol and its derivatives were formulated into pesticide/herbicide products. Disposal sites for such wastes are included in Tier 3.

Tier 4: Combustion sources such as industrial and municipal incinerators, home heating units, and PCB transformer fires. Hundreds of Tier 4 sites are believed to exist.

Tier 5: Sites where potentially dioxin-contaminated pesticides have been applied, such as certain rangelands, forests, and agricultural areas. Sites where pesticides/herbicides derived from 2,4,5-trichlorophenol have been used on a commercial basis. Hundreds of Tier 5 sites are known to exist.

Tier 6: Sites where inadequate quality control could result in formation of dioxin-contaminated chemical products. Hundreds of such sites probably exist.

Tier 7: Control sites, believed not to be contaminated with dioxins or related chemicals of interest to the EPA under the National Dioxin Strategy. Samples from these sites would be used to establish "background" levels of dioxins.

Sampling and analytical efforts for Tiers 3–6 were initiated through a special funding appropriation referred to as the "National Dioxin Study." Three EPA Laboratories participating in this effort have been referred to as "the Troika." Those laboratories (Environmental Research Laboratory-Duluth, MN, Environmental Monitoring Systems Laboratory-Research Triangle Park, NC, and Environmental Chemistry Laboratory-Bay St. Louis, MO) are expected to phase out their activities during 1986, with the EPA Contract Laboratory Program providing future analytical support. Troika methodology and quality assurance/quality control procedures have recently been published.[2]

Research efforts under the National Dioxin Strategy include five major topics of interest:

(1) Assessment of toxicities – most available data are based on the 2,3,7,8-TCDD isomer alone, but the EPA recognizes that other dioxin isomers and related chemicals are also of concern.
(2) Determination of sources – these dioxins and related compounds are often unwanted by-products, and may be found in relatively high concentrations in still bottoms which have been stored in hazardous waste sites. Secondly, areas which have been sprayed with contaminated pesticides/herbicides (e.g. 2,4,5-T) may be contaminated to significant extents. Thirdly, combustion sources may result in exposure to populations located nearby.
(3) Environmental fate and transport properties of dioxins and related compounds are of interest. Leaching through soil, transport through groundwater or sediment migration, and travel via airborne particulates are examples.
(4) Levels of exposure and risk assessment for a given level of contamination in a certain medium, and the resultant level of exposure to the population, need to be determined. The risk factor resulting from that exposure also needs to be calculated so that scientifically valid action levels can be determined.

(5) Legislative control recommendations can be provided once the above four topics have been addressed adequately.

In order to support EPA analytical efforts via the Contract Laboratory Program (CLP), a single, well detailed analytical method with quality assurance and reporting requirements was needed. Such a method must be rugged, and provide high-quality analytical results meeting the goals of the National Dioxin Strategy. The method was designed for all TCDD analytes, at 1–10 ppt detection limits. Initially, soil, sediment, and water matrices were targeted, but fish and other tissue samples would be analyzed, particularly at Tier 7 sites. Fish may serve as bioaccumulators, and the U.S. Food and Drug Administration has provided the guideline of 25–50 ppt maximum concentration of TCDD for edible fish tissue. Therefore, analysis of fish tissues could serve as indicators of low-level dioxin concentrations at a site, and such analyses would also be important to determine suitability of fish from such sites for human consumption.

The program required that a suitable analytical method needed to be identified, tested, and validated without extensive method comparison studies. Use of existing Contract Laboratory Program (CLP) procedures would be desirable for facile implementation. If existing quality control and reporting requirements could be adapted from another analytical method and proven successful for similar efforts, much time and effort could be saved. A literature search and evaluation were performed to assess the probable suitability of existing methods amenable to low-ppt determinations of TCDDs.

SURVEY OF ANALYTICAL METHODS

Numerous methods were located in the literature; using the criteria of (1) low-ppt detection limits, (2) confirmatory quality method, not a screening procedure, and (3) soil, sediment, water, and fish matrices, the following methods are noted here for background information.

1. EPA method of Harless and Dupuy,[2,3] and related Wright State University[4-7] and Battelle-Columbus Laboratories[8] methods
2. FDA (U.S. Food and Drug Administration) method of Firestone[9-11]
3. New York State method of O'Keefe[12,13]
4. Fish and Wildlife Service (FWS) method of Stalling and Smith[14-20] and related Rappe (University of Umeå, Sweden) method[21,22]
5. Dow Chemical Company method of Lamparski and co-workers[23-26]
6. U.S. Food and Drug Administration – National Center for Toxicological Research (NCTR) method of Mitchum and co-workers[27-31]

In general, interlaboratory validation of these methods has been performed only to the extent that:

1. quality assurance samples using appropriate matrices and spiked with known amounts of the analytes were analyzed precisely and accurately (±30% of known values).
2. environmental samples (fish tissue) with "naturally" incorporated analytes were analyzed with results consistent among different labs, each using its own methods. At low-ppt levels, relative standard deviations of 13–25% have been obtained across eight laboratories.[32]

The method of Tiernan and colleagues at Wright State University[4-7] is similar to the Harless and Dupuy method. Battelle-Columbus scientists[8] have performed analyses of the required quality; their method is also similar to the Harless/Dupuy method. The U.S. EPA Method 613 (using LRMS) has been validated for 2,3,7,8-TCDD and a method detection limit of 1.6 ng/L has been established.[33,34] The NCTR method involves oxygen-mediated negative ion atmospheric pressure ionization mass spectrometry (NIAPI-MS) for detection and quantification. This technique has not been applied (due partly to equipment and specialized experience requirements) by other laboratories, although it has been validated for that laboratory (NCTR), for 2,3,7,8-TCDD in a variety of matrices, with low-ppt detection limits. Three of these methods are discussed below as examples of current methodology directly applicable to EPA needs.

EPA METHOD OF HARLESS AND DUPUY

This method has been validated for that laboratory against other methods for a variety of matrices at detection limits required, including those of interest for this effort.* The method is applicable to low-ppt analyses for chlorinated dioxins and dibenzofurans (PCDDs and PCDFs). While the extracts are relatively more complex than those generated using other methods chosen, the resolving power of high-resolution mass spectrometry (HRMS) compensates, for most samples encountered to date.** Analysis time per sample is relatively short. For example, 20 samples per week can be analyzed per pair of workers using this method. A highly qualified HRMS operator is needed to achieve low-ppt detection limits. This method is being used in the National Dioxin Study, and has been described[2] in some detail, along with confirmation criteria, reporting requirements, and quality assurance procedures. Interlaboratory validation of this method was attempted

---

*In this context, validated means that analytical results are ±20 percent of true values on "known," spiked samples, and ±20 percent of the mean value obtained on homogenized environmental "field" samples analyzed by several laboratories experienced in such efforts.

** Extract "cleanliness" for different cleanup methods has been described.[23-26] Typically encountered analytical interferences have been reported.[14-20,36-39]

in 1982–83, and results have been reported.[35] The results indicated that the Harless and Dupuy method is technically difficult to apply for routine analysis.

FWS METHOD OF STALLING AND SMITH

This method has been validated in two laboratories (FWS, and University of Umeå) on fish tissue to low-ppt levels, for most PCDDs and for tetra-through octa-PCDFs. As a result of added cleanup steps, quadrupole low-resolution mass spectrometry (LRMS) in both electron ionization (EI) and negative ion chemical ionization (NICI) modes has been employed successfully. Detection limits are mostly a function of the type of GC/MS employed, with the LRMS quadrupole instrument being the limiting factor. Use of HRMS would be expected to improve the detection limits. Crucial elements of this method, which allow use of LRMS for ppt-level determinations, are the extra cleanup (afforded by GPC or cesium silicate, and carbon/foam, carbon/glass fiber, or carbon/silica gel) and the concentration enrichment afforded concomitantly by the carbon cleanup step. Many features of this method are very similar to the U.S. EPA Region 7 Method (IFB-WA84-A002)[40-43] which afforded ppb-level determinations of 2,3,7,8-TCDD in soil and sediment.

DOW METHOD

This method has been validated by that laboratory, against other laboratories' methods and results. With highly skilled operators, ppt and ppq detection limits are achievable. Extensive HPLC cleanup results in very clean extracts, with separation of analytes (rather than generation of a single extract containing all the analytes). GC/MS analysis time of the several extracts is therefore longer, but detection limits are improved for those situations where not all analytes can be measured during a single analytical determination (mass spectrometric "run"). With the extensive cleanup utilized, up to one week may be required to analyze one to five samples. This method, with the highly sensitive and selective HRMS technique, would provide excellent detection limits, but with high cost and long analysis time.

EVALUATION AND SELECTION OF METHODOLOGY

The literature survey and limited laboratory testing demonstrated several problems and limitations, with respect to use of existing methods in a Contract Laboratory Program-type application.

(1) The CLP-LRMS method developed by U.S. EPA Region 7, and applied under IFB Contract WA84-A002,[40-43] was rugged and dependable for ppb-level determinations of 2,3,7,8-TCDD, in certain soil-type matrices. Performance in other matrices, at ppt-level concentrations, has not been validated.

(2) Those methods which were proven applicable at ppt concentration levels had not been tested extensively on an interlaboratory basis. Indications were (at least for the Harless and Dupuy Method[3]) that additional ruggedness testing documentation, elaboration, and possibly modifications would be needed[35] for application to the CLP. While each method worked well at the "parent" laboratory, the requirement for this effort would be a method that could be expected to work well at many laboratories.

The existing methodologies have several common factors in cleanup, detection limits, and quality assurance requirements.

(1) With the exception of the Dow Chemical Company's extensive high-performance liquid chromatographic sample cleanup scheme, sample preparation methodology was similar throughout those methods proven applicable to ppb- and ppt-level dioxin determinations.

(2) Experience of the EPA, and other investigators as well, showed that carbon adsorption cleanup was efficacious.

(3) Cleanup schemes such as used in IFB WA84-A002 provided samples which were generally quite "clean," and probably the major limitation regarding detection limits for TCDDs would be instrumental rather than chemical interference related.

(4) Quality assurance/quality control procedures, and reporting requirements outlined in IFB WA84-A002, were adequate and could be modified to fit current National Dioxin Strategy goals.

(5) Routine achievement of low-ppt detection limits would necessitate use of instrumentation such as negative ion chemical ionization/low-resolution mass spectrometry (vacuum or atmospheric pressure), or electron ionization (EI) high-resolution mass spectrometry. The EPA has amassed considerable data demonstrating the success of the latter technique for a great number of samples in various matrices.[3] This technique (HRMS) was also being used for ppt-level determinations by most laboratories engaged in such work.

The method detection limit (MDL) for Method 613 was reported as 1.6 ppt.[33,34] The MDL for soil determinations using IFB WA84-A002 of 1.0 ppb is undergoing assessment, and efforts are underway to validate a detection limit in the 300-ppt range. Thus the analytical ranges chosen for the new HRMS method (1 ppt to 1.2 ppb in soil; 10 ppq to 12 ppt in water) will provide some overlap with the LRMS methods already in use by the EPA at the upper concentration end, and provide the best routinely attainable detection limits.

Using these criteria, the most efficient way to produce an analytical method for contract laboratory usage, suitable for testing (single laboratory or multiple laboratory), would be to adapt IFB WA84-A002 to high-resolution mass spectrometry. While a number of technical issues would have to be addressed, the amount of effort involved in method development, contract format, contract implementation, and laboratory cleanup experience levels would be minimal, thereby favoring such an approach. Further, EPA experience with HRMS determinations of ppt levels of dioxins could be applied to modify the methodology as required for different matrices. EPA experience could also be used to estimate detection limits, choose instrument parameters, modify reporting requirements, and adapt existing quality assurance/quality control requirements from the LRMS IFB method.

In order to provide a protocol for high-resolution mass spectrometric determination of TCDDs in the extract prepared as described in this chapter, a survey was taken informally of some laboratories engaged in such determinations using HRMS in electron impact (EI) mode. In summary, those laboratories agreed on the following items: perfluorokerosene (PFK) or FC-43 (perfluorotributylamine) as calibrant; peak top measurement (i.e., traditional selected ion monitoring at the mass centroid) — except for one laboratory using a 40-mμ window due to requirements of the data system; and splitless GC injection. The laboratories polled differed, however, in types of instrumentation and data systems: MAT 311A, MAT 8200, MAT 8230, VG ZAB-2F, VG 7070, and Kratos MS-80 instrumentation were employed with INCOS, SS-300, VG-2035, VG-11/250, DS-55, and analog data-handling systems. Additionally, instrument parameters differed considerably; resolutions of 5000, 8000, 10,000, and 15,000 were used. Lock masses m/z 255, 319, 331 (PFK), and 314 (FC-43) were used. In addition to m/z 320 and 322, certain laboratories used different combinations of the following: m/z 257, 259, 285, 287, 324, 328, 332, 336. Dwell times in milliseconds were reported as: 30/ion; 500/ion with 20 for lock mass and 50 for m/z 334; 150 (m/z 320 and 322) and 20 (m/z 332, 334); 400 (m/z 320, 322) with 250 (lock mass) and 100 (m/z 336). Interchannel delays of 12, 20, 30 msec were reported. Total cycle times of 0.4, 0.5, 1.23, 3.31, 1.2, 1.26, and 0.6 sec were used. Electron energies of 70, 30, and 65 eV were employed. Detection limits of 0.5–10 pg were attained. The amount injected into the GC and final extract volume are therefore of considerable importance.

This limited survey demonstrated that a number of instrumental parameters were greatly varied according to operator preference, and that some reasonable compromise value would have to be specified for the contractual method.

PROPOSED HRGC/HRMS METHOD: SUMMARY FOR SOIL AND WATER MATRICES

Soil Sample Extraction

The soil sample is centrifuged to remove gross amounts of water, and then air dried. A sample size of 20 g (± 0.5 g) is chosen, and 2.0 ng $^{13}C_{12}$-2,3,7,8-TCDD is added as a surrogate/internal standard for quantification. After adding 20 g sodium sulfate drying agent, the sample is extracted with 200–250 mL benzene using Soxhlet apparatus. This extraction procedure is used in SW-846 Method 3540,[44] and in RCRA Method 8280,[45] and also was employed in early EPA Region VII efforts for Missouri soil analyses.[40-43] It has been shown that this rigorous extraction procedure is more suitable than a stirred or shaken jar/hexane procedure for the variety of soil types to be encountered. Following extraction, the extract may be concentrated with Kuderna-Danish and Snyder apparatus, or by rotary evaporation under water-aspirator vacuum.

Water Sample Extraction

The water sample (2 L) is spiked with 2.0 ng $^{13}C_{12}$-2,3,7,8-TCDD, and extracted with a separatory funnel or continuous liquid-liquid extractor, using three 60-mL portions of methylene chloride. The extract is dried and concentrated with Kuderna-Danish and Snyder apparatus. This extraction method parallels EPA Method 613.

Extract Cleanup

Extract cleanup is performed as in the LRMS-IFB method. The extract is passed through a column containing 1 g silica gel, 4 g 40% (w/w) sulfuric acid/silica gel, and 1 cm of 1:1 Na_2SO_4/K_2CO_3, followed by a second column containing 1 cm sodium sulfate and 6 g acidic alumina. The extract is eluted from the first column, concentrated with 90 mL hexane, and eluted from the second column with 20 mL hexane followed by 30 mL 20% methylene chloride/hexane.

Final cleanup is performed with the Amoco PX-21 type carbon adsorption column. The column is conditioned with 2 mL toluene, 1 mL 75:20:5 methylene chloride/methanol/benzene, 1 mL 1:1 cyclohexane/methylene chloride, and 2 mL hexane. The extract is then eluted from the column at controlled flow rate with 2 × 1 mL hexane, 1 mL 1:1 cyclohexane/methylene chloride, and 1 mL 75:20:5 methylene chloride/methanol/benzene. The analytes are eluted (reversed flow) from the column with 6 mL toluene at a carefully controlled flow rate.

As a deviation from the LRMS IFB method, instead of adding

$^{37}Cl_4$-2,3,7,8-TCDD to the original sample (in part due to potential interference from PCBs), a recovery internal standard, $^{13}C_{12}$-1,2,3,4-TCDD in tridecane, is added with nitrogen blow/down to bring the extract to a final volume of 10 μL. This recovery standard allows improved accuracy in the calculation of percent recovery of the surrogate/internal standard.

HRMS Analysis

Performance of the HRGC/HRMS measurement, for purposes of the new protocol, would have to meet the following criteria and employ these stated procedures.

(1) Operate the HRMS in EI mode with static resolution of at least 10,000 (10% valley). Document attainment of that resolution at the beginning and end of each 12-hr period of operation (recommended after every run). To monitor and correct for mass drifts, a reference compound (high-boiling PFK is recommended) is introduced via molecular leak. Tune for 10,000 resolution at m/z 254.986 (or other mass close to m/z 259). Calibrate the voltage sweep at least across the range m/z 259 to m/z 334 (see Figure 1 for example) and verify that m/z 330.979 from PFK (or other mass close to 334) is measured within ± 5 ppm (i.e., 1.7 mμ) using m/z 254.986 as a reference, as shown in Figure 2. Documentation of the mass resolution must be provided by recording the peak profile of the PFK reference m/z 318.979 (see Figure 3 for example). This documentation must allow manual verification of the resolution attained. The result of the peak width measurement (performed at 5% of the maximum) must appear on the hard copy, and cannot exceed 31.9 mμ or 100 ppm.

(2) Use a 60-m SP-2330 or 50-m CP-Sil-88 (or equivalent) GC column, and demonstrate resolution of 2,3,7,8-TCDD from the other TCDD isomers in the EPA-supplied check mixture. This check mixture will contain 2,3,7,8-TCDD and closely eluting TCDDs, and also the first- and last-eluting TCDDs. Demonstrate GC resolution at the beginning and end of each 12-hr period of operation.

(3) Calibration for instrument response vs concentration will be performed with EPA-supplied solutions, containing native TCDD at concentrations of 2, 10, 50, and 200 pg/uL. These solutions will provide mass spectrometric calibration for the range 1 ppt to 100 ppt (soil) and 10 ppq to 1 ppt (water). In order that samples from 100 ppt to 1.2 ppb (soil) and 1 to 12 ppt (water) can be analyzed using that mass spectrometric calibration range, the extract is diluted 1:12 with recovery standard/tridacane solution.

(4) Recommended techniques will involve final extract volume of 10 μL, with 2 μL injections onto the GC with splitless injection. As an option, the laboratory may employ 1 or 3 μL injections, but such a technique must be used for all injections.

(5) SIM data must be acquired for the following ions, with total cycle

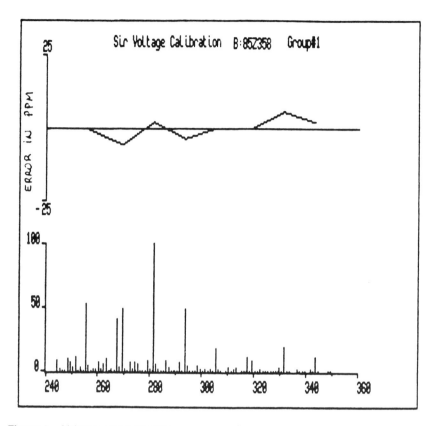

Figure 1. Voltage scan calibration.

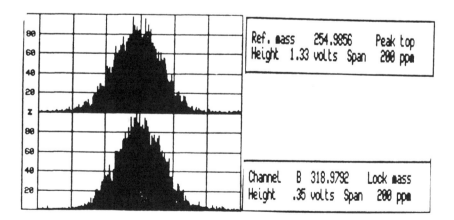

Figure 2. Peak matching m/z 319 using m/z 255 from PFK.

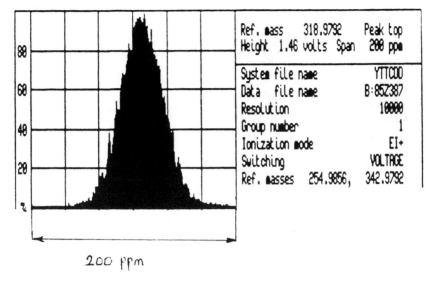

Ref. mass   318.9792   Peak top
Height 1.46 volts Span   200 ppm

System file name   YTTCDD
Data   file name   B:85Z387
Resolution   10000
Group number   1
Ionization mode   EI+
Switching   VOLTAGE
Ref. masses   254.9856,   342.9792

200 ppm

Figure 3. Peak profile display of PFK m/z 319.

time not exceeding 1 sec: 258.930 (TCDD−COCl); 313.984 (FC-43 lock mass) or 318.979 (PFK lock mass); 319.897 and 321.894 (unlabeled TCDD); 331.937 and 333.934 ($^{13}C_{12}$-TCDD).

(6) Quantifications will be performed using isotope-dilution methodology, based upon the $^{13}C_{12}$-2,3,7,8-TCDD which was added to the original sample. Calculation of percent recovery of that isotope diluent ("surrogate/internal standard") is facilitated by addition of a second isotopically labeled TCDD ($^{13}C_{12}$-1,2,3,4-TCDD) to the final extract as a "recovery internal standard."

(7) The following identification criteria must be met.

   a. Retention time for unlabeled 2,3,7,8-TCDD must be within −1 to +3 sec of that for the isotopically labeled surrogate/internal standard. Retention times of other TCDDs must fall within the RT window established by the GC column performance check mixture.

   b. Ion current responses for 258.930, 319.897, and 321.894 must maximize simultaneously (± 1 scan), and all ion current intensities must be at least 2.5 times noise level.

   c. The response ratios for 319.897/321.894 and for 331.937/333.934 must be between 0.67 and 0.90 (0.77 theoretical).

   d. Integrated ion currents for m/z 331.937 and 333.934 must maximize simultaneously (± 1 second).

   e. Recovery of the surrogate/internal standard ($^{13}C_{12}$-2,3,7,8-TCDD) must be between 40 and 120%.

Table 1.   Interferences from PFK.[a]

m/z from PFK	m/z from TCDD	mµ	Required Resolution
258.9805	258.9298	50.7	5,100
258.9586	258.9298	28.8	9,000
319.9845	319.8965	88.0	3,600
321.9869	321.8936	93.3	3,500
331.9844	331.9368	47.6	7,000
333.9891	333.9339	55.2	6,000

[a]Two batches, peak match at 10,000 resolution.

(8) The analytical report form will list sample number; dry weight, or volume if water sample; percent moisture in original soil sample; observed gas chromatographic retention times for unlabeled and labeled 2,3,7,8-TCDD; measured levels (ppt) of TCDDs and detection limits for samples which do not contain 2,3,7,8-TCDD; ion abundance ratios (320/322 and 332/334); percent recovery; signal/noise for 258.930, 321.894, and 333.934; date and time of analysis; and comments.

SOME ANALYTICAL CONSIDERATIONS

While much of this method borrows from proven technology, several analytical considerations and issues will be tested by this method.

(1) Possible streamlining of the extract cleanup, and determination of its performance on TCDD isomers other than 2,3,7,8-TCDD. For example, different types (acidic, neutral, basic; activity levels) of alumina may be tested, along with different eluent mixtures.

(2) Comparison of Carbopack C/Celite 545 to Amoco PX-21 on silica gel or some other support with respect to adsorption/desorption properties of each TCDD isomer.

(3) Performance of PFK and FC-43 calibrants. The former appears to have a significant level of potential interferences close to masses of interest (see Table 1 and Figure 4).

(4) Determintion of m/z 258.930 suitability (vs 256.933); calculations and experiments show (see Table 2) that the former is less subject to interference from PCBs; 258.930 also provides essentially the same relative response as 256.933 (259/257 = 0.975).

(5) $^{37}Cl_4$–2,3,7,8-TCDD is omitted from this HRMS protocol. Interference from PCBs is judged likely, making its value uncertain, based upon prior EPA HRMS experience in ppt-level dioxin analyses.

(6) Benzene is used for Soxhlet extraction instead of toluene, due to ease of obtaining the former in pure form, and ease of evaporation.

(7) Validation of the new method is needed, especially for TCDDs other than 2,3,7,8-TCDD, and for matrices other than relatively simple, clay-type soils.

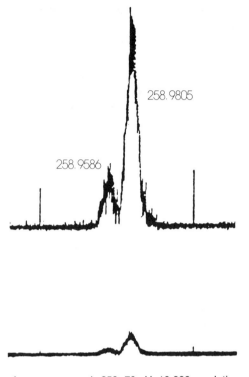

258.9805

258.9586

Figure 4.   PFK interferences near m/z 259; 70 eV; 10,000 resolution.

(8) Mass resolution of 10,000 is specified as a compromise between sensitivity and specificity.

(9) Control limits for 320/322 and 332/334 ratio and percent recovery were chosen based upon EPA experience in such analyses, but may be updated later based upon actual method performance.

(10) The retention time control limits for unlabeled and labeled TCDDs reflect EPA and contract laboratory experience that the labeled TCDD elutes from the GC column 1 second earlier than the corresponding unlabeled TCDD.

(11) The retention time window for all TCDDs will be established by the same GC column performance check mixture as is used to demonstrate adequate resolution of 2,3,7,8-TCDD from the other TCDD isomers which elute close to it on the allowed (polar) types of GC columns.

(12) A fortified field blank pair will be analyzed to track method performance and contamination. One blank is spiked with $^{13}C_{12}$-1,2,3,4-TCDD and the other with $^{13}C_{12}$-2,3,7,8-TCDD.

Table 2. Potential Interferences from PCBs Near m/z 257,259.

$$\text{TCDD (M}^{\cdot\,+} - \text{COCl}^{\cdot})^+ \quad 257 = C_{11}H_4O^{35}Cl_3 = 256.9328$$
$$259 = C_{11}H_4O^{35}Cl_2{}^{37}Cl_1 = 258.9298$$

Assignment of Fragment Ion	Accurate Mass	Required Resolution	Relative Abundance
Tetrachlorobiphenyls (M-Cl)$^+$			
$C_{12}H_6{}^{35}Cl_2{}^{37}Cl_1$	256.9506	14,400	1.00
$C_{12}H_6{}^{35}Cl^{37}Cl_2$	258.9476	14,500	0.33
Pentachlorobiphenyls (M-2C1)$^{\cdot\,+}$,^{13}C			
$^{12}C_{11}{}^{13}C_1H_5{}^{35}Cl_2{}^{37}Cl$	256.9461	19,200	1.00
$^{12}C_{11}{}^{13}C_1H_5{}^{35}Cl_1{}^{37}Cl_2$	258.9432	19,300	0.33
Hexachlorobiphenyls (M-3Cl)$^{+}$ [a]			
$C_{12}H_4{}^{35}Cl^{37}Cl_2$	256.9320	321,100	1.00
$C_{12}H_4{}^{37}Cl_3$	258.9290	323,600	0.11
Heptachlorobiphenyls (M-4Cl)$^{\cdot\,+}$,^{13}C [a]			
$^{12}C_{11}{}^{13}C_1H_3{}^{35}Cl_1{}^{37}Cl_2$	256.9275	48.700	1.00
$^{12}C_{11}{}^{13}C_1H_3{}^{37}Cl_3$	258.9245	48,800	0.11
Octachlorobiphenyls (M-5Cl)$^+$ or			
(M-H, 5Cl)$^+$			
$C_{12}H_2{}^{37}Cl_3$	256.9134	13,200[b]	
$^{12}C_{11}{}^{13}C_1H^{37}Cl_3$	256.9089	10,800[b]	

[a]PCBs most likely to coelute with TCDDs.
[b]No fragment near m/g 259.

SUMMARY

As a part of the National Dioxin Strategy, the U.S. EPA is investigating sites where contamination of the environment by dioxins and related compounds could occur. Currently, three EPA laboratories ("the Troika") are providing analytical support for this effort. In the near future, the EPA plans to obtain analytical efforts through contractual arrangement with private sector laboratories.

This chapter described efforts to incorporate proven analytical principles while providing state-of-the-art methodology suitable for parts-per-trillion level determinations in soil, sediment, and water matrices. Where possible, innovative thinking was used to provide rugged methodology which would afford very high-quality data. Current EPA plans are to use this method in a contractual situation after laboratory evaluation of the method, and testing of analytical considerations such as those discussed in this chapter.

ACKNOWLEDGMENTS

We wish to thank K. S. Kumar for assistance in preparing the first draft of the method contract protocol, and W. F. Beckert for assistance in preparing later drafts and coordinating the single laboratory study of the method described in this chapter.

J. R. Donnelly acknowledges funding through U.S. EPA contracts 68-03-3050 and 68-03-3249 to Lockheed Engineering and Management Services Co., Inc.

Y. Tondeur acknowledges funding through U.S. EPA Cooperative Agreement CR-809706-01 with Environmental Research Center, University of Nevada–Las Vegas.

NOTICE

Although the research described in this article has been funded by the U.S. Environmental Protection Agency, it has not been subjected to Agency policy review and therefore does not necessarily reflect the views of the Agency. Mention of trade names or commercial products does not constitute endorsement or recommendation for use.

REFERENCES

1. U.S. Environmental Protection Agency, Office of Water Regulations and Standards and the Office of Solid Waste and Emergency Response in conjunction with the Dioxin Strategy Task Force. "Dioxin Strategy," internal report dated November 28, 1983, Washington, DC 20460.
2. U.S. Environmental Protection Agency. "National Dioxin Study," 600/3-85/019, Environmental Research Laboratory, Duluth, MN 55840.
3. Harless, R., E. Oswald, M. Wilkinson, A. Dupuy, D. McDaniel, and H. Tai. *Anal. Chem.* 52:1239 (1980).
4. Taylor, M., T. Tiernan, J. Garrett, G. Van Ness, and J. Solch. "Assessment of Incineration Processes as Sources of Supertoxic Chlorinated Hydrocarbons: Concentrations of Polychlorinated Dibenzo-p-dioxins/Dibenzofurans and Possible Precursor Compounds in Incinerator Effluents," in *Chlorinated Dioxins and Dibenzofurans in the Total Environment*, G. Choudhary, L. Keith, and C. Rappe, Eds. (Stoneham, MA: Butterworth Publishers, 1983).
5. Tiernan, T., M. Taylor, J. Garrett, J. Solch, D. Wagel, and G. Van Ness. "Analytical Protocol for Quantitation of 2,3,7,8-Tetrachlorodibenzo-p-dioxin (TCDD) and Total TCDD Present in Soil using High Resolution Gas Chromatography-High Resolution Mass Spectrometry," Wright State University, Brehm Laboratory. Report to S. Billets, U.S. EPA, Environmental Monitoring and Systems Laboratory, Las Vegas, Nevada (January 1984).
6. Tiernan, T., M. Taylor, G. Van Ness, J. Garrett, D. Wagel, J. Solch, and T. Mazer. "Method of Analysis for Tetra-, Penta-, Hexa-, Hepta-, and Octa Chlorinated Dibenzo-p-dioxins and Dibenzofurans by Total Chlorinated Class and for 2,3,7,8-Tetrachlorodibenzo-p-dioxin (TCDD) and 2,3,7,8-Tetrachlorodibenzofuran (TCDF) in Chemical Wastes and Soils; Revised Method 8280," Wright State University, Brehm Laboratory. Report to S. Billets, U.S. Environmental Protec-

tion Agency, Environmental Monitoring and Systems Laboratory, Las Vegas, Nevada (May 1984).

7. Solch, J. G., G. L. Ferguson, T. O. Tiernan, B. F. Van Ness, J. H. Garrett, D. J. Wagel, and M. L. Taylor. "Analytical Methodology for Determination of 2,3,7,8-Tetrachlorodibenzo-p-dioxin in Soils," in *Chlorinated Dioxins and Dibenzofurans in the Total Environment II*, L. Keith, C. Rappe, and G. Choudhary, Eds. (Stoneham, MA: Butterworth Publishers, 1985).

8. Peterson, B. A. "Prospectus on Procedures for Identification and Quantification of Polychlorinated Dibenzo-p-dioxins and Furans using High Resolution Mass Spectrometry," Battelle-Columbus Laboratories report to Fred Haeberer, U.S. EPA, Chemical Analysis Programs, Washington, DC (November 1, 1983).

9. Firestone, D. *J. Agric. Food Chem.*, 25:1274 (1977).

10. Firestone, D., M. Clower, A. P. Borsetti, R. H. Teske, and P. E. Long. *J. Agric. Food Chem.* 27:1171 (1979).

11. Brumley, W. C., J. A. G. Roach, J. A. Sphon, P. A. Dreifuss, D. Andrzejewski, R. A. Niemann, and D. Firestone. *J. Agric. Food Chem.* 29:1040 (1981).

12. O'Keefe, P. W., M. S. Meselson, and R. W. Baughman. *J. Assoc. Off., Anal. Chem.* 61:621 (1978).

13. O'Keefe, P. W., R. M. Smith, D. R. Hilker, K. M. Aldous, and W. Gilday. "A Semiautomated Cleanup Method for Polychlorinated Dibenzo-p-dioxins and Polychlorinated Dibenzofurans in Environmental Samples," in *Chlorinated Dioxins and Dibenzofurans in the Total Environment II*. L. Keith, C. Rappe and G. Choudhary, Eds. (Stoneham, MA: Butterworth Publishers, 1985).

14. Smith, L., and J. Johnson. "Evaluation of Interferences from Seven Series of Polychlorinated Aromatic Compounds in an Analytical Method for Polychlorinated Dibenzofurans and Dibenzo-p-dioxins in Environmental Samples," in *Chlorinated Dioxins and Dibenzofurans in the Total Environment*, G. Choudhary, L. Keith, and C. Rappe, Eds. (Stoneham, MA: Butterworth Publishers, 1983).

15. Petty, J. D., M. Smith, P. A. Berqvist, J. L. Johnson, D. L. Stalling, and C. Rappe. "Composition of Polychlorinated Dibenzofuran and Dibenzo-p-dioxin Residues in Sediments of the Hudson and Housatonic Rivers," in *Chlorinated Dioxin and Dibenzofurans in the Total Environment*, G. Choudhary, L. Keith, and C. Rappe, Eds. (Stoneham, MA: Butterworth Publishers, 1983).

16. Stalling, D., J. Petty, L. Smith, C. Rappe, and H. Buser. "Isolation and Analysis of Polychlorinated Dibenzofurans in Aquatic Samples," in *Chlorinated Dioxins and Related Compounds*, O. Hutzinger, Ed. (Elmsford, NY: Pergamon Press, Inc. 1982).

17. Kuehl, D., R. Dougherty, Y. Tondeur, D. Stalling, L. Smith, and C. Rappe. "Negative Chemical Ionization Studies of Polychlorinated Dibenzo-p-dioxins, Dibenzofurans, and Naphthalenes in Environmental Samples," in *Environmental Health Chemistry*, J. McKinney, Ed. (Stoneham, MA: Butterworth Publishers, 1981).

18. Smith, L. *Anal. Chem.* 53:2151 (1981).
19. Stalling, D., J. Petty, and L. Smith. *J. Chrom. Sci.* 19:18 (1981).
20. Smith, L., D. Stalling, and J. Johnson. *Anal. Chem.* 56:1830 (1984).
21. Rappe, C., S. Marklund, M. Nygren, and A. Gara. "Parmeters for Identification and Confirmation in Trace Analyses of Polychlorinated Dibenzo-p-dioxins and Dibenzofurans," in *Chlorinated Dioxins and Dibenzofurans in the Total Environment*, G. Choudhary, L. Keith, and C. Rappe, Eds. (Stoneham, MA: Butterworth Publishers, 1983).
22. Rappe, C., P. A. Bergqvist, and S. Marklund. "Analysis of Polychlorinated Dibenzofurans and Dioxins in Ecological Samples," in *Chlorinated Dioxins and Dibenzofurans in the Total Environment II*, L. Keith, C. Rappe, and G. Choudhary, Eds. (Stoneham, MA: Butterworth Publishers, 1985).
23. Lamparski, L. L., T. J. Nestrick, and R. H. Stehl. *Anal. Chem.* 51:1453 (1979).
24. Nestrick, T. J., L. L. Lamparski, and R. H. Stehl. *Anal. Chem.* 51:2273 (1979).
25. Lamparski, L. L., and T. J. Nestrick. *Anal. Chem.* 52:2045 (1980).
26. Nestrick. T. J., and L. L. Lamparski. *Anal. Chem.* 54:2292 (1982); Crummett, W. B. *Ann. N.Y. Acad. Sci.* 320:43 (1979).
27. Mitchum, R. K., W. A. Korfmacher, G. F. Moler, and D. L. Stalling. *Anal. Chem.* 54:719 (1982).
28. Mitchum, R. K., G. F. Moler, and W. A. Korfmacher. *Anal. Chem.* 52:2278 (1980).
29. Korfmacher, W. A., and R. K. Mitchum. *J. High Res. Chrom. Chrom. Commun.* 4:294 (1981).
30. Mitchum, R. K., W. A. Korfmacher, and G. F. Moler. "Validation Study for the Gas Chromatography/Atmospheric Pressure Ionization/ Mass Spectrometry Method for Isomer-Specific Determination of 2,3,7,8-Tetrachlorodibenzo-p-dioxin," in *Chlorinated Dioxins and Dibenzofurans in the Total Environment*, G. Choudhary, L. Keith, and C. Rappe, Eds. (Stoneham, MA: Butterworth Publishers, 1983).
31. Korfmacher, W., L. Rushing, D. Nestorick, H. Thompson, R. Mitchum, and J. Kominsky. *J. High Res. Chrom. Chrom. Commun.* 8:12 (1985).
32. Ryan, J., C. Pilon, B. S. Conacher, and D. Firestone. *J. Assoc. Off. Anal. Chem.* 66:700 (1983).
33. Wong, A. S., M. W. Orbanosky, P. A. Taylor, C. P. McMillin, R. W. Noble, D. Wood, J. E. Longbottom, D. L. Foerst, and R. J. Wesselman. "Determination of 2,3,7,8-Tetrachlorodibenzo-p-dioxin in Industrial and Municipal Wastewaters, Method 613: Development and Detection Limits," in *Chlorinated Dioxins and Dibenzofurans in the Total Environment*, G. Choudhary, L. Keith, and C. Rappe, Eds. (Stoneham, MA: Butterworth Publishers, 1983).
34. McMillin, C. R., F. D. Hileman, D. E. Kirk, T. Mazer, B. J. Warner, R. J. Wesselman, and J. E. Longbottom. "Determination of 2,3,7,8-Tetrachlorodibenzo-p-dioxin in Industrial and Municipal Wastewaters, Method 613: Performance Evaluation and Preliminary Method Study

Results," in *Chlorinated Dioxins and Dibenzofurans in the Total Environment*, G. Choudhary, L. Keith, and C. Rappe, Eds. (Stoneham, MA: Butterworth Publishers, 1983).

35. Donnelly, J. R., A. E. Dupuy, Jr., D. D. McDaniel, R. L. Harless, and R. K. Robeson. "Quality Assurance Samples for the Dioxin Monitoring Program—An Interlaboratory Study," in *Chlorinated Dioxins and Dibenzofurans in the Total Environment II*, L. Keith, C. Rappe, and G. Choudhary, Eds. (Stoneham, MA: Butterworth Publishers, 1985).

36. Shadoff, L., W. Blaser, C. Kocher, and H. Fravel. *Anal. Chem.* 50:1586 (1978).

37. Phillipson, D., and B. Puma. *Anal. Chem.* 52:2328 (1980).

38. Deleon, I., E. Overton, and J. Laseter. "The Role of Analytical Chemistry in a Toxic Substance Spill into the Aquatic Environment," in *Analytical Techniques in Environmental Chemistry 2*, J. Albaiges, Ed. (Elmsford, NY: Pergamon Press, 1981).

39. National Research Council Canada, Associate Committee on Scientific Criteria for Environmental Quality. *Polychlorinated Dibenzo-p-dioxins: Limitations to the Current Analytical Techniques*, NRCC No. 18576, ISSN 0316–0114, Publications NRCC/CNRC, Ottawa, Canada, 1981.

40. U.S. Environmental Protection Agency, "Determination of 2,3,7,8-TCDD in Soil and Sediment," February 1983, Revised May 1983. Invitation for Bid Contract IFB-WA-82-A002.

41. Kleopfer, R. D., W. W. Bunn, K. T. Yue, and D. J. Harris. "Occurrence of Tetrachlorodibenzo-p-dioxin in Environmental Samples from Southwest Missouri," in *Chlorinated Dioxins and Dibenzofurans in the Total Environment*, G. Choudhary, L. Keith, and C. Rappe, Eds. (Stoneham, MA: Butterworth Publishers, 1983).

42. Kleopfer, R. D., K. T. Yue, and W. Bunn. "Determination of 2,3,7,8-Tetrachlorodibenzo-p-dioxin in Soil," in *Chlorinated Dioxins and Dibenzofurans in the Total Environment II*, L. Keith, C. Rappe, and G. Choudhary. Eds. (Stoneham, MA: Butterworth Publishers, 1985).

43. Kleopfer, R. D., and C. J. Kirchmer. "Quality Assurance Plan for 2,3,7,8-Tetrachlorodibenzo-p-dioxin Monitoring in Missouri," in *Chlorinated Dioxins and Dibenzofurans in the Total Environment II*, L. Keith, C. Rappe, and G. Choudhary, Eds. (Stoneham, MA: Butterworth Publishers, 1985).

44. "Test Methods for Evaluating Solid Waste—Physical/Chemical Methods," Government Printing Office, 1982, stock #055-002-81001-2.

45. Federal Register 40CFR261: 1978 (January 14, 1985).

CHAPTER 27

Evaluation of RCRA Method 8280 for Analysis of Dioxins and Dibenzofurans

J. R. Donnelly, T. L. Vonnahme, C. M. Hedin, and W. J. Niederhut

INTRODUCTION

Largely as a result of finding trace levels of 2,3,7,8-tetrachlorodibenzo-p-dioxin (2,3,7,8-TCDD) as a contaminant in commercial preparations of chlorophenol-based herbicides, the U.S. EPA initiated (in 1973) monitoring efforts for 2,3,7,8-TCDD in environmental samples. Later findings of contamination by 2,3,7,8-TCDD in soil samples from Niagara Falls, New York, and various sites in Missouri led to extensive sampling and analysis efforts. It is now known that many, if not all, of the 75 possible chlorinated dioxins and 135 structurally related chlorinated dibenzofurans possess relatively high toxicities to man and certain animal species. Most available acute and chronic toxicological data for chlorinated dioxins and dibenzofurans are based upon the 2,3,7,8-TCDD isomer. In certain animal species (notably, the guinea pig), extraordinarily low doses may be lethal. A wide range of systemic effects, including hepatic disorders, carcinoma, and teratogenicity, have been observed in animal species, although the major documented effect upon humans has been chloracne.

In 1983[1] the EPA proposed a ruling affecting disposal of hazardous wastes containing tetra-, penta-, and hexachlorinated dioxins and dibenzofurans. The EPA has determined[2] that enhanced toxicities are likely to be observed with samples containing tetra-, penta-, and hexachlorinated dioxins and dibenzofurans. These wastes would be managed under the Resource Conservation and Recovery Act (RCRA) and would be analyzed for the target chlorinated dioxins and dibenzofurans using an analytical method which was included as an appendix to the proposed rule.

In order to manage these wastes effectively, it is necessary to obtain data regarding the performance of the method included in the Federal Register on hazardous waste samples. This study was intended to determine and elaborate any changes needed for satisfactory method performance and to provide data showing satisfactory method performance after revisions had been made. Hazardous waste sample types/matrices which have been subjected to analysis under that method to date include soils, carbonaceous material (fly ash), and a chlorophenol production still bottom.

The RCRA Method 8280 consists of four major sections: 1. extraction of the analytes from the environmental sample 2. "open" column chromatographic cleanup with alumina, using methylene chloride/hexane eluent 3. HPLC cleanup 4. analysis by high-resolution column gas chromatography/ low-resolution mass spectrometry (HRGC/LRMS). In order to test the method efficiently and to develop appropriate modifications with minimal lost effort, each section of the methodology was tested separately. Initial tests were performed on a simple (pottery clay soil) sample matrix and upon standard solutions. Necessarily, the first step to be elaborated was the measurement technique. Both GC/MS and GC/EC (electron capture detection) were tested, using guidelines from the published RCRA method. Since the analytes could not be measured at the published m/z values by GC/MS, these values were corrected immediately. Corrected values are presented in Appendix A, Table 3A. Development of the other sections of the method is also presented in this chapter. Thus, method performance data are described separately in this chapter for each section of the RCRA method.

Performance data on the unrevised Method 8280, with the exception that the corrected m/z values were monitored for the analytes, were calculated using the recoveries obtained on the extraction procedures, and on the alumina cleanup procedure. These data are presented in Table 1.

EXTRACTION METHOD DEVELOPMENT AND REVISIONS

As a model for determining extraction efficiency on a simple soil sample, pottery clay was selected. A 50-ng spike of 2,3,7,8-TCDD was added, and extraction was performed as specified in the unrevised Method 8280. This experiment was performed in triplicate, and a similar experiment was performed in triplicate on wet pottery clay (i.e., 5.0 g clay, and 5.0 g water). Results are presented in Table 1. A modified extraction solvent system and procedure (similar to that specified in the published RCRA Method 8280 for chemical wastes) was developed, drawing also upon other EPA experience in dioxin analysis for guidance. This modified procedure adds methanol and sodium sulfate to the petroleum ether solvent specified in the

Table 1.   Performance of Unrevised[a] Method 8280.

Analyte	Matrix	Percent Recovery Extraction[b]	Percent Recovery Cleanup	Total Calculated Percent Recovery
2,3,7,8-TCDD	Clay	44	42	18
	Fly ash	56	42	23
1,2,3,4-TCDD	Clay	44	15	7
	Fly ash	56	15	8
1,2,3,4,7-PeCDD	Clay	44	0	0
	Fly ash	56	0	0
1,2,3,4,7,8- HxCDD	Clay	44	11	5
	Fly ash	56	11	6
2,3,7,8-TCDF	Clay	44	48	21
	Fly ash	56	48	27

[a]Corrected m/z values were used.
[b]Extraction recoveries estimated for other analytes are based upon actual recovery of $^{13}C_{12}$-2,3,7,8-TCDD.

RCRA method as published, and uses a Kuderna-Danish concentration technique. The revised procedure is presented in Appendix A of this chapter. Results are shown in Table 2; performance is improved on both dry and wet samples. As shown in the pottery clay and Missouri soil sections of this report, the revised extraction procedure resulted in satisfactory overall method performance (e.g., dioxin surrogate spike recovery values).

The Method 8280 specified a benzene/Soxhlet extraction method (SW-846 Method 3540)[3-5] for carbonaceous materials. These results are presented in the fly ash sample analysis section of this chapter.

DEVELOPMENT OF ALUMINA CLEANUP PROCEDURE

Experimental conditions for elution of the analytes from the "open" alumina liquid chromatographic column were determined to be crucial to method performance. Using conditions as specified in the Federal Register,[1] the elution profile shown in Table 3 was obtained. This profile was unsatisfactory, since many analytes of interest eluted in the 20% CH_2Cl_2/hexane fraction, which was to be discarded. Furthermore, over half of the 2,3,7,8-TCDD failed to elute, until an added step was performed, with 60% CH_2Cl_2 in hexane. Use of 10% and 60% CH_2Cl_2 in hexane eluents caused all analytes to appear in the 60% fraction, as shown in Table 4. Elution volume of 15 mL for each fraction was used consistently. Results obtained on oily samples indicated 8% and 60% CH_2Cl_2/hexane gave equal or improved separation.

Readily available Woelm Super 1 neutral alumina was also tested. This material provided elution patterns as shown in Tables 5 and 6. Since this material was more convenient to use, avoided exposure of the sample to basic materials, and provided fully satisfactory elution patterns, it was

Table 2.   Soil Extraction Method Development: RCRA Method as Published Vs Revised Extraction Procedure.

Sample ID	Weight Clay (g)	Weight Water (g)	Method	Percent Recovery of Spiked 2,3,7,8-TCDD[a]
RCRA-D1	10.0	---	8280[b]	37.5
RCRA-D2	10.0	---	8280[b]	55.0
RCRA-D3	10.0	---	8280[b]	38.7
RCRA-W1	5.0	5.0	8280[b]	ND[d]
RCRA-W2	5.0	5.0	8280[b]	ND
RCRA-W3	5.0	5.0	8280[b]	ND
D-1	10.0	---	Rev. 8280[c]	96
D-2	10.0	---	Rev. 8280[c]	72
W-1	5.0	5.0	Rev. 8280[c]	68
W-2	5.0	5.0	Rev. 8280[c]	96

[a]TCDD spiked at 50 ng/10 g sample; DL 75 pg.
[b]RCRA Method 8280 was used without any drying of sample, but with corrected m/z values being monitored.
[c]Revised Method 8280 as presented in this report, Appendix A.
[d]ND, none detected.

Table 3.   Elution of Analytes from Basic Alumina (Per RCRA Method 8280).[a]

Analyte	% CH$_2$Cl$_2$ in Hexane			
	3%	20%	50%	60%
2,3,7,8-TCDF		52	48	
1,2,3,4-TCDD		85	15	
2,3,7,8-TCDD			42	58
1,2,3,4,7-PeCDD		100		
1,2,3,4,7,8-HxCDD		89	11	
1,2,3,4,6,7,8-HpCDD		25	75	
OCDD			100	

[a]Amount of analyte in eluate expressed in percentage of total recovered from column.

Table 4.   Elution of Analytes from Basic Alumina (10%, 60% CH$_2$Cl$_2$/Hexane).[a]

Analyte	% CH$_2$Cl$_2$ in Hexane	
	10%	60%
2,3,7,8-TCDF		100
1,2,3,4-TCDD		100
2,3,7,8-TCDD		100
1,2,3,4,7-PeCDD		100
1,2,3,4,7,8-HxCDD		100
1,2,3,4,6,7,8-HpCDD		100
OCDD		100

[a]Amount of analyte in eluate expressed in percentage of total recovered from column.

specified in the revised method. To demonstrate the effect of deactivating this Woelm Super 1 alumina by addition of water (via wet solvents, wet sample, or intentional deactivation), elution profiles were determined on Woelm Super 1 with 1% and with 4% added water. These profiles are presented in Tables 7 and 8, respectively.

Note that the bulk of interferences elute in the 5%, 10%, and/or 20% methylene chloride/hexane fraction in all cases. These elution experiments were performed with the analytes spiked at 50 ng each, plus 2.5 µg each of Aroclor 1016®, Aroclor 1254®, Aroclor 1260®, pentachlorophenol, dibenzofuran, Silvex®, 2,4,5-T, 2,4,5-trichlorophenol, 2,4,6-trichlorophenol, and

Table 5.   Elution of Analytes from Neutral Alumina (Woelm Super 1) (5%, 20%, 60% CH$_2$Cl$_2$/Hexane).[a]

Analyte	% CH$_2$Cl$_2$ in Hexane		
	5%	20%	60%
2,3,7,8-TCDF			100
1,2,3,4-TCDD		95	5
2,3,7,8-TCDD			100
1,2,3,4,7-PeCDD		94	6
1,2,3,4,7,8-HxCDD		80	20
1,2,3,4,6,7,8-HpCDD		29	71
OCDD		4	96

[a]Amount of analyte in eluate expressed in percentage of total recovered from column.

Table 6.   Elution of Analytes from Neutral Alumina (Woelm Super 1) (10%, 60% CH$_2$Cl$_2$/Hexane).[a]

	% CH$_2$Cl$_2$ in Hexane	
Analyte	10%	60%
2,3,7,8-TCDF		100
1,2,3,4-TCDD		100
2,3,7,8-TCDD		100
1,2,3,4,7-PeCDD		100
1,2,3,4,7,8-HxCDD		100
1,2,3,4,6,7,8-HpCDD		100
OCDD		100

[a]Amount of analyte in eluate expressed in percentage of total recovered from column.

Table 7.   Elution of Analytes from Woelm Super 1 Neutral Alumina Deactivated with 1% Water.[a]

	% CH$_2$Cl$_2$ in Hexane	
Analyte	10%	60%
2,3,7,8-TCDF	2	98
1,2,3,4-TCDD	59.5	40.5
2,3,7,8-TCDD		100
1,2,3,4,7-PeCDD	64	36
1,2,3,4,7,8-HxCDD	74.5	25.5
1,2,3,4,6,7,8-HpCDD	46	54

[a]Amount of analyte in eluate expressed in percentage of total recovered from column.

Table 8.   Elution of Analytes from Woelm Super 1 Neutral Alumina Deactivated with 4% Water.[a]

	% CH$_2$Cl$_2$ in Hexane	
Analyte	10%	60%
2,3,7,8-TCDF	96	4
1,2,3,4-TCDD	97.5	2.5
2,3,7,8-TCDD	85	15
1,2,3,4,7-PeCDD	97.5	2.5
1,2,3,4,7,8-HxCDD	100	
1,2,3,4,6,7,8-HpCDD	99	1

[a]Amount of analyte in eluate expressed in percentage of total recovered from column.

p,p'-DDE in the spiking solution. Note that while the elution percentages in Tables 3–8 represent percentage of total recovered from the column, all columns yielded essentially quantitative recoveries.

ELABORATION OF HPLC CLEANUP PROCEDURE

Since the published method provided little experimental detail regarding HPLC cleanup, appropriate HPLC reverse phase operating parameters were developed for extract cleanup. Two problems to be solved were: (1) to select a suitable mobile solvent that would be safe to work with as well as

Table 9. **Typical HPLC Retention Times of 2,3,7,8-TCDD Interferences.**[a]

Compound	Amount (ng)	Retention Time (min)
Pentachlorophenol	50	3.81
2,4,6-Trichlorophenol	50	5.80
2,4,5-Trichlorophenol	100	6.30
Silvex	250	6.77
Dibenzofuran	50	8.21
p,p'-DDE	100	9.22
Aroclor 1016	100	6.91 – 9.52
Aroclor 1254	100	6.91 – 10.30
Aroclor 1260	200	6.75 – 15.70
2,3,7,8-TCDD	25	16.6

[a]two 4.6-mm × 25-cm octadecylsilyl columns, 5-μm particle size. Solvent at ambient temperature; column and detector maintained at 30°C. Flow 1 mL/min methanol (172 atm). UV detector at 235 nm, 0.02 AUFS. 100-μL sample loop.

provide adequate separation of compounds of interest, and (2) to find a fast but effective procedure to concentrate the effluent fraction from the HPLC.

Methanol performed satisfactorily as the mobile phase, and following the operating parameters (described later) produced a retention time of 16.6 min for 2,3,7,8-TCDD. Interferences commonly associated with 2,3,7,8-TCDD were also separated (Table 9 and Figures 1–3). Evaporating the HPLC effluent fraction using a 2-ball micro-Synder column and a 90°C water bath provided an efficient extract concentration procedure.

As an additional cleanup procedure, the charcoal adsorption methods of Mitchum and colleagues[6] and Smith et al.[7] were applied to more complex samples, particularly the still bottom discussed later in this chapter. For the HPLC procedure, a disposable 4.6-mm-i.d. by 7-cm-long column was employed with 5% (w/w) Amoco® PX-21 on 20-μ particle size Spherisorb-silica.

ANALYSIS OF POTTERY CLAY SAMPLES

While Method 8280 specified drying the samples, a simple change in the extraction solvent system resulted in the ability to handle wet or dry samples of soil with good results.

Contamination-free pottery clay was spiked with suitable interferences and with analytes of interest (CDD's and TCDF) to determine method performance on a clean soil matrix. Interferences such as Silvex®, DDE, and Aroclor 1260® were used at concentrations from 10 to 200 times that of each target analyte. Separation of interferences in clay samples was discussed in the previous sections. Recoveries of analytes from pottery clay samples using the revised Method 8280 are summarized in Table 10, using GC/EC measurement.

Performance Evaluation samples from the EPA Contract Laboratory

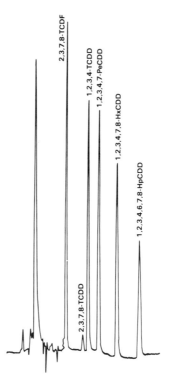

Figure 1. Analyte standard mixture by HPLC/UV.

Program (CLP) were analyzed by the revised RCRA Method 8280, with results shown in Table 11. These pottery clay samples had been spiked with nominal levels of 1 ppb TCDD and with 10–50 ppb levels of chlordane, DDT, DDE, and Aroclor 1260. Acceptance windows had been determined for the analysis of these PE samples under a CLP protocol.[8] A mean concentration of 0.84 ppb with standard deviation 18% was calculated for 21 sample analyses (interlaboratory); a 90% window (i.e., 90% of available data excluding outliers will fall within that window) was calculated at 0.53–1.2 ppb. The data gathered by revised Method 8280 using HRGC/LRMS fall within this acceptance window.

ANALYSIS OF NATIVE MISSOURI SOIL SAMPLES

Native Missouri soil samples, environmentally contaminated with 2,3,7,8-TCDD and interferences such as PCB's, were obtained from the EPA Contract Laboratory Program. These samples had been blended and aliquots of 20 g each prepared and bottled. The mean TCDD concentration of this soil reference material has been determined by interlaboratory study

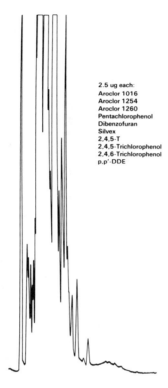

2.5 ug each:
Aroclor 1016
Aroclor 1254
Aroclor 1260
Pentachlorophenol
Dibenzofuran
Silvex
2,4,5-T
2,4,5-Trichlorophenol
2,4,6-Trichlorophenol
p.p'-DDE

Figure 2. Interference mixture eluted by 10% CH_2Cl_2/hexane from alumina column; HPLC/UV detection.

under the EPA Contract Laboratory Program to be 383 ppb. Upon analysis of this material, the revised RCRA Method 8280 provided the results shown in Table 12. Aliquots of this soil were spiked with other dioxin and dibenzofuran analytes, and percent recoveries were determined (Table 13).

ANALYSIS OF FLY ASH SAMPLES

A portion of RCRA municipal incinerator fly ash material was obtained in order to provide an environmental carbonaceous type matrix for testing. This type of sample historically has required toluene or benzene extraction with Soxhlet apparatus to remove adsorbed dioxins and dibenzofurans. These analytes are often found in such samples at high-ppt to low-ppb levels.[9-17] Concentrations of environmentally incorporated dioxins and dibenzofurans using revised Method 8280 (toluene/Soxhlet extraction) for analysis are shown in Tables 14 and 15.

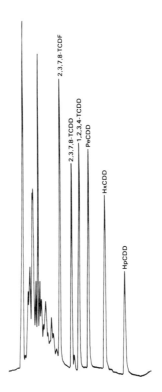

Figure 3.   60% CH$_2$Cl$_2$ eluate from alumina column; HPLC/UV detection.

ANALYSIS OF CHLOROPHENOL STILL BOTTOM SAMPLES

A still bottom sample from a chlorophenol production process was obtained from the EPA. Historically, these samples have been difficult to analyze for dioxins and dibenzofurans. Due to presence of massive amounts of analytical interferences, poor reproducibility, low recovery of spiked PCDD's and PCDF's, and relatively poor detection limits may be obtained.[18-21] However, this challenging sample represents a sample type likely to be encountered as a RCRA waste. The sample dissolved in toluene readily. Full-scan GC/MS using a 30-m DB-5 fused-silica capillary column was performed for matrix characterization. Results are shown in Figure 4. Characterization of an extract subjected to revised Method 8280 cleanup, excluding carbon adsorption HPLC, is presented in Figure 5.

The HPLC cleanup, using a silica Spherisorb®/PX-21 charcoal column, has been described.[6] A similar open column procedure has also been reported.[7]

A list of analytes quantitated is presented in Table 16. Recovery of additional analytes spiked into still bottom samples is shown in Table 17.

Table 10. Spike Percentage Recoveries from Pottery Clay Samples.[a]

	Samples								n = 7
Analyte	A	B	C	D	E	F	G	X̄	% Std. Dev.
1,3,6,8-TCDD	66.1	48.6	63.8	48.1	53.3	54.0	40.1	53.4	9.11
1,3,7,9-TCDD	100.7	94.7	97.0	92.6	96.8	101.4	86.1	95.6	5.21
1,2,7,8-TCDF	86.8	90.7	94.7	94.7	99.0	96.8	95.1	94.0	4.04
1,3,7,8-TCDD	85.6	80.2	90.8	89.1	96.2	94.9	93.2	90.0	5.62
1,2,7,8-TCDD	87.8	80.8	91.9	91.8	99.1	98.7	91.8	91.7	6.29
1,2,8,9-TCDD	86.5	80.5	91.5	92.3	100.0	98.7	92.7	91.7	6.72
1,2,3,7,8-PeCDD	87.6	80.6	89.4	95.9	103.4	100.7	95.6	93.3	7.93
1,2,3,4,7-PeCDD	88.2	81.4	88.2	97.8	102.6	105.6	96.6	94.3	8.70
1,2,3,7,8-PeCDF	84.6	81.4	89.6	93.7	92.7	93.0	93.0	89.7	4.86
1,2,3,4,7,8-HxCDF	81.7	80.0	85.4	94.2	89.9	92.9	90.5	87.8	5.51
1,2,3,4,7,8-HxCDD	82.0	79.6	83.1	97.0	93.9	93.7	93.8	89.0	7.13
1,2,3,4,6,7,8-HpCDD	72.1	77.7	81.4	105.8	96.3	84.0	99.4	88.1	12.48
OCDF	74.2	79.3	73.0	102.2	83.2	85.3	83.3	82.9	9.70
$^{13}C_{12}$-2,3,7,8-TCDD	82.6	81.8	90.8	90.5	96.8	98.0	91.0	90.2	6.24

[a]Spikes at 50 ng each component (50 pg/µL in extract). Analyses using Revised Method 8280 procedures, GC/EC.

Table 11. Analysis of Performance Evaluation Samples Obtained from Contract Laboratory Program, Using Revised RCRA Method 8280.

Sample ID	Nominal Conc. 2,3,7,8-TCDD	Exptl. Conc. Detd. (ppb)	% Recovery of $^{13}C_{12}$-2,3,7,8-TCDD
PE-1	1 ppb	1.00	62
PE-2	1 ppb	0.81	72
PE-3	1 ppb	1.13	90
PE-4	1 ppb	0.94	74

Table 12. Analysis of Missouri Soil Samples.

Analyte	(ppb)	% RSD
2,3,7,8-TCDD	372.2	2.77
2,3,7,8-TCDD	351.1	
2,3,7,8-TCDD	353.4	
2,3,7,8-TCDD	353.2	
Mean	357.5	

Table 13. Spike Recoveries from Native Missouri Soil.

Analyte	Average % Recovery	% RSD	n
1,3,6,8-TCDD	64.0	13.9	4
1,3,7,9-TCDD	61.3	10.8	4
1,3,7,8-TCDD	79.3	9.89	4
1,2,7,8-TCDF	71.1	8.40	4
1,2,3,4-TCDD	60.3	9.03	4
1,2,7,8-TCDD	75.3	12.2	4
1,2,8,9-TCDD	60.3	9.28	4
1,2,3,7,8-PeCDF	64.4	6.77	4
1,2,3,4,7-PeCDD	62.2	8.92	4
1,2,3,7,8-PeCDD	68.4	10.8	4
1,2,3,4,7,8-HxCDF	68.5	10.1	4
1,2,3,4,7,8-HxCDD	65.0	12.9	4
1,2,3,4,6,7,8-HpCDD	82.0	—	2
2,3,7,8-TCDF	81.5	—	2
OCDD	76.0	—	2
$^{13}C_{12}$-2,3,7,8-TCDD	78.8	9.14	4

Table 14. Analytes Found in Fly Ash Sample.

Analyte	Conc.[a] (ppb)	% RSD[b]
Total TCDD	200	--
Total PeCDD	150	4.2
Total HxCDD	262	4.7
Total TCDF	87.0	5.0
Total PeCDF	118	6.8
Total HxCDF	72.2	11.2

[a]Concentrations estimated using average response factor of available standards within a congener group, for others of that group.
[b]n = 4, except TCDD's (n = 2).

Table 15.   Analytes Found in Fly Ash Sample.

Analyte[a]	x,ppb	% # RSD[b]
1,3,6,8-TCDD	37.4	—
1,3,7,9-TCDD	99.3	—
1,3,7,8-TCDD	15.1	5.28
1,2,7,8-TCDF	6.7	11.1
1,2,3,4-TCDD	33.1	4.67
1,2,7,8-TCDD	2.10	19.7
1,2,3,4,7-PeCDD	25.0	10.3
1,2,3,4,7,8-HxCDF	31.7	8.71
$^{13}C_{12}$-2,3,7,8-TCDD[c]	55.6%	6.74

[a]Isomer identifications based on RT vs available standards. Other unidentified isomers might coelute, increasing measured responses.
[b]% RSD based on n = 4, except 1,2,7,8-TCDD (n = 3); — means n = 2.
[c]Recovery of isotopically labeled surrogate spiked into samples.

RATIONALE FOR SOME PROPOSED METHOD MODIFICATIONS

The changes needed for satisfactory method performance are shown in the revised RCRA Method 8280 (Appendix A). Each change is indicated by placing the original text as published in the Federal Register into square brackets, with the revision following, italicized.

4.3.2   Fused-silica columns are recommended as being chromatographically equivalent to glass, and easier to use. As shown in Table 1A, four columns were evaluated using the Contract Laboratory Program (CLP) GC column performance check mixture, one TCDF, and a commercially available mixture containing 1,2,3,4-TCDD, 1,2,3,4,7-PeCDD, 1,2,3,4,7,8-HxCDD, and 1,2,3,4,6,7,8-HpCDD. The GC column performance check mixture contains 1,4,7,8-TCDD, 2,3,7,8-TCDD, 1,2,3,4-TCDD, 1,2,3,7-TCDD, 1,2,3,8-TCDD, 1,2,7,8-TCDD, and 1,2,6,7-TCDD. Both CP-Sil-88 and SP-2250 gave reasonable separation of 2,3,7,8-TCDD from the other TCDD's. Separations of 2,3,7,8-TCDD from the closest eluting isomer using the CLP GC Column Performance Check Mixture were as follows: 21% valley on 30-m SP-2250; 16% valley on 50-m CP-Sil-88; 30% valley on 30-m DB-225; 64% valley on 30-m DB-5 column. Separation on all four columns was adequate to observe whether any close-eluting isomers (to 2,3,7,8-TCDD) were present in a given sample.

The temperature program included in Table 3A (developed by Wright State University) is used with a 60-m DB-5 column. The $^{37}Cl_4$-TCDD ion (327.885) may be omitted if this standard is not used. It may be desirable to add this standard, or $^{13}C_{12}$-1,2,3,4-TCDD, to the final extract for improved percentage recovery calculation accuracy.

4.3.3   As shown in the recommended list of ions to monitor, the MS system should be capable of monitoring ten (not just four) ions simultaneously.

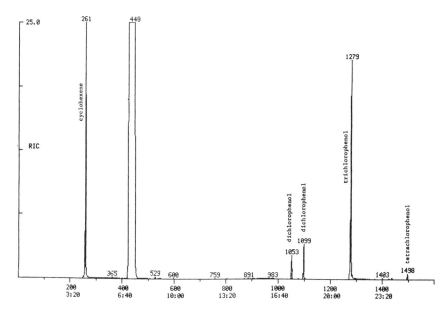

Figure 4. Still bottom sample in toluene, without cleanup. Tentative component identifications; GC/MS full-scan mode, library search. 30-m DB-5 column.

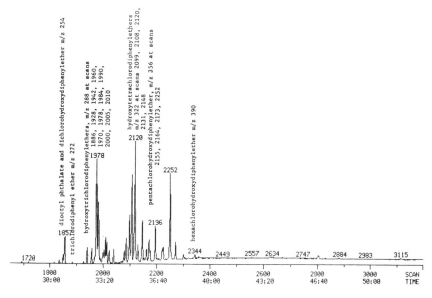

Figure 4, contd. Still bottom sample in toluene, without cleanup. Tentative component identifications; GC/MS full-scan mode, library search. 30-m DB-5 column.

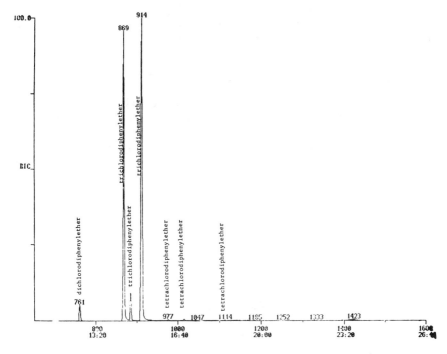

Figure 5.   Still bottom sample after alumina cleanup, GC/MS full-scan.

Table 16.   Analytes Found in Still Bottom Sample.

Analyte	Conc. (ppm)
2,3,7,8-TCDD + Coeluter	1.02
Total TCDD	2.65
Total TCDF	200
Total PeCDF	234
Total HxCDF	20.4

Table 17.   Spike Recoveries from Still Bottom Sample.

Analyte	% Recovery		
1,3,6,8-TCDD	76.4	83.9	
1,3,7,9-TCDD	81.8	102	
1,3,7,8-TCDD	84.0	96.9	
1,2,7,8-TCDF	104	104	
1,2,3,4-TCDD	92.5	95.1	
1,2,7,8-TCDD	81.7	95.2	
1,2,8,9-TCDD	80.3	82.1	
1,2,3,4,7-PeCDD	82.7	81.0	
1,2,3,7,8-PeCDD	85.5	82.6	
1,2,3,4,7,8-HxCDF	95.9	90.0	
1,2,3,4,7,8-HxCDD	96.8	81.5	
$^{13}C_{12}$-2,3,7,8-TCDD	68.5	65.2	75.4

Sensitivity of quadrupole low-resolution systems is consistent with the revised specification of 5:1 signal to noise on 0.15 ng of native TCDD.

4.3.4   Directly interfacing the column into the MS source is preferable.

4.3.6   This section reflects the specifications of the HPLC system used in developing the revised Method 8280 Section 10.

4.4   Equipment added reflects modifications in extraction/cleanup methodology for soil samples.

5.1   The 2% is a typographical error; 20% is correct.

5.4 $^{13}C_{12}$-TCDD is more readily available and cheaper by about a factor of 1/2 compared to $^{37}Cl_4$-TCDD. M/z 332 and 334 are suitable for $^{13}C_{12}$-TCDD monitoring. Additionally, m/z 328 is usually more subject to analytical interference.

5.5   It is suggested that Drierite® or equivalent ($CaSO_4$) be used as a drying agent in the desiccator.

6.1   Elution patterns/retention times on both HPLC and "open" (alumina) columns should be verified. It is a worthwhile quality assurance measure to check the 10% methylene chloride/hexane fraction to ensure no analytes are lost in that fraction.

6.3   Use of 0.15 ng native TCDD, monitored at m/z 320, reflects reasonable sensitivity requirements for quadrupole GC/MS systems. It is considered worthwhile to check GC column performance daily, as most columns which have been used for these types of analytes are subject to rapid performance degradation.

7.2   Incorporation of field blanks into the quality assurance "package" is a standard procedure.

8.2   The sentence on sample preservation is removed, since it is generally considered unnecessary on these types of samples, as the analytes are exceptionally stable.

9.1   The revised extraction procedure eliminates the need for drying the sample. Samples may be dried if analytical results based on dry sample weight are desired.

9.2.2   Comparison data presented in this chapter show significantly improved recovery of isotopically labeled TCDD using the revised soil sample extraction procedure on a pottery clay soil-type sample set. The published extraction method proved adequate for the fly ash sample type.

10.1−10.15   This revised procedure for HPLC provides useful details lacking in the published RCRA 8280 method.

11.3   For good quality assurance, monitoring the ions shown in Table 3A (Appendix A) is important.

11.3.4   This requirement has been discussed.[22,23]

11.7   Another technique which has been shown (under EPA Special Analytical Services contract laboratory efforts) to provide answers when interferences are present is HRGC/MS/MS, so this technique could be included.

CONCLUSIONS

In summary, major types of revisions which were made to the published method include:

1. Correct m/z values were substituted for those in the published method so that the mass spectrometer could detect analytes and standards introduced via the interfaced gas chromatograph. 2. The extract cleanup with an alumina column was revised so that all desired analytes eluted in a single fraction, with the bulk of the analytical interferences removed. 3. The method was elaborated so that wet samples could be accommodated. 4. HPLC procedures that could be reproducibly and effectively performed were developed.

The extraction revision which allows accommodation of wet samples also improves recovery of spiked analytes in dry soil samples.

The RCRA method, with revisions discussed above, was subjected to performance tests which included (1) analysis of reference materials containing 2,3,7,8-TCDD and interferences; (2) precision and accuracy determinations on samples having known composition through spiking the sample matrix at this laboratory. This revised method is included as Appendix A. Each change made to the published method from the Federal Register is indicated by placing the original text in square brackets and placing the revision in italics. Rationale for each change is provided in the preceding section of this chapter.

Performance of the revised method was investigated by precision and accuracy determinations (standard deviation of results), by recovery of spiked analytes and isotopically labeled standards, and by using two independent teams of analysts. Effects of experimental parameters, such as GC column type (coating) and alumina activation level, are noted in this chapter.

After incorporating necessary revisions, satisfactory method performance has been demonstrated on soil-type samples. Preliminary method performance data on fly ash and still bottom samples are acceptable.

ACKNOWLEDGMENTS

This method evaluation was funded by the U.S. Environmental Protection Agency, EMSL-LV, under contracts 68-03-3050 and 68-03-3249. Support and guidance by R. K. Robeson and S. Billets, Task Monitors, U.S. EPA, are gratefully acknowledged. Support and guidance by S. J. Simon and F. L. Shore, Managers, Lockheed-EMSCO, are gratefully acknowledged. Thanks are also extended to D. R. Youngman, W. J. Verret, and M. C. Doubrava for expert technical assistance.

Stimulating technical discussions, and technical management by R. K. Mitchum, U.S. EPA, are gratefully acknowledged. R. K. Mitchum also provided materials for the HPLC/PX-21/Spherisorb column.

NOTICE

REFERENCES

1. Federal Register 40 CFR 65: 14514, April 4, 1983.
2. Federal Register 40 CFR 261: 1978, January 14, 1985.
3. "Test Methods for Evaluating Solid Waste – Physical/Chemical Methods," G.P.O. Stock No. 055-022-81001-2.
4. Gray, A., V. Dipinto, and I. Solomon. *J. Org. Chem.* 41:2428 (1976).
5. Gray, A., S. Cepa, I. Solomon, and O. Aniline. *J. Org. Chem.* 41:2435 (1976).
6. Korfmacher, W., L. Rushing, D. Nestorick, H. Thompson, R. Mitchum, and J. Kominsky. *J. High Res. Chrom. Chrom. Commun.* 8:12 (1985).
7. Smith, L., D. Stalling, and J. Johnson. *Anal. Chem.* 56:1830 (1984).
8. U.S. Environmental Protection Agency, Procurement Section A.

"Chemical Analytical Services for 2,3,7,8-Tetrachlorodibenzo-p-dioxin." Invitation for Bid Solicitation Number WA 84-A002, December 9, 1983.

9. Olie, K., P. Vermeulen, and O. Hutzinger. *Chemosphere* 6:455 (1977).
10. Eiceman, G., R. Clement, and F. Karasek. *Anal. Chem.* 51:2343 (1979).
11. Eiceman, G., A. Viau, and F. Karasek. *Anal. Chem.* 52:1492 (1980).
12. Lustenhouwer, J., K. Olie, and O. Hutzinger. *Chemosphere* 9:501 (1980).
13. Kooke, R., J. Lustenhouwer, K. Olie, and O. Hutzinger. *Anal. Chem.* 53: 461 (1981).
14. Eiceman, G., R. Clement, and F. Karasek. *Anal. Chem.* 53:955 (1981).
15. Nestrick, T., L. Lamparski, W. Crummett, and L. Shadoff. *Anal. Chem.* 54: 824 (1982).
16. Karasek, F., R. Clement, and A. Viau. *J. Chromatogr.* 239:173 (1982).
17. Buser, H., H. Bosshardt, and C. Rappe. *Chemosphere* 7:165 (1978).
18. Jensen, S., and L. Renberg. *Ambio* 1:62 (1972).
19. Tulp, M., and L. Hutzinger. *Biomed. Mass Spectrom.* 5:224 (1978).
20. Shadoff, L., W. Blaser, C. Kocher, and H. Fravel. *Anal. Chem.* 50:1586 (1978).
21. Phillipson, D., and B. Puma. *Anal. Chem.* 52:2328 (1980).
22. Kormacher, W., R. Mitchum, F. Hileman, and T. Mazer. *Chemosphere* 12:1243 (1983).
23. Harless, R. "Analysis for Tetrachlorodibenzofurans," paper presented at Thirty-First Annual Conference of Mass Spectrometry and Allied Topics, American Society for Mass Spectrometry, Boston, Massachusetts, 1983, p. 221.

The changes needed for satisfactory method performance are indicated by placing the original text as published in the Federal Register into square brackets, with the revision following, italicized.

APPENDIX A: RCRA METHOD 8280 WITH PROPOSED REVISIONS BASED ON SINGLE LABORATORY TESTING. METHOD OF ANALYSIS FOR CHLORINATED DIBENZO-P-DIOXINS AND DIBENZOFURANS[1,2,3,4] 40 CFR 65: 14525-7 APRIL 4, 1983

Method 8280

1. Scope and Application

1.1   This method covers the determination of chlorinated dibenzo-p-dioxins and chlorinated dibenzofurans in chemical wastes including still bottoms, filter aids, sludges, spent carbon, and reactor residues, and in soils.

1.2   The sensitivity of this method is dependent upon the level of interferences.

1.3   This method is recommended for use only by analysts experienced with residue analysis and skilled in mass spectral analytical techniques.

1.4   Because of the extreme toxicity of these compounds, the analyst must take necessary precautions to prevent exposure to himself, or to others, of materials known or believed to contain CDD's or CDF's.

Infectious waste incinerators are probably not satisfactory devices for disposal of materials highly contaminated with CDD's or CDF's. A laboratory planning to use these compounds should prepare a disposal plan to be

---

[1]This method is appropriate for the analysis of tetra-, penta-, and hexachlorinated dibenzo-p-dioxins and dibenzofurans.

[2]Analytical protocol for determination of TCDD's in phenolic chemical wastes and soil samples obtained from the proximity of chemical dumps. T. O. Tiernan and M. Taylor. Brehm Laboratory. Wright State University. Dayton, OH 45435.

[3]Analytical protocol for determination of chlorinated dibenzo-p-dioxins and chlorinated dibenzofurans in river water. T. O. Tiernan and M. Taylor. Brehm Laboratory. Wright State University. Dayton, OH 45435.

[4]In general, the techniques that should be used to handle these materials are those which are followed for radioactive or infectious laboratory materials. Assistance in evaluating laboratory practices may be obtained from industrial hygienists and persons specializing in safe laboratory practice.

reviewed and approved by EPA's Dioxin Task Force (Contact Conrad Kleveno, WH-548A, U.S. EPA, 401 M Street, S. W., Washington, D.C. 20450).

2. Summary of the Method

2.1 This method is an analytical extraction cleanup procedure and capillary column gas chromatography/low-resolution mass spectrometry method, using capillary column GC/MS conditions and internal standard techniques which allow for the measurement of PCDD's and PCDF's in the extract.

2.2 If interferences are encountered, the method provides selected general purpose cleanup procedures to aid the analyst in their elimination.

3. Interferences

3.1 Solvents, reagents, glassware, and other sample processing hardware may yield discrete artifacts and/or elevated baselines causing misinterpretation of gas chromatograms. All of these materials must be demonstrated to be free from interferences under the conditions of the analysis by running method blanks. Specific selection of reagents and purification of solvents by distillation in all-glass systems may be required.

3.2 Interferences co-extracted from the samples will vary considerably from source to source, depending upon the diversity of the industry being sampled. PCDD is often associated with other interfering chorinated compounds such as PCB's which may be at concentrations several orders of magnitude higher than that of PCDD. While general cleanup techniques are provided as part of this method, unique samples may require additional cleanup approaches to achieve the sensitivity stated in Table 1.

3.3 The other isomers of tetrachlorodibenzo-p-dioxin may interfere with the measurement of 2,3,7,8-TCDD. Capillary column gas chromatography is required to resolve those isomers that yield virtually identical mass fragmentation patterns.

4. Apparatus and Materials

4.1 Sampling equipment for discrete or composite sampling.

4.1.1 Grab sample bottle—amber glass, 1-L or 1-qt volume. French or Boston Round design is recommended. The container must be washed and solvent rinsed before use to minimize interferences.

4.1.2 Bottle caps—threaded to screw on to the sample bottles. Caps must be lined with Teflon®. Solvent washed foil, used with the shiny side towards the sample, may be substituted for the Teflon if sample is not corrosive.

4.1.3   Compositing equipment — automatic or manual compositing system. No tygon or rubber tubing may be used, and the system must incorporate glass sample containers for the collection of a minimum of 250 mL. Sample containers must be kept refrigerated after sampling.

4.2   Water bath — heated, with concentric ring cover, capable of tempera-ture control ($\pm$ 2°C). The bath should be used in a hood.

4.3   Gas chromatograph/mass spectrometer data system.

4.3.1   Gas chromatograph: An analytical system with a temperature-pro-grammable gas chromatograph and all required accessories including syringes, analytical columns, and gases.

4.3.2   [Column: SP-2250 coated on a 30-m long × 0.25-mm I. D. glass column (Supelco No. 2-3714 or equivalent).] [Glass] capillary column con-ditions: Helium carrier gas at 30 cm/sec linear velocity run splitless. [Column temperature is 210°C.]

Fused silica capillary columns are recommended as chromatographically equivalent to glass and easier to use. Useful columns include the following: (a) 50-m CP-Sil-88 programmed 60°–190° at 20°/min, then 190°–240° at 5°/min; (b) 30-m DB-5 programmed 70°–225° at 20°/min; (c) 30 m SP-2250 programmed 70°–320° at 10°/min. Column/conditions (a) provide good separation of 2,3,7,8-TCDD from the other TCDD's at the expense of longer retention times for higher homologs. Column/conditions (b) and (c) provide some separation of 2,3,7,8-TCDD. Resolution of 2,3,7,8-TCDD from the other TCDD's is better on column (b), but column (c) is more rugged, and may provide better separation of certain classes of interferences from the analytes. An example temperature program is shown in Table 3 along with the m/z to monitor during each part of that program, in order to detect tetra- through octachlorodioxins and dibenzofurans.

4.3.3   Mass spectrometer: Capable of scanning from [35 to 450] *55 to 550* amu every 1 sec or less, utilizing 70 volts (nominal) electron energy in the electron impact ionization mode [and producing a mass spectrum which meets all the criteria in Table 2 when 50 ng of decafluorotriphenylphosphine (DFTPP) is injected through the GC inlet.] The system must also be capable of selected ion monitoring (SIM) for at least [4] *10* ions simultaneously, with a cycle time of 1 sec or less. Minimum integration time for SIM is 100 ms. Selected ion monitoring is verified by injecting [0.015 ng of TCDD Cl^{37}]

0.15 ng of native TCDD to give a minimum signal to noise ratio of 5 to 1 at mass 320.

4.3.4   GC/MS interface; Any GC-to-MS interface that gives acceptable calibration points at 50 ng per injection for each compound of interest and achieves acceptable tuning performance criteria (see Sections 6.1–6.3) may be used. GC-to-MS interfaces constructed of all glass or glass-lined materials are recommended. Glass can be deactivated by silanizing with dichlorodimethylsilane. The interface must be capable of transporting at least 10 ng of the components of interest from the GC to the MS. *Fitting the GC column directly into the MS source is recommended.*

4.3.5   Data system: A computer system must be interfaced to the mass spectrometer. The system must allow the continuous acquisition and storage on machine-readable media of all mass spectra obtained throughout the duration of the chromatographic program. The computer must have software that can search any GC/MS data file for ions of a specific mass and that can plot such ion abundances versus time or scan number. This type of plot is defined as an Extracted Ion Current Profile (EICP). Software must also be able to integrate the abundance, in any EICP, between specified time or scan number limits.

4.3.6   High-Performance Liquid Chromatography: Varian 5000 HPLC with Valco C6U loop valve injector, or equivalent. Equipped with a 5-μm particle size, 4.6-mm ID, 4.0-cm long Altex Ultrasphere ODS precolumn coupled with two 5-μm particle size 4.6-mm ID, 25-cm long Altex Ultrasphere ODS columns in series. Varian Model 50 UV variable-wavelength detector, or equivalent.

4.4   Apparatus Pipettes-Disposable, Pasteur, 150-mm long × 5-mm ID (Fisher Scientific Co., No. 13-678-6A or equivalent).

4.5   Flint glass bottle (Teflon®-lined screw cap).

4.6   Reacti-vial (silanized) (Pierce Chemical Co.).

4.7   500-mL Erlenmeyer flask (American Scientific Products cat #F4295-500FO) fitted with Teflon® stoppers (ASP #S9058-8 or equivalent).

4.8   Wrist Action Shaker (VWR #57040-049 or eq.).

4.9   125-mL Separatory Funnels (Fisher #10-437-5b or eq.).

4.10 500-mL Kuderna-Danish fitted with a 10-m concentrator tube and 3-ball Synder column (ACE Glass #6707-02, 6707-12, 6575-02 or eq.).

4.11 Teflon® boiling chips (Berghof America #15021-450 or eq.).

4.12 300-mm × 10.5-mm glass chromatographic column fitted with Teflon® stopcocks (ASP #C4669-1 or eq.).

4.13 15-mL conical concentrator tubes (Kontes #K-288250 or eq.).

4.14 Adaptors for concentrator tubes (14/20 to 19/22)(Ace Glass #9092-20 or eq.).

4.15 2-Ball micro-Snyder columns (Ace Glass #6709-24 or eq.).

4.16 Nitrogen evaporator (n-Evap #1156 or eq.).

4.17 Microflex conical vials (Kontes K-749000 or eq.).

4.18 Filter paper (Whatman #54 or eq.).

5. Reagents

5.1 Potassium hydroxide-(ACS), [2%] *20%* in distilled water.

5.2 Sulfuric acid-(ACS), concentrated.

5.3 Methylene chloride, hexane, benzene, petroleum ether, methanol, tetradecane, isooctane, toluene – pesticide quality or equivalent.

5.4 Stock standards in a glovebox, prepare stock standard solutions of TCDD and [Cl]$^{13}C_{12}$-TCDD (molecular weight [328] *332*). The stock solutions are stored in a [*refrigerator*] and checked frequently for signs of degradation or evaporation, especially just prior to the preparation of working standards.

5.5 Alumina-[basic], *neutral, Super 1*, Woelm; 80/200 mesh. [Before use activate overnight at 600°C, cool to] *Store at* room temperature in a desiccator *with CaSO$_4$ drying agent. Oven drying at 600°C overnight is acceptable, but alumina so processed should be checked for contamination by solvent rinsing and GC/EC analysis.*

5.6 Prepurified nitrogen gas.

6.0   Calibration

6.1   Before using any cleanup procedure, the analyst must process a series of calibration standards through the procedure to validate elution patterns and the absence of interferences from reagents. *Both open column and HPLC column performance must be checked.* Routinely check the 10% CH_2CL_2/hexane eluate of environmental extracts from the alumina column for presence of analytes.

6.2   Prepare GC/MS calibration standards for the internal standard technique that will allow for measurement of relative response factors of at least [three] *one* TCDD/$^{13}C_{12}$-TCDD and TCDF/[^{37}Cl]$^{13}C_{12}$-TCDF ratios.[5] The [^{37}Cl] $^{13}C_{12}$-TCDD/F concentration in the standard should be fixed and selected to yield a reproducible response at the most sensitive setting of the mass spectrometer.

6.3   Assemble the necessary GC/MS apparatus and establish operating parameters equivalent to those indicated in Section 11.1 of this method. [Calibrate the GC/MS system according to Eichelberger et al. (1975) by the use of decafluorotriphenyl phosphine (DFTPP).] By injecting calibration standards, establish the response factors for CDD's *vs* $^{13}C_{12}$-TCDD and for CDF's *vs* $^{13}C_{12}$-TCDF. [The] *An adequate instrument* detection limit [provided in Table 1] should be verified by injecting [0.015 ng of ^{37}Cl-TCDD] *0.15 ng of* $^{13}C_{12}$-TCDD which should give a minimum signal to noise ratio of 5 to 1 at mass [328] *320. GC column performance should be checked for resolution and peak shape daily using a mixed standard such as the GC column performance check mixture used in the EPA Contract Laboratory Program.*

7.   Quality control

7.1   Before processing any samples, the analyst should demonstrate through the analysis of a distilled water method blank that all glassware and reagents are interference-free. Each time a set of samples is extracted, or there is a change in reagents, a method blank should be processed as a safeguard against laboratory contamination.

7.2   Standard quality assurance practices must be used with this method.

---

5 - [^{37}Cl]-$^{13}C_{12}$-labeled TCDD and TCDF are available from K.O.R. Isotopes, Cambridge, MA, and *from Cambridge Isotope Laboratory, Incorporated, Woburn, MA.* Proper standardization requires the use of a specific labeled isomer for each congener to be determined. [However, the only labeled isomers readily available are ^{37}Cl-2,3,7,8-TCDD and ^{37}Cl-2,3,7,8-TCDF. This method, uses these isomers as surrogates for the CDD's and CDF's.] When labeled CDD's and CDF's of each homolog are available, their use will be required.

Field replicates must be collected to validate the precision of the sampling technique. Laboratory replicates must be analyzed to [validate] *determine* the precision of the analysis. Fortified samples must be analyzed to establish the accuracy of the analysis. *Field blanks must be collected to verify that sample collection processes are free from cross-contamination.*

8. Sample Collection, Preservation, and Handling

8.1   Grab and composite samples must be collected in glass containers. Conventional sampling practices should be followed, except that the bottle must not be prewashed with sample before collection. Composite samples should be collected in glass containers in accordance with the requirements of the RCRA program. Sampling equipment must be free of tygon and other potential sources of contamination.

8.2   The samples must be iced or refrigerated from the time of collection until extraction. Chemical preservatives should not be used in the field unless more than 24 hr will elapse before delivery to the laboratory. [If an aqueous sample is taken and the sample will not be extracted within 48 hr of collection, the sample should be adjusted to a pH range of 6.0–8.0 with sodium hydroxide or sulfuric acid.]

8.3   All samples must be extracted within 7 days and completely analyzed within 30 days of collection.

9. Extraction and Cleanup Procedures

9.1   Use an aliquot of 1 to 10-g sample of the chemical waste or soil to be analyzed. [Soils should be dried using a stream of prepurified nitrogen and pulverized in a ball-mill or similar device.] Transfer the sample to a tared 125 mL flint glass bottle (Teflon-lined screw cap) and determine the weight of the sample. Add an appropriate quantity of $^{13}C_{12}$-labeled 2,3,7,8-TCDD (adjust the quantity according to the required minimum detectable concentration), which is employed as an *surrogate/* internal standard.

9.2   Extraction

9.2.1   [Extract chemical waste samples by adding 10mL methanol, 40 mL petroleum ether, 50 mL doubly distilled water, and then shaking the mixture for 2 min.] Tars should be completely dissolved in any of the recommended neat solvents. Activated carbon samples must be extracted with benzene using Method 3540 in SW-846 (Test Methods for Evaluating Solid Waste — Physical/Chemical Methods, available from G. P. O. Stock #055-002-81001-2). Quantitatively transfer the organic extract or dissolved sample to a clean 125 mL separatory funnel (Teflon® stopper cap), add 50 mL doubly distilled water and shake for 2 min. Discard the aqueous layer and proceed with Step 9.3. *After Step 9.6, evaporate the benzene with a rotary evaporator (35°C water bath) and add 2 ml hexane to dissolve the residue. Rinse the flask with 2 × 1 mL hexane, and add to the above.*

[9.2.2   Extract soil samples by adding 40 mL of petroleum ether to the sample and then shake for 20 min. Quantitatively transfer the organic extract to a clean 250 mL flint glass bottle (Teflon®-lined screw cap), add 50 mL doubly distilled water, and shake for 2 min. Discard the aqueous layer and proceed with Step 9.3.]

[9.3   Wash the organic layer with 50 mL of 20% aqueous potassium hydroxide by shaking for 10 min and then remove and discard the aqueous layer.]

[9.4   Wash the organic layer with 50 mL of doubly distilled water by shaking for 2 min and discard the aqueous layer.]

[9.5   Cautiously add 50 mL concentrated sulfuric acid and shake for 10 min. Allow the mixture to stand until layers separate (approximately 10 min.), and remove and discard the acid layer. Repeat acid washing until no color is visible in the acid layer.]

[9.6   Add 50 mL of doubly distilled water to the organic extract and shake for 2 min. Remove and discard the aqueous layer and dry the organic layer by adding 10 g of anhydrous sodium sulfate.]

[9.7   Concentrate the extract to incipient dryness by heating in a 55°C water bath and simultaneously flowing a stream of prepurified nitrogen over the extract . . . Quantitatively transfer the residue to an alumina microcolumn fabricated as follows:]

[9.7.1   Cut off the top section of a 10 mL disposable Pyrex pipette at the 4.0 mL mark and insert a plug of silanized glass wool into the tip of the lower portion of the pipette.]

[9.7.2   Add 2.8 g of Woelm basic alumina (previously activated at 600°C overnight and then cooled to room temperature in a desiccator just prior to use).]

[9.8   Elute the microcolumn with 10 m of 3% methylene chloride-in-hexane followed by 15 mL of 20% methylene chloride-in-hexane and discard these effluents. Elute the column with 15 mL of 50% methylene chloride-in-hexane and concentrate this effluent (55°C water bath, stream of prepuri-fied nitrogen) to about 0.3–0.5 mL.]

[9.9   Quantitatively transfer the residue (using methylene chloride to rinse the container) to a silanized Reacti-Vial (Pierce Chemical Company). Evap-orate, using a stream of prepurified nitrogen, almost to dryness, rinse the walls of the vessel with approximately 0.5 mL methylene chloride, evapo-rate just to dryness, and tightly cap the vial. Store the vial at 5°C until analysis, at which time the sample is reconstituted by the addition of tride-cane.]

9.2.2   Extract soil samples by placing 10 g of sample and 10 g of anhydrous sodium sulfate in a 500-mL Erlenmeyer flask fitted with a Teflon® stopper. Add 50 ng of $^{13}C_{12}2,3,7,8$-TCDD to sample. Add 70 mL of petroleum ether and 30 mL of methanol, in that order, to the Erlenmeyer flask. Shake on a wrist-action shaker for 2 hr. You should be able to observe the solid portion turning over. If a smaller soil aliquot is used, scale down the amount of methanol proportionally.

9.2.2.1   Into a 500-mL Kuderna-Danish (KD) concentrator fitted with a 10 mL concentrator tube pour off the extract by filtering it through a glass funnel fitted with filter paper (Whatman #54 or equivalent) and filled with sodium sulfate. Add 50 mL of petroleum ether to the Erlenmeyer flask, restopper the flask and swirl the sample gently, remove the stopper carefully and decant the solvent as above. Repeat twice, for a total extract of 150 mL. Wash the sodium sulfate with two 10 mL portions of petroleum ether.

9.2.2.2   To the KD add a Teflon® boiling chip and a three-ball Snyder column. Concentrate in a 70°C steam bath to an apparent volume of 10 mL. Remove the apparatus from the steam bath and allow it to cool for 5 min.

9.2.2.3   To the KD add 50 mL of hexane and a new boiling chip. Concen-trate in a 100°C steam bath to an apparent volume of 10 mL. Remove the apparatus from the steam bath and allow to cool for 5 min.

9.2.2.4   Remove and invert the Synder column and rinse it down into the KD with two 1-mL portions of hexane. Pour off the KD and concentrator tube into a 125-mL separatory funnel. Rinse and pour off the KD with two additional 5-mL portions of hexane.

9.3   Partition the solvent against 40 mL of 20% w/w potassium hydroxide. Shake for 2 min. Remove and discard the aqueous layer (bottom).

9.4   Partition the solvent against 40-mL of distilled water. Shake for 2 min. Remove and discard aqueous layer (bottom).

9.5   Partition the solvent against 40 mL of concentrated sulfuric acid. Shake for 2 min. Remove and discard the aqueous layer (bottom).

9.6   Partition the solvent against 40 mL of distilled water. Shake for 2 min. Remove and discard aqueous layer (bottom).

9.7   Pack a gravity column (glass 300 mm × 10.5 mm), fitted with a Teflon® stopcock, in the following manner:

Insert a glass-wool plug into the bottom of the column. Add a 4-g layer of sodium sulfate. Add a 3.6-g layer of Woelm Super 1 neutral alumina. Tap the top of the column gently. Woelm Super 1 neutral alumina need not be activated or cleaned prior to use but should be stored and sealed in a desiccator. Add a 4-g layer of sodium sulfate. Elute with 20-mL of hexane and close the stopcock just prior to the exposure of the sodium sulfate layer to air. Discard the effluent. Check the column for channeling. If channeling is present discard the column. Do not tap on a wetted column.

9.8   Elute with the solvent directly from the separatory funnel; run the solvent extract through a sodium sulfate funnel and then onto the column. It is important that the solvent be dry by this step. Discard the effluent.

9.8.1   Elute with 10 mL of 10% methylene chloride by volume in hexane. As a quality assurance measure, check that no CDD's or CDF's eluted in this fraction.

9.8.2   Elute the CDD's and CDF's from the column with 15 mL of 60% methylene chloride by volume in hexane and collect this fraction in a conical shaped (15 mL) concentrator tube.

9.9   Carefully concentrate this fraction in an analytical evaporator (utilizing prepurified nitrogen gas) to almost dryness. Utilizing sand as a heat carrier in the evaporator will eliminate the possibility of water contamination. The walls of the concentrator tube should be rinsed down at least 3 times during this concentration step. When the tube is almost dry add 100 µL of isooctane and concentrate to approximately 50 µL. Quantitatively transfer the extract to a 1-mL conical gc vial and bring to a volume of 100 µL with rinsings from the concentrator tube. Store the gc vial fitted with a Teflon®-coated septum in a refrigerator until analysis.

9.10   Approximately 1 hr before GC/MS (HRGC/LRMS) analysis, dilute the residue in the micro-reaction vessel with an appropriate quantity of *isooctane or* tridecane. Gently swirl the tridecane on the lower portion of the vessel to ensure dissolution of the CDD's and CDF's. Analyze a sample by GC/EC to provide insight into the complexity of the problem and to determine the manner in which the mass spectrometer should be used. Inject an appropriate aliquot of the sample into the GC MS instrument, using a syringe.

9.11   If, upon preliminary GC/MS analysis, the sample appears to contain interfering substances which obscure the analyses for CDD's and CDF's, high performance liquid chromatographic (HPLC) cleanup of the extract is accomplished, prior to further GC/MS analysis.

10.0   HPLC Cleanup Procedure

[10.1   Place approximately 2mL of hexane in a 50-mL flint glass sample bottle fitted with a Teflon®-lined cap.]

10.1   Operating Parameters: Column temperature: 30°C, Column pressure: 172 atm., Flow rate: 1.0 mL/min., UV detector: 235 nm, Detector sensitivity: 0.02 AUFS, Isocratic operating mode, Injection via 100 µL sample loop.

10.2   Elution Profile for 2,3,7,8-TCDD: An authentic standard of 2,3,7,8-TCDD demonstrated a retention time of 16.0 min under the above conditions. Typically, collect the entire column effluent eluting between 15.7 and 22.0 min following introduction of the sample extract onto the HPLC columns into a 12.0-mL concentrator tube. The retention time of 2,3,7,8-TCDD must be determined by each analyst for the particular system in use by injecting 2,3,7,8-TCDD standards.

10.3   Solvent Exchange, Instrument Analysis, and Sample Concentration: After the methylene chloride/hexane effluent is taken to dryness, add 0.5 mL of methanol to the concentrator tube.

10.4   Using a gentle stream of prepurified nitrogen and a 55°C sand bath evaporate the effluent to 40 µL.

10.5   Using a 100-µL syringe, draw the residue solution carefully into the syringe until an air bubble appears in the barrel. Place a second 60-µL. aliquot of methanol into the concentrator tube, rotate quickly for rinsing and draw this into the syringe.

10.6   Inject the total volume into the HPLC.

[10.2] *10.7*   At the appropriate retention time, position sample bottle to collect the required fraction. *Add 5 mL hexane to the sample bottle.*

[10.3]*10.8*   Add 2 mL of 5% (w/v) sodium carbonate to the sample fraction collected and shake for 1 min.

[10.4]*10.9*   Quantitatively remove the hexane layer (top layer) and transfer to a micro-reaction vessel. *Add an additional 5 mL of hexane and shake for 1 min. Transfer to a 15-mL concentrator tube.*

10.10   A Teflon® boiling chip is added and the concentrator tube is fitted with a 2-ball Snyder column.

[10.5   Concentrate the fraction to dryness and retain for further analysis.]

10.11   The sample is concentrated to near dryness in a 90°C water bath.

10.12   After cooling for 10 min, the sample is evaporated just to dryness using a gentle stream of prepurified nitrogen and a 55°C sand bath.

10.13   Transfer the residue to a 1.0-mL microflex vial with a small amount of toluene. Be very careful to transfer the entire rinse solution.

10.14   Again concentrate the residue to 100 μL using a gentle stream of prepurified nitrogen and a 55°C sand bath.

10.15   Store the microflex vial under refrigeration until just prior to GC/ MS analyses.

11.   GC/MS Analysis

11.1   *Use a GC column and operating conditions described in Section 4.3.2.* [The following column conditions are recommended: Glass capillary column conditions: SP-2250 coated on a 30-m long × 0.25-mm i. d. glass column (Supelco No. 2-3714, or equivalent) with helium carrier gas at 30 cm/sec linear velocity, run splitless. Column temperature is 210°C. Under these conditions the retention time for TCDD's is about 9.5 min.] Calibrate the system daily with a minimum of [*one*] injection of standard mixtures.

11.2   Calculate response factors for standards relative to $[^{37}Cl]^{13}C_{12}$-TCDD/F (see Section 12).

11.3   Analyze samples with selected ion monitoring [of at least two ions from Table 3.] *using ions listed in Table 3A*. Proof of the presence of CDD or CDF exists if the following conditions are met:

11.3.1   The retention time of the peak in the sample must match that in the standard, within the performance specifications of the analytical system.

11.3.2   The ratio of ions must agree within 10% with that of the standard.

11.3.3   The retention time of the peak maximum for the ions of interest must exactly match [that of the peak].

11.3.4   For confirmation of a CDF, molecular ion response (m/z 374, 376, for example) for diphenyl ethers (see Table 3A) must be absent at that retention time.

11.4   Quantitate the CDD and CDF peaks from the response relative to the ^{37}Cl-TCDD/F internal standards. Recovery of the internal standard should be greater than 50%. *An additional, different recovery internal standard (e.g. $^{13}C_{12}$-1,2,3,4-TCDD) should be added to the final extract to improve the accuracy of percent recovery calculations.*

11.5   If a response is obtained for the appropriate set of ions, but is outside the expected ratio, a co-eluting impurity may be suspected. In this case, [another set of ions characteristic of the CDD/CDF molecules should be analyzed. For TCDD a good choice of ions is m/e 257 and m/e 259. For TCDF a good choice of ions is m/e 241 and 243. These ions are useful in characterizing the molecular structure of TCDD or TCDF.] *further sample cleanup should be performed and the sample re-analyzed. As an example using ions in Table 3A*, for analysis of TCDD good analytical technique would require using [all four ions] m/e [257] *259*, 320, 322, 328, *and/or 332* to verify detection and signal to noise ratio of [5] *2.5* to 1. Suspected impurities such as DDE, DDD, or PCB residues can be confirmed by checking for their major fragments. These materials can be removed by the cleanup columns. Failure to meet criteria *even after addition cleanup and analysis* should be explained in the report [or the sample reanalyzed].

11.6   If broad background interference restricts the sensitivity of the GC/MS analysis, the analyst [should] *must* employ additional cleanup procedures such as carbon adsorption HPLC and reanalyze by GC/MS.

11.7   In those circumstances where these procedures do not yield a definitive conclusion, the use of high resolution mass spectrometry *or HRGC/MS/MS* is suggested.

12.   Calculations

12.1   Determine the concentration of individual compounds according to the formula:

$$\text{Concentration, } \mu g/gm = \frac{A \times A_s}{G \times A_{is} \times R_f}$$

Where:
A   = μg of *surrogate*/internal standard added to the sample.[6]
G   = gm of sample extracted
A_s = area of characteristic ion of the compound being quantified
A_{is} = area of characteristic ion of the *surrogate*/internal standard
R_f = response factor

[6] The proper amount of standard to be used is determined from the calibration curve (See Section 6.0).

Response factors are calculated using data obtained from the analysis of standards according to the formula:

$$Rf = \frac{A_s \times C_{is}}{A_{is} \times C_s}$$

Where:
C_{is} = concentration of the *surrogate*/internal standard
C_s = concentration of the standard compound

12.2   Report results in micrograms per gram with[out] correction for recovery data. When duplicate and spiked samples are analyzed, all data obtained should be reported.

[12.3   Accuracy and Precision. No data are available at this time.]

Table 1.   Gas Chromatography of TCDD.

	Column	Retention time (min)	Detection limit (mg/kg)[a]
Glass Capillary		9.5	0.003

[a]Detection limit for liquid samples is 0.003 μg/L. This is calculated from the minimum detectable GC response being equal to five times the GC background noise assuming a 1 mL effective final volume of the ato both electron capture and GC/MS detection. For further details, see 1-L sample extract, and a GC injection of 5 microliters. Detection levels 44 *FR* 69526 (December 3, 1979).

Replace this table with Table 1A.

Table 1A.   Representative Gas Chromatographic Retention Times[a] of Analytes.

Analyte	50-m CP-Sil-88	30-m DB-5	30-m SP-2250	30-m DB-225
2,3,7,8-TCDD	23.6	22.8	26.7	37.3
1,2,3,4-TCDD	24.1	22.7	26.5	37.6
1,2,3,4,7-PeCDD	30.0	25.2	28.1	NM[b]
1,2,3,4,7,8-HxCDD	39.5	28.0	30.6	NM
1,2,3,4,6,7,8-HpCDD	57.0	34.2	33.7	NM
2,3,7,8-TCDF	25.2	22.5	26.7	42.5

Temperature Programs
CP-Sil-88   60°–190° at 20°/min; 190°–240° at 5°/min
30-m DB-5   10 min at 170°, 8°/min to 280°
SP-2250   70°–320° at 10°/min
DB-225   70°–230° at 10°/min
Column Manufacturers
CP-Sil-88   Chrompack Incorporated, Bridgewater, New Jersey
DB-5, DB-225   J&W Scientific, Incorporated, Rancho Cordova, California
SP 2250   Supelco, Incorporated, Bellefonte, Pennsylvania

[a]Retention time in minutes, using temperature programs shown above.
[b]NM, not measured.

Table 2. DFTPP Key Ions and Ion Abundance Criteria.[a,b]

Mass	Ion Abundance Criteria
51	30–60% of mass 198
68	Less than 2% of mass 69
70	Less than 2% of mass 69
127	40–60% of mass 198
197	Less than 1% of mass 198
198	Base peak, 100% relative abundance
199	5–9% of mass 198
275	10–30% of mass 198
365	Greater than 1% of mass 198
441	Present but less than mass 443
442	Greater than 40% of mass 198
443	17–23% of mass 442

[a]J.W. Eichelberger, L.E. Harris, and W.L. Budde. Reference compound to calibrate ion abundance measurement in gas chromatography mass spectrometry. *Analytical Chemistry* 47:995 (1975).
[b]*Use of this calibration technique was found to be optional for purposes of this monitoring method.*

Table 3. List of Accurate Masses Monitored Using GC Selected-Ion Monitoring, Low Resolution, Mass Spectrometry for Simultaneous Determination of Tetra-, Penta-, and Hexachlorinated Dibenzo-P-Dioxins and Dibenzofurans.

Class of Chlorinated Dibenzodioxin or Dibenzofuran	Number of Chlorine Substituents (x)	Monitored m/z for Dibenzodioxins $C_{12}H_{8-x}OCl_x$	Monitored m/z for Dibenzofurans $C_{12}H_{8-x}O_2Cl_x$	Approximate Theoretical Ratio Expected on Basis or Isotopic Abundance
Tetra	4	303.902[a]	319.897[a]	0.74
		305.899	321.894	1.00
			327.885[b]	—
			256.933[c]	0.21
			258.930[c]	0.20
Penta	5	337.863[a]	353.858[a]	0.57
		339.860	355.855	1.00
Hexa	6	373.821	389.816	1.00
		375.818	391.813	0.87

[a]Molecular ion peak
[b]Cl$_4$-labeled standard peaks
[c]Ions which can be monitored in TCDD analyses for confirmation purposes.

NOTE: Replace this table with new Table 3A, next 2 pages.

Table 3A. Sequence of Operations in GC/MS/DS Quantitation of CDD's/CDF's in Extracts of Environmental Samples.[a]

Elapsed Time (min)	Event	GC Column Temperature (°C)	Temperature Program Rate (°C/min)	Ions Monitored by Mass Spectrometer (m/z)	Identity of Fragment Ion	Compounds Monitored	Approximate Theoretical Ratio of $[M]^+:[M+2]^+$
0.00	Injection, splitless	190					
1.00	Turn on split valve	190					
1.00	Begin temperature program to 220°C	190	5				
6.00	Open column flow to mass spectrometer	215	5				
7.00	Column temperature hold	220					
14.00	Start tetra program; sweep = 350 ppm; time/mass = 0.08 sec	220					
				240.938	$[M-COCl]^+$	TCDF	
				258.930	$[M-COCl]^+$	TCDD	
				303.902	$[M]^+$	TCDF	
				305.899	$[M+2]^+$	TCDF	
				315.942	$[M]^+$	$^{13}C_{12}$-TCDF	
				319.897	$[M]^+$	TCDD	
				321.894	$[M+2]^+$	TCDD	0.77
				327.885	$[M]^+$	$^{37}Cl_4$-TCDD	
				331.937	$[M]^+$	$^{13}C_{12}$-TCDD	1.00
				373.840	$[M]^+$	HxDPE[b]	
22.00	Stop tetra program	220					
22.50	Start penta program; sweep = 350 ppm; time/mass = 0.12 sec						
				274.899	$[M-COCl]^+$	PeCDF	
				290.894	$[M-COCl]^+$	PeCDD	
				337.863	$[M]^+$	PeCDF	
				339.860	$[M+2]^+$	PeCDF	
				353.858	$[M]^+$	PeCDD	
				355.855	$[M+2]^+$	PeCDD	0.57
				407.801	$[M]^+$	HpDPE[b]	1.00
23.00	Begin temperature program to 235°	220	5				
26.00	Column temperature hold	235					
32.00	Stop penta program	235					
32.50	Start hexa program; sweep = 350 ppm; time/mass = 0.20 sec						
				310.857	$[M-COCl]^+$	HxCDF	
				326.852	$[M-COCl]^+$	HxCDD	
				373.821	$[M]^+$	HxCDF	
				375.821	$[M+2]^+$	HxCDF	
				385.861	$[M]^+$	$^{13}C_{12}$-HxCDF	
				389.816	$[M]^+$	HxCDD	
				391.813	$[M+2]^+$	HxCDD	1.00
				411.856	$[M]^+$	$^{13}C_{12}$-HxCDD	
				443.759	$[M]^+$	ODPE[b]	0.87

Table 3A. (contd.)

Elapsed Time (min)	Event	GC Column Temperature (°C)	Temperature Program Rate (°C/min)	Ions Monitored by Mass Spectrometer (m/z)	Identity of Fragment Ion	Compounds Monitored	Approximate Theoretical Ratio of $[M]^+:[M+2]^+$
33.00	Begin temperature program to 250°C	235	5				
36.00	Column Temperature hold	250					
42.50	Stop hexa program						
43.00	Start hepta program; sweep = 350 ppm; time/mass = 0.30 sec	250		344.818	$[M\text{-}COCl]^+$	HpCDF	
				360.813	$[M\text{-}COCl]^+$	HpCDD	
				407.782	$[M]^+$	HpCDF	
				409.779	$[M+2]^+$	HpCDF	
				423.777	$[M]^+$	HpCDD	1.00
				425.774	$[M+2]^+$	HpCDD	1.00
				477.720	$[M^+]$	NDPE[b]	
53.00	Stop hepta program	250					
53.50	Start octa program; sweep = 350 ppm; time/mass = 0.30 sec	250		378.768	$[M\text{-}COCl]^+$	OCDF	
				394.774	$[M\text{-}COCl]^+$	OCDD	
				441.732	$[M]^+$	OCDF	
				443.740	$[M+2]^+$	OCDF	
				453.772	$[M]^+$	$^{13}C_{12}$-OCDF	
				457.738	$[M]^+$	OCDD	
				459.735	$[M+2]^+$	OCDD	0.86
				469.779	$[M]^+$	$^{13}C_{12}$-OCDD	1.00
				471.776	$[M+2]^+$	$^{13}C_{12}$-OCDD	
				511.681	$[M]^+$	DDPE[b]	
54.00	Begin temperature program to 270°	250°	5				
58.00	Column temperature hold	270°					
65.00	Stop octa program	270°	5				
65.00	Begin temperature program to 300°	270°					
71.00	Column temperature hold	300°					
75.00	Cool column to 190°						

[a]This table was provided by Wright State University, Brehm Laboratory, under Cooperative Agreement Number CR 809972-04-0 with U.S. EPA. The GC temperature program was designed for a 60-m DB-5 column. Ions monitored for HxCDD, HxCDF, HpCDD, HpCDF, OCDD, and OCDF are actually $(M+2)^{\bullet+}$ and $(M+4)^{\bullet+}$ for better sensitivity.
[b]HxDFE, HpDPE, ODPE, NDPE, DDPE are abbreviations which designate (respectively) hexachloro-, heptachloro-, octachloro-, nonachloro-, and decachlorodiphenyl ethers. Use of $[M+2]^{\bullet+}$ for these DPE's would result in better sensitivity.

CHAPTER 28

Approaches to the Determination of Limit of Detection in 2,3,7,8-TCDD Analyses

Cliff J. Kirchmer, Tenkasi S. Viswanathan, and Robert D. Kleopfer

INTRODUCTION

In defining the limit of detection we are inherently dealing with uncertainty. The purpose of the limit of detection is to help us make a decision regarding whether or not the analytical response we observe indicates the presence of analyte. The theory of limit of detection, as developed over the last 20–25 years, tells us that the variability of blank responses is the preferred basis for determining limit of detection.

If the blanks are analyzed by exactly the same procedure as that used for samples, the limit of detection determined is for a "complete analytical procedure." The variability of the blanks should be measured in terms of the statistical probability of errors of the first and second kinds. The error of the first kind is the error of accepting an apparent effect arising by chance as a real effect (i.e., false positives). The probability of this is denoted by α, the significance level of the test used. The error of the second kind is the error of failing to recognize a real effect (i.e., false negatives). The probability of this is denoted by β.

Instrumental detection limits, on the other hand, have often been defined in terms of signal-to-noise ratio. These do not pretend to measure the limit of detection of a complete analytical procedure, but only one part of the procedure, namely the instrumental part. In the case of analysis of 2,3,7,8-

tetrachlorodibenzo-p-dioxin (2,3,7,8-TCDD) by isotope dilution gas chromatography/mass spectrometry (IDMS) a clear distinction between these two approaches to limit of detection becomes somewhat less clear.

In the following section, the procedures for calibration and analysis in IDMS are briefly reviewed. This is followed by a discussion of various procedures for determining limit of detection, with particular emphasis on the analysis of 2,3,7,8-TCDD by IDMS. Finally, recommendations are given regarding the best procedure to follow in establishing the limit of detection for 2,3,7,8-TCDD analyses, and suggestions are made for further study.

The examples cited in this article deal with data obtained using low-resolution mass spectrometry (LRMS). Lower detection limits may be achieved by the use of high-resolution mass spectrometry (HRMS) or tandem (MS/MS) mass spectrometry.[1] Although the emphasis is on LRMS, the principles discussed should also apply to HRMS and MS/MS.

CALIBRATION AND ANALYSIS

Isotope dilution mass spectrometry is a type of internal standard calibration and analysis, in which the internal standard is an isotopically labeled analog of the analyte. When added initially to the sample, the internal standard serves to correct for losses during the processing of samples, and to compensate for errors owing to differences in injected volume and unnoticed variations in instrumental sensitivity. A stable isotope-labeled analog of the analyte is an ideal internal standard, since its chemical and physical properties can be expected to be almost identical to the analyte, thus assuring negligible differences in extraction, cleanup and chromatographic properties during sample processing.[2] The carbon-13-labeled isotope, $^{13}C_{12}$2,3,7,8-TCDD, has been most commonly used as an internal standard in the analysis of 2,3,7,8-TCDD.

Internal standard calibration requires both a calibration solution and a spiking solution. The calibration solution is used to determine the relative responses of analyte and internal standard, while the spiking solution is used to add a known amount of internal standard to each sample.

The calibration curve for IDMS analysis of 2,3,7,8-TCDD is obtained by plotting ratios of the GC peak areas (or heights) vs ratios of the weights (or concentrations), the ratios corresponding to those of the unlabeled 2,3,7,8-TCDD and internal standard, as illustrated in Figure 1. The slope of the calibration curve is equal to the relative response factor. Analytical responses can be converted to concentrations using a calibration curve, or by use of Equation 1, in which the relative response factor appears.

$$C = (As/Ais) \cdot (1/RRF) \cdot (Qis/W) \qquad (1)$$

As = response for ions characteristic of analyte

CALIBRATION CURVE FOR
ISOTOPE DILUTION MASS SPECTROMETRY

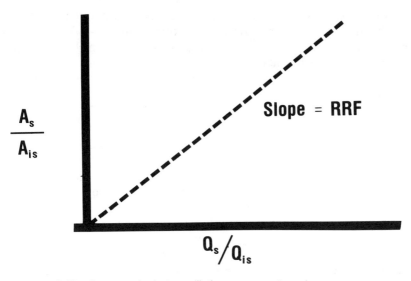

Figure 1. Calibration curve for isotope dilution mass spectrometry.

Ais = response for ions characteristic of the isotopically labeled internal standard
RRF = relative response factor
Qis = weight of internal standard added to the sample
W   = weight of the sample

Calibration is related to the limit of detection because a calibration factor must be used to convert response to concentration units.

Note in Equation 1 that the error in estimating concentration is principally governed by the error in the ratio As/Ais, since it should be possible to accurately measure the amount of internal standard added (Qis) and the weight of the sample (W). Equation 1 also indicates that one way to measure lower concentrations, and therefore to achieve lower limits of detection, is to use a larger weight of sample, W. While this is true in theory, there are practical reasons for not handling large amounts of hazardous samples. Also, the gain may not be directly proportional to W, since interferences may increase with increased sample size.

LIMIT OF DETECTION PROCEDURES BASED ON SIGNAL-TO-NOISE RATIO

Procedures based on signal-to-noise ratio are most commonly used in determining limit of detection for the analysis of 2,3,7,8-TCDD. The expression "signal-to-noise" implies that only instrumental noise is considered. In fact, a more appropriate description would be "signal-to-background response" since the responses for samples may include chemical noise as well as instrumental noise. A characteristic of these procedures as applied in the analysis of 2,3,7,8-TCDD is that the limit of detection is calculated specific to each sample.

EPA IFB Procedure

For the IFB Method (i.e., the EPA Invitation for Bid method for dioxin in a soil/sediment matrix) there are two different procedures for determining the lower concentration limit above which the analyte is not considered to be present.[3] The decision regarding which procedure to use depends on whether the selected ion responses (m/z 257, m/z 320, and m/z 322) are less than or greater than 2.5 times the background signal. The choice of a factor of 2.5 is somewhat arbitrary, and is not based on any statistical considerations.

If the response is determined to be less than 2.5 times the background signal, the limit of detection (LOD) is calculated according to Equation 2.

$$LOD = \frac{2.5 \cdot Ax \cdot Qis}{Ais \cdot RRF \cdot W} \tag{2}$$

where Ax = peak height or integrated ion abundance for either m/z 320 or m/z 322

If the response is determined to be greater than 2.5 times the background signal, but qualitative identification criteria are not met, an "estimated maximum possible concentration" is calculated according to Equation 1. This is the same way positive samples are calculated, except in this case the result is not identified as being positive, because the qualitative identification criteria are not met.

This case illustrates a general problem of defining a lower reporting limit when criteria are specified for qualitative identification. When qualitative ion ratio criteria and minimum ion responses are specified, the decision regarding detection becomes more complex. Not only must analytical responses for selected ions (m/z 257, m/z 320 and m/z 322) exceed a defined level, but the qualitative criteria must also be met (i.e., the ion abundances of the ions 257, 320 and 322 should have definite ratios; 257/322 = 0.25 to 0.45 and 320/322 = 0.67 to 0.87). Note that in these procedures, the limit of detection is not a tool for deciding whether dioxin has been detected.

These procedures assume that background signal can be accurately measured. The procedure specified for determining background signal is to integrate ion abundances (area) for either m/z 320 or 322 in the appropriate region of the selected ion chromatogram profile (SICP). Sometimes it is more convenient to measure background signal as peak height.

In Equation 2, the 320 or 322 ion is specified for estimating the limit of detection, but for qualitative identification the 257 ion (with an abundance of only a third of the abundance of the 322 ion) is also required to be greater than 2.5 times the background signal. In applying this definition of limit of detection to a particular sample, it is probably correct to conclude that the concentration of dioxin is less than the calculated LOD. However, one should not interpret the LOD as implying that it would be possible to identify dioxin at a concentration equal to the LOD, since the response for the 257 ion would probably not be greater than 2.5 times the background signal.

As an illustration of the above, the analyses of 12 method blanks and background uncontaminated soil samples over a 3-month period at the EPA Region VII Laboratory gave significant responses for the m/z 320 and 322 ions with a calculated (Equation 1) mean concentration of 32 ppt (ng/kg) and a standard deviation of 5.5 ppt. (Note—As discussed later, the cause of these responses was contamination of the spiking solution with unlabeled 2,3,7,8-TCDD, but this was not known initially). Since none of the 12 samples had a S/N of 2.5 or greater for the m/z 257 ion, and most of them failed to meet the 320/322 ratio criteria, the samples were judged to be "nondetects," with an "estimated maximum possible concentration" of approximately 32 ppt.

For comparative purposes, it was assumed that no significant responses were obtained for the m/z 320 and 322 ions, and the background response was used to calculate limits of detection using Equation 2. This gave a mean limit of detection of 14 ppt, with a standard deviation of 6.7 ppt. This example illustrates that qualitative identification according to the criteria specified is unlikely at concentrations several times greater than the LOD as calculated based on the signal-to-noise ratio (Equation 2). This situation is the reverse of that usually encountered when calculating the limit of detection based on the variability of the blank (see the section on variability of blank response) in which qualitative identification is possible at low concentrations, but a good quantitative estimate is not. Here, a quantitative estimate can be made, but a qualitative identification cannot.

The EPA has recommended elsewhere that "if the criteria for qualitative identification of the analyte is [sic] based upon pattern recognition techniques, the least abundant signal necessary to achieve identification must be considered in making the estimate."[4] If it were desired to follow this recommendation, then the limit of detection calculation should be based on the 257 ion.

Finally, in the IFB procedure the use of the surrogate standard $^{37}Cl_4 2,3,7,8$-TCDD plays an important indirect role in assuring that one can analyze samples at the required limit of detection. The surrogate is added to each sample at a concentration of 1 µg/kg. The fact that one is able to detect and accurately measure the presence of the surrogate at the 1 µg/kg level provides indirect support that one can also detect unlabeled 2,3,7,8-TCDD at that concentration level. Of course the ions monitored are different for the two compounds, so the relationship is not perfect. Also, one must assume that both the unlabeled 2,3,7,8-TCDD and the surrogate are extracted completely from the sample.

EPA National Dioxin Study Procedure

A second procedure which utilizes signal-to-noise ratio for estimating the limit of detection is the "surrogate analyte present" technique, as proposed in the "Analytical Procedures and QA Plan for the National Dioxin Study."[5] The procedure states that if 2,3,7,8-TCDD is not detected, then a minimum detection limit (MDL) should be calculated using Equation 3. As for the IFB procedure, a decision regarding detection must be made prior to calculating the limit of detection. The limit of detection is therefore not a tool for deciding whether dioxin has been detected.

$$MDL = \frac{(2.5 \text{ N})(Qss)}{Ass} \tag{3}$$

where MDL = minimum detection limit
  N = 1 standard deviation of noise (at analyte elution time)
  Qss = concentration of surrogate standard in the samples
  Ass = response for ion characteristic of surrogate standard

Figure 2 illustrates how measurements are obtained from a chromatogram.

Note that the standard deviation of noise is used as a measure of background signal. This parameter is a computer-generated number whose value will depend on the software of the instrument being used.

The surrogate standard is 1,2,3,4-TCDD, while the internal standard is $^{13}C_{12}$ 1,2,3,4-TCDD. Both are added to the sample prior to processing. Normally, the surrogate concentration is measured with reference to the internal standard for the purpose of evaluating performance. However, in this procedure, the surrogate is, in effect, used as a second internal standard, as indicated in Equation 3. The advantage of this procedure is to have a reference internal standard at a low concentration, so that its presence indicates that low concentrations can indeed be measured. Of course if the ion responses for 1,2,3,4-TCDD are not detectable, it will not be possible to get an estimate of the MDL.

IF 2,3,7,8 - TCDD IS **NOT** DETECTED THEN:

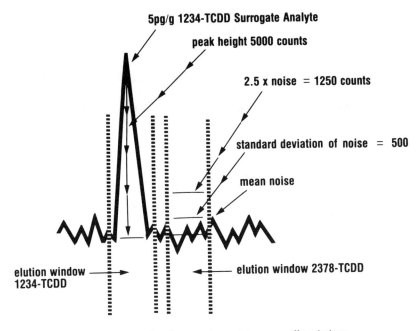

Figure 2.   National dioxin study—"surrogate analyte present" technique.

Another disadvantage is that at low concentrations the response of the 1,2,3,4-TCDD will be variable due to random error, giving rise to an uncertain estimate of the limit of detection. In effect, calibration and analysis are combined into one step, and the uncertainty in the relative response factor is therefore greater than one would have by preparing a calibration curve from multiple determinations of a calibration standard solution. Finally, the proportional losses of 1,2,3,4-TCDD and 2,3,7,8-TCDD may differ during extraction, cleanup, and chromatography; therefore the 1,2,3,4-TCDD may not fulfill the requirements of an ideal internal standard as outlined previously.

If 2,3,7,8-TCDD is judged to be present in the sample, a MDL is still calculated. However, in this second case, the concentration of 2,3,7,8-TCDD is first calculated and then used as an internal standard to calculate the equivalent concentration corresponding to 2.5 times the standard deviation of the noise. There is a question as to whether a MDL is needed in this second case, since the value of the MDL is not used in deciding whether 2,3,7,8-TCDD is present. Since the concentration of 2,3,7,8-TCDD is estimated from measurement, its use as an internal standard may result in propagation of error in the estimated MDL.

Table 1.   Concentrations of Unlabeled 2,3,7,8-TCDD in Spike Solutions.

Date Issued	Batch No.	Concentration (ng/mL)
9/83	C–057	0.7
5/84	C–073	2.9
11/84	C–086	2.9
3/85	C–096	0.9

LIMIT OF DETECTION PROCEDURES BASED ON THE VARIABILITY OF RESPONSE

Variability of Blank Response

This approach to limit of detection has a solid theoretical foundation. It was first developed by Kaiser,[6] and later more fully developed by Currie.[7] Wilson[8] applied the procedure to water analyses. Kaiser compared the limit of detection problem to that of one trying to find a ship in a stormy sea. Is it a ship or a larger wave than usual? The waves correspond to blank responses. The ship corresponds to the response of a sample. Kaiser, and later Currie, provided a statistical decision-making basis upon which to decide whether the signal for a sample was larger than that which one might obtain for a blank.

An analytical result (R) always corresponds to the sample response (S) minus the blank response (B) (i.e., R = S–B). Currie treated two possible cases for the blank: (1) paired observations and (2) "well-known" blank. In the first case, one obtains a result by individually blank correcting every sample response. In the second case, one obtains a result by subtracting a constant "well-known" blank response from each sample response.

Taking into account both errors of the first and second kinds, Currie defined limit of detection for the "well known" blank as $2t_{0.1}s$ and for the "paired comparison" as $2t_{0.1}(\sqrt{2})$ s, where $t_{0.1}$ is the 10% point of the t-statistic for double-sided tests, and s is the estimated within-run standard deviation of the blank.[7]

A need for application of this theory presented itself in the EPA Region VII laboratory when it was discovered that the spiking solution contained a small amount of unlabeled 2,3,7,8-TCDD as an impurity (Table 1). Blank responses of 6.2, 5.3, 6.2, 5.8 and 5.5 ng/kg were obtained when analyzing fish samples, with a mean blank value of 5.8 ng/kg and a standard deviation of 0.41 ng/kg. Blanks were analyzed by the same procedure used for samples, which involved the spiking of 50-g samples with internal standard at a level of 1000 ng/kg.

The limit of detection was thus calculated on the basis of a "well known" blank, i.e.,

$$LOD = 2(t_{0.1})s$$
$$= 1.8 \text{ ng/kg}$$

Note that this LOD corresponds to the analytical result obtained after the blank response is subtracted from the sample response. For example, if the sample response is 7.2 ng/kg, then the analytical result would be equal to the sample response minus the blank response (i.e., 7.2–5.8 = 1.4). Since 1.4 is less than 1.8, the limit of detection, the result is expressed as less than 1.8 ng/kg.

The preceeding illustrates that when blank correction is necessary, the procedures based on variability of the blank provide a better estimate of the LOD. Approaches based on signal-to-noise ratios do not allow for blank correction, and can give biased results when blank correction is required, such as in the case of the fish analyses described above.

The isotopic impurity of the spiking solution was normally not noticed for soil samples. This was because the required detection limit on soil samples was 1 ppb (1000 ng/kg), and therefore analyses were not done at the level of sensitivity necessary to detect the impurity. However, the impurity of the spiking solution was found to be the cause of the m/z 320 and 322 responses for soil blanks mentioned in the section on the IFB procedure. Application of the procedure for calculation of limit of detection based on the variability of the blank to those results gives an LOD of 20 ppt. Blank correction in this case allows one to measure concentrations below the levels found in the blanks.

Ironically, the statistically based calculation of LOD becomes easier when analyte impurity is present in the spiking solution. This is because one can give an exact statistical treatment to the variability of blank response. When only background signal or noise is present, an exact statistical treatment is more difficult.

Variability of the Analyte Response

This approach to limit of detection was developed by the Environmental Protection Agency[9] and has been included as an appendix to the 600 series methods.[4] The approach was developed for the determination of detection limits in wastewater, and has not to our knowledge been applied to the analysis of dioxin in soils.

The procedure specifies that one determine the standard deviation of the analytical response for low concentrations of analyte. Blank correction based on a "well-known" blank is permitted, but the blank is not used for estimating the limit of detection, referred to as the method detection limit (MDL). Instead, the MDL is calculated from the variability of responses for samples that either contain or have been spiked with a low concentration of the analyte. Equation 4 is the result of the theoretical treatment.

$$MDL = t_{0.02}\, s \qquad\qquad (4)$$

where $t_{0.02}$ = the 2% value of the t-statistic for a two-sided table

s = the standard deviation of the replicate analyses

This equation shows some similarity to those based on the variability of blank response. The primary differences are the values chosen for the t-statistic and the use of the standard deviation of sample responses instead of the standard deviation of blank responses. If one assumes that blank and sample responses are equal for low concentrations of analyte, the only difference in the two approaches is the value of the t-statistic that is chosen, which in turn reflects a difference in the acceptable level of error of the first kind (i.e., Currie[7] and Wilson[8] chose α = 0.05 and Glaser et al.[9] chose α = 0.01).

The major difficulty in application of this procedure is to choose the concentration level at which to perform the replicate analyses. It is known that standard deviation generally increases with concentration, so if the concentration level is too high, the estimate of the method detection limit will also be too high.

As a quality control measure, EPA has required contract laboratories to analyze background soil samples spiked with 2,3,7,8-TCDD at a level of 1 μg/kg. The following is a statistical summary of these analyses.

Number of analyses, n = 228
Mean spike amount = 1.06 μg/kg
Mean percent recovery = 104
Standard deviation of the percent recovery = 25.8

Based on these results, the overall detection limit for these laboratories can be calculated as

$$
\begin{aligned}
MDL &= t_{0.02}s \\
&= 2.33 \times 0.28 \\
&= 0.66 \ \mu g/kg
\end{aligned}
$$

A similar analysis done at EPA Region VII Laboratory on 18 spiked samples during the 18-month period ending June, 1985, gave a mean of 1.01 μg/kg with a standard deviation of 0.078 μg/kg and a calculated MDL of 0.20 μg/kg.

In estimating the LOD based on the variability of blank responses, the need for blank correction provides a solid rationale for following this approach. No such rationale exists for calculating the LOD based on the variability of analyte responses. This again raises the question as to whether a single limit of detection should apply to all samples, or if a limit of detection should be determined for each sample, and leads us to search for an alternative approach which combines the best elements of the approaches based on signal-to-noise ratio and variability of response.

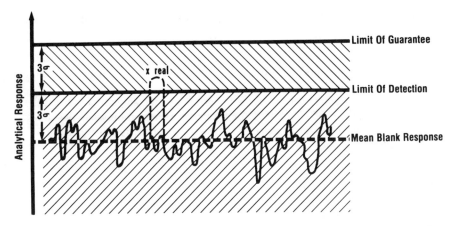

Figure 3.   Definitions of limit of detection and limit of guarantee (adapted from Kaiser[6])

LIMIT OF DETECTION BASED ON THE VARIABILITY OF THE SIGNAL-TO-NOISE RATIO FOR INDIVIDUAL SAMPLE RESPONSES

The two approaches to limit of detection discussed thus far differ in several ways. One of the more significant differences is that of defining a limit of detection for individual samples or for a "complete analytical procedure."

A synthesis of these two ideas may be possible since in isotope dilution mass spectrometry, the internal standard allows one to analyze and provide a reference for calibration at the same time. Within each chromatographic run therefore, there is a variability of background response which can be evaluated statistically to determine if the response(s) for the selected ion(s) of interest exceed the background signal.

In his description of limit of detection based on the variability of blank responses, Kaiser used a continuous curve (Figure 3) to illustrate the concept of variability of the blank.[6] In subsequent analyses of samples, detection is based on whether or not some multiple of the standard deviation of blank response has been exceeded. Three standard deviations ($\alpha = 0.0014$) was chosen by Kaiser for the definition of limit of detection. Although Kaiser was not referring to internal standardization, one might extend his reasoning to apply to the case of internal standardization such that the variability of blank response is determined from a single chromatogram.

The use of three standard deviations for the variability of background response takes into account only errors of the first kind (i.e., false positives). Kaiser referred to this as the limit of detection. Kaiser also defined another term, limit of guarantee, as equal to six standard deviations of blank response to take into account both errors of the first and second kinds ($\alpha = \beta = 0.0014$). The two terms were defined by Kaiser to answer two different questions.

1. What is the least amount that is 'certainly' present as the result of an analytical procedure? Answer: A concentration equal to or greater than the limit of detection.
2. What purity (characterized by an upper limit to the amount possibly still contained) can be guaranteed for a sample with "certainty," if the substance sought cannot be detected? Answer: A concentration equal to the limit of guarantee.

Qualitative criteria for identification complicate the application of this approach described by Kaiser.

An extension of Kaiser's arguments to the case of isotope dilution mass spectrometry leads us to conclude that background response might be determined from the chromatograms of individual sample analyses. The same chromatograms can then be examined to determine if the response(s) at retention time(s) corresponding to the compound of interest (i.e., 2,3,7,8-TCDD) exceed three (or six) standard deviations of the variability of background response. Note that other multiples of standard deviation might be used, depending on the significance levels chosen for α and β.

The data systems of modern mass spectrometers allow us to take this approach. It is important that the algorithm incorporated into the software measure the standard deviation of background noise accurately. It must have a criterion for distinguishing between background noise and "significant" peaks. The standard deviation of background noise should have as its reference the mean background noise.

We have done a comparison of the variability of background response for a sample using normal statistical techniques (Equation 5) and the Finnigan INCOS data system (Equation 6). For a chromatogram, $X(I)$, $I = 1, \ldots, n$, the two methods would compute the variability as:

$$S = \left\{ \frac{1}{n-1}\left[\sum_{i=1}^{n} x^2_i - \left(\sum_{i=1}^{n} x_i \right)^2 / n \right]^{1/2} \right\} \tag{5}$$

$$\text{Noise} = N = \frac{1}{n-2}\left(\sum_{i=2}^{n-1} D_i \right) \tag{6}$$

$$\text{where } D_i = |X_{i-1} - 2X_i + X_{i+1}| \tag{7}$$

A sample was chosen for which only high-frequency instrumental noise appeared to contribute to the background response so that no points in the chromatogram were rejected as too large to be noise using the test

$$D_i < 8[\text{Min}(X_{i-1}, X_i, X_{i+1})]^{1/2} \tag{8}$$

Table 2. A Comparison of the Standard Deviation of Noise Calculated Manually (Equation 5) and by Computer (Equation 6).

	Selected Ions (m/z)			
	320	322	323	320 + 322
Manual	85	85	106	123
Computer	145	166	194	252

Table 2 presents the results of the comparison of standard deviation as calculated manually and by the computer, using 61 data points.

While there may be a need to improve the software for defining the standard deviation of background response, the principle of using computerized calculation of the standard deviation of background response to determine a sample specific limit of detection would appear to be a valid application of the general principle of defining the limit of detection based on the variability of background or blank response.

The above-outlined approach would appear to offer important advantages when blank correction is not necessary, that is, for those cases in which background responses for selected ions in the region of interest are of the same order of magnitude for both samples and blanks. In these cases, the physical recovery of the internal standard is the key variable in determining the limit of detection. One is in effect comparing a relatively constant background response against a variable internal standard response. This variability in internal standard recovery is increased by the fact that the method includes several cleanup options[3] each one of which is also affected by the sample matrix. In effect, there is not one method but rather several, each of which has a different mean physical recovery of the internal standard. Determining a limit of detection of a "complete analytical procedure" based on the variability of background response from replicate blanks or samples would lead to underestimates or overestimates of limit of detection when applied to specific samples.

CONCLUSIONS AND RECOMMENDATIONS

An overall approach to the subject of limit of detection is summarized in Figure 4. The first step should be to define a required limit of detection based on data needs. The required limit of detection is an important factor in selecting a method. A recommended rule of thumb is that the required LOD should be 10–20% of the concentration of the environmental criteria or standard to be enforced.[10] This is because random error at a concentration corresponding to the limit of detection is high. By requiring a lower limit of detection, we assure that measurements done at higher concentrations will be precise enough to determine whether or not the environmental criteria or standard has been exceeded.

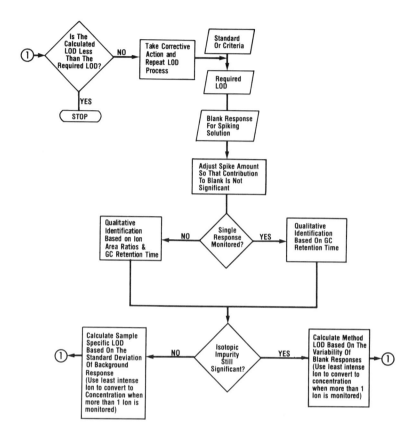

Figure 4.   Flow chart for approach to determining the limit of detection.

The next step is to determine the blank response for the spiking solution. It is best to analyze the concentrated spike solution to initially determine whether blank response might be significant. Our ultimate interest is in the blank response for the complete analytical procedure. Our experience has shown that the presence of 2,3,7,8-TCDD as an isotopic impurity in the internal or surrogate standards is the principal cause of significant blank responses. If possible, the amount of the spike solution added to a sample should be adjusted so that the concentration equivalent contribution to the blank response is not significant. If blank response is significant, blank correction as described earlier will be necessary.

A decision must then be made to monitor a single ion or multiple ions. (One might also monitor several ions, but program the computer to sum these ions and treat them as a single response.) If only a single ion is monitored, then qualitative identification is based solely on GC retention time. If a significant response was found for the blank, then the limit of detection should be calculated based on the variability of the blank. If blank response was not significant in comparison with background noise, then a limit of detection should be calculated as the concentration equal to three times the standard deviation of background response for each sample, as determined by the computer data system. Only if the response for the selected ion is greater than three standard deviations from the mean will 2,3,7,8-TCDD be regarded as "detected." Finally, for nondetects, the calculated LOD should be compared with the required LOD. If the calculated LOD is greater than the required LOD, the sample needs to be reanalyzed.

If more than one ion is monitored, then qualitative identification requires meeting ion ratio criteria as well as GC retention time criteria. The remaining procedures for determining LOD are then the same as for monitoring a single ion, with one exception. One must decide which of the ions to use in converting background response to an equivalent concentration. Consistency would suggest that the least intense ion should be used so that the calculated LOD would correspond to a concentration for which both qualitative and quantitative determination would be possible.

Improvement in the detection of significant responses might be achieved with improved data system software for measuring the variability of background response, and for distinguishing between background response and "significant" peaks. Further advances might also come from the application of pattern recognition techniques to the selected ion chromatogram profiles.

REFERENCES

1. Johnson, J. V., and R. A. Yost. "Tandem Mass Spectrometry for Trace Analysis," *Anal. Chem.* 57(7):739A (1985).
2. Watson, J. T. *Introduction to Mass Spectrometry: Biomedical, and Forensic Applications.* (New York: Raven Press, 1976).
3. "Selected Ion Monitoring (SIM) GC/MS Analysis with Jar Extraction Procedure," U.S. EPA Invitation for Bid IFB WA 84-A002 (September 15, 1983).
4. "Methods for Organic Chemical Analysis of Municipal and Industrial Wastewater," U.S. EPA Report—600/4-82-057 (1982).
5. "Analytical Procedures and Quality Assurance Plan for the National Dioxin Study—Study Tiers 3-7 of the National Dioxin Strategy," U.S. EPA Environmental Research Laboratory—Duluth, MN (November, 1984).
6. Kaiser, H. *Two Papers on the Limit of Detection of a Complete Analytical Procedure* (London: Adam Hilger Ltd., 1968), Part 2.
7. Currie, L. A. "Limits for Qualitative Detection and Quantitative Determination," *Anal. Chem.* 40(3):586–593 (1968).
8. Wilson, A. L. "The Performance Characteristics of Analytical Methods—II," *Talanta* 20:725–732 (1973).

9. Glaser, J. A., D. L. Foerst, G. D. McKee, S. A. Quave, and W. L. Budde. "Trace Analyses for Wastewaters," *Environ. Sci. Technol.* 15:1426–1435 (1981).
10. Gardiner, J., and A. L. Wilson. "Accuracy Required of Analytical Results for Water Quality Data Banks," Water Research Centre (England) Technical Report TR 34 (1976).

CHAPTER 29

Characterization of Polychlorinated Dibenzo-p-Dioxins and Dibenzofurans by Powder X-Ray Diffraction

Joseph S. Cantrell, Patrick J. Slonecker, and Thomas A. Beiter

INTRODUCTION

The mathematical relationship between known PCDD and PCDF single-crystal structure determinations and their powder X-ray diffraction (PXRD) patterns provides a basis to make rapid and conclusive isomeric assignments by PXRD.[1-5] In cases where the X-ray structure determinations have not been completed, characterization by gas chromatography/mass spectrometry (GC/MS), electron capture gas chromatography (EC/GC) and differential scanning calorimetry are required to establish the isomer assignment and purity.

This chapter contains a set of powder X-ray diffraction patterns of well characterized polychlorinated dibenzo-p-dioxins (PCDD's) and polychlorinated dibenzofurans (PCDF's). They have been determined to provide unique analytical "fingerprints" for cases where suspect PCDD or PCDF standards need to be validated in their isomeric assignment, in the analysis of unknown PCDD and PCDF mixtures, and in laboratories where these standards are not available.

EXPERIMENTAL

The EC/GC consisted of a Perkin Elmer Sigma 1 equipped with a nickel-63 ECD and a 30-m x 0.25-mm i.d. DB-5 fused-silica capillary column. The oven temperature was programmed from 180°C to 280°C at 5°C/min and held at 280°C for 15 min. The injection port and detector temperatures were maintained at 280°C. A split mode of injection (30:1) was used in all EC/GC determinations. The carrier gas was helium at ca. 1.0 mL/min and the makeup gas was nitrogen at ca. 30 mL/min. Quantities (2 μL) of 1.0-ppm benzene solutions of the samples were injected to purposefully overload the electron capture detector and thus allow detection of potential low-level impurities.

The GC/MS characterizations were carried out on a Hewlett Packard 5993B GC/MS equipped with a 30-m x 0.32-mm i.d. DB-5 fused-silica capillary column. Hydrogen gas at a linear velocity of ca. 30 cm/sec was used as the carrier gas. All determinations via GC/MS were made with a splitless liner. The oven temperature was programmed from 200°C to 290°C at 5°C/min. The samples for GC/MS characterization were diluted in benzene. Perfluorokerosene was used as the mass spectrometer calibrant.

The powder X-ray diffraction patterns were recorded by a Rigaku D/Max automated wide-angle diffractometer with a 12-kW rotating anode and a Phillips Norelco (XRG-3000) generator and goniometer. The samples for the PXRD work were mounted on a microscope slide via double-stick tape. Silicon, NBS SRM 640A, was used as an internal standard. PXRD analysis required 3 to 5 mg of each sample.

The differential scanning calorimetry work was carried out on a Perkin Elmer Model DSC-2. Approximately 1 to 10 mg of each sample was sealed in a gold boat prior to analysis. Indium, lead and potassium chromate were used as DSC standards. The method of purity calculation was based on that reported by Driscoll et al.[6] All samples were screened for gross amounts of impurities (>6%) by EC/GC prior to analysis and purity determination by DSC.

For further details on the EC/GC, GC/MS and DSC studies, refer to the work of Slonecker and Cantrell.[4-5]

RESULTS

We have characterized four PCDF and twelve PCDD standards by PXRD. Sources for these compounds, and their relative purities, are given in Table 1. To verify their purity, they have been screened by EC/GC and checked by DSC. Their isomer chlorine substitution has been verified by GC/MS, as outlined by Rappe et al.[7] In several cases, such as the 1,2,3,7,8,9-CDD, the PXRD patterns are in good agreement with patterns

Table 1.   Sources and Relative Purities of PCDD and PCDF Standards.[a]

Standard	Purity (%)	Where Available
1,2,3,4,6,7,8,9-CDD	>99.5	Dominion Rubber Co., Canada M. Kulka
1,2,3,4,6,7,8-CDD	>99	Wright State University T. Tiernan
1,2,3,4,6,7,9-CDD	>95	Illinois Institute Of Technology J. Uchic
1,2,3,7,8,9-CDD	>99	Purification of Dow Chemical Material J. Cantrell
1,2,3,4,7,8-CDD	>96	Food and Drug Administration D. Firestone
1,2,3,6,7,8-CDD	>96	Purification of Dow Chemical Material J. Cantrell
1,2,4,6,7,9-CDD	>95	Food and Drug Administration D. Firestone
1,2,3,4-CDD	>95	Analabs, Inc., Foxboro Labs North Haven, CN
1,2,4-CDD	>95	Analabs, Inc., Foxboro Labs North Haven, CN
2,7-CDD	>95	Analabs, Inc., Foxboro Labs North Haven, CN
2,3-CDD	>95	Analabs, Inc., Foxboro Labs North Haven, CN
1-CDD	>95	Analabs, Inc., Foxboro Labs North Haven, CN
1,2,3,4,6,7,8,9-CDF	>99.9	Analabs, Inc., Foxboro Labs North Haven, CN
2,3,7,8-CDF	>95	Wright State University T. Tiernan
3,6-CDF	>95	Analabs, Inc., Foxboro Labs North Haven, CN
CDF	>95	Analabs, Inc., Foxboro Labs North Haven, CN

[a]Purities were determined by DSC, FID/GC and EC/GC.

calculated from known PCDD crystal structure determinations and other reported PXRD patterns.[1-3]

The individual PXRD patterns are tabulated in Tables 2–7. Both the 2Θ and d-spacings are presented with the relative intensities. In Figures 1–4, the PXRD patterns of the PCDF and PCDD standards are depicted in plots of relative intensity versus 2Θ in degrees. The uniqueness of these "fingerprints" is illustrated in composite PXRD plots of closely related isomers in Figure 5.

Plot A of Figure 5 compares the PXRD patterns of the 1-CDD, 2-CDD, 2,3-CDD, and 2,7-CDD. The 1-CDD can easily be differentiated in the presence of the other two via a single line at 13.6 2Θ. Also, the 2-CDD may be distinguished from the 2,3-CDD and 2,7-CDD by the lines at 10.63 2Θ and 21.31 2Θ. The 18.00 and 32.18 2Θ lines of the 2-CDD are too closely spaced to the lines of the 2,3-CDD and the 2,7-CDD to be of value in its

Table 2. Powder X-Ray Diffraction Data for CDF, 2-CDD and 1-CDD.

CDF			2-CDD			1-CDD		
2Θ	dÅ	I/Io	2Θ	dÅ	I/Io	2Θ	dÅ	I/Io
9.15	9.66	100.	8.92	9.91	4.0	10.3	8.59	0.2
15.84	5.59	0.2	10.63	8.322	100.	11.6	7.63	3.0
18.40	4.82	20.	12.40	7.138	2.0	13.6	6.51	100.
19.00	4.67	4.0	12.78	6.927	13.	17.9	4.96	0.1
19.55	4.54	4.0	14.09	6.285	3.0	20.25	4.39	0.1
20.57	4.32	0.3	14.43	6.138	7.0	20.78	4.27	0.1
23.25	3.83	0.9	15.42	5.746	20.	22.35	3.98	0.1
23.65	3.76	12.	17.55	5.053	5.0	23.40	3.80	1.0
25.00	3.56	0.3	18.00	4.928	49.	24.35	3.66	0.1
27.10	3.29	0.3	21.31	4.169	78.	25.90	3.44	0.1
27.80	3.21	8.0	22.55	3.943	6.0	27.43	3.25	9.0
28.17	3.17	4.0	23.40	3.801	3.0	29.90	2.99	0.2
30.20	2.96	0.3	24.30	3.663	2.0	30.25	2.95	0.1
32.00	2.80	0.3	25.00	3.562	3.0	31.44	2.85	0.1
36.10	2.49	0.4	25.83	3.449	5.0	35.45	2.53	0.3
36.55	2.46	0.4	26.35	3.382	3.0	36.50	2.46	0.2
37.33	2.41	1.0	26.81	3.325	2.0	41.70	2.17	0.2
38.45	2.34	0.5	27.32	3.246	2.0	52.75	1.74	0.1
39.90	2.26	0.2	28.52	3.130	7.0	56.68	1.62	0.3
42.30	2.14	0.3	28.94	3.085	2.0			
45.15	2.01	0.4	29.68	3.010	1.0			
46.80	1.94	0.6	31.13	2.873	17.			
47.20	1.93	0.7	32.18	2.782	47.			
48.80	1.87	0.3	33.00	2.714	1.0			
50.25	1.82	0.4	33.43	2.608	4.0			
54.90	1.67	0.2	34.05	2.633	2.0			
57.50	1.60	0.3	35.44	2.533	3.0			
			36.38	2.469	2.0			
			38.45	2.341	6.0			
			42.18	2.142	1.0			
			42.59	2.123	2.0			
			43.50	2.080	13.			
			44.03	2.057	3.0			
			44.90	2.019	2.0			
			46.10	1.969	2.0			
			46.82	1.940	2.0			
			48.25	1.886	2.0			

Table 3. Powder X-Ray Diffraction Data for 3,6-CDF 1,2,3,4-CDD, and 1,2,4-CDD.

3,6-CDF			1,2,3,4-CDD			1,2,4-CDD		
2θ	dÅ	I/Io	2θ	dÅ	I/Io	2θ	dÅ	I/Io
4.89	18.09	0.7	5.01	17.64	0.9	2.40	36.81	15.
8.10	10.92	16.	7.90	11.19	3.0	8.40	10.53	21.
14.30	6.196	1.0	10.39	8.514	100.	10.45	8.465	100.
15.05	5.889	1.0	11.95	7.406	47.	11.08	7.985	77.
16.22	5.466	40.	15.80	5.609	2.0	13.42	6.598	84.
19.45	4.565	4.0	16.74	5.296	2.0	15.45	5.735	15.
24.35	3.656	100.	20.87	4.256	39.	16.29	5.441	15.
25.45	3.500	3.0	21.73	4.090	28.	18.62	4.765	19.
26.30	3.389	2.0	24.15	3.685	2.0	20.18	4.400	25.
32.75	2.735	65.	24.30	3.663	2.0	21.03	4.224	41.
41.21	2.191	18.	26.35	3.382	1.0	21.43	4.146	26.
44.15	2.051	4.0	26.95	3.308	1.0	22.01	4.038	12.
50.00	1.824	13.	27.93	3.194	2.0	22.38	3.972	22.
54.65	1.679	1.0	28.90	3.089	2.0	23.15	3.842	18.
59.18	1.561	1.0	31.55	2.836	84.	23.70	3.754	8.0
			32.16	2.783	16.	24.40	3.648	12.
			33.87	2.646	4.0	25.05	3.555	32.
			36.60	2.455	5.0	25.65	3.473	23.
			37.58	2.393	14.	26.00	3.427	16.
			39.10	2.304	2.0	27.50	3.243	34.
			40.10	2.249	2.0	28.20	3.164	27.
			41.45	2.178	0.9	28.85	3.095	40.
			42.55	2.125	5.0	29.65	3.013	45.
			43.00	2.103	3.0	30.52	2.929	12.
			44.40	2.040	2.0	30.84	2.899	12.
			44.75	2.025	2.0	31.25	2.862	23.
			46.60	1.949	5.0	31.82	2.812	12.
			53.15	1.723	1.0	32.33	2.769	11.
			53.75	1.699	9.0	33.36	2.686	11.
			55.50	1.656	0.8	34.05	2.633	8.0
			57.40	1.605	1.0	34.54	2.597	15.
			62.00	1.497	0.9	36.40	2.468	18.
			63.17	1.472	2.0	37.10	2.423	10.
			66.35	1.409	1.0	37.75	2.383	14.
			72.62	1.302	0.9	39.05	2.307	14.
						41.05	2.199	18.
						42.07	2.148	15.
						42.44	2.130	19.
						47.85	1.901	10.

Table 4.   Powder X-Ray Diffraction Data for 2,3-CDD, 2,7-CDD, and 2,3,7,8-CDF.

2,3-CDD			2,7-CDD			2,3,7,8-CDF		
2Θ	dÅ	I/Io	2Θ	dÅ	I/Io	2Θ	dÅ	I/Io
5.00	17.67	3.0	4.95	17.85	2.0	5.03	17.57	15.
7.68	11.51	4.0	8.90	9.94	51.	12.50	7.081	18.
15.70	5.64	100.	13.40	6.607	10.	15.20	5.829	100.
19.65	4.52	3.0	17.70	5.011	49.	15.80	5.609	50.
21.70	4.10	2.0	20.20	4.396	4.0	21.08	4.214	20.
22.10	4.02	2.0	23.25	3.826	2.0	22.00	4.040	23.
23.10	3.85	5.0	24.62	3.616	13.	22.75	3.909	25.
23.70	3.75	5.0	27.00	3.302	100.	23.50	3.786	33.
24.26	3.67	3.0	27.70	3.220	4.0	25.25	3.527	23.
25.38	3.51	3.0	29.45	3.033	3.0	26.30	3.389	25.
25.70	3.47	3.0	30.45	3.935	8.0	28.00	3.187	50.
26.96	3.31	11.	33.37	2.685	2.0	28.70	3.110	53.
28.00	3.19	7.0	36.28	2.476	46.	31.00	2.885	25.
28.25	3.16	6.0	41.13	2.195	3.0	34.05	2.663	38.
30.15	2.96	5.0	42.37	2.133	6.0	35.15	2.553	20.
31.80	2.81	36.	46.17	1.966	3.0	40.05	2.251	20.
33.28	2.69	3.0	48.48	1.878	3.0	48.75	1.868	20.
35.78	2.51	3.0	50.80	1.797	13.	59.60	1.551	23.
36.50	2.46	2.0	56.14	1.638	3.0			
38.47	2.34	2.0						
38.90	2.32	3.0						
40.50	2.23	2.0						
42.74	2.12	3.0						
43.77	2.07	2.0						
46.60	1.95	3.0						
48.60	1.87	4.0						
49.65	1.84	2.0						
50.01	1.82	3.0						

Table 5. Powder X-Ray Diffraction Data for 1,2,3,4,7,8-CDD, 1,2,3,6,7,8-CDD, and 1,2,3,7,8,9-CDD.

1,2,3,4,7,8-CDD			1,2,3,6,7,8-CDD			1,2,3,7,8,9-CDD		
2Θ	dÅ	I/Io	2Θ	dÅ	I/Io	2Θ	dÅ	I/Io
6.58	13.43	42.	4.80	18.4	15.	4.40	20.08	3.0
9.25	9.560	21.	6.19	14.3	8.0	4.95	17.85	4.0
10.40	8.506	65.	9.90	8.930	12.	11.87	7.455	100.
13.18	6.717	23.	10.50	8.425	81.	15.21	5.825	3.0
13.95	6.348	23.	13.48	6.568	79.	17.35	5.111	3.0
14.72	6.018	56.	17.85	4.969	15.	23.10	3.850	6.0
16.80	5.277	21.	18.80	4.720	15.	23.96	3.714	26.
19.23	4.615	49.	19.98	4.444	54.	25.16	3.539	5.0
20.90	4.250	51.	21.21	4.189	85.	26.50	3.363	8.0
23.45	3.793	47.	23.75	3.746	19.	27.80	3.209	8.0
24.60	3.619	53.	26.06	3.419	23.	28.53	3.129	5.0
26.30	3.389	98.	27.30	3.267	56.	30.50	2.931	4.0
26.40	3.376	100.	27.71	3.219	42.	31.85	2.810	4.0
27.10	3.290	30.	29.12	3.066	17.	33.13	2.704	8.0
28.27	3.157	53.	30.17	2.962	56.	35.05	2.560	8.0
28.80	3.100	58.	32.10	2.788	100.	36.26	2.477	5.0
29.78	3.000	65.	33.23	2.696	50.	37.80	2.380	4.0
31.60	2.831	33.	33.39	2.683	52.	38.89	2.316	4.0
33.00	2.714	65.	34.75	2.581	40.	41.76	2.163	4.0
34.05	2.633	56.	35.99	2.495	42.	46.57	1.950	4.0
36.00	2.495	33.	37.20	2.417	15.	47.40	1.918	3.0
38.75	2.353	42.	38.61	2.332	42.	49.40	1.845	3.0
39.05	2.307	21.	40.68	2.218	17.	54.60	1.681	3.0
44.20	2.049	23.	41.55	2.173	15.			
45.00	2.014	28.	42.20	2.141	15.			
			44.15	2.051	27.			
			46.34	1.959	17.			
			47.60	1.910	17.			
			50.50	1.807	13.			
			51.90	1.762	17.			
			54.95	1.671	13.			
			55.68	1.651	29.			

Table 6.   Powder X-Ray Diffraction Data for 1,2,3,4,6,7,9-CDD and
1,2,3,4,6,7,8-CDD.

1,2,3,4,6,7,9-CDD			1,2,3,4,6,7,8-CDD		
2Θ	dÅ	I/Io	2Θ	dÅ	I/Io
7.49	11.80	100.	7.79	11.35	16.
11.95	7.406	14.	10.85	8.154	10.
15.09	5.871	74.	11.31	7.823	76.
18.52	4.791	10.	11.63	7.609	10.
21.85	4.068	11.	15.20	5.829	8.0
22.64	3.927	11.	15.69	5.648	100.
23.35	3.809	28.	17.75	4.997	5.0
24.75	3.597	16.	20.92	4.246	9.0
26.18	3.404	19.	21.65	4.105	7.0
26.46	3.368	13.	22.80	3.900	19.
27.25	3.272	16.	23.61	3.768	40.
28.80	3.100	16.	25.09	3.549	7.0
29.45	3.033	15.	25.46	3.498	7.0
30.40	2.940	20.	25.71	3.465	7.0
30.56	2.925	21.	26.23	3.397	41.
30.90	2.894	15.	26.95	3.308	10.
31.60	2.831	10.	28.99	3.080	29.
34.20	2.622	26.	30.24	2.955	9.0
34.45	2.603	34.	30.75	2.908	6.0
34.58	2.594	31.	31.63	2.829	10.
38.60	2.332	14.	32.17	2.782	6.0
39.65	2.273	10.	32.70	2.738	22.
41.95	2.154	13.	33.15	2.702	14.
42.68	2.118	10.	36.74	2.446	31.
46.65	1.947	16.	39.93	2.258	12.
			46.16	1.966	16.
			48.43	1.879	26.

Table 7. Powder X-Ray Diffraction Data for 1,2,3,4,6,7,8,9-CDF and
1,2,3,4,6,7,8,9-CDD.

1,2,3,4,6,7,8,9-CDF			1,2,3,4,6,7,8,9-CDD		
2Θ	dÅ	I/Io	2Θ	dÅ	I/Io
9.22	9.591	40.	7.49	11.80	38.
11.05	8.007	25.	11.77	7.519	27.
11.66	7.589	49.	12.04	7.351	26.
13.35	6.632	12.	14.48	6.117	10.
14.85	5.965	52.	14.96	5.922	74.
15.75	5.626	21.	16.81	5.274	3.0
17.03	5.206	19.	23.15	3.842	36.
18.23	4.866	13.	24.82	3.587	11.
18.90	4.695	10.	26.32	3.386	24.
19.48	4.557	19.	27.06	3.295	38.
20.05	4.428	12.	29.25	3.053	7.0
22.39	3.971	28.	30.18	2.961	50.
23.08	3.853	72.	32.85	2.726	7.0
23.48	3.789	43.	33.49	2.676	15.
24.27	3.667	16.	34.06	2.632	16.
25.13	3.544	19.	34.32	2.613	100.
26.14	3.409	27.	35.98	2.496	4.0
26.93	3.311	27.	37.55	2.395	4.0
28.25	3.159	94.	38.14	2.359	22.
28.54	3.127	93.	41.33	2.184	7.0
28.95	3.084	31.	41.92	2.155	5.0
30.06	2.973	37.	44.51	2.035	10.
31.24	2.863	60.	45.30	2.002	42.
31.81	2.813	90.	46.17	1.966	60.
32.13	2.786	100.	46.66	1.947	8.0
33.92	2.643	63.	49.83	1.830	4.0
34.57	2.594	96.	50.68	1.801	4.0
35.80	2.508	69.	53.32	1.718	9.0
38.67	2.328	21.	54.01	1.698	6.0
40.90	2.206	12.	54.42	1.686	8.0
42.21	2.141	15.	56.03	1.641	4.0
43.20	2.094	19.	58.42	1.580	4.0
44.05	2.056	25.	60.06	1.540	4.0
46.12	1.968	24.	60.80	1.523	5.0
47.75	1.905	18.	61.78	1.502	11.
50.60	1.804	15.	63.02	1.475	13.
51.25	1.782	18.	63.91	1.457	8.0
51.55	1.773	18.	68.03	1.378	3.0
52.29	1.749	12.			
54.57	1.682	12.			
55.29	1.661	19.			
58.50	1.578	18.			

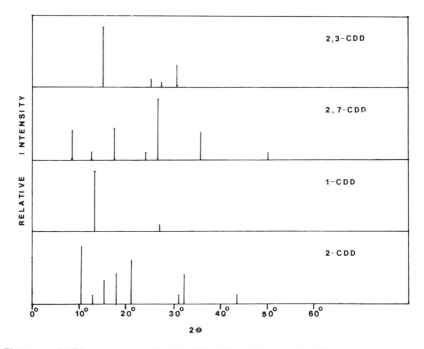

Figure 1. PXRD patterns for 2,3-CDD, 2,7-CDD, 1-CDD, and 2-CDD.

analysis. However, the 2,3-CDD and the 2,7-CDD possess resolved, unique lines at 15.70 2Θ and at 8.9, 27.0 and 36.28 2Θ, respectively, which may be used in their analysis.

The 1,2,4-CDD and 1,2,3,4-CDD PXRD patterns are superimposed in plot B of Figure 5. It can be seen that the tetra-CDD can be easily identified in the presence of the tri-CDD by the line at 31.55 2Θ. In addition, two lines of the 1,2,4-CDD that distinguish it from the tetra-CDD are at 2.40 2Θ and 13.42 2Θ. The line at 2.40 2Θ of the 1,2,4-CDD PXRD pattern is unique in the sense that it can serve as its marker in mixtures of all compounds investigated; however, this is true only at high concentrations because of its low relative intensity.

In plot C of Figure 5, the PXRD patterns of the 1,2,3,4,7,8-CDD, the 1,2,3,7,8,9-CDD and the 1,2,3,6,7,8-CDD are compared. They have been designated by Mh, Mv and C to denote their respective molecular horizontal mirror plane, vertical mirror plane or center of symmetry. The 1,2,3,6,7,8-CDD is easily distinguished from the 1,2,3,4-CDD and its Smiles rearrangement partner, the 1,2,3,7,8,9-CDD, by its intense lines at 13.48, 21.21 and 32.10 2Θ. The 1,2,3,4,7,8-CDD is marked by its strong lines at 26.30 and 26.40 2Θ. The 1,2,3,7,8,9-CDD can be identified in the presence of the other two hexa-CDD's, with care, using the lines at 11.87 and 23.96 2Θ.

Plot E of Figure 5 exhibits the composite of the 3,6-CDF, the 2,3,7,8-

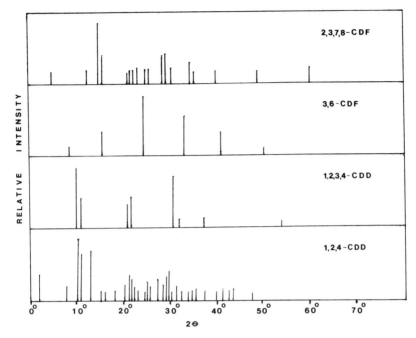

Figure 2.   PXRD patterns for 2,3,7,8-CDF, 3,6-CDF, 1,2,3,4-CDD, and 1,2,4-CDD.

CDF and the 1,2,3,4,6,7,8,9-CDF PXRD patterns. The di-CDF is easily distinguished from the other two CDF's by its intense line at 24.35 2Θ. With care, the 2,3,7,8-CDF can be identified in this mixture by its strong line at 15.20 2Θ. The identification of the Octa-CDF in this mixture is difficult because of interfering lines of the 2,3,7,8-CDF at 28.25, 28.54, 28.00 and 28.70 2Θ; however, with care, the octa-CDF may be differentiated from the other CDF's of this mixture by using its intense line at 31.81 2Θ.

CONCLUSIONS

We have presented in this chapter the powder X-ray diffraction patterns of sixteen samples of PCDF's and PCDD's which have been carefully characterized by EC/GC, GC/MS and DSC to establish their purity and verify their chlorine substitution.

In the absence of "validated" PCDF and PCDD standards, PXRD patterns of unknown or suspect PCDF's or PCDD's may be compared to the reported PXRD patterns. In addition, mixtures of these compounds may be qualitatively analyzed in the milligram range by PXRD.

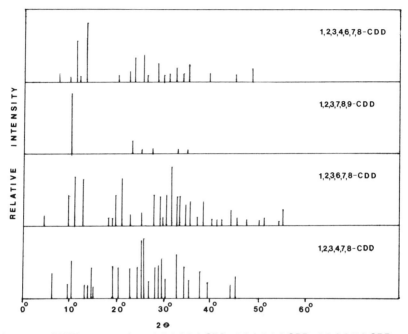

Figure 3. PXRD patterns for 1,2,3,4,6,7,8-CDD, 1,2,3,7,8,9-CDD, 1,2,3,6,7,8-CDD and 1,2,3,4,7,8-CDD.

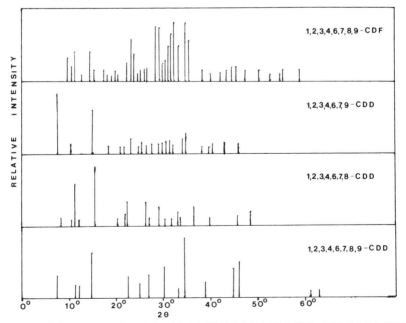

Figure 4. PXRD patterns for 1,2,3,4,6,7,8,9-CDF, 1,2,3,4,6,7,9-CDD, 1,2,3,4,6,7,8-CDD, and 1,2,3,4,6,7,8,9-CDD.

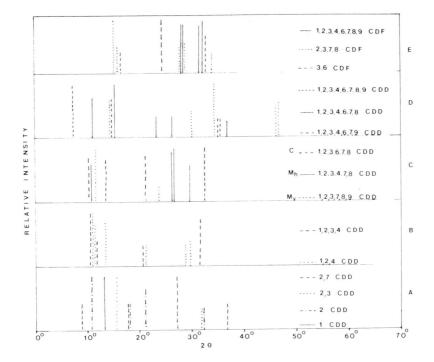

RELATIVE INTENSITY

— 1,2,3,4,6,7,8,9 C D F
····· 2,3,7,8 C D F   E
- - - 3,6 C D F

····· 1,2,3,4,6,7,8,9 C D D
— 1,2,3,4,6,7,8 C D D   D
- - - 1,2,3,4,6,7,9 C D D

C - - - 1,2,3,6,7,8 C D D
M$_h$ — 1,2,3,4,7,8 C D D   C
M$_v$ ····· 1,2,3,7,8,9 C D D

- - - 1,2,3,4 C D D   B
····· 1,2,4 C D D

- - - 2,7 C D D
····· 2,3 C D D   A
- · - · 2 C D D
— 1 C D D

0°   10°   20°   30°   40°   50°   60°   70°
2 θ

Figure 5. Composite PXRD patterns.

REFERENCES

1. Cantrell, J. S., N. C. Webb, and A. J. Mabis. "The Identification and Crystal Structure of a Hydropericardium—Producing Factor: 1,2,3,7,8,9-Hexachlorodibenzo-p-dioxin," *Acta Cryst.* B(1):150–156 (1969).

2. Boer, F. P., M. A. Neuman, F. P. Van Remoortere, P. P. North, and H. W. Rinn. "X-Ray Diffraction Studies of Chlorinated Dibenzo-p-dioxins," in *Chlorodioxins Origin and Fate*, E. H. Blair, Ed., (Washington, DC: American Chemical Society, 1973) Adv. Chem. Ser. 120, p. 33.

3. Gray, A. P., S. P. Cepa, and J. S. Cantrell. "Intervention of the Smiles Rearrangement in Syntheses of Dibenzo-p-dioxins. 1,2,3,6,7,8- and 1,2,3,7,8,9-Hexachlorodibenzo-p-dioxin (HCDD)," *Tetrahedron Lett.* 33:1872–6 (1975).

4. Slonecker, P. J., and J. S. Cantrell. "Identification of Polychlorinated Dibenzo-p-dioxin Isomers by Powder X-Ray Diffraction with Electron Capture Gas Chromatography," *Anal. Chem.* 54:1543–1547 (1983).

5. Slonecker, P. J. "Characterization of Polychlorinated Dibenzo-p-dioxins," MS Thesis, Miami University, Oxford, OH (1982).

6. Driscoll, G. L., I. N. Dulling, and F. Magnotta. "Purity Determination by Differential Scanning Calorimetry," in *Proceedings of the American*

Chemical Society Symposium on Analytical Chemistry (New York: Plenum Press, 1968), p. 271.

7. Buser, H.-R., and C. Rappe. "Identification of Substitution Patterns in Polychlorinated Dibenzo-p-dioxins (PCDD's) by Mass Spectrometry," *Chemosphere* 2:199–211 (1978).

CHAPTER 30

Mathematical and Statistical Methods in Modeling the Seveso Dioxin Episode

Sergio P. Ratti, Giuseppe Belli, Agostino Lanza, and Silvia Cerlesi

INTRODUCTION

On July 10, 1976, tetrachlorodibenzo-p-dioxin (TCDD) was accidentally released from a smokestack of the ICMESA plant near Seveso, Italy. Many aspects of this incident, known as the Seveso episode, have been investigated.[1,2]

Under a contract between Regione Lombardia, University of Pavia and the Commission of the European Community, our group elaborated the possible "rationale" for the interpretation of the data collected in different campaigns from 1976 through 1984.

Considering the accident as an historical fact, we extracted the maximum possible statistical information from every bit of measurement performed in the territory around the ICMESA factory.

Figure 1a shows the location of the chemical factory and a map of the surrounding territory. TCDD was spread primarily in a region of about 87 ha (Zone A) still closed to the public, but also partially in an area of about 270 ha along the direction of the blowing wind (Zone B). An additional area of approximately 1430 ha (Zone R) has been kept under surveillance and control for safety purposes.

In several campaigns, samples were extracted and analyzed following preset mapping grids. As an example, Figure 1b shows the location of the

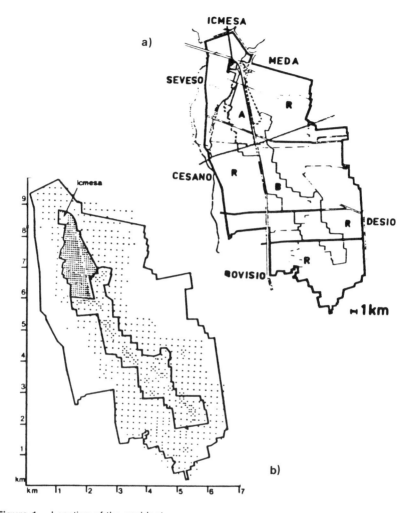

Figure 1. Location of the accident.
a) Topography of the site, where zones A, B and R are indicated.
b) Mapping grids of the 1976/1977 campaigns reproducing the (x,y) coordinates of the sampling locations.

measurements made in 1976. As can be seen, the steps were different in the different regions (one can also notice minor regions where collection of the samples was impossible, such as the "Milan-Como highway," well visible in Zone A).

OVERALL STATISTICAL SIGNIFICANCE OF THE ANALYSIS

The difficulties encountered in giving a full description of the contamination of the territory arise mainly from the following facts:

1. The total area to be considered, $S_{(tot)}$, is much larger than the area $S_{(an)}$ (sum of all collected samples "submitted to chemical analysis"). For instance, in the 1976 campaign, $S_{(tot)} = 1.79 \times 10^7$ m² and $S_{(an)} = 7.65$ m².
2. Local measurements fluctuate widely.
3. Data collected in different years may have been chemically analyzed with different methods.

Point 1 imposes a severe warning on the extrapolation of the data, and can be overcome by using nontrivial techniques suggested by numerical analysis. Indeed the "magnification factor" between "actual information" S(an) and "representation" S(tot) is

$$R_{[A+B+R]} = S_{(tot)}/S_{(an)} = (1.79 \times 10^7)/7.65 = 2.34 \times 10^6 \qquad (1)$$

Even limited to Zone A, the magnification factor is extremely large, i.e.

$$R_{[A]} = S'_{(an)} = (8.73 \times 10^5)/2.87 = 3.04 \times 10^5 \qquad (2)$$

Point 2 can be overcome by verifying the nature of the phenomenon and by estimating the expected fluctuations.

Point 3 requires a systematic investigation of the possiblities to "translate" numerical values "from one method into another," via a proper renormalization.

A first determination of the maximum contamination regression line on the data from the campaign of 1976 has suggested an "octet" procedure for a checkup mapping in 1979,[3] which was able to save about 70% of the chemical analytical effort.

RISK MAPPING AND RELATED PROBLEMS

The possibility of drawing "risk maps" is bound to the statistical handling of numerical values of the contaminant (e.g., in micrograms per square meter) close to the "detection limit" of the apparati used in the chemical analysis. The information collected in the region of high contamination (H.C.) can be properly transferred to the low contamination (L.C.) region.

Improving the Data Analysis

In particular, in the highly contaminated region the fluctuations of the values in the TCDD content ($\mu g/m^2$) around the average value is lognormal;[4] therefore, the adoption of the variable $w = \ln(z) - z$ being the amount of TCDD in $\mu g/m^2$—is most suitable for data handling.

Figure 2 shows the data collected in the campaigns of 1976 in Zone A. It is obvious how much easier it is to reproduce the numerical values of w

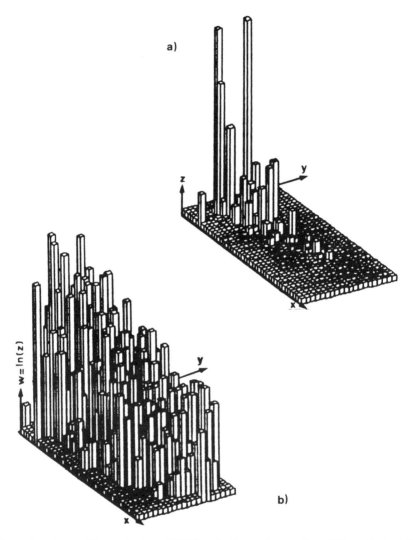

Figure 2.a Lego-plot reproducing z (TCDD) vs (x,y) sampling locations [(50m x 50m) grid] in zone A.
2.b Lego-plot reproducing w = ln[z (TCDD)] vs (x,y) in zone A.

(Figure 2b) rather than the values of z (Figure 2a). In an earlier study[3] the lognormal nature of the phenomenon was verified experimentally, and it is consistent with a reasonable model of toxic powder release from a smoke-stack plus fallout and gravity deposition on the ground.

From Figure 3 we learn that we cannot do better than $\sigma = 0$ (i.e., zero variance!). Actually, at the point where σ goes to zero, the real value of w

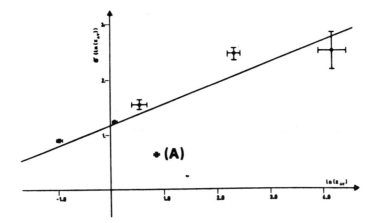

Figure 3. Fitted straight line to the measured correlation between standard deviations and average values for five independent campaigns in different regions.

does not go to zero, but rather the value $w_{(meas)}$, which is measured by the analytical method.

Therefore, the value $w_{(meas)}$ corresponding to $\sigma = 0$ is the minimum possible detectable amount of contaminant with the method actually employed in that particular investigation, and it indicates the overall maximum possible sensitivity of the complex "instrument plus analytical method," (number of extractions, etc.).

The point labeled (A) in Figure 3 indeed corresponds to the improved chemical analytical method, adopted after 1980, which is clearly outside the linear trend.

The best fitted linear interpolation among the five points of Figure 3 gives a value

$$w = \ln(z) = -(3.11 \pm 0.51)$$

or   (3)

$$x = (0.045 \pm 0.007) \, \mu g/m^2$$

This result is relevant in view of the following facts: a) It is obtained and inferred on a purely statistical ground, and ignores what method is involved. It has, therefore, a rather general validity, provided the "nature" of the phenomenon is known. The numerical value,[4] when converted into the proper units,[5] corresponds to a theoretical limit $z_{(th)} = 1$ ppt, consistent with the best estimates on the chemical method used.[5] b) It can give a reasonable estimate of the dispersion even in "quick" mapping campaigns in which, for whatever reason, σ cannot be determined.

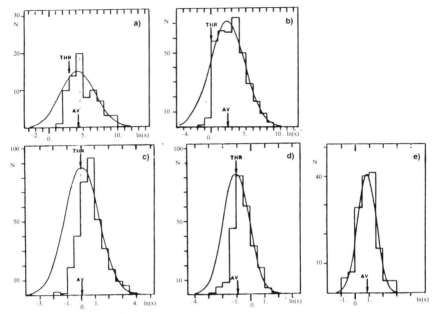

Figure 4.   Distribution of the values w = ln (TCDD), in different zones and periods.
a) zone A 1976;
b) zone A 1976/77;
c) zone (B + R) 1976/77;
d) zone R 1980;
e) zone B 1980/81 (new chemical method).

The Final Risk Mapping

On the basis of the analysis performed above it is possible to draw a properly prepared risk map.[4]

First of all, if, in a given zone around Seveso a campaign of N measurements gives (N-K) values below the sensitivity of the instrument and K "measured" quantities of TCDD with average value $z_{(av)}$, the lognormal assumption allows a correction of $\ln[z(av)]$ into $\ln[z(corr)]$, to which a given value of σ $[\ln[z_{(corr)}]]$ corresponds.

Some examples from an earlier analysis[4] are shown in Figure 4. The parameters $w_{(av)} = [\ln(z)_{av}]$ and σ (w) of the gaussians (solid lines) are also taken from this earlier study.[3]

For instance, in Figure 4a the "missing events" on the left-hand side of the distribution (53 $\pm$ 8 events, estimated as difference between gaussian curve and histogram) are consistent with the number of samples in which the amount of TCDD could not be detected (38 ± 6 samples) in the highly contaminated [$w_{(av)} = 2.5$] area. (The difference A = (53-38) = (15 ± 10) is consistent with zero, within 1.5 standard deviations).

On the other hand, Figure 4e shows that with the improved analytical method, no losses are seen in the data of the campaign of 1980/81, in a moderately contaminated area [$z_{(av)} = 0.85$].

Given these corrections, it is possible to define the following factors:

F_{poll} = pollution factor: the ratio between the number of contaminated sampling points and the total number of samples [this ratio is corrected for the estimated lost events resulting from measurements below the sensitivity of the instruments]

$P_{poll}(z')$ = the maximum pollution probability: the probability that a given random sample value exceeds the specified quantity z' [this probability can be immediately calculated, given the graph of Figure 3].

On the basis of these assumptions we may define a "risk probability" as

$$P_{risk}(z') = F_{poll} \times P_{poll\,(z')} \tag{4}$$

$P_{risk(z')}$ defines the absolute probability that any sample will show a contamination larger than z'. It is worth noting here that $P_{risk}(...)$ can be estimated for any given arbitrary preset value z_{preset}, on the basis of the lognormal distribution.

CONTINUOUS REPRESENTATION OF THE GROUND CONTAMINATION

One of the appealing problems in environmental science, in the case of a chemical accident, may be to describe the contaminant distribution in the geographical area implied in the phenomenon. Indeed, it is not a trivial task.

The problem of finding a continuous mathematical function, representing

$$z = f(x,y) \tag{5}$$

in an area which is over a million times larger than the area in which the variable z is known, is similar to the problem of finding a mathematical function $z = f(x,y)$, "measured" in a very limited number N of points [x(i),y(i)] so that N values are known:

$$z(i) = f[x(i),y(i)] \qquad i = 1,2,.....,N. \tag{6}$$

THE SEVESO CASE (ZONE A)

This problem has been tackled[6] for Zone A around Seveso. From numerical analysis, we learn that it is possible to approximate even an "unknown" function to any degree of accuracy, provided the numerical value of the unknown function is known in a given number of points.[7]

ICMESA

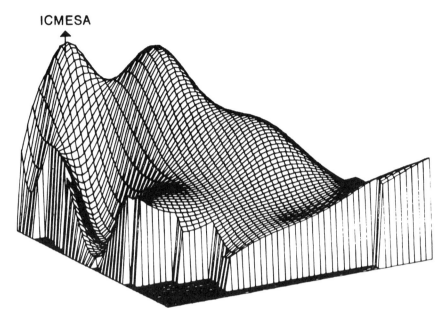

Figure 5. Graphic display of the approximate function fitted to the samples of 431 measurements (enriched by the Shepard method) in zone A. Campaign of year
(1976). View rotated by 30° with respect to the x-axis.

Without entering into the details of the approach used, for which the
reader can refer to Belli et al.,[6] we simply mention that, due to the severe
amplification factors discussed earlier, we must increase artificially the
number of measured points. In order to increase the number of points
[x(k),y(k)] in which z can be assumed to be known, the method of "pseudo-
measurements" suggested by Shepard and Gustafson et al.[8] has been invalu-
able.

Furthermore, the approximate functions used have been the linear combination of an orthonormal set of bidimensional Tchebychev polynomials.[7]
The result of the approximation is shown in Figures 5 and 6 in three-
dimensional representations.

The vertical extension of the figure (vertical axis) is proportional to $w =$
$\ln(x)$, the value of the logarithm of the TCDD density in $\mu g/m^2$. The two
inclined axes are geometrical coordinates, respectively parallel to the east-
west and the north-south directions shown for instance in Figure 1b.

In Figures 5 and 6 one can immediately notice the "floating effect" typical
of the release of powder from a smokestack; the height of the three-dimen-
sional curve [proportional to the logarithm of the amount of TCDD deposited on the ground] decreases with increasing distance from the ICMESA
factory, and follows preferentially the direction of the wind blowing at the
time of the accident.

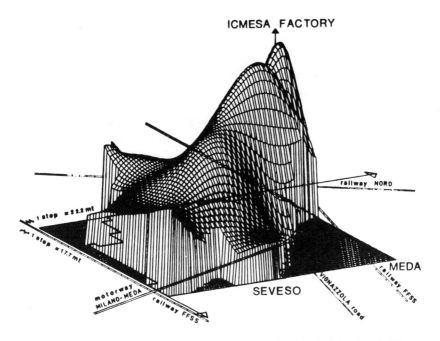

Figure 6.   Graphic display of the approximant function fitted to the samples of 431 measurements (enriched by the Shepard method) in zone A. Campaign of year (1976). View from the direction of the blowing wind (–60° rotation).

As a matter of practice, the indication of a geometrical path in the (x,y) plane will immediately lead to an easy calculation of the integral amount of TCDD deposited along the particular path.

In spite of the very severe limitations of the original data, we consider the present result to be very satisfactory.

The graphical representation given in Figures 5 and 6 corresponds to a set of numerical paramethers so that the "probable value" of $w = \ln[z_{TCDD}]$ can be numerically estimated at any given point of Zone A.

Extention to All the Seveso Area

In order to apply the method to all the area (Zones $A + B + R$) we have first to recover all possible measurements, even if performed using different analytical methods, in different periods of time.

The comparison of two different chemical analytical methods and the indication of a possible strategy to approach the problem, ignoring the "chemical" content of the matter, will not be discussed here. It will be explicitly presented at the next Symposium on Dioxins (Bayreuth, West Germany). Here we can only state that we have not yet succeeded in giving a full representation of the phenomenon, although we have indeed been able to reach acceptable results if we "cut" the surface into smaller portions, within which also small variations (in the low-contamination regions) can be seen.

The failure (which is only of a pictorial nature) of the final application seems to be mostly due to two reasons: a) the very limited number of data points. As concluded in Equation 1, they represent only one part per 2.3 million parts of the area we want to be described. b) There is a large variation of the values of $w = \ln(z)$ along about 7 units from values around 4 down to values around –3. This fact reduces dramatically the geometrical sensitivity. As a consequence, the surface in Zone R is reduced to a very flat, anonymous surface.

REFERENCES

1. Hutzinger, O., R. W. Frei, E. Merian, and F. Pocchiari, Eds. *Chlorinated Dioxins and Related Compounds: Impact on the Environment.* Pergamon Series on Environmental Science, Vol. 5 (Elmsford, NY: Pergamon Press, 1982).
2. *Chemosphere* 12(4,5) (1983).
3. Belli, G., G. Bressi, E. Calligarich, S. Cerlesi, and S. Ratti. In: *Chlorinated Dioxins and Related Compounds: Impact on the Environment.* Pergamon Series on Environmental Science, Vol. 5 (Elmsford, NY: Pergamon Press, 1982), p. 155.
4. Belli, G., G. Bressi, S. Cerlesi, and S. Ratti. *Chemosphere* 12:517 (1983).
5. Facchetti, S. Commission of the European Communities, Joint Research Centre, Ispa (VA), Italy. Personal communication (1985).
6. Belli, G., S. Cerlesi, E. Milani, and S. Ratti. "Statistical Interpolation Model for the Description of Ground Pollution due to the TCDD Produced in the 1976 Chemical Accident at Seveso in the Heavily Contaminated Zone A," report IFNUP/RL 14–83, University of Pavia preprint, Pavia, Italy.
7. Vitali, G. and G. Sansone, "Series Development of Orthogonal Functions," C. N. R. Monographies in Applied Matematics. (Zanichelli, Bologna, 1946).
8. Shepard, D. *Proceedings of the 23rd ACM National Conference*, Las Vegas, NV (1968), p. 517.
9. Gustafson, S. A., K. O. Kortaned, and J. R. Sweigart. *J. Appl. Meteorol.* 16:1234 (1977).

SECTION VI

Synthesis and Destruction

CHAPTER 31

Preparation of Dioxin Standards for Chemical Analysis

Leslie T. Gelbaum, Donald G. Patterson, and Donald F. Groce

INTRODUCTION

The chlorinated dibenzo-p-dioxins (CDDs) have been shown to be highly toxic trace contaminants in the environment. Since the toxicity of these compounds is highly dependent on the number and substitution pattern of the chlorine, one must quantify the presence of specific toxic isomers to determine the true toxicity of an environmental sample. We have, therefore, begun a program to synthesize and purify specific toxic isomers so that quantitative standards can be prepared.

EXPERIMENTAL

The synthesis of CDDs has been carried out by the self-condensation of alkali metal salts of orthochlorinated phenols[1] or the more facile condensation of the alkali metal salts of chlorinated catechols with chlorinated benzenes[2] or orthochlorinated nitrobenzenes.[3] The chlorinated benzenes and most of the orthochlorinated nitrobenzenes needed for preparing the polychlorinated dibenzo-p-dioxins (PCDDs) are commercially available; however, none of the chlorinated catechols or 2,3,6– and 2,4,6–trichloronitrobenzene are commercially available.

The 4–monochlorocatechol, 4,5–dichlorocatechol, and 3,4,5–trichlorocatechol were prepared by the direct chlorination of catechol with sulfuryl chloride.[3] The 3-monochlorocatechol and the 3,4-, 3,5- and 3,6-dichlorocatechols were synthesized by using the Reimer-Tiemann reaction[4] on the appropriate chlorinated phenol, followed by the Dakin oxidation reaction[5] to give the chlorinated catechol. 2,4,6-Trichloronitrobenzene was prepared by nitration of 2,4,6-trichlorobenzene.[6] 2,3,6-Trichloronitrobenzene was synthesized from p-dichlorobenzene by initial dinitration followed by amination using ethanolic ammonia.[7] The isolated 3,6-dichloro-2-nitroaniline was converted to 2,3,6-trichloronitrobenzene using standard Sandmeyer reaction conditions.

PCDDs were prepared using both the catechol chlorobenzene and the catechol orthochloronitrobenzene reactions. The reaction of chlorinated catechols with chlorinated benzenes required approximately one week to reach completion. During this time, the PCDDs of one less chlorine were produced as the main impurity. Purification of these materials was difficult because of the presence of these additional PCDDs. When the PCDDs were prepared using chlorinated catechols with orthochlorinated nitrobenzenes, the reaction was complete in less than three hours and the main impurity was the nitro CCDs. This impurity was easily removed by silica gel chromatography of the reaction mixture, with hexane used as the eluting solvent. Further purification, if necessary, was carried out with the use of a 1-in. Dynamax C-18 column with aqueous acetonitrile as the eluting solvent.

Using the above procedures, we prepared the 22 tetrachlorinated, one pentachlorinated, and three hexachlorinated dibenzo-p-dioxins in milligram quantities. When two or more isomers were formed, they were not separated. Table 1 lists the dioxins prepared, the starting materials and the yields of purified PCDDs. The purity of the samples was determined by gas chromatography/electron capture detection (GC/ECD), gas chromatography/Fourier infrared detection (GC/FID), and high-resolution GC/MS; purity was determined to be greater than 95% for the specific isomers prepared.

For the GC analysis we first used a 30-m Supelco SP2330 column; however, this was unable to separate the 1,2,3,8- from the 1,2,3,7-tetra CDD (TCDD), the 1,2,4,6- from the 1,2,4,9-TCDD, and the 1,2,4,7- from the 1,2,4,8-TCDD. To determine that the reaction to prepare these isomers did indeed give two products, we had to develop a GC analysis to separate these isomers. We found that a 50-m CP SiL 88 column from Chrompack separated the 1,2,3,7- from the 1,2,3,8-TCDD and the 1,2,4,6- from the 1,2,4,9-TCDD; however, it did not separate the 1,2,4,7- from the 1,2,4,8-TCDD. The relative retention times for the 22 TCDD isomers on both SP2330 and CP SiL 88 are given in Table 2.

Table 1. PCDDs Prepared and Reactants.

Isomer	Chlorocatechols Catechol	Chloronitrobenzenes Hexachlorobenzene	% Yield
1234			55
1378[a]			
2378	4,5-di	1,2,4,5-tetrachlorobenzene	8
1236,1239	3,4,5-tri	2,3-di	5
1247,1248	4-mono	2,3,5,6-tetra	25
1237,1238	4-mono	2,3,4,5-tetra	20
1246,1249	3-mono	2,3,5,6-tetra	42
1368,1379	3,5-di	2,4,5-tri	37
1478,1278	4,5-di	2,3,6-tri	41
1469,1269	3,6-di	2,3,6-tri	5
1268,1279	3,5-di	2,3,4-tri	33
1369,1268,1279	3,5-di	2,3,6-tri	12
1267,1269,1289	3,4-di	1,2,3,4-tetrachlorobenzene	7
1289,1267	3,4-di	2,3,4-tri	20
12378	4,5-di	2,3,4,5-tetra	5
123478	4,5-di	hexachlorobenzene	4
123678,123789	3,4,5-tri	2,3,4,5-tetra	14

[a]Purchased from KOR Incorporated, Cambridge, MA, Catalog No. DD–42, Lot No. AMA-IX-21, 98% pure.

Table 2. Relative Retention Time of 22 TCDDS.

Isomer	Relative RT[a,b] SP2330	Relative RT[a] CP SiL 88
1368	0.773	0.732
1379	0.825	0793
1378	0.882	0.860
1369	0.915	0.904
1247	0.922	0.908
1248	0.922	0.908
1268	0.957	0.949
1478	0.988	0.987
2378	1.002	1.003
1237	1.014	1.016
1246	1.014	1.023
1234	1.014	1.027
1249	1.014	1.028
1238	1.014	1.028
1236	1.042	1.055
1279	1.042	1.054
1278	1.110	1.141
1469	1.117	1.149
1239	1.142	1.178
1269	1.166	1.211
1267	1.211	1.256
1289	1.362	1.450

[a]Retention times are calculated relative to internal, $^{13}C_{12}2,3,7,8$-TCDD.

[b]30 = m Supelco SP2330, GC program; 100° for 2 min, then 20°/min to 180°, then 30°/min to 200°, hold for 40 min. Elution order determined on a commercial mix of 22 TCDDs on the basis of Supelco data.

[c]50-m CP SiL GC program; 170° for 3 min, then 20°/min to 200°, hold for 40 min. Elution order determined by chromatography of isomers shown in Table 1.

Table 3. Concentration of 2,3,7,8-TCDD in Human Omental and Breast Adipose Samples.

Area	Sex	Age	Concentration[a] (ppt)
Atlanta	M	16	6.4/5.8
Atlanta	F	35	7.6/8.1
Atlanta	M	45	4.8/5.0
Atlanta	M	69	6.0/6.5
Atlanta	F	57	9.6
Atlanta	M	48	3.8/3.2
Atlanta	M	71	8.6
Atlanta	M	71	15.2
Atlanta	M	28	6.7/6.4
Atlanta	M	19	7.6/7.2
Atlanta[b]	F	—	6.1
Atlanta[b]	F	—	7.2
Atlanta[b]	F	—	8.9
Atlanta	F	39	3.3
Atlanta	M	78	8.9
Atlanta	F	46	8.6
Atlanta	M	67	6.5
Atlanta	F	56	8.2
Utah	M	57	7.1
Utah	M	51	3.2/3.8
Utah	M	59	6.7
Utah	M	52	5.6
Utah	M	64	2.7
Utah	F	76	11.3/12.9
Utah	F	41	7.2
Utah	F	18	3.8

[a]More than one result indicates separate 10-g samples from the same large piece of adipose tissue. All samples are ~ 10 g wet weight (weighed to the nearest tenth of a gram).
[b]Breast adipose tissue.

APPLICATION TO CHEMICAL SAMPLES

The analytical standards generated by the synthetic program have been used to prepare quantitative standards for the analysis of the 22 TCDDs in human adipose tissue in the low-part-per-trillion range.[8] We have analyzed a number of human adipose tissue samples collected from local and regional centers for TCDDs. A summary of the results is presented in Table 3, along with the information we have concerning these samples (age, sex, location). It should be pointed out that the method[8] allows for the quantitation of all 22 TCDDs; however, only the 2,3,7,8–TCDD isomer was detected in the samples. In addition, we have not found a human omental adipose sample in which 2,3,7,8–TCDD could not be detected. Very limited data are available on the levels of 2,3,7,8–TCDD that may be found in human adipose tissue. The data critically needed by researchers to define expected levels of 2,3,7,8–TCDD in the general population or in selected cohorts do not exist. Many biological, environmental and other factors may influence the 2,3,7,8–TCDD levels we observed, and these factors must be carefully described and defined before any valid interpretation of our data can be made.

DISCLAIMER

Use of trade names is for identification only and does not constitute endorsement by the Public Health Service or the U.S. Department of Health and Human Services.

REFERENCES

1. Nestrick, T. J., L. L. Lamparski, and R. H. Stehl. *Anal. Chem.* 51(13):2273 (1979).
2. Kende, A. S., J. J. Wade, D. Ridge, and A. Poland. *J. Org. Chem.* 39(7):931 (1974).
3. Gray, A. P., S. P. Cepa, I. J. Solomon, and O. Aniline. *J. Org. Chem.* 41(14):2435 (1976).
4. Vogel, A. I. *Elementary Practical Organic Chemistry Part I; Small Scale Preparation* (London:Longmans, 1966), p. 316.
5. Dakin, H. D. *American Chemical Journal* 42(6):477 (1909).
6. Huntress, E. H., and F. H. Carten. *J. Amer. Chem. Soc.* 62:511 (1940).
7. Nason, E.J. *Amer. Chem. Soc.* 40:1602 (1918).
8. Patterson, D. G., J. S. Holler, C. R. Lapeza, Jr., L. R. Alexander, D. F. Groce, R. C. O'Connor, S. J. Smith, J. A. Liddle, and L. L. Needham. "A High-Resolution Gas Chromatographic/High-Resolution Mass Spectroscopic Method for the Analysis of 2,3,7,8–TCDD in Human Adipose Tissue," Submitted to *Anal. Chem.* (1985)

CHAPTER 32

Synthesis and Characterization of Polybromo- and Polybromochloro-Dibenzo-p-Dioxins and Dibenzofurans

B. Ramalingam, T. Mazer, D. J. Wagel, C. M. Malloy,
M. L. Taylor, T. O. Tiernan, J. H. Garrett, and Arleen B. Rifkind

INTRODUCTION

Since 1970 considerable attention has been focused on the environmental distribution and the potential hazards of the polychlorinated dibenzo-p-dioxins (PCDDs) and dibenzofurans (PCDFs). Considerably less is known about the formation and toxicology of other halogenated dibenzo-p-dioxins and dibenzofurans. Reports on the possible introduction of the brominated dibenzo-p-dioxins and dibenzofurans into the environment have appeared from studies of certain fire retardants.[1,2] With the increasing acceptance of plastics in all phases of modern construction, flammability of plastics has become an increasingly important issue. Several brominated organic compounds have become commercially important fire retardants. The phenomenal growth that can be expected for these products is illustrated by the market for fire retardant-reinforced polyester laminates, consumption of which was about 100 million pounds in 1969. An increase to an estimated consumption of 1.5 billion pounds of these materials in recent times has been noted — a phenomenal fifteenfold increase over this period.[3]

Widespread usage of products containing brominated organic fire retardants provides an impetus for increased research regarding the normal stabil-

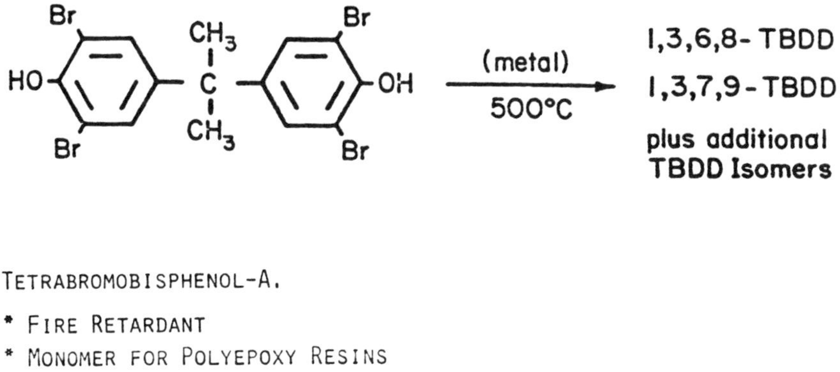

TETRABROMOBISPHENOL-A.

* FIRE RETARDANT
* MONOMER FOR POLYEPOXY RESINS

Figure 1.   Formation of TBDDs from tetrabromobisphenol-A.

ity of such fire retardants and the possible thermal decomposition products arising from these materials. In recent investigations in our laboratory,[1] the thermal decomposition of 4,4'-isopropylidene-bis-(2,6-dibromophenol), more commonly known as tetrabromobisphenol-A or Firemaster BP-4A, was demonstrated to produce 1,3,6,8- and 1,3,7,9-tetrabromodibenzo-p-dioxins (1,3,6,8- and 1,3,7,9-TBDD), as shown by the reactions in Figure 1.

Studies[2] on another important fire retardant, Firemaster BP-6, the major component of which is 2,2',4,4',5,5'-hexabromobiphenyl shows that 3,3'4,4'-tetrabromobiphenyl can arise from this bromobiphenyl through photolysis. While 3,3',4,4'-tetrabromobiphenyl in itself exhibits significant biological activity, this compound could also react further to form 2,3,7,8-tetrabromodibenzofuran (TBDF), as shown in Figure 2, since it is known that the corresponding tetrachlorobiphenyl has been shown to produce

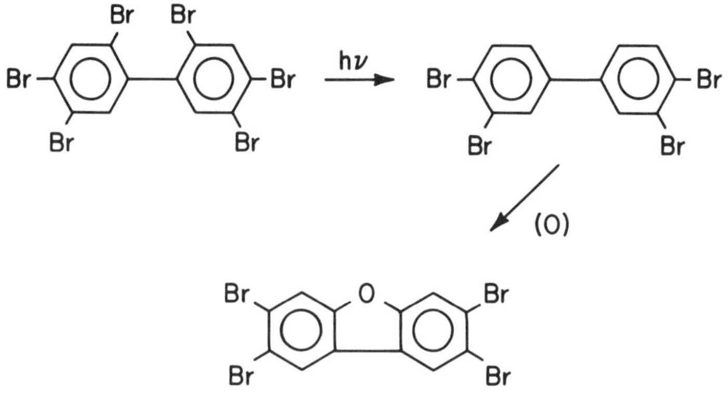

Figure 2.   Formation of 2,3,7,8-TBDF from Firemaster BP-6.

2,3,7,8-tetrachlorodibenzofuran (TCDF) via the reaction shown in Figure 3.[4]

It has also been reported on the basis of the enzyme-inductive effects seen in the chick embryo assay[5] that bromo-substituted dioxins are as potent as the corresponding chlorinated isosteres. Kende et al.[5] observed that 2,3,7,8-TCDD and 2,3,7,8-TBDD exhibited equal potencies when induction of aryl hydrocarbon hydroxylase (AHH) was evaluated in the chick embryo assay. Interestingly, Kende et al.[5] also found that 2,3-dichloro-7,8-dibromodibenzo-p-dioxin is a more potent inducer of AHH than either 2,3,7,8-TCDD or 2,3,7,8-TBDD. In addition, results of preliminary investigations accomplished in our laboratory and reported previously[1] indicate that a 1,3,6,8-/1,3,7,9-TBDD mixture is a more potent inducer of AHH than 1,3,6,8-TCDD.

In view of the widespread usage of brominated organic chemicals as fire retardants and the possibility that potentially toxic brominated organics such as brominated dioxins and dibenzofurans can arise as a result of pyrolytic degradation of these fire retardants, further investigations into the chemistry and biological activity of these compounds is warranted. Reported herein are the initial results of studies performed in this laboratory in which a series of brominated dibenzo-p-dioxins, dibenzofurans and certain chlorobromodioxins were synthesized and subsequently evaluated for biological activity using the chick embryo model.

SAFETY CONSIDERATIONS

Since the bromo-substituted dibenzo-p-dioxins and dibenzofurans are expected to be at least as potent biologically as the chloro-substituted dibenzo-p-dioxins and dibenzofurans, extreme caution is warranted in their synthesis. The syntheses reported herein were carried out in a limited access laboratory. The procedures employed to ensure the safety of personnel involved in this effort are identical to those employed in our previous programs for synthesis of TCCD isomers.[7]

SYNTHESIS OF POLYBROMODIBENZO-P-DIOXINS (PBDDs), POLYBROMOCHLORODIBENZO-P-DIOXINS (PBCDDs) AND POLYBROMODIBENZOFURANS (PBDFs)

Previous reports on the synthesis[5] of 2,3,7,8-TBDD and 2,3-dibromo-7,8-dichlorodibenzo-p-dioxins described methods entailing the condensation of a substituted catechol with the appropriate halobenzene to obtain the desired product. Alternatively, direct halogenation of dibenzo-p-dioxin was employed. Such syntheses yield various products, including meager quanti-

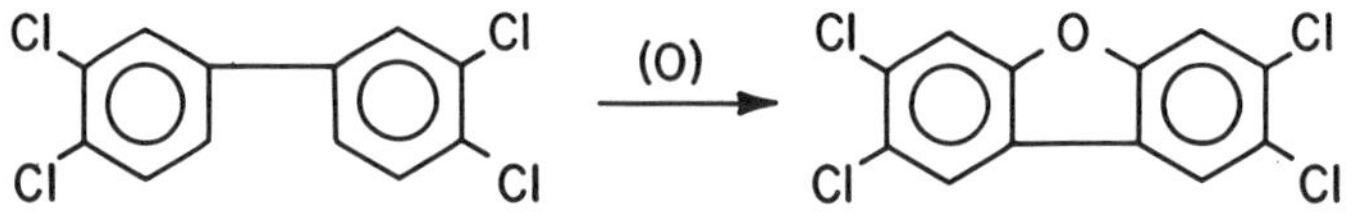

Figure 3. Formation of 2,3,7,8-TCDF from 3,3′,4,4′-Tetrachlorobiphenyl.

ties of several different brominated isomers. Based upon investigations in this laboratory, the optimal method for the synthesis of halodibenzo-p-dioxins seems to be the condensation of an appropriate halosubstituted catechol with an appropriate halosubstituted nitrobenzene in the presence of potassium carbonate. This general method, described in detail below, has been employed for the synthesis of halodibenzo-p-dioxins or mixed halodibenzo-p-dioxins.

In the synthesis of polychlorodibenzo-p-dioxins (PCDD), we have employed the halogenated catechol/halogenated nitrobenzene condensation procedure to produce various tetra- through heptachlorinated dibenzo-p-dioxins in very high-purity ($>90\%$) and good yields ($>40\%$). However, it was found that reactions involving chloronitrobenzenes having less than three chlorine substituents required higher temperatures and resulted in poorer yields. This latter difficulty was particularly pronounced in reactions involving catechols which were orthochloro-substituted. On the other hand, reactions of bromocatechols with bromonitrobenzenes were expected to occur readily, even under milder conditions, because of steric and electronic considerations. Indeed, this was found to be the case in the present study using the general procedure outlined above, bromo- and bromochlorodibenzo-p-dioxins were synthesized in 100- to 500-mg quantities. The details of the method of synthesis of these halodibenzo-p-dioxins are discussed below.

Synthesis of Polybromodibenzo-P-Dioxins

One equivalent of bromocatechol and one equivalent of polybromonitrobenzene was placed, along with 2.5 equivalents of anhydrous K_2CO_3, in 70–100 mL of acetone (Figure 4). The mixture was refluxed for 4 to 6 hr and subsequently the reaction products were precipitated by the addition of water. The precipitate was separated by filtration and, after washing and drying, was recrystallized from chloroform or anisole. The products so obtained were very pure (typically $>90\%$) and yields of the desired bromo-dibenzo-p-dioxin were as shown in Table 1, 50% or greater.

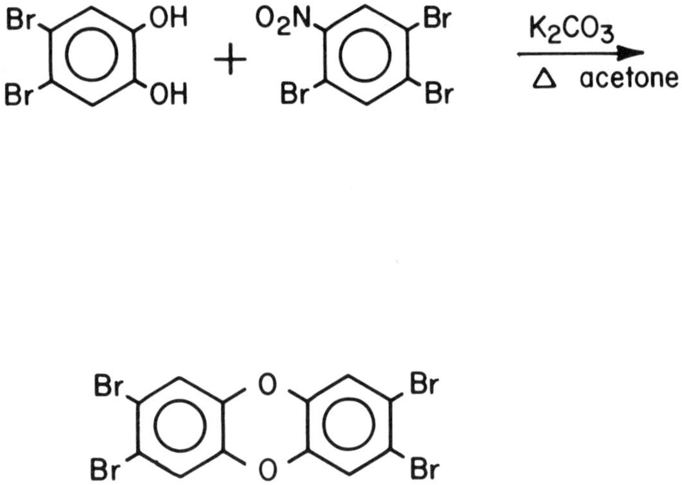

Figure 4. Synthesis of PBDDs.

Synthesis of Polybromochlorodibenzo-P-Dioxins

One equivalent of a bromocatechol or a chlorocatechol and one equivalent of a polychloronitrobenzene or a polybromonitrobenzene were placed, along with 2.5 equivalents of anhydrous K_2CO_3, in 70–100 mL of acetone and refluxed for 4–6 hr (Figure 5). The procedure utilized for isolation and purification of the resulting chlorobromodibenzo-p-dioxin was the same as described above for bromodibenzo-p-dioxins. The purity and percent yields of the isomers obtained are shown in Table 2.

SYNTHESIS OF BROMOCATECHOLS

Bromocatechols needed as intermediates in the synthesis of halodibenzo-p-dioxins were synthesized either by (a) direct bromination of catechol or by (b) the Dakin reaction of bromo-substituted salicylaldehydes, as shown in Figure 6. A listing of the bromocatechols synthesized in the present study is given in Table 3.

Table 1. Polybromodibenzo-p-Dioxins Synthesized.

Polybromodibenzo-p-Dioxin	Yield (%)	Purity[a] (%)
2,3,7/2,3,8-Tribromodibenzo-p-dioxin mixture	60	>95
2,3,7,8-Tetrabromodibenzo-p-dioxin	75	>98
1,3,7,8-Tetrabromodibenzo-p-dioxin	67	>97
1,3,6,8/1,3,7,9-Tetrabromodibenzo-p-dioxim mixture	47	>97

[a]Based upon EC/GC analysis of recrystallized material.

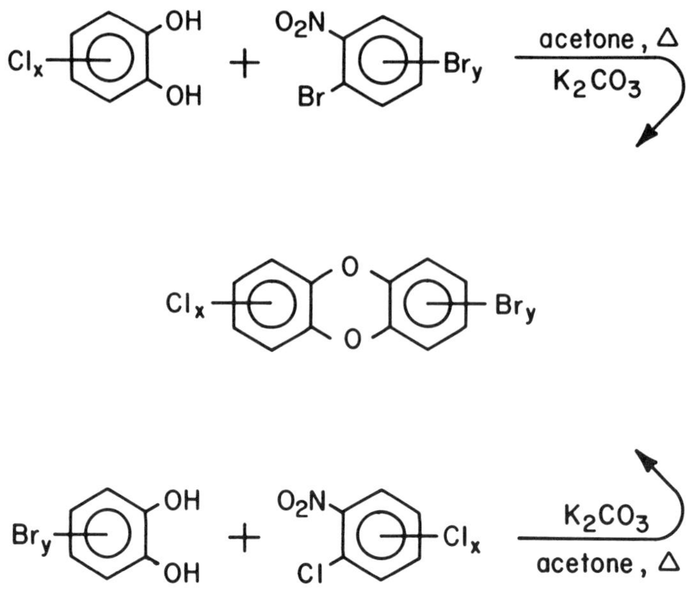

Figure 5. Synthesis of PBCDDs.

SYNTHESIS OF CHLOROCATECHOLS

The synthesis of chlorocatechols used in this study has been described previously.[7,8]

SYNTHESIS OF BROMONITROBENZENES

The bromonitrobenzene intermediates required in this study were not available commercially; therefore, these compounds were synthesized using methods published in the literature as outlined below. The two bromonitrobenzene intermediates used in the study are listed in Table 4 along with the melting points determined in the present study and those reported previously. Direct nitration of 1,2,4-tribromobenzene using a mixture of concentrated nitric acid and sulfuric acid yielded 2,4,5-tribromonitrobenzene. The latter compound was obtained by hydrogen peroxide oxidation of 2,4,6-tribromoanaline.

Table 2.   Bromochlorodibenzo-p-Dioxins Synthesized.

Bromochlorodibenzo-p-Dioxin	Yield (%)	Purity (%)
2,3-Dibromo-7,8-dichlorodibenzo-p-dioxin	55	>97
1,3-Dichloro-7,8-dibromodibenzo-p-dioxin	50	>97

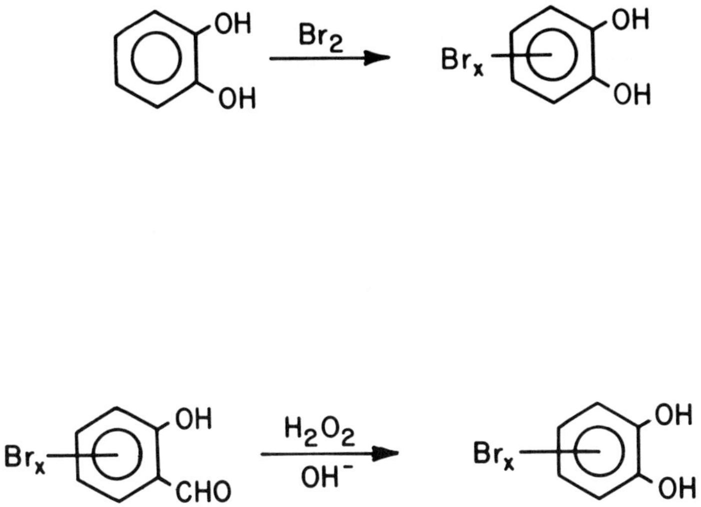

Figure 6. Synthesis of bromocatechols.

SYNTHESIS OF TBDFs

The TBDFs were synthesized by the palladium-catalyzed cyclization of bromo-substituted diphenyl ethers (Figure 7). The synthesis procedure is similar to the procedure employed previously for the synthesis of TCDFs and entailed refluxing a mixture of 1 equivalent of bromodiphenyl ether, 2 equivalents of palladium, and 7 mL of acetic acid containing about 0.5 mL of methanesulfonic acid. The reaction mixture was cooled after 2–3 hr and extracted with methylene chloride. The methylene chloride layer was washed with aqueous base, then dried over anhydrous sodium sulfate, and

Table 3. Bromocatechols Synthesized.

		Melting Point (°C)		
Catechol	Method[a]	Reported[8]	Found	Yield (%)
4-Bromocatechol	b	86–87	85	80
4,5-Dibromocatechol	a	122–124	117	55
3,5-Dibromocatechol	b	58–60	60	80

[a]a, direct bromination; b, dakin reaction.

Table 4. Bromonitrobenzenes Synthesized.

	Literature	Melting point (°C)	
Bromonitrobenzene	Citation	Reported[8]	Found
2,4,5-Tribromonitrobenzene	10	85.4	85
2,4,6-Tribromonitrobenzene	11	127	124–125

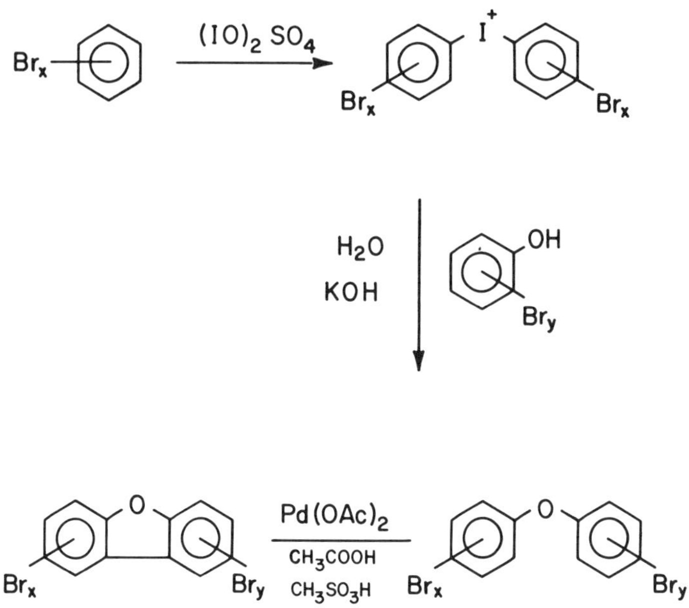

Figure 7. Synthesis of TBDF.

concentrated to remove the solvent. Finally, the crude tetrabromofuran was recrystalized from chloroform.

ELECTRON CAPTURE GAS CHROMATOGRAPHIC ANALYSIS OF PBDDs

The electron capture gas chromatographic (EC/GC) system used to ana-lyze the PBDDs was a Varian 3740 gas chromatograph equipped with an electron capture detector (GC/ECD). A 60-m DB-5 (0.25-μm coating thick-ness) column was employed. The column was maintained at 180°C for 1 min and then raised to 300°C at 10°C/min. The GC/ECD analysis of the PBDDs synthesized as described above revealed that the isomer products were extremely pure. The chromatogram obtained for 2,3,7,8-TBDD is shown in Figure 8.

Table 1 summarizes all the bromodibenzo-p-dioxins synthesized in this study, and the purity of these as determined by GC/ECD analysis.

Figure 9 shows the chromatograms obtained for various TBDD isomers as well as a mixture containing all of the isomers. It is interesting to note that the elution pattern observed for the TBDDs parallels the well estab-lished elution pattern of the corresponding TCDDs. For instance, 1,3,6,8/1,3,7,9-TBDDs elute before 1,3,7,8-TBDD, which in turn, elutes before 2,3,7,8-TBDD.

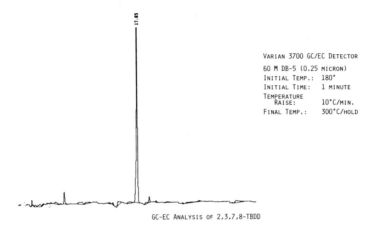

GC-EC Analysis of 2,3,7,8-TBDD

Figure 8. Electron capture gas chromatogram obtained in the analysis of 2,3,7,8-TBDD.

The mixed bromochlorodibenzo-p-dioxins synthesized in this study are listed in Table 2. The GC/ECD analysis of these isomers and the conditions used for the analysis are shown in Figure 10.

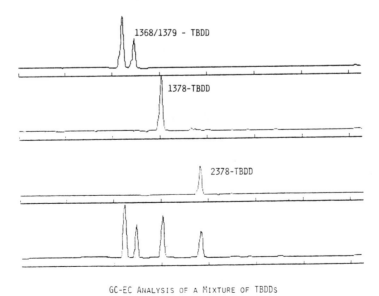

GC-EC Analysis of a Mixture of TBDDs

Figure 9. Comparison of Chromatograms showing the order of elution of TBDDs.

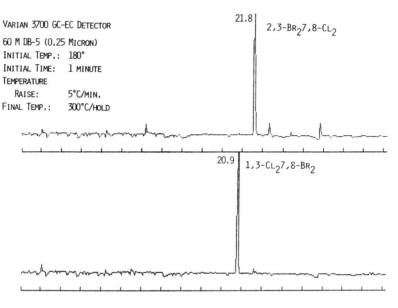

Figure 10. Comparison of chromatograms showing the order of elution of bromochlorodibenzo-p-dioxins.

GAS CHROMATOGRAPHIC/MASS SPECTROMETRIC CHARACTERIZATION OF PBDDs, PBC/DDs AND PBDFs

Final confirmation of the identity of the bromo-substituted dioxins and furans synthesized in this study was obtained from gas chromatographic/ mass spectrometric (GC/MS) analysis of the isomers. The GC/MS system consisted of a Perkin-Elmer Sigma III GC coupled through a custom fabricated glass interface to a Kratos MS-25 double-focusing mass spectrometer equipped with a Nova 3/DS-55 data system. The GC conditions employed were identical to the GC/EC analysis described above. The conditions for the mass spectrometer were as below:

- ionizing voltage: 70 eV
- accelerating voltage: 4 kV
- source temperature: 250°C

Tetrabromodibenzo-p-Dioxins (TBDDs)

The mass chromatogram obtained for 2,3,7,8-TBDD is shown in Figure 11. Octachlorodibenzo-p-dioxin was employed as an internal standard. The ion masses (m/z) monitored were 498, 500, 502, 423 and 421. The m/z 423 and m/z 421 ion masses correspond to the species resulting from the loss of the [Br] moiety. The mass chromatograms obtained for the other TBDDs showed similar results.

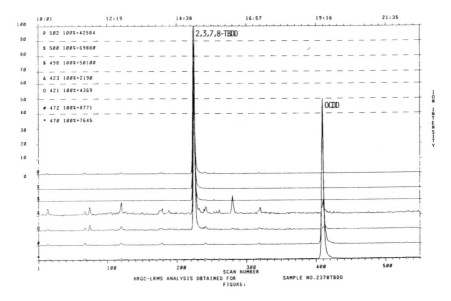

Figure 11. Mass Chromatograms obtained in the analysis of 2,3,7,8-TBDD.

Tetrahalogenated Dibenzo-p-Dioxins (TBD/DDs)

The mixed tetrahalogenated dibenzo-p-dioxins synthesized were also characterized by GC/MS. The mass chromatogram obtained for 1,3-dichloro-7,8-dibromodibenzo-p-dioxin is shown in Figure 12. The following ion masses (m/z) were monitored: 408, 410, 412, 414, 303, 347 and 349. The ion masses m/z 347 and m/z 349 represent species resulting from a loss of the [COCl] moiety, while m/z 303 results from a loss of [COBr].

Tetrabromodibenzofurans (TBDFs)

The mass chromatograms obtained for 2,3,7,8-TBDF are shown in Figure 13. The ion masses m/z 480, m/z 482, m/z 484, and m/z 486 were monitored, along with m/z 375 and m/z 377. The latter two ion masses arise from loss the of [COBr] moiety. As indicated in the mass chromatogram, there are at least two isomers in this synthesis product mixture.

When 3,3', 4,4'-tetrabromobiphenyl is cyclized, it is theoretically possible that the cyclization can proceed to yield 2,3,7,8-TBDF, 1,2,7,8-TBDF and 1,2,8,9-TBDF. Based upon careful study of the results of cyclization of the corresponding tetrachlorobiphenyl (unpublished results obtained in the Brehm Laboratory), it is reasonable to expect that the corresponding brominated isomers would be obtained as diagrammed in Figure 14. However, isolation of the reaction products indicated that 1,2,8,9-TBDF must have been either formed in very low amounts or was removed in the process of purification. Even though it is reasonable to assume that the two products

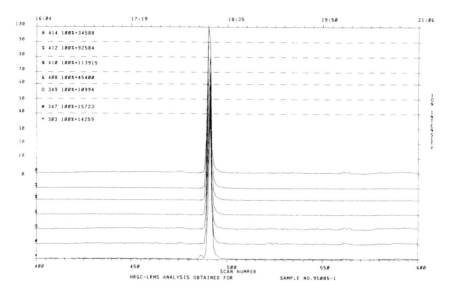

Figure 12. Mass Chromatograms obtained in the analysis of 1,3-dichloro-7,8-dibromodibenzo-p-dioxin.

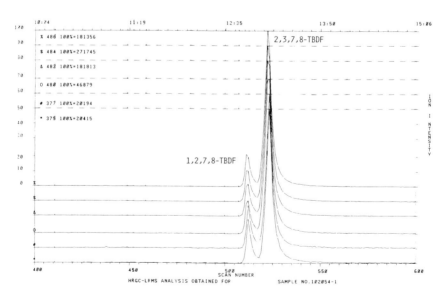

Figure 13. Mass Chromatograms obtained in the analysis of 2,3,7,8-TBDF.

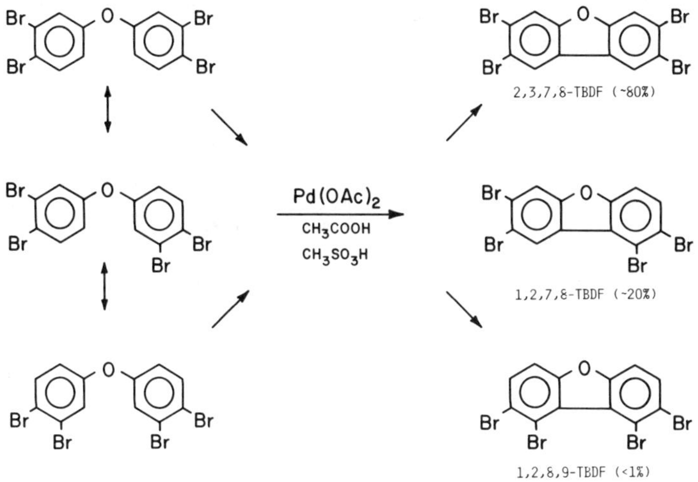

Figure 14. Possible reaction products arising from reductive cyclization of 3,3',4,4'-tetrachlorobiphenyl.

obtained in the synthesis are 2,3,7,8-TBDF (80%) and 1,2,7,8-TBDF (20%), it is not possible at this time to distinguish one from the other. Further studies are in progress.

PRELIMINARY RESULTS OF ASSAYS TO DETERMINE THE BIOLOGICAL ACTIVITY OF THE HALOGENATED DIBENZO-P-DIOXINS SYNTHESIZED IN THE PRESENT STUDY

As mentioned earlier, Kende et al.[5] utilized a chick embryo model to evaluate the enzyme inductive effects (AHH) of halogenated dibenzo-p-dioxins, including 2,3,7,8-TCDD, 2,3,7,8-TBDD and 2,3-dichloro-7,8-dibromodibenzo-p-dioxin (2,3-DC-7,8-DBDD). A similar assay system was implemented to evaluate and further characterize the enzyme inductive effects of the compounds synthesized in the present study. However, in the assay implemented in the present work, the levels of both aryl hydrocarbon hydroxylase (AHH) and 7-ethoxyresorufin deethylase (7-ER) were quantitated as a measure of induction capability.[6] An outline of the chick embryo assay method is given below, and some of the results obtained using this assay are shown in Table 5.

1. Inject 17-day-old chick embryos with 10 μL of dioxane containing 0.003 to 0.3 nmol 2,3,7,8-TCDD or 0.003 to 2500 nmol TBDDs. One group of six eggs was employed for each dose. Two groups of six eggs served as the controls; one control group was injected with 10 μL dioxane, the other control group with 20 μL dioxane.

Table 5. Relative Potencies of 2,3,7,8-Tetrabromodibenzo-p-Dioxin and 2,3-Dichloro-7,8-Dibromodibenzo-p-Dioxin Based upon Induction of 7-Ethoxyresorufin Deethylase (7-ER) and Aryl Hydrocarbon Hydroxylase (AHH).

| | 7-ER | | |
| | Dose Producing 1/2 Maximal Induction (pmol/egg) | Relative Potency | |
		Present Study	Previous Study[5]
2,3,7,8-TCDD	11.1	1.0	—
2,3,-Dichloro-7,8-Dibromo DD	7.35	1.51	—
2,3,7,8-TBDD	9.40	1.18	—
	AHH		
2,3,7,8-TCDD	9.70	1.0	1.0
2,3-Dichloro-7,8-Dibromo DD	6.50	1.49	1.1
2,3,7,8-TBDD	9.00	1.08	1.0

2. Incubate for 24 hr post injection.
3. Excise livers, pool the six livers from each dosage group, homogenize and prepare 9000 xg supernatant.
4. Perform triplicate assays for AHH and 7-ER.

As seen in Table 5, both 2,3,7,8-TBDD and 2,3,-DC-7,8-DBDD are, on the basis of these results, more potent inducers of 7-ER and AHH than 2,3,7,8-TCDD. Also indicated in Table 5 are the results previously reported by Kende et al.[5] The results of the present study indicate that 2,3,7,8-TBDD and 2,3-DC-7,8-DBDD may be even more potent relative to 2,3,7,8-TCDD as inducers of the enzymes studied than the earlier work suggested. More extensive studies of the biological activities of the halogenated dioxins and dibenzofurans are currently underway in our laboratory.

SUMMARY

Methodologies for the synthesis, isolation, purification and characterization of brominated dibenzo-p-dioxins and dibenzofurans, as well as bromochlorodibenzo-p-dioxins, have been developed and applied in the present research effort. The gas chromatographic and gas chromatographic/mass spectrometric characterizations of these compounds revealed that the desired isomers were obtained in very good yields (>50%) and were sometimes >90% pure. Preliminary evaluation of the biological activities of the halogenated dioxins was accomplished using a chick embryo model, and the results of this study generally agree with the conclusions of Kende et al.[5] — that is, the brominated and chlorinated-brominated dioxins are as potent or more potent than the corresponding chlorinated isosteres.

REFERENCES

1. Tiernan, T. O., M. L. Taylor, B. Ramalingam, A. B. Rifkind, J. H. Garrett, and D. J. Wagel. "Formation of Tetrabrominated Dibenzo-p-dioxins (TBDD) by Pyrolysis of Brominated Phenolic Compounds and Assessment of the Biological Activity of Several TBDD Isomers," Paper presented at the *Fourth International Symposium on Chlorinated Dioxins and Related Compounds*, Ottawa, Ontario, Canada (October 16–18, 1984).
2. Robertson, L. W., B. Chittim, S. H. Safe, M. O. Mullin, and C. M. Pochini. *J. Agric. Food Chem.* 31:454 (1983).
3. Encyclopedia of Chemical Technology, 2nd Ed. (New York: Interscience, 1970), p. 468.
4. Mazer, T., F. D. Hileman, R. W. Nobel, and J. J. Brooks. *Anal. Chem.* 55:104 (1983).
5. Kende, A. S., J. J. Wade, D. Ridge, and A. Poland. *J. Org. Chem.* 39:931 (1974).
6. Rifkind, A. B., A. J. Firpo, and D. R. Alonso, *Toxicol. Appl. Pharmacol.* 72:343–354 (1984).
7. Taylor, M. L., T. O. Tiernan, B. Ramalingam, D. J. Wagel, J. H. Garrett, J. G. Solch, and G. L. Ferguson. "Synthesis, Isolation, and Characterization of the Tetrachlorinated Dibenzo-p-dioxins and Other Related Compounds," in *Chlorinated Dioxins and Dibenzofurans in the Total Environment II*, H. Keith, C. Rappe and G. Choudhary, Eds. (Stoneham, MA: Butterworth Publishers, 1985) p. 17.
8. Gray, A. P., S. P. Copa, L. J. Solomon, and O. Aniline, *J. Org. Chem.* 41:2435 (1976).
9. *Dictionary of Organic Compounds, 5th Ed.* (New York: Chapman and Hall, 1982).
10. Davis, F., and N. Cook. *Chem. Tech.* 7:626–629 (1977).
11. Emmons, W. D. *J. Am. Chem. Soc.* 76:3470–3472 (1954).

CHAPTER **33**

Polychlorinated Biphenylene Production for Qualitative Reference Material

E.R. Barnhart, D.G. Patterson, Jr., J.A.H. MacBride, L.R. Alexander,
C.A. Alley, and W.E. Turner

INTRODUCTION

Polychlorinated dibenzo-p-dioxins (PCDDs) and furans (PCDFs) are
well known environmental contaminants. A number of laboratories have
synthesized many of the PCDD and PCDF congeners for analytical reference material. Some of these congeners are also available commercially. A
review of the environmental chemistry literature, as well as publications of
commercial firms, showed that another class of potentially toxic chlorinated aromatic compounds is not available: the polychlorinated biphenylenes
(PCBPs). The PCBPs are closely related to PCDDs and PCDFs. The
2,3,6,7-TCBP isomer has been studied by Poland et al.[1] and found to have
about the same degree of toxicity as 2,3,7,8-TCDD. Rappe et al.[2] reported
on two accidental PCB fires in Sweden. In the fire where oil-filled capacitors burned, PCDFs were the main polychlorinated products formed. In
another capacitor fire (possible electrical malfunction), PCBPs were the
main products. The only PCBP isomer available as a reference for this
study was the 2,3,6,7-TCBP. Williams et al.[3] described results from four
PCB fires in which the PCBPs were present at concentrations 10 times
higher than the PCDFs. In fact, these researchers found that PCBPs were
the main chlorinated species present, excluding the PCBs. Smith et al.[4] also

reported finding a large number of PCBPs in soot produced during an electrical accident that involved the pyrolysis of PCBs in a state office building in Binghamton, New York, in 1982. Since analytical standards for the chlorinated biphenylene congeners are not available, we have prepared them by either photolysis of octachlorobiphenylene (OCBP)[5] or chlorination of biphenylene.

DECHLORINATION

Photodechlorination of OCBP was performed by weighing an aliquot of OCBP and dissolving it in 10% benzene-hexane (50–400 μg/mL). Quartz tubes, 1 × 13 cm, each containing about 5 mL of the solution, were placed in a carousel to revolve around a 450-watt ultraviolet (UV) lamp. After timed periods of illumination, the tubes containing the photolysis products were placed on a rack in the dark. In some cases, aliquots were removed for GC/ECD analysis. In other instances, the tube contents were evaporated to dryness and resuspended in toluene for GC/MS or in DMSO for microbial sensitivity tests. After the 75-min illumination of a solution at an original OCBP concentration of 56 μg/mL, chromatographic analysis yielded the tracing in Figure 1. The six relatively intense peaks, correlated with those generated by negative chemical ionization-mass spectrometry (NCI-MS), led to the assignments shown in Figure 1. The simple pattern of major products suggests that only one of two series of dechlorination products (α or β type) is occurring preferentially, producing one Cl_7, three Cl_6, one Cl_5, and one Cl_4 congener. The α-attack series is depicted in Figure 2, with arrows indicating sites of dechlorination and hydrogen replacement; attack meta to the bridge (β attack series) is depicted in Figure 3. The β series seems to fit the chromatogram best, since the two lower Cl_6 isomers (Figure 3) would have similar dipole moments and the top Cl_6 isomer would be more polar. Peaks for the two less polar Cl_6BPs would be expected to be very close together or superimposed at a retention time less than that of the more polar Cl_6BP. The α-attack series could be expected to generate Cl_6 isomers resolved into three major peaks, since polarity differences appear to be greater. Even if the two polar compounds (top and bottom structures, Cl_6 Figure 2) were superimposed, the double-intensity peak would presumably have a longer retention time than that for the middle, nonpolar structure. We plan to prepare specific isomers for GC/MS reference compounds for identifying these products. This preferential β series dechlorination would allow predictions of dechlorination products of any congener and therefore facilitate the preparation of specific congeners for reference material.

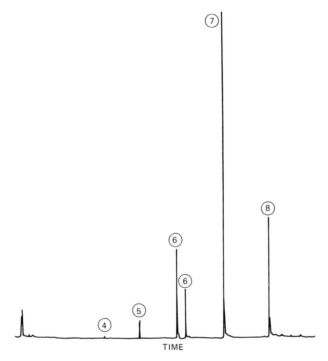

Figure 1. Chromatogram (ECD) of products of ultraviolet irradiation of octachlorobiphenylene in 10% benzene-hexane after 75 min exposure to A 450-W lamp. Numbers
refer to chlorine atoms per biphenylene congener (MS). Chromatographic conditions: column, 30-m DB-5 at 170°C programmed to 290°C at 2°/min and held
at 290°C for 30 min; injector, 290°C; detector, 305°C; He flow, 1.5 mL/min.

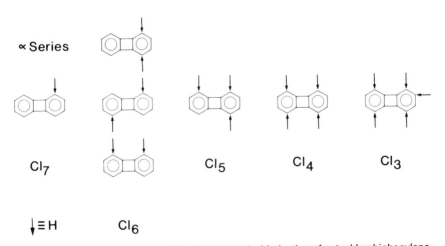

Figure 2. Series of preferential ultraviolet photodechlorination of octachlorobiphenylene.
Arrows indicate possible sites of dechlorination and replacement with H.

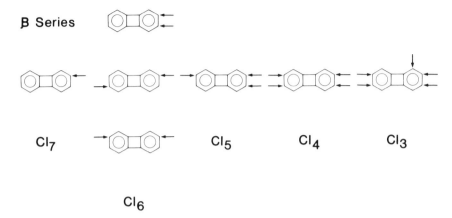

Figure 3. Products of ultraviolet photolysis of octachlorobiphenylene. Arrows indicate position of dechlorination of OCBP, β series.

MICROBIAL TOXICITY

The OCBP and photolysis products were dissolved in DMSO after the solvent evaporated, and we tested 10-μL aliquots for relative toxicity using the bacterial agar-plate inhibition assay (bioassay) described by Turner et al.,[6] which is similar in principle to other plate assays used for measuring bactericidal and bacteriostatic effects of antibiotics. Ten strains were used: *Photobacterium phosphoreum* (NRRL B-11177), *Bacillus subtilis* (ATCC 4925), *Escherichia coli* K12 (HB 101 and 185NX), *E. coli* (ATCC 25922), *Salmonella typhi* (ATTCC 6539), and *Salmonella typhimurium* (Ames TA 98, TA100, TA1537, and TA1538). *B. subtilis, P. phosphoreum*, and *S. typhimurium* (TA 98) showed sensitivity or growth inhibition (over solvent blanks) with the application of OCBP photolysis products after as little as 5 min of illumination, but showed no toxic effects to OCBP (360 μg/10-μL aliquot). Replicative experiments have produced variable results with these three strains; however, a trend was observed. Overall, growth inhibition (measured as zone sizes) with these strains occurred after 5 min of irradiation; it tended to increase after longer exposure times (< 30 min), and tended to decrease upon extended exposure (> 30 min). Interpretation beyond 45 min of irradiation was confounded because in some instances the residue of the irradiated solvent blanks also produced zones of inhibition. Computer-assisted profiling of the microbial toxicity of the chlorinated aromatic congeners is being developed, and experiments are being designed for evaluating the growth inhibition properties of individual congeners using a modification of Lab Systems' Mutascreen[7] toxicity testing protocol.

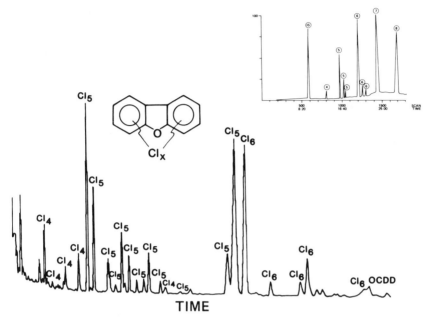

Figure 4. Chromatogram (MS-TIC) of the products of ultraviolet irradiation of octochlorodi-
benzofuran in 10% benzene-hexane after 75 min. exposure to A 450W lamp.
Numbers refer to the number of chlorine atoms per biphenylene congener.
(Insert: Under a faster program on a different capillary column the Cl_7 and Cl_8
isomers are more readily detected.) Chromatographic conditions: GC/
MS,VG2AB2F, Column 30-m SP2330 at 100°C for 2 min, then programmed to
180°C at 20°/min, and then to 220°C at 3°/min at held at 228°C for 30 min.
Insert conditions: GC/MS, Finnigan 4500, Column, 30m DB-5 at 100°C for 2
min, programmed to 200°C at 20°/min and then to 300°C for 4°/min and held
for 30 min.

COMPARISON OF PCDF AND PCBP

Photodechlorination of PCDF was reported by Hutzinger et al.[8] and by
Buser,[9] who found that all products were due to dechlorination. A simultaneous illumination of OCBP and OCDF was conducted for a comparison of
the products. The dechlorination of OCDF produced products that were
resolved by GC into the peaks seen in Figure 4. The numbers of chlorine
atoms per dibenzofuran congener are labeled on the basis of analysis by
GC/MS. Congeners with 4, 5, and 6 chlorine atoms appear to be abundant,
especially in comparison with the OCBP photodechlorination products
(Figure 1). To be useful as a reference material for method development
(recovery studies, retention time windows, etc.), these congeners (Cl_4-Cl_6)
need to be in the mixtures. Even after very long (8 hr) periods of UV
exposure, OCBP did not produce the Cl_4 isomers of PCBP in any significant concentration.

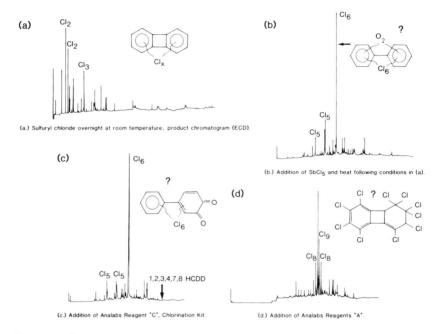

Figure 5. Chromatograms of biphenylene chlorination reaction products. Chromatographic conditions: column, 30-m DB-5 at 130°C for 4 min, then programmed to 16°/min to 290°C and held for 30 min.

CHLORINATION

Experience with chlorination of a fluorobiphenyl with Analabs perchlorination kit, Reagent B (SO_2Cl_2, I_2 and $SbCl_5$), had demonstrated that synthesis according to the manufacturer's direction produced a spectrum of partially chlorinated congeners rather than the perchlorinated product sought; therefore, we attempted to chlorinate biphenylene using several reagents and conditions as shown in Figure 5 (a, b, c, and d). Conditions for chlorination of BP to give di- and trichlorobiphenylenes and their chromatographic separation are shown in Figure 5a.

In chromatograms b and c (Figure 5), the major peak was attributed by MS analysis to a hexachloro compound with the molecular weight of the hexachlorodibenzo-p-dioxin (HCDD). However, the mass fragmentation pattern (Figure 6) differs from that characteristic of HCDDs (Figure 7) in that the molecular ion with the loss of a chlorine atom is the base peak. This (M-Cl) ion is of very low intensity for HCDDs. Although efforts to isolate this compound by reverse phase HPLC (RP-HPLC) have not been successful, the very early elution by RP-HPLC relative to the other HCDDs suggests a much more polar and/or less planar molecule.

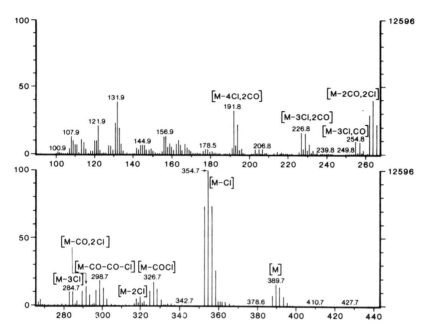

Figure 6. Mass spectrum of a chlorinated derivative of biphenylene.

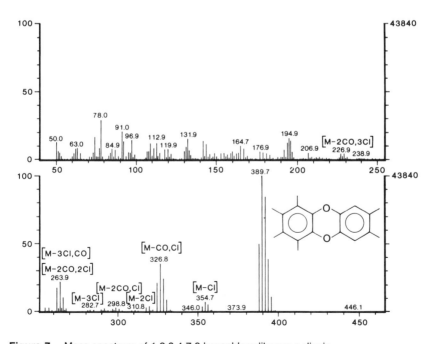

Figure 7. Mass spectrum of 1,2,3,4,7,8-hexachlorodibenzo-p-dioxin.

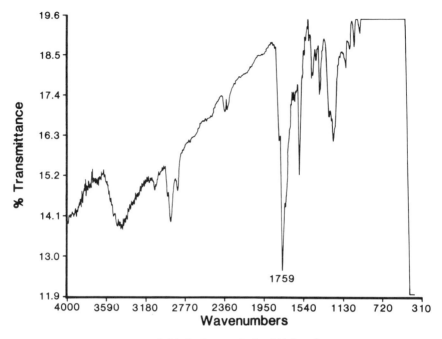

Figure 8.   Infrared spectrum of chlorination products of biphenylene.

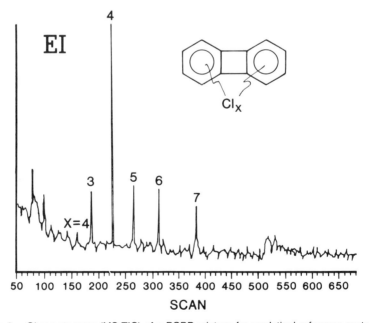

Figure 9.   Chromatogram (MS-TIC) of a PCBP mixture for analytical reference material.

INFRARED SPECTROMETRY

We subjected the reaction mixture to infrared spectral analysis in a KBr pellet to further characterize this potential interference in dioxin analyses. The large 1759-cm^{-1} band (Figure 8) suggests the presence of one or more carbonyl groups in the molecule. Kuboyama,[10] in 1981, prepared and studied the electronic absorption and emission of biphenylene-2,3-dione, and reported that it was stable and unique in this regard with its four-membered ring. All the reaction products represented in Figure 5 were precipitated with aqueous HCl, which may have oxidized the PCBPs to enols, or to keto- or diketo-PCBs.

The products shown in Figure 5d are evidently addition products, because several minor peaks have the molecular weight and chlorine clusters characteristic of decachlorobiphenyl, of which there is only one possible isomer. However, loss of a double bond by chlorination of biphenylene seems to be one possible explanation. We hope to isolate these various major compounds for NMR analyses to aid in our understanding of biphenylene chemistry.

REFERENCE MIXTURE

Chlorination of biphenylene at less than 70°C and for less than 2 hr produced material of greatest use to us. By combining reaction products from photolysis of OCBP and chlorination of biphenylene, we produced material that generated the chromatogram shown in Figure 9. Separation of the compounds detected here and characterization of the individual congeners will be included in a future publication.

ACKNOWLEDGMENT

The authors are indebted to Leslie T. Gelbaum, PhD, Toxicology Branch, Clinical Chemistry Division, Center for Environmental Health, Centers for Disease Control (CDC), for his interest and suggestions. Use of trade names is for identification only and does not constitute endorsement by the Public Health Service or the U.S. Department of Health and Human Services.

REFERENCES

1. Poland, A., E. Glover, and A. I. Kende. *J. Biol. Chem.* 251:4936 (1976).
2. Rappe, C., S. Marklund, P. -A. Bergquist, and M. Hansson. *Chemica Scripta* 20:56 (1982).
3. Williams, C. H., C. L. Prescott, P. B. Stewart, and G. Choudhary. "Formation of Polychlorinated Dibenzofurans and Other Potentially

Toxic Chlorinated Pyrolysis Products in PCB Fires," in *Chlorinated Dioxins and Dibenzofurans in the Total Environment II*, L. H. Keith, C. Rappe, and G. Choudhary, Eds. (Stoneham, MA: Butterworth Publishers, 1985).

4. Smith, L. W., D. L. Stalling, and L. Johnson. *Anal. Chem.* 56:1830 (1984).
5. Kanoktanaporn S., and J. A. Hugh MacBride. *J. Chem. Research* (M):2901 (1980).
6. Turner, W. E., R. H. Hill, Jr., W. H. Hannon, J. T. Bernert, Jr., E. M. Kilbourne, and D. D. Bayse. *Arch. Environ. Contam. Toxicol.* 14:261 (1985).
7. *Mutascreen User's Manual*, Lab Systems, Inc. Morton Grove, Illinois (1984).
8. Hutzinger, O., S. Safe, B. R. Wentzell, and V. Zitko. *Envir. Health Pers.* (Sept) 267 (1973).
9. Buser, H. R. *J. Chromat.* 129:303–307 (1976).
10. Kuboyama, A. *Bull. Chem. Soc. Jpn* 54:873 (1981).

CHAPTER 34

Destruction of Dioxin-Containing Wastes in a Mobile Incineration System

Robert D. Kleopfer, Ralph H. Hazel, Frank J. Freestone, and Paul E. des Rosiers

INTRODUCTION

Over 40 sites have been identified in the state of Missouri which are contaminated with 2,3,7,8-tetrachlorodibenzo-p-dioxin at levels above 1.0 part per billion.[1] Many of these sites resulted from waste disposal practices of a now defunct producer of hexachlorophene. Most of these sites were identified more than 10 years after the initial contamination. The high levels found at these sites attest to the remarkable stability and persistance of 2,3,7,8-TCDD in the soil environment.

A number of different options for remedial action have been proposed.[2] These include stabilization in place, onsite storage, transportion to secure landfill, solvent extraction, biodegradation, chemical degradation in place, and incineration. Each of these proposals has limitations regarding the economic, technical and political feasibility/acceptability. This chapter presents the results of the evaluation and demonstration of a transportable rotary kiln incineration system.

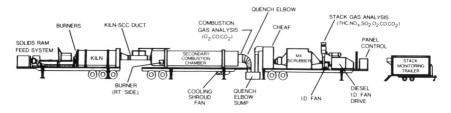

Figure 1. EPA Mobile Incinerator System.

SYSTEM DESCRIPTION

The Mobile Incineration System was designed and built to provide a mobile facility for onsite thermal destruction/detoxification of hazardous and toxic organic substances. The total system consists of: (1) major incineration and air pollution control equipment mounted on three heavy-duty, over-the-road semi-trailers; (2) combustion and stack gas monitoring equipment housed within a fourth trailer; and (3) ancillary support equipment. As illustrated in Figure 1, the Mobile Incineration System consists principally of (1) a rotary kiln, (2) a secondary combustion chamber (SCC), (3) a wetted-throat quench elbow with sump, (4) a cleanable high-efficiency air filter (CHEAF), (5) a mass transfer (MX) scrubber, and (6) an induced-draft fan. Ancillary support equipment consists of bulk fuel storage, waste blending, and feed equipment for both liquids and solids; scrubber solution feed equipment; ash receiving drums; and an auxiliary diesel power generator.

The Mobile Incineration System is controlled and monitored via electrical relay logic and conventional industrial process instrumentation and hardware. Safety interlocks and shutdown features comprise a major portion of the control system. Fuel, waste, and combustion air feed rates, combustion temperatures, and stack gas and SCC fuel gas concentration of carbon monoxide, carbon dioxide, oxygen, and nitric oxides are continuously monitored, thus assuring compliance with Resource Conservation and Recovery Act (RCRA) and Toxic Substances Control Act (TSCA) regulatory requirements. In addition, sulfur dioxide and total hydrocarbons can also be monitored as needed.

The design and procurement of the Mobile Incineration System was sponsored by the Oil and Hazardous Material Spills Branch of the EPA's Cincinnati Hazardous Waste Engineering Research Laboratory. The design and assembly of major equipment items were performed by MB Associates (San Ramon, CA) who delivered the trailers in September 1980. Subsequent redesign, assembly, and modification were performed by EPA's operating contractors—Mason and Hanger-Silas Mason Company (Louisville, KY), and IT Corporation (Edison, N.J.). IT Corporation was the operating contractor for EPA's Mobile Incineration System, and was responsible for modification, inspection, and testing the system. IT Corporation was responsible for its subsequent field operation during the trail burns described in this chapter.

The rotary kiln is designed to operate at temperatures between 1400° and 1900°F, with a residence times of up to 30 min. The secondary combustion chamber operates at 2000° to 2400°F, with a residence time of about 2.2 sec. The three primary "waste streams" produced by the system are the stack emissions, the kiln ash, and wastewater (primarily from the MX scrubber system).

OBJECTIVES

In 1982 the incineration system successfully destroyed liquid organic wastes containing polychlorinated biphenyls (PCBs), ortho-dichloro-benzene, and carbon tetrachloride.[3] Consequently, the system was granted TSCA and RCRA permits to burn liquid PCBs and RCRA-designated liquids having heats of combustion equal to or greater than that of carbon tetrachloride. The primary purpose of the trial burn described here is to obtain data which verify (1) that chlorinated dioxins and other hazardous organic liquid and solid materials are adequately destroyed by incineration in the EPA system, and (2) that the resulting stack emissions do not pose an unacceptable risk to the health or safety of the surrounding communities. The trial burn data also support the issuance of the federal and state permits required for extended use of the system at the Denney Farm site in southwestern Missouri.

EPA and state regulatory offices currently require trial burn data on both liquids and solids because of the processing differences between liquids and solids in the kiln of a kiln-plus-secondary combustion chamber type incinerator. In part, this is because organic liquids can be passed through a burner and directly destroyed, whereas solids are heated by thermal radiation and contact with a refractory in a kiln. Further, because of the political sensitivity of dioxins, a special dioxin trial burn—as applied to liquids and solids—is also required by the EPA's Office of Solid Waste and Emergency Response (OSWER) in support of the regulations (July 15, 1985) relating to the dioxin listing rule. The RCRA and TSCA regulations both address the measurement of incinerator stack emissions as the primary performance criterion. The required destruction and removal efficiency (DRE) achieved by an incineration system under TSCA for PCBs is at least 99.9999%; under RCRA for other hazardous materials, it is at least 99.99%. A DRE of 99.9999% is being required by OSWER for dioxins. The DRE is computed from the emission rate of hazardous materials in the incinerator stack and the feed rate of the material to the incinerator.

This chapter describes the results of four trial burns of about 6 hr each, in which 2,3,7,8-TCDD-contaminated liquids and soils were fed simultaneously.

SAMPLING/ANALYTICAL PROCEDURES

A number of different sample types were collected and analyzed for a variety of parameters. Key samples included the input waste feed (solids and liquids), the residual kiln ash, the CHEAF, the stack gas, and the wastewater.

The solid waste feed material consisted of soil taken from the Denney Farm site during excavation activities in 1981.[4] Samples were analyzed for tetrachlorodibenzo-p-dioxin (2,3,7,8-TCDD) by standard procedures based on a benzene-Soxhlet[5] jar extraction with isotope dilution HRGC/HRMS.[6,7] Concentrations ranged up to 1010 ppb.

The liquid waste feed material consisted of 2,4,5-trichlorophenol still bottom waste,[8] which was diluted with solvent to achieve concentrations around 300 ppm (mg/kg) of 2,3,7,8-TCDD. The wastes also contained 2,4,5-trichlorophenol (50%), hexachloroxanthene,[9] and a complex mixture of chlorinated aromatic hydrocarbons.[8] These oils were analyzed for 2,3,7,8-TCDD by HRGC/LRMS following dilution with toluene and liquid chromatographic cleanups based on silica gel and alumina.[10]

The residual kiln ash was analyzed for 2,3,7,8-TCDD using HRGC/LRMS and HRGC/HRMS following benzene/Soxhlet extraction and silica gel/alumina cleanup.[4,10] TCDD was not detected in any kiln ash sample at detection limits as low as 1 ppt. The ash was also analyzed for a number of other organics (PCDDs/PCDFs, PCBs, PNAs, chlorophenols, chlorobenzenes, and hexachlorophene) for RCRA delisting purposes.

The CHEAF samples were analyzed for PCDDs and PCDFs using HRGC/LRMS methods. No TCDD was observed at a detection limit of 0.05 ppb.

The purge water was analyzed for PCDDs and PCDFs using HRGC/LRMS-based methods. The water was also analyzed for other organics (PCBs, PNAs, chlorophenols, chlorobenzenes, and hexachlorophene) for RCRA delisting purposes. No detectable levels were found for any organic constituent.

The EPA Modified Method 5 (MM5) sampling train was used for collection of particulates, HCl, and the semivolatile organics (PCDDs/PCDFs) in the stack gas. Figure 2 presents a schematic of the sampling device. Analytical results are summarized in Tables 1, 2, and 3.

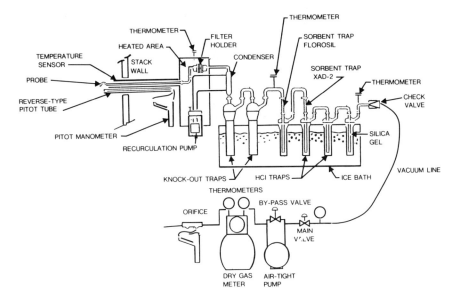

Figure 2. Modified Method 5 sampling train.

QUALITY ASSURANCE

The accurate measurement of 2,3,7,8-TCDD and related chemicals was crucial to the success of this project. The primary contractor (IT Corporation) was responsible for all sampling and analytical activities. This included the preparation of a detailed quality assurance project plan as

Table 1. Isomer-Specific 2,3,7,8-TCDD Measurements for Four Test Burns.

Sample Type	2,3,7,8-TCDD Concentration
Soil (Waste Feed)	101–1010 ppm
Oil (Waste Feed)	179–357 ppm
CHEAF	ND[a] (0.05 ppb)
Kiln Ash	ND (0001 ppb)
Waste Water	ND (0.001 ppb)
Stack Gas	ND (4.0 ng)

[a]ND, none detected; detection limits in parentheses.

Table 2. PCDD Measurements for Typical Test Burn.

Sample Type	TCDD	PCDD	HxCDD	HpCDD	OCDD	Units
Soil (Waste Feed)	274	ND[a] (0.4)	ND (0.3)	ND (1.6)	17	ppb
Oil (Waste Feed)	240	ND (7)	ND (1)	ND (0.6)	ND (2)	ppm
CHEAF	ND (0.03)	ND (0.01)	ND (0.03)	ND (0.01)	0.05	ppb
Kiln Ash	ND (0.06)	ND (0.1)	ND (0.1)	ND (0.02)	ND (0.09)	ppb
Waste Water	ND (0.001)	ND (0.001)	ND (0.001)	ND (0.0004)	ND (0.001)	ppb
Stack Gas	ND (4)	ND (3)	ND (3)	ND (7)	25	ng

[a]NC, none detected; detection limits in parentheses.

Table 3.   PCDF Measurements for Typical Test Burn.

Sample Type	TCDD	PCDF	HxCDF	HpCDF	OCDF	Units
Soil (Waste Feed)	33	ND[a] (0.2)	ND (0.2)	ND (0.7)	ND (0.4)	ppb
Oil (Waste Feed)	ND (32)	ND (9)	ND (1)	ND (29)	ND (4)	ppm
CHEAF	ND (0.08)	ND (0.04)	ND (0.05)	ND (0.07)	ND (0.01)	ppb
Kiln Ash	ND (0.3)	ND (0.02)	ND (0.03)	ND (0.02)	ND (0.05)	ppb
Waste Water	ND (0.001)	ND (0.0003)	ND (0.002)	ND (0.001)	ND (0.001)	ppb
Stack Gas	ND (5)	ND (5)	ND (0.5)	ND (3)	ND (3)	ng

[a]NC, none detected; detection limits in parentheses.

required for all EPA monitoring activities.[11] EPA maintained a quality assurance overview role which included the following:

1. Analyses of "split" samples
2. Inclusion of quality control (performance evaluation) samples
3. Review of all analytical data
4. Provision of certified analytical standards
5. Onsite technical evaluations (audits) of the participating laboratories
6. Observation of the test burns and associated sample collection activities

RESULTS AND CONCLUSIONS

Table 4 summarizes the results of the four trial burns. During the trials, 1750 gal of chlorinated dioxin-contaminated liquid and more than 40 tons of chlorinated dioxin-contaminated soils were burned in the incinerator. These materials contained up to 357 ppm of the highly toxic 2,3,7,8-TCDD isomer. In each case the 2,3,7,8-TCDD destruction and removal efficiency was demonstrated to be greater than 99.9999%. The measurement of the actual DRE was limited by the amount of 2,3,7,8-TCDD which could be introduced into the kiln and the limit of detection for TCDD in the stack gas. TCDD was not detected in any stack gas sample. In addition, 2,3,7,8-TCDD was not detected in any of the other major waste streams (kiln ash, CHEAF, and wastewater).

The data produced during this project are considered highly reliable because of the comprehensive quality assurance program. For example, analytical results of the split sample demonstrated substantial agreement for all 2,3,7,8-TCDD measurements. The results of this study demonstrate conclusively that the rotary kiln/secondary combustion technology can safely

Table 4.   Dioxin Test Burn Results Summary.

Test Run	TCDD Destruction Efficiency (%)
1	>99.999973
2	>99.999986
3	>99.999995
4	>99.999989

Table 5. Material Incinerated During Field Demonstration.

Material	Estimated Quantity	2,3,7,8-TCDD Concentration
Denny Farm		
MDB soil	210 yd^3	500 ppb
Mixed solvents and water	2590 gal	low
Chemical solids and soils	31,150 lb	1–2 ppm
Drum remnants and trash	84 85-gal overpack drum	unknown
Verona		
Hexane/isopropanol	10,000 gal	0.2 ppm
Methanol	5,000 gal	ppt
Extracted still bottoms	5,000 gal	0.2 ppm
Activated carbon	5,000 gal	unknown
Decontamination solvents	1,000 gal	unknown
Sodium sulfate salt cake	23 yd^3	1 ppb
Miscellaneous trash	84 55-gal drums	unknown
Neosho		
Spill area soil	25 yd^3	60 ppb
Bunker soil/residue	15 drums	2 ppm
Tank asphaltic material	75 gal	2 ppm
Erwin Farm		
Empty drums with residue	30 drums	8 ppb
Rusha Farm		
Spill area soil	10 yd^3	unknown
Talley Farm		
Spill area soil	10 yd^3	6 ppb
Eastern Missouri		
Times Beach soil sample	3 yd^3	500 ppb
Piazza Road soil sample	3 yd^3	1600 ppb

and effectively be utilized to detoxify both solid and liquid wastes contaminated with 2,3,7,8-TCDD.

Future activities with the EPA incineration system will include a field demonstration at the Denney Farm site in which a wide variety of dioxin-contaminated solids and liquids will be fed (Table 5). This demonstration began in July 1985. When the demonstration is finished, most of the chlorinated dioxin-contaminated material in southwestern Missouri will have been destroyed.

ACKNOWLEDGMENTS

We wish to acknowledge the substantial contributions of numerous individuals and organizations. These include EPA Region VII (Charles Hensley and lab staff), EPA/EMSL–Las Vegas (Ron Mitchum and staff), IT Analytical Services (Barry Stevenson and staff), Midwest Research Institute (Jim Spigarelli and staff), and Roy F. Weston, Inc. (Cliff Kirchmer and staff).

REFERENCES

1. Kleopfer, R. D. "2,3,7,8-TCDD Contamination in Missouri," paper presented at 4th International Symposium on Chlorinated Dioxins and Related Compounds, Ottawa, Canada, October 1984.
2. "Interim Report of the Missouri Dioxin Task Force," submitted to Governor Christopher S. Bond, June 1, 1983.
3. "1982 PCB Study," Internal EPA Report.
4. Harris, D. J. Paper presented at the National Conference of Uncontrolled Hazardous Waste Sites, Washington, DC (October 1981).
5. "Consolidated Trial Burn Plan, Quality Assurance Project Plan, and Detailed Sampling/Analytical Procedures," IT Corporation, Edison, NJ, December 6, 1984.
6. Kleopfer, R. D., K. T. Yue, and W. W. Bunn. In *Chlorinated Dioxins and Dibenzofurans in the Total Environment II* L. H. Keith, C. Rappe, and G. Choudhary, Eds. (Stoneham, MA: Butterworth Publishers, 1985), p. 355.
7. "*Statement of Work*: Dioxin Analysis; Soil/Sediment Matrix Multi-Concentration; Selected Ion Monitoring (SIM) GC/MS Analysis with Jar Extraction Procedure," EPA Office of Emergency and Remedial Response (1983).
8. Exner, J. H., J. D. Johnson, O. D. Ivins, M. N. Wass, and R. A. Miller, *Detoxication of Hazardous Waste* (Ann Arbor, MI: Ann Arbor Science Publishers, 1982), p. 269.
9. Viswanathan, T. S., and R. D. Kleopfer. Chapter 14, this volume.
10. "Final Dioxin and Dibenzofuran Report for EPA Mobile Incinerator Missouri Trial Burn," Oil and Hazardous Materials Spills Branch, EPA Contract No. 68-03-3069 (May 1985).
11. "Guidance for Preparation of Combined Work/Quality Assurance Project Plans for Environmental Monitoring," Office of Water Regulations and Standards, EPA (May 1984).

CHAPTER 35

Preparation and Characterization of Clay-Based Sorbents for the Removal of Trace Levels of 2,3,7,8-TCDD from Industrial Wastewaters

Keeran R. Srinivasan and H. Scott Fogler

INTRODUCTION

Occurrence of polychlorinated dibenzo-p-dioxins (PCDDs), especially 2,3,7,8-TCDD, as unwanted contaminants in the synthesis of several useful commercial products such as 2,4,5-trichlorophenol, 2,4,5-trichloro-phenoxyacetic acid (2,4,5-T), pentachlorophenol, and hexachlorobenzene has been well documented.[1,2] Similarly, commercial polychlorinated biphenyl mixtures (PCBs) have been implicated in the presence of polychlorinated dibenzofurans (PCDFs) in industrial wastewater.[2] Removal or detoxification of these pollutants is an ideal compromise between a clean environment and the continued use of the above-mentioned commercial products. Although several methods of detoxification, such as (1) photolytic dechlorination, (2) microbial degradation, and (3) thermal decomposition, are currently available or being developed,[3] there is no alternative to adsorption for the removal of trace levels of organic pollutants from very large volumes of wastewater. For example, a typical presence of small amounts of 2,3,7,8-TCDD (3.5 mg/day) in very large quantities of industrial effluent water (20×10^6 gal/day) can be most efficiently removed by adsorption on a suitable sorbent.

In the following two chapters, we report on the progress of work done to evolve a scheme to reduce the concentration of 2,3,7,8-TCDD in industrial effluent water. In this chapter, we describe the methods for the preparation and characterization of suitable sorbents and in Chapter 36, we report on the binding of 1,2,3,4,6,7,8,9-octachlorodibenzo-p-dioxin (OCDD), 2,3, 7,8-tetrachlorodibenzo-p-dioxin (2,3,7,8-TCDD) and 2,4,5,2',4',5'-hexachlorobiphenyl (HCB) to these sorbents.

DESIGN OF SORBENTS

An ideal sorbent must be cost-effective, easy to prepare, abundant in supply, amenable to easy characterization, and finally, and most importantly, it should have high binding affinity and sorption potential for the target organic molecules.

Binding of organic pollutants to natural organic sediments, suspended particles containing humic material and even microorganisms has been studied.[4-6] Use of these materials in the sorption of TCDD is impractical because (a) there is a considerable variation in the binding constants from source to source, except in the case of bacteria, and (b) they are poorly characterized, heterogeneous materials, and it would be quite difficult to draw any conclusions regarding the binding mechanism. It should be noted that the binding affinities are fixed and cannot be varied as is possible with synthetic sorbents.

Clay, on the other hand, is readily available, is chemically uniform, has a high surface area, and is cost-effective. In addition, clay is being used in many instances as a liner for toxic waste dumps, and any improvements achieved as a part of this study can be extended to these systems as well.

However, clay is reported to have a low binding affinity for polycyclic aromatics such as Chlorinated dioxins.[4,5] A high binding constant is essential for the successful use of sorption to remove 2,3,7,8-TCDD. Preliminary calculations have revealed that a minimum value of the surface excess of the sorbate at equilibrium should be 10,000 for the economic viability of the sorption process. These calculations assumed rapid equilibrium between the free and the bound states of the sorbate and a series of at least three continuously stirred tank adsorbers (CSTAs) to carry out the operation. On the basis of the above considerations, we chose montmorillonite as the clay, which, when suitably chemically modified will have a high binding affinity for 2,3,7,8-TCDD. Montmorillonite, being a smectite with a high surface area/charge ratio,[7] swells extensively in water. This property can be taken advantage of in constructing a sorbent of high surface area with a micropore structure capable of binding dioxin molecules very strongly.

There are two effective methods for modifying clay, the first of which takes into account the chemical nature of the sorbate, namely 2,3,7,8-TCDD in the present case. The low aqueous solubility of 2,3,7,8-TCDD

(0.32 ppb) combined with its high octanol-water partition coefficient (log K_{OW} = 5.38) and the large negative standard free energy change for transfer of 2,3,7,8-TCDD from water to nonpolar solvents[8] ($\triangle G^0_{298}$ = –8.5 to –11.3 kcal/mol) suggest hydrophobic surface sites as most likely to increase binding affinity. Soil humic acid (HA), which is reported to have high binding affinity for polycyclic aromatics in solution,[9] was selected as one of the modifiers of clay. When bound irreversibly to clay, humic acid was expected to provide high binding affinity for 2,3,7,8-TCDD. In other words, a high concentration of humic acid on clay surface in conjunction with the intrinsic high surface area of clay were considered important for rapid and efficient removal of 2,3,7,8-TCDD from water. Since humic acid is a polyanionic heteropolymer at near neutral pH, hydroxy aluminum (HYDAL), a polycationic, polynuclear complex of aluminum hydroxide at low pH, was used as a "cement" to bind humic acid irreversibly to the clay surface. The advantage of hydroxy aluminum lies in its ability to bind nonexchangeably to the cationic sites on clay, thereby reducing the chance of humic acid being leached out from the surface of the clay.

Hydroxy aluminum polymers are formed in aqueous solution by the reaction of Al^{3+} ions in solution with OH^- (from NaOH) successively to yield hydroxy aluminum as follows:

$$Al^{3+} \quad + \quad OH^- \quad \rightarrow \quad Al(OH)^{2+} \qquad (1)$$
$$Al(OH)^{2+} \quad + \quad OH^- \quad \rightarrow \quad Al(OH)_2^{1+} \qquad (2)$$
$$Al(OH)_2^{1+} \quad + \quad OH^- \quad \rightarrow \quad Al(OH)_3 \text{ (s)} \qquad (3)$$

If the addition of OH^- is regulated to give OH^-/Al ratio between 2 and 3 (Equation 2), then unstable $Al(OH)_2^{1+}$ polymerizes spontaneously in solution as shown in Equation 4:[10]

$$6 \, Al(OH)_2^{1+} \quad \rightarrow \quad [Al_6(OH)_{12} \cdot n \, H_2O]^{6+} \qquad (4)$$

The above description of the hydroxy aluminum polymers in solution is necessarily concise, and we may point out that by varying the preparation conditions, it is possible to produce a range of polynuclear complexes with a variety of cationic charges, shapes, and sizes.

The second approach to the design of a suitable sorbent would be to modify the structure of the clay by cross-linking repeating units of clay layers with hydroxy aluminum polymer solution described above. It has been shown that such an intercalation increases the surface area of the resultant clay and imparts to it a "zeolite-like" microporous structure.[11] A schematic representation of a cross-linked clay is shown in Figure 1. It was anticipated that, due to the higher surface area and the unique micropore structure, hydroxy aluminum modified clay (HYDAL-clay) would have a high sorption potential and an increased binding affinity for toxic organic molecules such as chlorinated dioxins.

As described in Chapter 36, hydroxy aluminum-clay and humic acid-

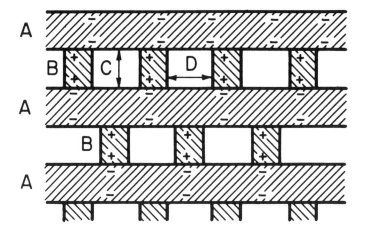

A: CLAY LAYER
B: CROSS-LINKING HYDROXY-ALUMINUM
C: HEIGHT OF CROSS-LINK
D: LATERAL SPACING

Figure 1.   Schematic representation of cross-linked clay.

hydroxy aluminum-clay were used as sorbents before attempting further modifications of clay. The stepwise approach in the development of the sorbents is illustrated schematically in Figure 2.

PREPARATION OF SORBENTS

Na-montmorillonite was swollen in water and treated twice with a commercially available solution of hydroxy aluminum. The total equivalents of aluminum used was about 30 times the cation exchange capacity of the clay. Following the treatment, the hydroxy aluminum-clay was extensively washed with distilled water, dried at 60°C, and used in the sorption experiments.

Isotherms of humic acid binding directly to Na-montmorillonite and to hydroxy aluminum-clay were performed in a batch adsorber. Concentrations of humic acid were measured as dissolved organic carbon (DOC) or by absorbance measurements at 465 nm using DOC analysis to prepare a calibration curve for each batch of humic acid stock solution. Ash-free humic acid stock solution was prepared from Aldrich humic acid by established procedures[12] followed by extensive dialysis against distilled water. Humic

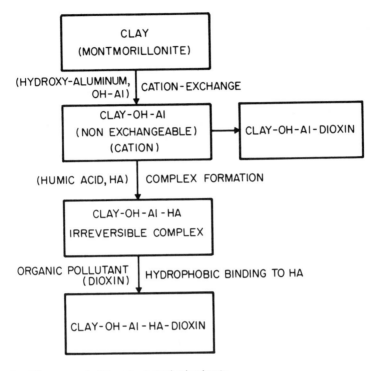

Figure 2.   The concept of the clay-based adsorbents.

acid at pH 3.0 was obtained from the stock solution by cation exchange. Further details are given elsewhere.[13]

The binding of humic acid was studied as a function of pH (3.0, 4.5, and 6.3) and ionic strength (0 and 25 mM NaCl at the designated pH). Humic acid bound to unmodified clay or hydroxy aluminum-clay was assayed as organic carbon after extensive washing to remove loosely bound humic acid.

BET surface areas of clay, hydroxy aluminum-clay, humic acid-bound clay, and humic acid-bound hydroxy aluminum-clay were measured (Table 1).

Rows 2 and 3 in Table 1 reveal the interesting trend that the hydroxy aluminum treatment opens up the clay surface while the binding of large amounts of humic acid reduces the pore volume of the modified clay. As discussed in Chapter 36, these results have startling effects on the binding of dioxins and biphenyl to the respective sorbent. Ionic strength effect was significant in the case of unmodified clay-humic acid interaction and absent in the case of humic acid-hydroxy aluminum-clay.

Table 1. BET Surface Area of Clay-Based Adsorbents.

Clay-Compound	% Organic Carbon	Surface Area (m²/g)
Clay	None	27.6
Hydroxy aluminum-clay	None	114.0
Humic acid-HYDAL-clay (prepared at pH 3.0)	8.6	1.6
Humic acid-unmodified clay (prepared at pH 3.0)	1.8	78.2
Humic acid-HYDAL-clay (prepared at pH 6.3)	2.2	61.8

CHARACTERIZATION OF SORBENTS

Extensive washing of the humic acid-hydroxy aluminum-clay with water and treatments with 1 M phosphate buffer or 0.1 M NaOH indicated that hydroxy aluminum-clay binds humic acid strongly and irreversibly (85% of the initially bound humic acid is retained) while unmodified clay loses most of the initially bound humic acid, except at pH 3.0, where about 25% is retained.

Irreversible binding of humic acid is known to be mediated through an ion exchange type interaction between humic acid and the edge cations of the clay. Alternatively, chelate formation between the proximal COOH and OH groups on humic acid and the edge cations of the clay also need to be considered.[14,15] In the case of hydroxy aluminum-clay, an ion exchange type of interaction or chelate formation between humic acid and polycationic hydroxy aluminum polymer is also probable. Therefore, the addition of hydroxy aluminum to clay is expected to significantly stabilize the resultant humic acid-hydroxy aluminum-clay.

In the case of the second type of adsorbent, namely hydroxy aluminum-clay by itself, the extent of hydroxy aluminum in the interlayer space is of great importance, since this will have a direct effect on the surface area and the microporosity of the hydroxy aluminum-clay, which will in turn affect the sorption capacity and the binding potential of the resultant hydroxy aluminum-clay.

Fourier transform infrared (FTIR) difference spectra of humic acid bound to hydroxy aluminum-clay and of humic acid in solution have been recorded using a KBr pellet technique to help us understand the mode of binding of humic acid to hydroxy aluminum-clay. A new procedure was developed to generate a stable baseline for these spectra so that quantitative information can be obtained.[16] A set of FTIR difference spectra is shown in Figure 3. From bottom to top in the figure, the difference spectra are respectively those of humic acid in solution at pH 3.0, surface-bound humic acid in humic acid-hydroxy aluminum-clay prepared at pH 3.0 and containing 8.6% by weight organic carbon, and humic acid in solution at pH 6.3.

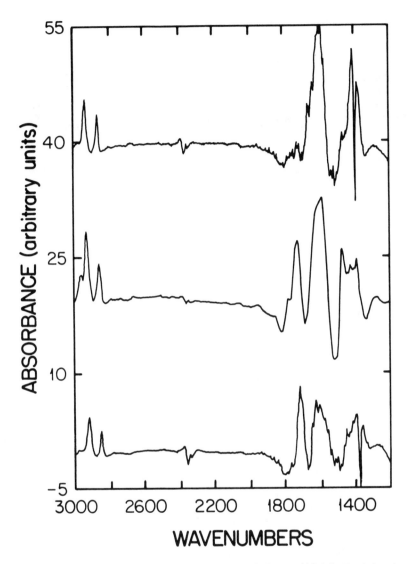

Figure 3.   FTIR difference spectra of humic acid in solution at pH 3.0 (bottom), humic acid on clay-hydroxy aluminum—humic acid prepared at pH 3.0 (middle) and humic acid in solution at pH 6.3 (top). The peak positions are given in the text.

The difference spectra were generated by subtracting the solvent IR spectrum from the solution IR spectrum using linear least square methods to minimize the residuals. For example, in the case of surface-bound humic acid, the "solvent" would be the hydroxy aluminum-clay containing no humic acid.

All spectra in Figure 3 display the prominent features of humic acid

Table 2.   Comparison of Peak Intensity Ratios of IR Bands of Different Forms of Humic Acid.

pH 3.0

Peaks Ratioed	HA (Solution)	HA-HYDAL Complex (Solution)	HA-HYDAL-Clay
2920/1460 cm^{-1}	1.36	1.30	1.30
1725/1620 cm^{-1}	1.60	0.60	0.58
1725/1400 cm^{-1}	2.0	1.0	0.93

pH 6.3

Peaks Ratioed	HA (Solution)	HA-HYDAL Complex (Solution)	HA-HYDAL-Clay
2920/1374 cm^{-1}	0.75	0.80	ND[a]
1725/1620 cm^{-1}	0.12	0.19	0.16
1725/1400 cm^{-1}	0.22	0.27	0.20

[a]ND, not detected.

infrared spectrum, such as the 2920 cm^{-1} and 2850 cm^{-1} bands (due to CH stretch), 1720 cm^{-1} band (due to C=O of COOH), and the companion bands of 1630 cm^{-1} and 1400 cm^{-1} (due to C=O of COO$^-$).[17] IR intensities from the hydrocarbon region of humic acid were used as fingerprints to verify that the surface-bound humic acid and the humic acid in humic acid-hydroxy aluminum complex in solution were chemically identical to humic acid in solution. We then made a comparison of the intensities of the C=O band in the protonated and deprotonated forms of COOH in the humic acid in the three cases. The results are summarized in Table 2. At pH 3.0, the surface-bound humic acid and the humic acid-hydroxy aluminum complex in solution have a much greater amount of the COO$^-$ form compared to humic acid in solution at the same pH, indicating strong cationic interaction resulting in a greater degree of dissociation of the COOH group involved in binding at this pH. This would confirm our observations that at pH 3.0, humic acid binds strongly and irreversibly to hydroxy aluminum-clay and that a large amount of humic acid can be bound to the surface at this pH. It can be seen from Reactions 1–4 above that low pH is expected to produce smaller hydroxy aluminum polymers with large net positive charge capable of shielding the humic acid partially, if not completely, from the negative charge of clay layers. This would tend to impart greater stability to the humic acid bound to the hydroxy aluminum-clay under conditions of low ionic strength and near neutral pH, as was experimentally observed in this study.

On the other hand, similar analysis of humic acid binding to hydroxy aluminum-clay at pH 6.3 (Table 2) revealed that the amount of the COO$^-$ form of humic acid on the surface or the complex in solution was not significantly different from that in the solution, verifying our finding that the binding of humic acid to hydroxy aluminum-clay at pH 6.3 was much weaker.

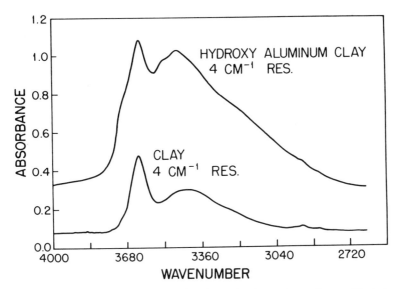

Figure 4.   FTIR absorbance spectra of hydroxy aluminum-clay and unmodified clay.

Similar quantitative studies on the mode of intercalation of hydroxy aluminum on the clay are under way. However, some preliminary qualitative understanding of the location of hydroxy aluminum polymers on the clay surface can be obtained from the FTIR spectra shown in Figures 4 and 5.

Figure 4 is the absorbance spectrum of unmodified clay (bottom) and the hydroxy aluminum-clay (top) in the spectral region of 4000–2600 cm^{-1}. The two spectra are nearly identical except for a prominent shoulder around 3700 cm^{-1} in the case of the hydroxy aluminum-clay spectrum. This band has been attributed to the O-H--O stretch of aluminum hydroxide[18] and has been shown to be characteristic of polymeric hydroxy aluminum in the interlayer space.[19] As pointed out earlier, a greater preponderance of hydroxy aluminum in the interlayer space is expected to be beneficial, and the existence of the 3700 cm^{-1} band in hydroxy aluminum-clay, shown by the difference spectrum in Figure 5, verifies that hydroxy aluminum is in the interlayer space.

It should be pointed out that there is a considerable upward drift in the baseline of the difference spectrum shown in Figure 5 compared to similar difference spectra in Figure 3. This is due to the procedure by which these difference spectra were obtained. Unlike the humic acid difference spectra shown in Figure 3, the difference spectrum of Figure 5 was obtained by subtracting incremental amounts of the unmodified clay infrared spectrum until the 3620 cm^{-1} band of the native clay had minimal intensity. The value of 1.7 shown at the top of Figure 5 represents the factor by which the

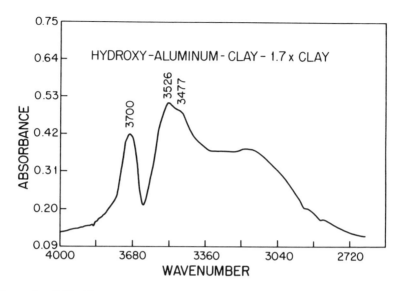

Figure 5. FTIR difference spectrum of hydroxy aluminum on hydroxy aluminum-clay. Refer to the text for details.

infrared intensity of the unmodified clay in the same spectral region had to be multiplied in order to minimize the 3620 cm^{-1} band.

Apart from the 3700 cm^{-1} band, two other bands at 3527 and 3477 cm^{-1} are also seen. The identity of the former is not known, but the 3477 cm^{-1} band has been seen in similar montmorillonite hydroxy aluminum complexes containing interlayer hydroxy aluminum.[19] Thus, the FTIR evidence indicates that a significant amount of interlayer hydroxy aluminum is present in hydroxy aluminum-clay. Such a result is not unexpected, since, as shown in Chapter 36, hydroxy aluminum-clay exhibits enormous sorption potential and a high binding affinity toward all the three sorbates tested in that study.

We are in the process of combining FTIR measurements with other physicochemical measurements to develop a structure-function relationship between sorbents and their sorption behavior toward toxic organic molecules in industrial wastewaters.

SUMMARY

The present study describes two procedures for modifying expanding clays like montmorillonite in order to enhance their binding affinity toward toxic organic molecules at trace levels in water. In the first of these, humic acid, is strongly and irreversibly bound to the clay through a sandwich layer of positive charge provided by hydroxy-aluminum to form humic acid-hydroxy aluminum-clay. The second approach is to use the hydroxy-aluminum clay itself, which is known to have a "zeolite-like" micropore structure and greatly enhanced surface area accessible to toxic organic molecules. FTIR studies confirm that humic acid is bound to hydroxy aluminum-clay through strong electrostatic interactions. Similar FTIR studies on hydroxy aluminum-clay alone reveal that the bulk of hydroxy-aluminum may be between clay lamellae.

ACKNOWLEDGMENTS

This study was supported by a grant from the Dow Chemical Company, Midland, Michigan. The authors wish to acknowledge the help of Prof. Phil Meyers, University of Michigan, Ann Arbor, Michigan, and of Dr. Peter F. Landrum, Great Lakes Environmental Research Labs., Ann Arbor, Michigan, in organic carbon determinations. The authors wish to thank Prof. Steve F. LeBlanc, University of Toledo, Toledo, Ohio, for his help with the BET surface area measurements. Many fruitful discussions held with Dr. Lanny Robbins, Dow Chemical Company, Midland, Michigan, are gratefully acknowledged. Finally, the authors would like to recognize the help of Mr. Timothy F. Nolan in all phases of this study.

REFERENCES

1. Poland, A., and J. C. Knutson. "2,3,7,8-Tetrachlorodibenzo-p-dioxin and Related Halogenated Aromatic Hydrocarbons: Examination of the Mechanism of Toxicity," *Ann. Rev. Pharmacol. Toxicol.* 22:517 (1982).
2. McConnell, E. E., J. A. Moore, J. K. Haseman, and M. W. Harris. "The Comparative Toxicology of Chlorinated Dibenzo-p-dioxins in Mice and Guinea Pigs," *Toxicol. Appl. Pharmacol.* 44:335 (1978).
3. Exner, J. H. "Detoxification of Chlorinated Dioxins," Abstract, ACS Symposium, 186th National Meeting, Division of Environmental Chemistry 23(2):30 (1983).
4. Weber, W. J., Jr., T. C. Voice, M. Pirbazari, G. E. Hunt, and D. M. Ulanoff. *Water Res.* 17(10):1443 (1983).
5. Di Toro, D. M., H. M. Lewis, and M. C. Casey. Final Report, Environmental Engineering and Science Program, Manhattan College, Bronx, NY (1982).
6. Nau-Ritter, G., C. F. Wurster, and R. G. Rowland. *Water Res.* 16:1616 (1982).
7. Barrer, R. M. *Zeolites and Clay Minerals as Sorbents and Molecular Sieves* (London: Academic Press, 1978), p. 430.
8. Robbins, L. Private Communication, Dow Chemical Company, Midland, MI (1983).
9. Landrum, P. F., S. R. Nihart, B. J. Eadie, and W. S. Gardner. *Environ. Sci. Technol.* 18:187 (1984).

10. Reed, M. G. *J. Pet. Tech., Trans. AIME* 253:860 (1972).
11. Pinnavaia, T. J. "Intercalated Clay Catalysts," *Science* 220:365 (1983).
12. Carter, C. W. "The Binding of Nonpolar Organic Compounds to Dissolved Humic Materials," PhD Thesis, Drexel University, 1982.
13. Srinivasan, K. R., H. S. Fogler, E. Gulari, T. F. Nolan, and J. S. Schultz. "Removal of Trace Levels of Dioxins from Water by Sorption on Modified Clay," *Environ. Progress*, Vol. 4, no. 4, 239 (1985).
14. Greenland, D. J. "Interaction between Humic and Fulvic Acids and Clays," *Soil Sci.* 111:34 (1971).
15. Theng, B. K. G. *The Chemistry of Clay-Organic Reactions*, (London: Adam Hilger, 1974), p. 291.
16. Oram, A. University of Michigan, Unpublished results (1985).
17. Schnitzer, M., and S. U. Khan. *Humic Substances in the Environment*, (New York: Marcel Dekker, 1972), p. 72.
18. Nakamoto, K., M. Margoshes, and R. E. Rundle. "Stretching Frequencies as a Function of Distance of Hydrogen Bonds," *J. Amer. Chem. Soc.* 77:6480 (1955).
19. Brydon, J. E., and H. Kodama. "The Nature of Aluminum Hydroxide-Montmorillonite Complexes," *The American Mineralogist* 51:875 (1966).

CHAPTER 36

Binding of OCDD, 2,3,7,8-TCDD and HCB to Clay-Based Sorbents

Keeran R. Srinivasan and H. Scott Fogler

INTRODUCTION

In Chapter 35, we described two procedures for modifying clay. The first of these was the use of humic acid (HA) as the modifier and hydroxy aluminum as the "cement" to hold humic acid on the clay surface so that humic acid is irreversibly bound to the surface in aqueous medium near neutral pH. The second was the intercalation of hydroxy aluminum polymers between clay layers to form a "zeolite-like" micropore structure.

It has been reported that sorption of hydrophobic molecules such as chlorinated dioxins and biphenyls by sorbents is mediated by the organic carbon content of the sorbent.[1] The applicability of this hypothesis to the experimental data of a given binding process is usually verified by examining the validity of the following predictive relationship to the binding process:[2]

$$\log K_{OC} = \log K_{OW} - 0.21 \tag{1}$$

Where K_{OC} is the organic carbon normalized linear partition coefficient, i.e.,

$$K_{OC} \ = \ K_p/f_{OC} \tag{2}$$

K_p is the apparent equilibrium constant for the partitioning of the sorbate between the solution and the sorbent, and f_{OC} is the organic carbon content of the sorbent expressed as the weight fraction. Finally, K_{OW} is the experimental octanol-water partition coefficient for the test sorbate. The basis for Equations 1 and 2 is the assumption that sorption of a hydrophobic molecule on an organic carbon-containing sorbent may be likened to the partitioning of the same molecule between water-saturated octanol and octanol-saturated water.

The sorption of toxic organic chemicals such as chlorinated dioxins and biphenyls by clay-based sorbents is also expected to occur through a molecular sieving process similar to the one between microporous carbons and silica gels and organic compounds.[3] In this case, the surface area and the microporosity are the critical factors in determining the strength and the extent of binding of organic molecules to the surface. Hydroxy aluminum-clay can be considered as an example of this class of sorbents.

In this study, an attempt has been made to evaluate the relative contributions of the above two factors to the sorption potential and binding affinity of the clay-based adsorbents toward chlorinated dioxins and biphenyls. It is our belief that extension of our analysis to similar toxic molecules such as chlorinated dibenzofurans and polynuclear aromatics would be quite fruitful.

The extent of binding of OCDD, 2,3,7,8-TCDD, and HCB to unmodified clay, hydroxy aluminum-clay, and humic acid-hydroxy aluminum-clay is described. The organic carbon content of humic acid-hydroxy aluminum-clay used in this study ranged from 2% to 8.6% by weight. Finally, humic acid-alumina containing 1.8% by weight organic carbon was also prepared at pH 6.3 and used in the present study.

MEASUREMENT OF PARTITION COEFFICIENTS

Binding of the three sorbates to the above sorbents was followed by a radiometric method. The ^{14}C-labeled OCDD (21 mCi/mmol, 96 dpm/ng), 2,3,7,8-TCDD (117 mCi/mmol, 789 dpm/ng) and HCB (12.6 mCi/mmol, 76 dpm/ng) were obtained from Pathfinder Labs (St. Louis, MO.). They were used without further purification, since all the isomers were rated at greater than 98% purity. The sorption experiments were carried out in glass vials containing a predetermined concentration of the sorbent and the sorbate in filtered (0.2-μm filter), deionized water. The vials were shaken in a constant-temperature water bath maintained at 25°C. Following a minimum of 24 hr equilibration (longer time periods showed virtually no change in the observed binding constants) the sorbent was separated by centrifugation in a horizontal rotor at 1600 g for 45 min and an aliquot of the super-natant was counted to give the "free" or solution phase concentration. The balance of the supernatant was carefully removed and the sorbent pellet was quantitatively transferred in a minimum volume of water to a filter pad and counted to yield the "bound" value. Finally, the scintillation fluid was directly added to the vial and the amount of sorbate bound to "glass" was measured. A mass balance was carried out for each experiment and if it was close to 100% but not less than 85%, and if the "glass" value did not exceed 10%, the linear partition coefficient (PC) was calculated from the following expression:

PC = (g bound sorbate/g sorbent)/(g sorbate in solution/mL solution)

The units of partition coefficient are mL/g. In most of the experiments we obtained a mass balance of 90% and the "glass" value did not exceed 6%.

The use of ^{14}C label precluded the use of initial sorbate concentrations below 1 ng/mL without an elaborate procedure to assay the amounts of sorbate in the three phases mentioned above: namely, the "free," the "bound," and the "glass." In the case of HCB, this posed no problem, since the reported aqueous solubility of HCB is 8.8 ng/mL.[4] In the case of 2,3,7,8-TCDD, the availability of very high specific radioactivity enabled us to use an initial concentration of 0.2 ng/mL, which is roughly 60% of the aqueous solubility of this isomer. However, OCDD is known to be extremely insoluble in water, with a reported aqueous solubility of 2×10^{-9} g/L.[5] Measurements in our laboratory indicate that the solubility may be closer to 1×10^{-7} g/L or 0.1 ng/mL. In any case, it was observed that addition of tetrahydrofuran (THF) to water at ppm levels increased the "solution" concentration of OCDD to 70×10^{-7} g/L at a THF volume fraction of 600 ppm.

All OCDD sorption experiments reported in this study contained THF in solution up to a volume fraction of 11,000 ppm. Control experiments in duplicate, i.e., water-THF solution without any sorbent, were performed with each sorption experiment to monitor OCDD concentration in the solution at the beginning and at the end. The purpose of the control experiments was to establish that there was minimal nonspecific adsorption or precipitation of OCDD on glass. These control measurements revealed that the solution concentration of OCDD was invariant with 5% during the course of the experiment. Centrifugation of control solutions at 1600 g showed no change in OCDD concentration, establishing that large microaggregates of the sorbate were not present. Finally, in all cases, the initial OCDD concentration never exceeded 50% of the highest solution concentration that could be maintained in water-THF binary mixture containing a given volume fraction of THF. Table 1 lists the relevant data for the highest OCDD concentration in water-THF binary solution obtained by above criteria.

Table 1. Data for the Highest OCDD Concentration Obtained in Water-THF Binary Solution.

Volume Fraction THF (ppm)	Mole Fraction ($\times 10^4$)	OCDD Concentration in Solution ($\times 10^7$ g/L)
0	0	1
120	0.3	40
300	0.7	62
600–3,000	1.4–6.8	>70
11,000	24.9	~250

Tetrahydrofuran (THF) was the carrier solvent in which OCDD was supplied to us. THF is miscible in water in the range of concentrations of interest and has a boiling point of 66°C.[6] Its vapor phase dipole moment of 1.70 Debye would mitigate against any tendency toward micelle formation at the mole fractions used in this study. Control experiments in water-THF solutions indicated that THF volatilization, if any, had very little effect on the solution concentration of OCDD for up to 48 hr, provided all the steps were carried out under tightly sealed conditions.

BINDING OF OCDD

The initial set of experiments was performed with OCDD in water-THF binary solution to establish the effects of (1) the amount of organic carbon on the sorbent; (2) the surface area of the sorbent; and (3) the microporosity of the sorbent on the value of the linear partition coefficient for different sorbents, and the results are summarized in Table 2.

A study of Table 2 reveals that the fractional organic carbon content of the sorbent has no significant effect on the binding, as has been previously stated.[1] In fact, hydroxy aluminum-clay with no organic carbon content has

Table 2. Relation of Amount of Organic Carbon and Surface Area of the Sorbent to the Linear Partition Coefficient for Different Sorbents.[a]

Clay-Compound[b]	% Organic Carbon[c]	Surface Area[d] (m^2/g)	PC $\times 10^3$ (mL/g)
Clay	none	27.6	2.0 ± [g]0.5
HA-HYDAL-Clay (6.3)[e]	2.2	61.8	7.8 ± 1.8
HA-HYDAL-Clay (3.0)	8.6	1.6	4.6 ± 0.5
HA-HYDAL-Clay (3.0)	3.9	ND[f]	10.6 ± 2.7
HYDAL-Clay(no HA)	none	114.0	171.6 ± 16.0
HA-Alumina	1.8	189.2	17.0 ± 1.0

[a]Sorbent concentration = 0.47 mg/mL; volume fraction THF = 600 ppm; OCDD concentration = 35 $\times 10^{-7}$ g/L.
[b]HA, humic acid; HYDAL, hydroxy aluminum.
[c]The error in organic carbon value is ± 0.02%.
[d]The error in surface area measurement in 5%.
[e]Indicates the pH at which HA was bound to clay-HYDAL.
[f]ND, not determined.
[g]Indicates the range of PC values from the mean value of quadruplicates.

the highest partition coefficient, and the corresponding K_{OC} value of infinity (according to Equation 2) is physically meaningless. It may be argued that the source of humic acid and/or its possible desorption from the surface during the partition experiments might have been responsible for the poor performance of humic acid-hydroxy aluminum-clay sorbents.

Ancillary experiments carried out in our laboratory show that humic acid in solution binds strongly to OCDD. Therefore, the source of humic acid (Aldrich Chemical Co.) may not be responsible for the poor binding affinity of OCDD to the surface-bound humic acid. Moreover, there was no indication of humic acid dissociating from the surface during the OCDD binding experiments. As pointed out earlier, humic acid-hydroxy aluminum-clay sorbents were extensively washed in distilled water prior to use to mitigate against humic acid desorbing during binding experiments. Secondly, we prepared, at pH 3.0, a series of humic acid-hydroxy aluminum-clay sorbents with the organic carbon content ranging from 1.65 to 8.6% by weight. As reported in Table 2, the highest partition coefficient value was obtained with the 3.9% sorbent, while the sorbents with higher and lower organic carbon contents had lower values of PC. If the leaching of humic acid from the surface of the sorbent was the contributory factor to the low PC values of humic acid-hydroxy aluminum-clay sorbents, then the highest PC would be expected with the sorbent having the lowest organic carbon content. However, this was not the case. Finally, the presence of 600 ppm of tetrahydrofuran in solution precluded the use of organic carbon analysis to detect the presence of humic acid in solution. It should be mentioned that the effect of organic carbon has been mostly seen with aquatic sediments and the chemical identity of the organic carbon is not well established. Using our synthetic procedure, we plan to prepare modified clays containing organic carbon, but with no concomitant loss of the available surface area. This would enable us to sort out the organic carbon effect in a more systematic way.

In any case, it is our belief that the dramatic loss of the surface area of humic acid-hydroxy aluminum-clay when its percent organic carbon content goes up to a value of 8.6% is due to the possibility that the binding of humic acid (a polymer of greater than 3500 daltons molecular weight) to hydroxy aluminum-clay blocks the surface pores. This is clearly seen to be detrimental to the sorbent's effectiveness.

A comparison between the binding affinity of humic acid-alumina and hydroxy aluminum-clay shows that the microporosity of the sorbent has the most profound effect on the binding affinity of the sorbent toward OCDD. Clearly, hydroxy aluminum-clay represents a new class of sorbents uniquely suited for the microseparation of toxic organic molecules from industrial wastewaters.

In a series of experiments, the volume fraction of THF was varied from 600 to 11,000 ppm to establish its role on the binding of OCDD to hydroxy

Table 3. Binding of 2,3,7,8-TCDD to Hydroxy Aluminum-Clay.[a]

Type	Value	% of Total Sorbate Added
Free TCDD in Solution	0.013 ng/mL	6.0 ± 0.5
TCDD Bound to the Sorbent	3.19 ng	88.6 ± 1.4
TCDD on the Glass Vial	0.22 ng	6.0 ± 1.0
Mass Balance		98.0
Partition Coefficient		67,100 ± 6,000 mL/g

[a]TCDD aqueous solubility = 0.32 ng/mL; TCDD concentration used = 0.20 ng/mL; TCDD specific radioactivity = 789 dpm/ng; sorbent concentration = 0.22 mg/mL; THF = not added.

aluminum-clay. The values of partition coefficient decreased from 171,000 to 45,000 mL/g at the two extrema of THF volume fractions.

Being a good solvent for OCDD, THF was expected to decrease the hydrophobicity of OCDD, which is reflected in a lowered value of the partition coefficient. THF could also reduce the values of partition coefficient by binding to the surface sites accessible to OCDD. In the former case, the decrease in the partition coefficient will be regulated by a corresponding increase in the "free" concentration of OCDD, while in the latter case, a decrease in the amount of "bound" sorbate can be expected.

A study of the raw data revealed that the "free" concentration of OCDD increased fourfold. Thus the decrease in hydrophobicity of OCDD due to the presence of THF accounts for the entire observed decrease in the partition coefficient. Obviously, the solvent properties of THF are significantly more important than its ability to act as a second ligand. A plot of the reciprocal of the observed PCs against the concentration of THF in water-THF solutions extrapolated to a THF free partition coefficient value of 220,000 mL/g for the OCDD binding to hydroxy aluminum-clay in pure water at the chosen initial OCDD concentration of 3.5 ng/mL.

BINDING OF 2,3,7,8-TCDD

These experiments were performed at Pathfinder Labs (St. Louis, MO) using the same procedure as described before. Stock solutions of the sorbate in water were prepared and used in the binding experiments. In the first set of experiments, hydroxy aluminum-clay was exclusively used as the sorbent, and in subsequent tests a comparison between different sorbents was attempted. The results are summarized in Tables 3 and 4.

The value of partition coefficient shown in Table 3 is at least 6 to 7 times higher than the minimum value required for economic viability. It is diffi-

Table 4.   Binding of TCDD to Different Sorbents.[a]

Sorbent	PC × 10^{-3} (mL/g)
Clay	2.8
HA-HYDAL-Clay	3.3
HYDAL-Clay	67.1
Activated Carbon (Calgon Filtrasorb 400)	24.0

[a]Experimental conditions same as in Table 3.

cult to account for the threefold lowering of the PC value between OCDD and 2,3,7,8-TCDD without similar experiments between hydroxy aluminum-clay and a series of polychlorinated dioxins (PCDDs). As in the case of OCDD, hydroxy aluminum-clay appears to have the highest binding affinity for 2,3,7,8-TCDD. An interesting finding is that hydroxy aluminum-clay is three times better than the activated carbon used in this study. It is our estimate that activated carbon is more expensive than the hydroxy aluminum-clay. Thus, *the results of OCDD and 2,3,7,8-TCDD experiments taken together clearly establish hydroxy aluminum-clay as the sorbent of choice for the removal of these and many other toxic organic molecules from industrial wastewaters.*

We have also carried out a few TCDD binding experiments in water-THF mixtures, and the lowering of the partition coefficient due to the presence of THF follows a trend identical to that of OCDD.

BINDING OF HCB

The hexachlorobiphenyl isomer used in our study was the 2,4,5,2',4',5'-HCB. Unlike the dioxin molecules, the biphenyl molecule is nonrigid and the angle between the planes of the phenyl rings can be greater than zero. The presence of chlorine atoms in the ortho positions (2 and 2') is expected to accentuate the tendency toward a nonzero dihedral angle.[7] This isomer of HCB was used to probe the topological nature of the hydroxy aluminum-clay's micropore structure. In other words, are the micropores on the modified clay more conducive to accommodating planar, rigid molecules as opposed to nonplanar ones? The results are shown in Table 5.

Even though hydroxy aluminum-clay has the highest binding affinity of all the adsorbents used, the absolute values of partition coefficients are lower compared to OCDD or 2,3,7,8-TCDD. They are still three times the minimum required. This behavior suggests that the micropore structure on the hydroxy aluminum-clay may have a bias toward rigid, planar molecules. It is interesting to note that THF has no observable effect on the binding of HCB to hydroxy aluminum-clay. Further experiments are clearly warranted to verify the generality of this effect.

Table 5.   Binding of HCB to Different Sorbents.[a]

Sorbent	PC × 10^{-3} (mL/g)
Clay	1.5 ± 0.2
HA-HYDAL-Clay	5.6 ± 1.0
HYDAL-Clay	32.0 ± 1.0
HYDAL-Clay[b]	30.0 ± 1.0

[a]Sorbent concentration = 0.25 mg/mL; initial HCB concentration = 3.1 ng/mL.
[b]Experiment carried out in presence of THF.

CONCLUSIONS AND FUTURE WORK

The experimental results described above demonstrate that hydroxy aluminum-clay has the potential to be a highly efficient removal agent for toxic organic molecules in industrial wastewaters. It is cost-effective and easy to prepare. Since hydroxy aluminum-clay is also known to be a good catalyst,[8] it is possible to degrade the adsorbed pollutant in situ and regenerate the sorbent.

Further work is required to refine the formulation of the hydroxy aluminum-clay so that its possible use as a general purpose wastewater treatment agent can be explored. Chromatography type experiments using packed-bed or continuously stirred slurry reactors need to be undertaken to establish optimal operating conditions. We have already carried out a few preliminary experiments on a packed-bed column using hydroxy aluminum-clay and 2,3,7,8-TCDD, and the results indicate that even under nonequilibrium flow conditions, the sorbent retains about 70–75% of the total adsorbate.

An area of research not yet explored is the potential use of hydroxy aluminum-clay as liner material for toxic waste dumps. Hydroxy aluminum has been used extensively to prevent dispersion of clay particles by water.[9] This property, combined with the ability of hydroxy aluminum-clay to strongly bind chlorinated dioxins and biphenyls, bodes well for its use as a liner material for toxic waste dumps.

ACKNOWLEDGMENTS

This study was funded by a grant from the Dow Chemical Company, Midland, MI. The authors wish to thank Dr. Lanny Robbins of the Dow Chemical Company, Midland, Michigan, for his help in arranging to have the 2,3,7,8-TCDD experiments performed by us at the Pathfinder Labs, St. Louis, Missouri. Finally, we would like to acknowledge the able assistance of Ms. Cathy Peters and Mr. Timothy F. Nolan.

REFERENCES

1. Voice, T. C., C. P. Rice, and W. J. Weber, Jr. "Effect of Solids Concentration on the Sorptive Partitioning of Hydrophobic Pollutants in Aquatic Systems," *Environ. Sci. Technol.* 17(9):513 (1983).
2. Karickoff, S. W., D. S. Brown, and T. A. Scott. *Water Res.* 13:241 (1979).
3. Barrer, R. M. *Zeolites and Clay Minerals as Sorbents and Molecular Sieves* (London: Academic Press, 1978), p. 6.
4. Hutzinger, O., S. Safe, and V. Zitko. *The Chemistry of PCBs* (Cleveland, OH: CRC Press, Inc., 1974).
5. Barrie, W. G. K., L. P. Sarna, and D. C. G. Muir. "K_{OW} of 1,3,6,8-T_4CDD and OCDD by Reverse Phase HPLC," Abstract, *ACS Symposium*, 186th National Meeting, Division of Environmental Chemistry 23(2):316 (1983).
6. *The Merck Index* 8th Edition, (Rahway, NJ: Merck & Co., Inc., 1968), p. 1026.
7. McKinney, J. D., and P. Singh. *Chem. Biol. Interact.* 33:271 (1983).
8. Shabtai, J. "Class of Cracking Catalysts: Acidic Forms of Cross-Linked Smectites," U. S. Patent No. 4,238,364 (1980).
9. Reed, M. G. *J. Pet. Tech, Trans. AIME* 253:860 (1972).

CHAPTER 37

The Seveso Lesson: Advances in Reclamation and Disposal Techniques

G. Umberto Fortunati

INTRODUCTION

As is well known,[1-8] on July 10, 1976 at 12:37 p.m., an accidental release in the atmosphere, deriving from the alkaline hydrolysis of 1,2,4,5-tetrachlorobenzene to Na-2,4,5-trichlorophenate, an intermediate for the processing of TCP (2,4,5-trichlorophenol), contaminated an area of 1810 ha in the municipalities of Seveso, Meda, Cesano Maderno, and Desio and, to a lesser degree, seven other municipalities all in the Milan (Italy) area.

The TCP Plant which contaminated the area belongs to ICMESA (Industrie Chimiche Meda Società Azionaria), an Italian Company of the Givaudan Hoffman-La Roche Multinational Concern.

The TCP was usually transported to another factory of the same group to manufacture herbicides and antiseptics.

SOIL RECLAMATION OPTIONS

Target Limits in Lombardy Region

Several methods of TCDD disposal have been listed in the special Law for Seveso voted by the Lombardy Regional Government on January 2, 1977.[9] The remedial action was aimed to reach the limits set forth as follows:

Interior of homes, schools, offices, factories	10 ng/m^2
Agricultural land	7 ppt or 750 ng/m^2
Nonagricultural land	45 ppt or 5000 ng/m^2

Incineration

The incineration method has been considered "preferential" for the detoxification of soil having a TCDD content higher than 45 ppt for an estimated total of 120,000 m^3, and, if necessary, also the less contaminated one with a TCDD content higher than 7 ppt, or an additional 200,000 m^3.[10] The characteristics foreseen for the incineration plant were:

Capacity	
soil	250 tons/day
vegetation	80 tons/day
dead animals	40 tons/day
Temperature	
of combustion	1000°C
in the post-combustion chamber	1200°C
ash discharge	1000°C

The dust in the fumes was to be minimized with a filter system completed by a scrubber. Two rotary kilns reached, according to the program, the long residence time required; the project considered the possibility of grinding the ashes to be recycled into the kilns to guarantee complete destruction of the contaminant. The fumes after the scrubber could be reheated to increase the emission height. The plant was planned to be inside the ICMESA property area, and a pilot unit was to have been built before finalizing the actual project. Later on, however, both the population and the local authorities showed concern not only for the danger of a TCDD fallout from the stack, but also because once a large and efficient incinerator was built in the area its utilization would not be limited to treating the TCDD-contaminated material but, eventually, any dangerous and toxic material produced in northern Italy.

In relation to the incineration option, it is worth mentioning that the OCS (Ocean Combustion Service), owner of the "VULCANUS," proposed to adapt another ship to incinerate at sea the Seveso solid material.[11] This possibility was discarded due to the risks and difficulties of transferring the several hundred thousand tons of soil, etc., to be loaded in a harbor into the incinerator ship through a heavily populated part of the country. The research programs listed in the law had the declared purpose of finding a more convenient method than incineration.

Photodegradation

Photodegradation in situ of TCDD trapped on the wall surfaces of contaminated homes was considered. The UV light option was proposed by Professor Liberti[12,13] of Rome as early as September 1976. The method comprises the spraying of a mixture of ethyl oleate/xylene (1:2 on the interior surfaces; 2:1 on outside walls). In the experiments carried out in the A2 subzone in Seveso the average degradation reached 82% with artificial

UV light (20 μW/cm^2) applied for three days at 20°C. A series of tests have also been conducted with highly contaminated soil (around 7 ppb) after having been treated with a 1:1 solution of ethyl oleate/xylene, reaching, after 5 days exposure:

79% degradation for the 1st cm of soil
63% degradation for the 2nd cm of soil
55% degradation for the third cm of soil

The scientific validity of this method was recognized by the Regional Research Committee. There were, however, no real chances to reduce the final contamination down to the target limits foreseen by the Lombardy Regional Government, especially if one considers the difficulty of spraying all the surfaces with the solvent mixture and of reaching many hidden parts of the contaminated homes with the UV light. The preparation of walls and surfaces for application of the UV method proved to be more difficult and expensive than the detoxification method normally followed: vacuum cleaning and washing with water and detergent as done in 1978–79 in the sub-zones A6 and A7.

Gamma-Ray Method

The gamma ray methodology was proposed in March 1978 by the Pharmacological Research Foundation "M. Negri" of Milan as a study contribution to solve the soil contamination problem.[14] By adding to the artificially contaminated soil different types of solvents (acetone, ethanol, etc.) and exposing it to gamma rays, the dechlorination of the TCDD was noted. The dechlorination was a function both of the quantity of gamma rays absorbed by the soil/solvent mixtures and of the type of solvent used. The best results were achieved with ethanol — more than 80% dechlorination. The difficulty of applying such a method in the field on a large scale convinced the Regional Committee not to pursue the initial experiments any further.

Ozone Method

The ozone method has twice been proposed to the Special Office for Seveso. The first time, one year after the accident, the University of Milan (Professor Cavalloni) offered to test the efficiency of blowing air enriched with ozone over soil and walls within the A zone, in the presence of carbon tetrachloride. Tests have been done only in the laboratory. The method was examined a second time in 1980 after the advice of the late Dean of the Organic Chemistry Faculty, Professor Canonica, to detoxify the solid residues contained inside the TCP reaction vessel where the anomalous reaction took place. It was necessary to use CCl_4 as a solvent to make the reaction effective. In both cases the ozone method was not used because:

1. The need to have aerosols containing CCl_4 in large quantities would have increased the risk for the workers, since CCl_4 is carcinogenic.
2. With the air/ozone-enriched gaseous mixtures technically available, it was not possible to go deep under the surface in the detoxification of the soil.
3. In the case of the solid mass contained in the reaction vessel, it would have been necessary first to take out the material in pieces and grind it finely before mixing with solvent. Therefore, the advantage of destroying the TCDD contained in the vessel residues in the vessel itself would have been lost, and the grinding operation would have been a potential dust source.

Extensive tests carried out both in Milan University and at ETH University in Zurich (Switzerland) proved the necessity of finely dispersing the solid in the solvent before applying the ozone in order to reach satisfactory results.

Solvent Extraction Method

This last method was suggested by analogy with the analytical procedure used for TCDD quantification. The plant to be used, according to DAM Progetti, the proposing Company of Ravenna, Italy,[15] would be similar to the hexane extraction of oil from previously pressed olives. Lab tests demonstrated that the yields claimed by the proposing party (96.5%–99.8%) could not be achieved (Table 1).

The feasibility of building a pilot plant to apply the extraction method was discussed thoroughly; however, after the laboratory test results were known the project was not carried out because of the unlikelihood of reaching the target limits for the recovery and purification of the solvent mixture (contaminated), and because of the risk of handling large amounts of materials in a rather complex chemical plant, which would not in fact be similar to the standard olive oil extraction process plant as proposed by the DAM Progetti Company. The development work had to be started de novo, from the bench-scale data. All of these problems considered, the solvent extraction proposal was abandoned.

SEA Marconi Dechlorination Method

Finally, a method proposed by the Italian Company SEA Marconi of Turin, to detoxify contaminated materials in situ, was tested. This dechlorination method uses a mixture of reagents, including peroxide and polyethylene glycol (PEG). Tests have been carried out both in the laboratory of the Joint Research Center of the European Community in Ispra (Varese, Italy) on behalf of the Special Office of Seveso, and inside the contaminated TCP plant of ICMESA.[16] The purpose of the experiments was the

Table 1.   Extraction Yields Experienced with Hexane and Hexane/Acetone Mixtures.

Solvent	Hexane	Hexane/Acetone
Soil with 17.5% Water	12.5%	87.5%
Soil with 1% Water (Dried)	47.7%	52.6%

decontamination of materials and equipment contained in the Reaction Department.

Different experiments were carried out: on a pure sample of TCDD present in a solvent; on a TCDD-contaminated glass surface; on a pipe of the plant; and on the soil of Seveso.

The results showed that, working at 85°C with TCDD present in n-decane, 99.9% of decomposition occurs after 2 hr, when the reagent is composed of PEG 6000, K_2CO_3 and Na_2O_2; at 50°C the reaction rate is lower, but high enough, since after 3 hr no TCDD was detected.

Because the reagent's formulation can produce a solid mixture at room temperature, it was possible to cover a TCDD-contaminated glass surface with such a solid and observe it without hazard of contamination: after 8 days its TCDD content decreased by 50%. Also, on contaminated pipe sections of the ICMESA plant and on a sample of contaminated soil from Seveso, this dechlorination method was judged satisfactory, but it was not used, as it appeared that the total contaminant in the equipment and pipes, excluding the A101 vessel, did not reach a total of 3 g, and therefore the use of the SEA Marconi method could not be justified. All the equipment of the TCP plant was emptied of residues and carefully sealed. Then the outside surfaces were vacuum-cleaned and painted before being buried in commercial containers in the very center of Basin No. 2. The TCDD-containing residues were incinerated (June 1985) at Ciba-Geigy's special oven in their main factory in Basel (Switzerland).

The reaction products of this dechlorination method were sent to the Swiss Toxicological Institute and were found to be nontoxic, as they did not affect the laboratory test animals. As found by GC/MS, the tetrachloro is progressively transformed into its tri-, di-, monochloro derivatives and, at the end, into dioxin which, in turn, polymerizes.

CONSIDERATION OF THE STUDIED OPTIONS

In general terms, the more complex the operations planned to handle the contaminated wastes, the higher the risk involved; consequently, more precautions must be taken to keep to a minimum the danger to the environment and to the cleanup workers. Furthermore, if a process plant is used, its decontamination is also complex. Moreover, most of the treatment processes required an intermediate removal step involving temporary storage.[16]

The "main" technical option, i.e., the incineration of the removed material in an ad hoc rotary kiln, was considered, but discarded on social and political grounds.

Alternative methods were considered; some were tested at the laboratory bench scale, and a few (three in all) were tested in the contaminated areas. However, of all the methods proposed and evaluated in 1977–1980 by the

Special Office, none was judged suitable to be transferred successfully to the "industrial" scale. In fact, this transfer implies the solution of many engineering, construction, operating, and safety problems posed by the fact that hundreds of thousands of metric tons of waste have to be treated. The dioxin concentration varied greatly from a few ng/kg of soil to a few mg/kg of soil in the same highly polluted land area, the "A" Zone.

All the decontamination programs and the subsequent management practice of the Special Regional Office in charge of fulfilling the Law were influenced by the hypothesis mentioned in the literature of a rapid natural decomposition of TCDD in the soil by light, microorganisms, or other factors. In the Seveso soil, the composition of which is largely alluvial and clayish, the reduction of TCDD concentration with time proved to be negligible.

Other possible methods of collecting and of disposing of the waste were evaluated:

    a) direct dumping of waste into the Atlantic Ocean
    b) disposal in a salt pit (in drums)
    c) disposal in situ in sealed concrete boxes
    d) disposal in situ in controlled basins

The International Conventions on the use of the seas[17,18] and the hostility of the population and local authorities to the disposal in salt pits made the projects listed under a) and b) impracticable. Therefore it became advisable to remove the toxic compound from the area in order to prevent any further dispersion into the surrounding territory.

CLEANUP TECHNIQUE: SOIL SCARIFICATION AND DISPOSAL INTO BASINS

A decision was made according to point d) above, i.e., to collect the contaminated materials (vegetation, rubble of demolished houses, and scarified soil) and dispose of them in two basins in the A zone itself.* TCDD is not expected to leak out from the basins[16] for the following reasons:

1. TCDD is not water soluble.
2. TCDD is bound to the clay, which is abundant in the Seveso soil.
3. In the disposal of the collected soil resulting from the reclamation work, the most-contaminated soil has been deposited in the central part of the basin, while the less-contaminated has been deposited all around the core.
4. A plastic sheet (high-density polyethylene) 2.5 mm thick is welded so as to constitute a unique blanket under and above the material disposed of.
5. A 15-cm-thick foundation made of sand and bentonite (reinforced con-

---

*The reclamation activity done during 4 years (1980–1984) is documented in an 18-hr video cassette available in Italian, English, and French.

crete) is also used; this conglomerate shows high impermeability as well as plasticity.

6. The deposit cap is protected by a layer of gunite (concrete slabs) to avoid any mechanical damage from above.
7. A 75-cm-thick clean soil layer finally covers the closed basins.

A well with a drainage pump allows extraction of the water which gathered at the bottom of the basin during the filling phase. The water extracted with the pump has been analyzed and found to be dioxin-free at the ppt detection level. An activated charcoal filter could be used, if the water should be found contaminated, before its discharge.

Basin Control Programs

The types of control foreseen for the basins are:

1) control of the contamination of the aquifier downstream
2) analytical monitoring of gaseous releases as well as liquid drainage
3) control of the mechanical deformation of the two hills resulting from the landscaping of the two closed basins
4) a search for new techniques to stop possible leakage of the high-density polyethylene (HDPE) membranes

The data collected so far (June 1985) do not indicate any problem or damage to the membranes' integrity.

COMPARISON BETWEEN 1st AND 2nd BASIN

The two basins were planned in 1980 and 1982 respectively. The experience gathered in the first project was useful in improving the second design. Whereas in the first basin the slope was 50%, in the second basin the slope was drastically reduced to 25%, thus making it possible to use compacting machines. The width of the 2.5-mm-thick HDPE used was also different in the two cases (12 m and 2.5 m, respectively); this difference increased the welding length in the second basin. The welding itself has been done with different techniques and different control methodologies: the first was done with the addition of new material, the second by parallel heating and welding the superimposed edges. Difference in size of the HDPE material, and consequently the less extended welding and the welding technique itself, recommend the method used in the first basin, but the second was considerably less expensive.

The drainage system is more complete in the second basin, as the water collection has been designed with two independent pipe systems: one inside at the bottom of the basin and one underneath the HDPE lining, thus making it possible to detect the origin of the leakage.

The mechanical protection with 10-m × 10-m concrete slabs of the first basin has been changed, in the second one, into a finely subdivided concrete

Table 2. Comparison of 1st and 2nd Basin Technical Characteristics.

		First Basin	Second Basin	Notes
Dimension				
Covered area,	m^2	9,300	20,150	The first basin had to be
Bottom area,	m^2	4,457	9,800	dug among the highway, the
Depth,	m	9	9	Certesa river and the Via
Filling height,	m	6	7	Vignazzola. The second basin
Slope,	%	50	25	was placed in the most con-
Capacity, 1000	m^3	80	200	venient location for soil
scarification and removal				
purposes.				
Lining				
Bentonite density				
bottom, kg/m^3		1.916	2.123	The second basin shows
slope, kg/m^3		1.826	2.082	improvement over the first.
HDPE				
thickness,	mm	2.62	2.55	
density, kg/m^3		943.7	946.7	The first basin is slightly
tensile strength, MPa		34.0	34.6	better, regarding liner quality,
stretch at breakage		935	980	than the second one.
Control of welding				
tensile strength, MPa		16.9	16.4	
Bentonite				
permeability		—	$>5 \times 10^{-10}$	Or 1.6 cm/yr. The 15-cm layer
could leak after 10 yr.				

Source: Lombardy Regional Government, Special Office for Seveso—Milan, September 20, 1984.[16]

tile cover sprayed in place, which better adapts itself to the deformations due to soil settlement. The differences in the technical characteristics of the two basins are shown in Table 2.

DISMANTLING OF THE ICMESA TCP PLANT (1983)

The Plant

The 230-m^2 plant consisted basically of three 10,000-L mixing vessels, various columns, condensers, pumps, connecting pipework, supporting constructions, and platforms. The heaviest single piece of equipment was a lead-lined vessel weighing over 7 tons. The total weight of the plant, including platforms and supporting constructions, was 75 tons.

Planning

A study prepared by the Ente Nazionale Energia Nucleare (ENEA) of Rome[19] (the Italian Nuclear Energy Authority) on behalf of Givaudan formed the essential basis on which the dismantling operations were planned. The methods proposed by ENEA represented the "nuclear solution," i.e., the work was to be carried out using safety precautions used in the nuclear field.

The first operation, carried out without difficulties in the summer of 1982, involved the removal of 2272 kg of contaminated residues from reaction vessel A 101. This was accomplished by erecting a decontamination unit which allowed safe access to the contaminated plant area.

The planning of the dismantling work could thus be based both on the ENEA study and on the experience gained in 1982 during the emptying operation.

The Operation Manual prepared by the Safety Department of Hoffmann-La Roche contained:

- additions to the preexisting infrastructure
- personnel, industrial hygiene, emergency measures
- work programs
- techniques for the dismantling of the individual parts of the plant
- tools, equipment, auxiliary aids
- packing materials, containers
- final removal by truck

Thanks to the Manual, the project management was able to discuss the individual steps in advance with the team and control operations.

Preparatory Work

Prior to the commencement of the dismantling work it was necessary to make additions to the existing infrastructure:

a) A large equipment decontamination steel frame building (SAS) was erected to decontaminate larger single equipment and containers used for the loading of small items and piping, and to remove these from the TCP plant area without the risk of contaminating the environment.

b) Two air exhaust units having a total capacity of 10,000 m^3/hr were fitted to the outer wall of the TCP plant. The exhaust air was cleaned using three series of filters.

Dismantling of the TCP Plant

The dismantling work on the TCP plant was divided into three phases:

1. marking of the pieces of equipment and piping
2. emptying of all residues within the plant
3. actual dismantling

After completing the extensive preparatory work, the operations began on April 18th, 1983.

Neither cutting torches nor discs could be used during this work. All flanged connections were undone, and cutting, where unavoidable, was done with a portable electric band saw.

Since the building structure did not allow the suspension of loads from the roof, mobile tubular scaffolding was used.

In order to keep the degree of contamination in the rooms down to the minimum, the noncontaminated equipment was dismantled before the contamined equipment.

Once all the preparatory work had been completed and the hoists anchored in the correct places, the most difficult part of the entire operation commenced. During hours of work, the A 101 reaction vessel was safely lowered to the ground.

After the last piece of equipment had been carried outside the building through the equipment decontamination building (SAS), the rooms themselves were thoroughly cleaned and the floor painted to fix any residual contaminated dust.

The dismantling was completed on August 12th, 1983. In all, 2200 hr had been spent inside the TCP plant in ventilated protective suits without any accidents.

Building Demolition

The demolition and disposal of the building itself began after the TCP plant had been totally emptied and the residual contamination fixed with paint. Having evaluated various alternatives, a bulldozer with cutting grab was finally chosen for the demolition work. Water was continuously sprayed on the areas where demolition was under way in order to keep dust formation down to the very minimum. A total of 6136 m^3 or 2260 truckloads of rubble were so deposited in Basin No. 2.

The TCP plant building was demolished, including floor slab, foundations, and sewage pipes. The subsoil was removed in its entirety down to ca. 1 m below ground level, analytically controlled, and then filled up with clean gravel.

The most important measures for the success of the TCP plant decontamination work were:

1. Detailed planning for:
   a. individual activities
   b. maintenance of the decontamination units (personnel and material)
   c. monitoring of the safety equipment
2. Analytical control in all areas
3. Use of ventilated protective "Mururoa" type suit

Consideration of the TCP Plant Dismantling Technique

The medical followup examinations of all personnel involved gave no indication of any toxic consequence whatsoever. It must be emphasized that the precautions were very strict: self-contained protective suits and 2-hr shifts only, and a maximum of 2 shifts per day.

Key factors for the success of the operation were:

• meticulous planning
• careful recruiting of personnel

Full credit is hereby given to the Safety Department of Hoffman-La Roche under the supervision of Mr. E. Uhlmann, who handled this complex decontamination problem with professional care and efficiency.

POST-RECLAMATION PROBLEMS

Post-reclamation problems include the use of the ex-A area as a park, the control of the two basins, and gradual removal of existing limitations in the use of the low-contaminated B Zone. A major risk assessment is under way by controlling TCDD accumulated in sample animals, in eggs, and in vegetation fed and grown in the area. The eating habits of the 4700 persons living in the B Zone are also being evaluated and quantified.

CONCLUSION

The low acceptable limits fixed for TCDD in the environment by the Lombardy Regional Government,[9] the early evacuation of 736 persons from the most contaminated area, and the confinement of 1/2 million metric tons of soil inside the two basins reduced to a very minimum the hazard for the population. The Mortality and Cancer Registries[20] to be kept for 12 years (up to 1997) will show whether the low level of TCDD in the Seveso area increased any pathology.

Remedial action is necessarily incomplete and may endanger the personnel engaged in it. Prevention is in fact the only real means to protect the environment. If in the future an emergency should happen or a contamination should be discovered, the Seveso experience, as well as the advances made since the start of the Special Office activity, will definitely prove

Table 3.   Zone A—Reclamation Unit Costs.

	$/m^{3a}	$/ton^b	% Total
Soil Preparation	3.7	2.3	8
Scarification	15.8	9.9	34
Building Dismantling	1.8	1.1	4
Transportation to Basins	7.5	4.7	16
Basin Lining	7.4	4.6	16
Filling with New Soil/Landscaping	5.6	3.5	12
Miscellaneous	4.6	2.9	10
Total (excl. analytical)	46.4	29.0	100

[a] 1 US $ = 2000 Italian lire.
[b] 1 m^3 = 1.6 ton of soil.
Source: Lombardy Regional Government—Special Office for Seveso—Milan, September 20, 1984.[16]

Table 4.   Zone B and R Reclamation Figures.

		Zone B	Zone R
Agricultural Areas	ha	143	776
Soil Treatment, No.		17	13
Cost of Agronomic Treatment	1000^a	1311	1311
Private Gardens, No.		780	3408
Indemnification	1000$	1056	3075
Cleanup Costs	1000$	1250	110
Indemnification to Farmers	1000$	1885	1885
Total Cost Zone B + R (excl. analytical)	1000$	8686	

[a] 1 US$ = 2,000 lt. lire
Source: Lombardy Regional Gov't—Special Offices for Seveso—Milano Sept. 20, 1984.[16]

useful in minimizing costs and risks. Better techniques such, as the in situ treatments now being developed, will likely be used in the future.

However, chemical engineers, chemists and politicians, the industry and the general public must first have reached consensus that it is a moral and social obligation to find solutions to the contamination problems. The cost borne by society for the Seveso case has been high indeed in terms of suffering and of resources (Tables 3 and 4).

Meetings on technical and scientific exchange, such as the ones organized by the ACS in Miami, Florida in April 1985, are efficient tools in reaching consensus in our society.

REFERENCES

1. Mazza, B., and V. Scatturin. "ICMESA: come e perchè," *Sapere* 10–36 (1976).

2. Marshall, V. C. "Seveso, an Analysis of the Official Report," *The Chemical Engineer*, 499–516 (1980).
3. Esposito, M. P., T. O. Tiernan, and F. E. Dryden. "Dioxin," EPA-600/2-80-197, Cincinnati, OH, 1980.
4. Pocchiari, F., A. de Domencio, V. Silvano, G. Viviano, and G. Zapponi. "Accidental Release of 2,3,7,8-Tetrachlorodibenzo-p-dioxin (TCDD) at Seveso, Italy," *Ecotoxicol. Environ. Safety* 4:282–356 (1980).
5. Cardillo, P., and A. Girelli. "Aspetti Chimico-Fisici dell'evento di Seveso," *Chimica e Industria*, September 1980, pp. 651–655.
6. Hay, A. *The Chemical Scythe*, (New York: Plenum Press, 1982).
7. Sambeth, J. "What Really Happened at Seveso," May 16, 44–47 (1983).
8. Theofanous, T. G. "The Physicochemical Origins of the Seveso Accident," *Chem. Eng. Sci.* 38:1615–1636 (1983).
9. "Bollettino Ufficiale della Regione Lombardia—Milano," 14.7.1977–Provvedimenti per Seveso, p. 3057.
10. "Bollettino Ufficiale della Regione Lombardia—Milano," 14.7.1977–Provvedimenti per Seveso, p. 3071.
11. " 'Neptunus' Incineration of Solid Waste at Sea—A Preliminary Study," October 1978. Attached to OCS letter addressed to Special Office for Seveso, Oct. 10, 1978.
12. Liberti, A., D. Brocco, I. Allegrini, and G. Bertoni. "Field Photodegradation of TCDD by Ultra-Violet Radiations," in *Dioxin: Toxicological and Chemical Aspects* F. Cattabeni, A. Cavallaro, and G. Galli, Eds., (1978) pp. 195–200.
13. Liberti, A., D. Brocco, I. Allegrini, A. Cecinato, and M. Possanzini. "Solar and UV Photodecomposition of 2,3,7,8-TCDD in the Environment," *Sci. Total Environ.* 10:97–104 (1978).
14. Fanelli, R., C. Chiabrando, M. Salmona, S. Garattini, and P. G. Caldera. "Degradation of 2,3,7,8-TCDD in Organic Solvents by Gamma Ray Irradiation," *Experientia* 34:1126–1127 (1978).
15. DAM Progetti SPA: "Proposta di Sperimentazione per la Bonifica del Terreno Inquinato a Seguito Incidente ICMESA," February 2nd, 1977.
16. "La Risposta Tecnologica agli Inquinamenti Chimici," Atti del Convegno tenutosi a Milano, Italia, September 20–22, 1984.
17. Barcelona Convention. Convention pour la Protection de la Mer Méditerranée contre la Pollution, February 16, 1966.
18. London Convention, December 29, 1972. Convention sur la Prévention de la Pollution des mers Resultant de l'Immersion de Déchets.
19. ENEA-Roma. Progetto Smantellamento Impianto Produzione TCF ICMESA, Meda (MI). Contract Givaudan/ENEA dated 10/01/81 attached to letter 06/02/82 to the Special Office.
20. Bollettino Ufficiale della Regione Lombardia-Milano 16.7.1799–Provvedimenti per Seveso, p. 3122. The expiry date of the two Registries was fixed by the Lombardy Regional Government in Spring 1984.

List of Authors

K. M. Aldous, Wadsworth Center for Laboratories and Research, New York State Department of Health, Empire State Plaza, Albany, New York 12201

L. R. Alexander, Toxicology Branch C-17/1814, Centers for Disease Control, 1600 Clifton Road, Atlanta, Georgia 30333

C. A. Alley, Toxicology Branch C-17/1814, Centers for Disease Control, 1600 Clifton Road, Atlanta, Georgia 30333

A. Balasso, Lombardy Region, Special Office for Seveso, Seveso (Milano) ITALY

Elizabeth R. Barnhardt, Toxicology Branch C-17/1814, Centers for Disease Control, 1600 Clifton Road, Atlanta, Georgia 30333

Thomas A. Beiter, Department of Chemistry, Miami University, Oxford, Ohio 45056

Giuseppe Belli, Department of Nuclear Theoretical Physics, University of Pavia - via Bassi, 6 , 127100, Pavia, ITALY

Per-Anders Bergqvist, Department of Chemistry, University of Umeå, SWEDEN

S. Billets, U.S. EPA, EMSL, P.O. Box 15027, Las Vegas, Nevada 89114

D. W. Brewster, Pesticide Research Center, Michigan State University, East Lansing, Michigan 48824

D. W. Brombick, Pesticide Research Center, Michigan State University, East Lansing, Michigan 48824

Darryl E. Brown, Food and Drug Administration, Dallas, Texas 75204

J. S. Cantrell, Department of Chemistry, Miami University, Oxford, Ohio 45056

Silvia Cerlesi, Regione Lombardia, Ufficio Speciale per Seveso, via S. Carlo 4, 120030 Seveso (Milano), ITALY

T. E. Clevenger, University of Missouri, Columbus, Missouri 65203

A. Columbis, New York State Department of Environmental of Conservation, Division Air Toxics, 50 Wolf Road, Albany, New York 12201

John Constable, Department of Surgery, Harvard Medical School, Massachusetts General Hospital, Boston, Massachusetts 01214

Deborah Crane, Wadsworth Center for Laboratories and Research, New York State Department of Health, Empire State Plaza, Albany, New York 12201

J. DeKanel, Department of Chemistry, Florida State University, Tallahassee, Florida 32306

Paul Desrosiers, U.S. EPA, 401 M Street, SW, Washington, DC 20460

Lennart Domellof, Department of Chemistry, University of Umeå, S-901 87, Umeå, SWEDEN

J. R. Donnelly, Lockheed-EMSCO, P.O. Box 15027, Las Vegas, Nevada 89114

R. Donnelly, Wadsworth Center for Laboratories and Research, New York State Department of Health, Empire State Plaza, Albany, New York 12201

R. Dougherty, Department of Chemistry, Florida State University, Tallahassee, Florida 32306

M. Erickson, Midwest Research Institute, 426 Bolder Blvd., Kansas City, Missouri 64110

Jurgen H. Exner, IT Enviroscience, 1815 Arnold Drive, Martinez, California 94553

S. Fachetti, Commission of the European Communities, Joint Research Centre, Ispa (VA), ITALY

Henry Falk, Centers for Disease Control, 1600 Clifton Road, Atlanta, Georgia 30333

C. Fichtner, Lombardy Region, Special Office for Seveso, Seveso (Milano), ITALY

D. Firestone, Food and Drug Administration, 200 C Street, SW, Washington, DC 20204

H. Scott Fogler, Department of Chemical Engineering, University of Michigan, Ann Arbor, Michigan 48109

G. U. Fortunati, Regione Lombardi, Ufficio Speciale per Seveso, via S. Carlo 4, Seveso (Milano), ITALY

G. Frare, Lombardy Region, Special Office for Seveso, Seveso (Milano), ITALY

R. A. Freeman, Monsanto Company, 800 N., Lindberg Blvd., St. Louis, Missouri 63167

Frank J. Freestone, U.S. EPA, Woodbridge Avenue, Raritan Depot, Bldg. 10, Edison, New Jersey 08837

John H. Garrett, The Brehm Laboratory and Department of Chemistry, Wright State University, Dayton, Ohio 45459

L. T. Gelbaum, Toxicology Branch, Clinical Chemistry Division, Center for Environmental Health, Centers for Disease Control. 1600 Clifton Road, Atlanta, Georgia 30333

J. F. Gierthy, Wadsworth Center for Laboratories and Research. New York State Department of Health, Empire State Plaza, Albany, New York 12201

Richard O. Gilbert, Energy Systems Department, Battelle Memorial Institute, Pacific Northwest Laboratories, P.O. Box 999, Richland, Washington 99352

George Gitlitz, Department of Surgery, State University of New York, Upstate Medical Center, Clinical Campus, Binghamton, New York 13901

D. F. Groce, Toxicology Branch, Clinical Chemistry Division, Center for Environmental Health, Centers for Disease Control 1600 Clifton Road, Atlanta, Georgia 30333

Michael Gross, Department of Chemistry, University of Nebraska, Lincoln, Nebraska 68588

H. M. Grotta, Center for Analytical and Structural Chemistry, Battelle Columbus Laboratory, 505 King Avenue, Columbus, Ohio 43201–2693

Annette Guiseppi-Elie, Department of Civil Engineering, University of Maryland, College Park, Maryland 20742

Marianne Hansson, Department of Chemistry, University of Umeå, S-901 87, Umeå, SWEDEN

Lennart Hardell, Department of Chemistry, University of Umeå, S-901 87, Umeå, SWEDEN

J. J. Harwood, University of Missouri, Columbia, Missouri 65203

Ralph Hazel, U.S. EPA, Region VIII, 726 Minnesota Avenue, Kansas City, Kansas 66101

C. M. Hedin, Lockheed-EMSCO, P.O. Box 15027, Las Vegas, Nevada 89114

Daniel Heggem, Field Studies Branch, TS 798, Office of Pesticides and Toxic Substances, U.S. EPA, 401 M Street SW, Washington, DC 20406

Fred D. Hilemen, Monsanto Company, 800 N. Lindberg Blvd., St. Louis, Missouri 63167

D. R. Hilker, Wadsworth Center for Laboratories and Research, New York State Department of Health, Empire State Plaza, Albany, New York 12201

David Ben Hur, Roy F. Weston, Inc., 7720 Lorraine Drive, Suite 102, Stockton, California 95210

Danny R. Jackson, Center for Analytical and Structural Chemistry, Battelle Columbus Laboratory, 505 King Avenue, Columbus, Ohio 43201-2693

S. Kapila, University of Missouri, Columbia, Missouri 65203

William D. Keffer, U.S. EPA, Region VIII, 25 Funston Road, Kansas City, Kansas 66115

D. C. Kendall, Division of Chemistry, FDA/National Center for Toxicological Research, Jefferson, Arkansas 72979

R. Kerr, New York State Department of Environmental Conservation, Division of Air Toxics, 50 Wolf Road, Albany, New York,

Margaret W. King, NIEHS/NIH, P.O. Box 12233, Research Triangle Park, North Carolina, 27709

Robert R. Kinnison, Energy Systems Department, Battelle Memorial Institute, Pacific Northwest Laboratories, P.O. Box 999, Richland, Washington 99352

Cliff J. Kirchmer, Roy F. Weston, Inc., Gateway Centre II, Suite 306, Kansas City, Kansas 66101

L.-O. Kjeller, Department of Organic Chemistry, University of Umeå, S-901 87, Umeå, SWEDEN

Robert D. Kleopfer, U.S. EPA, Region VIII. 25 Funston Road, Kansas City, Kansas 66115

J. R. Koninsky, NIOSH, Cincinnati, Ohio 45226

W. A. Korfmacker, Division of Chemistry, FDA/National Center for Toxicological Research, Jefferson, Arkansas 72079

Joseph P. La Fornara, U.S. EPA, Environmental Response Team, Edison, New Jersey 08837

Agostino Lanza, Department of Nuclear Theoretical Physics, University of Pavia - via Bassi, 6, 127100, Pavia, ITALY

R. C. Lao, Analytical Services Division, River Road Environmental Technology Center, Environment Canada, Ottawa Ontario, CANADA K1A 1CB

Ben P.-Y Lau, Food Research Division, Food Directorate, Health and Welfare Canada, Tunney's Pasture, Ottawa, Ontario, CANADA K1A 0L2

G. L. LeBel, Monitoring and Criteria Division, Environmental Health Directorate, Health and Welfare Canada, Tunney's Pasture, Ottawa, Ontario, CANADA K1A 0L2

A. Leono, Lombardy Region, Special Office for Seveso, Seveso (Milano), ITALY

Gunilla Linstrom, Department of Chemistry, University of Umeå, S-901 87, Umeå, SWEDEN

Raymonde Lizotte, Food Directorate, Health Protection Branch, Ottawa, Ontario, CANADA K1A 0L2

J. A. H. MacBride, Sunderland Polytechnic, United Kingdom

B.V. Madhukar, Pesticide Research Center, Michigan State University, East Lansing, Michigan 48824

C.M. Malloy, The Brehm Laboratory, and Department of Chemistry, Wright State University, Dayton, Ohio 45435

S. Marklund, Department of Organic Chemistry, University of Umeå, S-901 87, Umeå, SWEDEN

Fumio Matsumura, Pesticide Research Center, Michigan State University, East Lansing, Michigan 48824

C. Mauri, Lombardy Region, Special Office For Seveso, Seveso (Milano), ITALY

T. Mazer, The Brehm Laboratory and Department of Chemistry, Wright State University, Dayton, Ohio 45435

Jay C. Means, Chesapeake Biological Lab and Department of Chemistry, Solomons, Maryland 20688

R. K. Mitchum, U.S. EPA, EMSL, P.O. Box 15027, Las Vegas, Nevada 89114

C. Moody, Midwest Research Institute, 425 Volker Blvd., Kansas City, Missouri 64110

W. J. Niederhut, Quality Assurance Laboratory, University of Nevada, Las Vegas, Nevada 89114

Richard A. Niemann, Food and Drug Administration, 200 C Street SW, Washington, DC 20204

Roy W. Nobel, Monsanto Company, 800 N. Lindberg Blvd., St. Louis, Missouri 63167

Martin Nygren, Department of Chemistry, University of Umeå, S-901 87, Umeå, SWEDEN

P. W. O'Keefe, Wadsworth Center for Laboratories and Research, New York State Department of Health, Empire State Plaza, Albany, New York 12201

Mats Olsson, Swedish National Museum of Natural History, S-104 05, Stockholm, SWEDEN

J. Palausky, Environmental Trace Substances Research Center, University of Missouri, Route 23, Columbia, Missouri 65203

D. G. Patterson, Jr. Toxicology Branch C-17/1814, Centers for Disease Control, 1600 Clifton Road, Atlanta, Georgia 30333

Raphe Pavlick, Roy F. Weston, Inc., 7720 Lorraine Drive, Suite 102, Stockton, California 95210

B. Ramalingam, The Brehm Laboratory and Department of Chemistry, Wright State University, Dayton, Ohio 45435

M. M. Rao, Department of Civil Engineering, University of Maryland, College Park, Maryland 20742

C. Rappe, Department of Organic Chemistry, University of Umeå, S-901 87, Umeå, SWEDEN

S. P. Ratti, Department of Nuclear Theoretical Physics, University of Pavia - Bassi, 6, 127100, Pavia, ITALY

Arleen B. Rifkind, Department of Pharmacology, Cornell University Medical College, New York, New York 10021

M. H. Roulier, U.S. EPA, Cincinnati, Ohio 45268

K. L. Rowland, Division of Chemistry, FDA/National Center for Toxicological Research, Jefferson, Arkansas 72079

L. G. Rushing, Division of Chemistry, FDA/National Center for Toxicological Research, Jefferson, Arkansas 72079

S. W. Rust, Center for Analytical and Structural Chemistry, Battelle Columbus Laboratory, 505 King Avenue, Columbus, Ohio 43201-2693

C. Ryan, Department of Chemistry, Florida State University, Tallahassee, Florida, 32306

John J. Ryan, Food Directorate, Health Protection Branch, Ottawa, Ontario, CANADA K1A 0L2

J. J. Ryan, Health and Welfare Canada, Tunney's Pasture, Ottawa, Ontario, CANADA K1A 0L2

Eric Sampson, Centers for Disease Control, 1600 Clifton Road, Atlanta, Georgia 30333

Arnold Schecter, Department of Preventive Medicine, Upstate Medical Center, State University of New York, Clinical Campus. Binghamton, New York 13901

Carol M. Schiller, NIEHS/NIH, P.O. Box 12233, Research Triangle Park, North Carolina 27709

Louis F. Schneider, FDA, Detroit, Michigan 48207

Wayne Schramm, Missouri Division of Health, St. Louis, Missouri

Jerry M. Schroy, Monsanto Company, 800 N. Lindberg Blvd., St. Louis, Missouri 63167

Patrick J. Slonecker, Department of Chemistry, Miami University, Oxford, Ohio 45056

J. S. Smith, Walter B. Satterthwaite Assoc. Inc., 11 North Five Points Road, West Chester, Pennsylvania 19380

L. M. Smith, USDI, Fish and Wildlife Service, Columbia National Fisheries Research Laboratory, Columbia, Missouri 65201

R. M. Smith, Wadsworth Center for Laboratories and Research, New York State Department of Health, Empire State Plaza, Albany, New York 12201

W. Allen Smith, Roy F. Weston, Inc., Gateway Centre II, Suite 306, Kansas City, Kansas 66101

G. W. Sovocol, U.S. EPA, EMSL, P.O. Box 15027, Las Vegas, Nevada 89114

D. L. Stalling, USDI, Fish and Wildlife Service, Columbia National Fisheries Research Laboratory, Columbia, Missouri 65201

Keeran R. Srinivasan, Department of Chemical Engineering, University of Michigan, Ann Arbor, Michigan 48109

Paul Stehr, Centers for Disease Control, 1600 Clifton Road, Atlanta, Georgia 30333

Gary Stein, Centers for Disease Control, 1600 Clifton Road, Atlanta, Georgia 30333

Wing-F. Sun, Food Directorate, Health Protection Branch, Ottawa, Ontario, CANADA K1A 0L2

Stephen Swanson, Midwest Research Institute, 425 Volker Blvd., Kansas City, Missouri 64110

Michael L. Taylor, The Brehm Laboratory and Department of Chemistry, Wright State University, Dayton, Ohio 45459

R. S. Thomas, Analytical Services Division, River Road Environmental Technology Center, Environment Canada, Ottawa, Ontario, CANADA K1A 1C8

H. C. Thompson, Jr., Division of Chemistry, FDA/National Toxicological Research, Jefferson, Arkansas 72079

Thomas O. Tiernan, The Brehm Laboratory and Department of Chemistry, Wright State University, Dayton, Ohio 45459

Y. Tondeur, Environmental Research Center, University of Nevada, Las Vegas, Nevada 89154

W. E. Turner, Toxicology Branch C-17/1814, Centers for Disease Control, 1600 Clifton Road, Atlanta, Georgia 30333

M. Tysklind, Department of Organic Chemistry, University of Umeå, S-901 87, Umeå, SWEDEN

Michael J. Urban, US.S. EPA, Environmental Response Team, Edison, New Jersey 08837

M. Vasconi, Lombardy Region, Special Office for Seveso, Seveso (Milano), ITALY

Tenkasi S. Viswanathan, Jacobs Engineering Group, Roy F. Weston, Inc, TAT, 306 Gateway Center II, 4th Street and State Avenue, Kansas City, Kansas 66101

T. L. Vonnahme, Lockheed-EMSCO, P.O. Box 15027, Las Vegas, Nevada 89114

Wagel, The Brehm Laboratory and Department of Chemistry, Wright State University, Dayton, Ohio 45459

Ramsey Walden, NIEHS/NIH, P.O. Box 12233, Research Triangle Park, North Carolina 27709

Rick Walters, Department of Civil Engineering, University of Maryland, College Park, Maryland 20742

J. S. Warner, Center for Analytical and Structural Chemistry, Battelle Columbus Laboratory, 505 King Avenue, Columbus, Ohio 66101

Karen Webb, St. Louis School of Medicine, St. Louis, Missouri

N. C. A. Weerasinghe, Department of Chemistry, University of Nebraska, Lincoln, Nebraska 68588

David T. Williams, Monitoring and Criteria Division, Environmental Health Directorate, Health and Welfare Canada, Tunney's Pasture, Ottawa, Ontario, CANADA K1A 0L2

L. Wilson, Wadsworth Center for Laboratories and Research, New York State Department of Health, Empire State Plaza, Albany, New York 12201

Armon Yanders, Environmental Trace Substances Research Center, University of Missouri, Columbia, Missouri 65203

Index